Advances in Intelligent Systems and Computing

Volume 1108

The series "Advances in Intelligent Systems and Computing" contains publications on theory, applications, and design methods of Intelligent Systems and Intelligent Computing. Virtually all disciplines such as engineering, natural sciences, computer and information science, ICT, economics, business, e-commerce, environment, healthcare, life science are covered. The list of topics spans all the areas of modern intelligent systems and computing such as: computational intelligence, soft computing including neural networks, fuzzy systems, evolutionary computing and the fusion of these paradigms, social intelligence, ambient intelligence, computational neuroscience, artificial life, virtual worlds and society, cognitive science and systems, Perception and Vision, DNA and immune based systems, self-organizing and adaptive systems, e-Learning and teaching, human-centered and human-centric computing, recommender systems, intelligent control, robotics and mechatronics including human-machine teaming, knowledge-based paradigms, learning paradigms, machine ethics, intelligent data analysis, knowledge management, intelligent agents, intelligent decision making and support, intelligent network security, trust management, interactive entertainment, Web intelligence and multimedia.

The publications within "Advances in Intelligent Systems and Computing" are primarily proceedings of important conferences, symposia and congresses. They cover significant recent developments in the field, both of a foundational and applicable character. An important characteristic feature of the series is the short publication time and world-wide distribution. This permits a rapid and broad dissemination of research results.

** Indexing: The books of this series are submitted to ISI Proceedings, EI-Compendex, DBLP, SCOPUS, Google Scholar and Springerlink **

More information about this series at http://www.springer.com/series/11156

S. Smys · João Manuel R. S. Tavares ·
Valentina Emilia Balas · Abdullah M. Iliyasu
Editors

Computational Vision and Bio-Inspired Computing

ICCVBIC 2019

Set 2

 Springer

Editors
S. Smys
Department of CSE
RVS Technical Campus
Coimbatore, India

Valentina Emilia Balas
Faculty of Engineering
Aurel Vlaicu University of Arad
Arad, Romania

João Manuel R. S. Tavares ⓘ
Faculty of Engineering
Faculdade de Engenharia
da Universidade do Porto
Porto, Portugal

Abdullah M. Iliyasu
School of Computing
Tokyo Institute of Technology
Tokyo, Japan

ISSN 2194-5357   ISSN 2194-5365   (electronic)
Advances in Intelligent Systems and Computing
ISBN 978-3-030-37217-0   ISBN 978-3-030-37218-7   (eBook)
https://doi.org/10.1007/978-3-030-37218-7

This Springer imprint is published by the registered company Springer Nature Switzerland AG
The registered company address is: Gewerbestrasse 11, 6330 Cham, Switzerland

We are honored to dedicate the proceedings of ICCVBIC 2019 to all the participants and editors of ICCVBIC 2019.

Foreword

It is with deep satisfaction that I write this foreword to the proceedings of the ICCVBIC 2019 held in RVS Technical Campus, Coimbatore, Tamil Nadu, on September 25–26, 2019.

This conference was bringing together researchers, academics, and professionals from all over the world, experts in computational vision and bio-inspired computing.

This conference particularly encouraged the interaction of research students and developing academics with the more established academic community in an informal setting to present and to discuss new and current work. The papers contributed the most recent scientific knowledge known in the field of computational vision, soft computing, fuzzy, image processing, and bio-inspired computing. Their contributions helped to make the conference as outstanding as it has been. The local organizing committee members and their helpers put much effort into ensuring the success of the day-to-day operation of the meeting.

We hope that this program will further stimulate research in computational vision, soft computing, fuzzy, image processing, and bio-inspired computing and provide practitioners with better techniques, algorithms, and tools for deployment. We feel honored and privileged to serve the best recent developments to you through this exciting program.

We thank all authors and participants for their contributions.

S. Smys
Conference Chair

Preface

This conference proceedings volume contains the written versions of most of the contributions presented during the conference of ICCVBIC 2019. The conference provided a setting for discussing recent developments in a wide variety of topics including computational vision, fuzzy, image processing, and bio inspired computing. The conference has been a good opportunity for participants coming from various destinations to present and discuss topics in their respective research areas.

ICCVBIC 2019 conference tends to collect the latest research results and applications on computational vision and bio-inspired computing. It includes a selection of 147 papers from 397 papers submitted to the conference from universities and industries all over the world. All of accepted papers were subjected to strict peer-reviewing by 2–4 expert referees. The papers have been selected for this volume because of quality and the relevance to the conference.

ICCVBIC 2019 would like to express our sincere appreciation to all authors for their contributions to this book. We would like to extend our thanks to all the referees for their constructive comments on all papers; especially, we would like to thank to organizing committee for their hardworking. Finally, we would like to thank the Springer publications for producing this volume.

S. Smys
Conference Chair

Acknowledgments

ICCVBIC 2019 would like to acknowledge the excellent work of our conference organizing the committee, keynote speakers for their presentation on September 25–26, 2019. The organizers also wish to acknowledge publicly the valuable services provided by the reviewers.

On behalf of the editors, organizers, authors, and readers of this conference, we wish to thank the keynote speakers and the reviewers for their time, hard work, and dedication to this conference. The organizers wish to acknowledge Dr. Smys, Dr. Valentina Emilia Balas, Dr. Abdul M. Elias, and Dr. Joao Manuel R. S. Tavares for the discussion, suggestion, and cooperation to organize the keynote speakers of this conference. The organizers also wish to acknowledge for speakers and participants who attend this conference. Many thanks are given to all persons who help and support this conference. ICCVBIC 2019 would like to acknowledge the contribution made to the organization by its many volunteers. Members contribute their time, energy, and knowledge at local, regional, and international levels.

We also thank all the chair persons and conference committee members for their support.

Contents

Breast Cancer Detection Using CNN on Mammogram Images

Kushal Batra$^{(\boxtimes)}$, Sachin Sekhar, and R. Radha

SRM Institute of Science and Technology, Kattankulathur, India
kushalbatra_an@srmuniv.edu.in,
sachinsekhar1397@gmail.com,
radha.ra@ktr.srmuniv.ac.in

Abstract. The most commonly occurring form of cancer is breast cancer and thereby requires urgent attention. In our paper, we work on these CT images of breast cancer to detect cancer in early stages. These images are classified into benign (non-cancerous) and malignant (cancerous) using our proposed method of Convolution Neural Network (CNN) following which medical care can be given. In our paper, image segmentation along with the image preprocessing is used which provides enhanced images. This helps in increasing the overall accuracy of the system. To conclude, the paper compares the trade-off between accuracy and time to train the network obtained in different execution environment i.e. Tensorflow and Matlab. Tensorflow gave 87.98% accuracy with training time of 6 h and Matlab gave 84.02% accuracy with training time of 45 min when the neural network's architecture remains the same.

1 Introduction

Breast Cancer has become the most common type of cancer in women and second most common type of cancer among men [1]. Over 2 million new cases were recorded in the year 2018 itself. The factors governing the occurrences of breast cancer include age, family history, genetics, gender and many more [2].

Even with this increasing number of cases of breast cancer, the number of doctors and nurses available to diagnose or treat these cases are just not enough. Through a study conducted by Breast Cancer Care, statistics recorded implied that 42% of National Health Service (NHS) trusts admit that they do not possess the staff to assign to individuals who could be classified as a breast cancer specialist nurse. [3] According to a survey done for Cancer Search UK, it was found that for women who diagnosed with breast cancer at the earliest stage, 90% of these cases were able to survive their disease for at least 5 years. In contrast to this approximately 15% of the women diagnosed with the most advanced stage of disease were able to survive for the same period. [4] These numbers show the importance of detection of breast cancer as early as possible, and how it is not possible to achieve this with limited man-power and an increasing trend seen in new cases of breast cancer coming up.

To check for presence of breast cancer among patients medical imaging techniques are put to use such as CT scans, MRI, mammography, X-Ray etc. Among these methods, MRI and mammography are the safest and most reliable means. Mammograms generally

© Springer Nature Switzerland AG 2020
S. Smys et al. (Eds.): ICCVBIC 2019, AISC 1108, pp. 708–716, 2020.
https://doi.org/10.1007/978-3-030-37218-7_80

take 10 to 15 min. The breast of the patient is placed between two plates. X-Rays are used to check for radiations being expelled from the patient's breast, which are recorded, and an image is obtained to be used for the diagnosis process [5].

An image recognition system based on image segmentation and Convolutional Neural Network (CNN), which is a type of artificial neural networks, are used. The system is trained to detect and classify images based on the presence of abnormalities into malignant (cancerous abnormality), benign (non-cancerous abnormality), or normal. The system aims on achieving high levels of accuracy in the diagnosis of cases, allowing the process of reading mammogram images to be a process which is independent of human interference. This allows increasing the number of cases diagnosed, as computers can keep diagnosing images, and also eliminates the human error that may occur. CNN has proved to be dominant and effectual technique. For example, skin cancer, the most common occurring cancer in humans is now easily detectable with the introduction of CNN [6]. Another example is of lung cancer where it is very difficult to detect cancer in pre stages but with CNN it is now feasible [7]. These reasons have made CNN method gain popularity in biomedical science, especially in the field of cancer detection.

2 Methodology

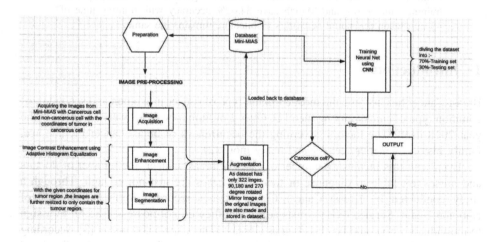

Fig. 1. Flowchart

Dataset

For our paper, mini-Mammographic Image Analysis Society (mini-MIAS) dataset has been used [8]. It is used to perform the analysis of mammogram images. Images stored as part of the Mini-MIAS database have their size reduced. It does so by decreasing the 200 micron pixel edge original image to a 50 micron pixel edge. The dataset comprises of mammogram images of 161 patients. Each patient's left and right mammogram images are included in the data set. Therefore, the total number of images adds up to 322. The dimensions of the images in the dataset are 1024 × 1024 pixels.

Mini MIAS dataset further provides some important information related to cancerous tissue. Every image has been tagged with class of abnormality, background tissue, radius and centre coordinates for the cancerous blob. The sample format of image in dataset (Figs. 1 and 2).

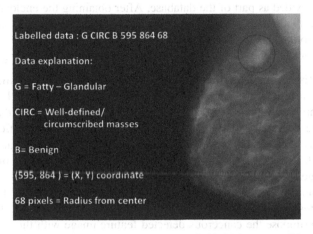

Labelled data : G CIRC B 595 864 68

Data explanation:

G = Fatty – Glandular

CIRC = Well-defined/
        circumscribed masses

B= Benign

(595, 864) = (X, Y) coordinate

68 pixels = Radius from center

Fig. 2. Breast Image Format from Mini-MIAS Dataset [2]

Image Pre-Processing
Pre-processing of image plays an eminent role in cleansing, filtering and extraction of feature point in images. This involves operations on images at lower levels of abstraction.

Data Augmentation
The dataset mini-MIAS has very small collection of mammogram images i.e. limited to only 322 images (including malignant and benign categories). This volume of images is not enough to train our CNN. To further improve the model's accuracy, sub-sampling of the dataset was done with the help of image transformations and data augmentation. Images rotated by an angle of 90, 180, 270° are included in the dataset. For every rotation, a horizontally reflected version is also included to further increase the dataset size. This makes the size of the dataset in terms of number of images to 2576 (322 * 4 * 2) images. The data augmentation has proved out to be of crucial importance in obtaining the higher accuracy rates with mini-MIAS [9].

Image Segmentation
Pixel size of the data stored in its raw form is 1024 × 1024 which is too large to feed into the CNN. Also, a lot of processing time is needed for this size of image. In most of the images, only 20% of the image consists of our ROI i.e. the cancer tissue region. Hence, using the entire 1024 × 1024 pixel sized image would be highly inefficient. Instead, resizing allows for greater accuracy and decrease in computation time [9].

In case of large-sized images being fed to a CNN classifier, a lot of computation time is spent on unnecessary information or noise which doesn't have much significance. Also, the accuracy and result of the classifier will be affected.

To achieve this, the image was cropped to a size of 48 × 48 pixels, containing of only the malignant portion of the breast tissue. The cropping is done using the x and y-coordinates given as mentioned and provided by mini-MIAS database. These coordinates provide the center of malignant tissue. For the circle enclosing the tissue, the radius is also provided as part of the database. After obtaining the enclosing circle, the image is resized to a size of just 48 × 48 pixels. Following these steps, the images are fed to the Convolution Network.

The two major techniques that have been used in our pre-processing phase are: -

1. **Adaptive Histogram Equalization** – Histogram Equalization augments the contrast of an image. In our case it helps in enhancing the contrast levels at cancerous region of the tissue for faster feature detection. All the images in dataset are not having same degree of contrast and hence for different images different contrast threshold has to be set. For this reason, Adaptive Histogram Equalization technique can be used. This technique is more efficient for the scenario as of ours where dataset images have different degree of contrast [10].

2. **Image Fusion** (Fuzzy Fusion technique) – Image fusion is the techniques of superimposing one image upon the second image. In our algorithm, this method is used to superimpose the cancerous detected feature image with the original image. After the cancerous blob detected is fused with the original 48 × 48 image and then fed to our CNN. This helps in improving the accuracy of the model. Fuzzy fusion technique was our choice of implementation as it gave outstanding results in detecting Head and Neck Cancer [11]. Hence also it has been found to be useful in case of detection of breast cancer as discussed further in the paper.

3 Convolutional Neural Network

We split the dataset images into 2 sections. One section is for training the CNN. The other section is used to test it. We split the dataset into a ratio of 7:3, 7 for training and 3 testing. Both cancerous and non-cancerous, referred to as malignant and benign respectively in dataset, are part of the two sections.

Convolutional Neural Networks or CNN is a class of Deep Neural Networks (DNN). CNN's are neural network consisting of many layers such that at each perceptron layer there is minimum requirement of image pre-processing to be done. CNN's are generally used in the studies which involves imagery like images and videos. The factor that distinguishes CNN from other class of neural network is the working dimension of neural net. As a result, CNN can work with 3D input data.i.e. - height, width and length. CNNs implement a local connectivity pattern between neurons of adjacent layer, exploiting spatial locality. Such features of CNN's, help greatly in vision or image based problems. CNN's are very good in detection of tumour as surveyed by researchers [13]. Recent works on brain tumour detection [12] and Lung cancer detection implemented using CNN is a sufficient statement to justify the above fact.

Implementation

The fused 48 × 48 images are now used for Breast Cancer detection using CNN (BCDCNN) [9]. The Fig. 3 illustrates the CNN structure's layout that is the used in our proposed system. The CNN implements 3 hidden layers and a final, fully connected layer. In our implementation, the predicting class is modified to have 2 output classes - cancerous and non-cancerous class. The learning rate for the CNN is set to 0.0001. The kernel sizes used in all the layers are of filter size 5 × 5. Epoch number implemented is 80 with testing set frequency is set to 100. After every convolution layer, a batch normalization layer is present which helps in normalizing the output obtained from the current layer. This normalized output is fed to the next output layer. Another max-pooling layer is also added of 2 × 2 filter size. Max Pooling is a down-sampling strategy in Convolutional Neural Networks. Firstly, it decreases parameters within the model. Secondly, it generalises the results obtained from the convolutional filter. This makes the detection of features invariant to scale or orientation changes. The sliding window's stride value for maxpooling layer is set to 2. The above mentioned 2 layers, i.e. - batch normalization layer and max-pooling layer have been implemented to improve the performance. The BCDCNN in [9] uses an adam optimizer in the final layer, but our implementation instead uses an stochastic gradient descent with momentum (sgdm) layer yielding better performance with our pre-processed dataset. The first of the activation layers uses a ReLu activation layer. The second layer implements a leaky ReLu layer. Leaky ReLu layer as shown in Fig. 5 helps in solving the problem of 'dead ReLu' by adding some gradient value in non- active region of normal ReLu layer as shown in Fig. 4. The last layer is a soft-max layer which provides us with our categorical output.

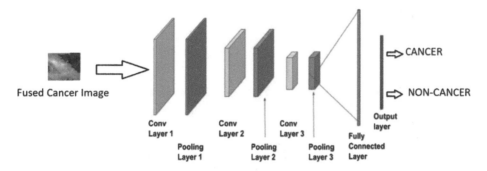

Fig. 3. CNN Layout

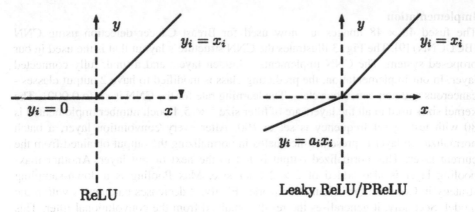

Fig. 4. ReLu function   Fig. 5. Leaky ReLu function

4   Results and Discussions

The above CNN implementation was implemented on both Matlab and Tensorflow environment. The results were calculated and plotted (Fig. 6).

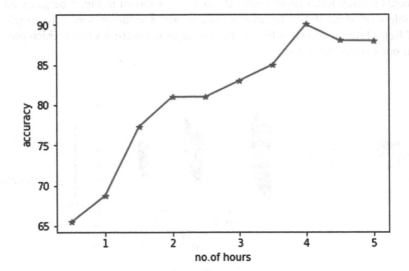

Fig. 6. Accuracy Plot for Tensorflow

From the final results, it could easily be deduced that CNN implementation on tensorflow outperforms CNN implementation on Matlab. On the other hand Matlab takes less time to train the dataset when compared to time taken to train the neural network in Tensorflow. This difference can be attributed to the time taken to train the dataset. The final accuracy rates achieved with Tensorflow was 87.98% with 5 h to train the dataset and for Matlab it is 84.02% with 30 min to train the dataset. The above results show the trade-off between accuracy rate and time taken to train the dataset for the Mini-MIAS Dataset.

The graph for tensorflow indicates that the accuracy rates increase steeply for 2 h under the training until for 4 h when the saturation point is reached. This saturation point is the point from where the neural network starts to overfit.

The Training progress graph obtained from Matlab provides information about the Training loss as well which indicates that with time the loss% for training and testing has both decreased with time. The accuracy plot shows that the initially the testing was low but after 400 iterations the testing accuracy line started to have a positive slope. The training accuracy for the above shows a exponential graph reaching the saturation point nearly at 1100 iterations (Fig. 7).

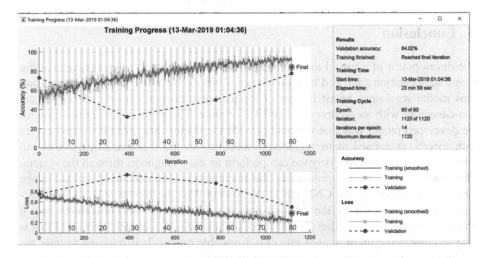

Fig. 7. Accuracy plot for Matlab

The Fig. 8 is the comparison graph comparing the accuracies of our proposed system with the existing system i.e. BCDCNN. The BCDCNN [7] showed accuracy rate of 82.71% whereas our proposed systems accuracy rates in both the environments yield better results −87.98% in Tensorflow and 84.02% in Matlab.

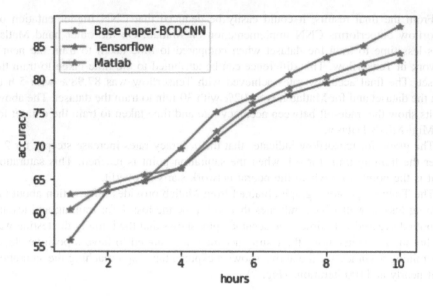

Fig. 8. Proposed System vs Existing System

5 Conclusion

The development of a model to detect cancerous tissues in mammogram images using image segmentation followed by Convolutional Neural Network has been carried out. This model segregates breast images of the patients into two classes: cancerous and non-cancerous with an accuracy rate of 84% or more. The most important use-case for the developed model was that it would provide the doctors and even dedicated nurses with speedy, diagnosed mammograms, to make up for the shortage of personnel and time consuming diagnosis. Anyone can perform an easy diagnosis through the system. The model begins by pre-processing the mammograms to obtain an output image which is processed by the CNN easily. Following this, the model will calculate differences between each of the labelled data by considering various features that it obtains from the images. After 20,000 times of comparison, a model is obtained that is able to classify the input images into closer possible output. It is required to train the model only once. To carry out evaluation on predict images, it would not take more than a few minutes to complete, unless a situation came up where a change in parameter is required. Following this, the model would require a few hours to complete training again. To conclude, detection of breast cancer by using CNN was successfully carried out and mammogram images of 166 patients' breasts were used to test it. The model brings forth a method to obtain a quick diagnosis and a system providing high accuracy.

The works presented by this paper can be used further in future to identify and design a new convolution neural network such that the trade-off between accuracy and training is reduced to minimal. Also the given results can be enhanced further by identifying a new feature extraction technique which helps in distinguishing between cancerous and non-cancerous tissue.

Compliance with Ethical Standards
   ✓ All authors declare that there is no conflict of interest.
   ✓ No humans/animals involved in this research work.
   ✓ We have used our own data.

References

1. World Cancer Research Fund. https://www.wcrf.org/dietandcancer/cancer-trends/breast-cancer-statistics
2. Breast Cancer Org. https://www.breastcancer.org/risk/factors
3. Breast Cancer Care. https://www.breastcancercare.org.uk/
4. Cancer Research UK. https://www.cancerresearchuk.org/about-cancer/cancer-symptoms/why-is-early-diagnosis-important
5. Very Well Health. https://www.verywellhealth.com/understanding-your-mammogram-report-430283
6. Esteva, A., Kuprel, B., Novoa, R., Ko, J., Susan, M.S., Blau, H.M., Thrun, S.: Dermatologist-level classification of skin cancer with deep neural networks. Nature, 542 (2017). https://doi.org/10.1038/nature21056
7. Sun, W., Zheng, B., Qian, W.: Computer aided lung cancer diagnosis with deep learning algorithms (2016)
8. Suckling, J., Parker, J., Dance, D., Astley, S., Hutt, I., Boggis, C., Taylor, P.: The mammographic image analysis society digital mammogram database. In: Exerpta Medica. International Congress Series, vol. 1069, pp. 375–378, July 1994
9. Tan, Y.J., Sim, K.S., Ting, F.F.: Breast cancer detection using convolutional neural networks for mammogram imaging system, pp. 1–5 (2017). https://doi.org/10.1109/icoras.2017.8308076
10. Pizer, S.M., Amburn, E.P., Austin, J.D., Cromartie, R., Geselowitz, A., Greer, T., Romeny, B.H., Zimmerman, J.B., Zuiderveld, K.: Adaptive histogram equalization and its variations. Comput. Vis. Graph. Image Process. **39**, 355–368 (1987). https://doi.org/10.1016/s0734-189x(87)80186-x
11. Schöder, H., Yeung, H.W., Gonen, M., Kraus, D., Larson, S.M.: Head and neck cancer: clinical usefulness and accuracy of PET/CT image fusion1. Radiology **231**, 65–72 (2004). https://doi.org/10.1148/radiol.2311030271
12. Joshi, D.M., Rana, N.K., Misra, V.M.: Classification of brain cancer using artificial neural network. In: ICECT 2010 Proceedings of the 2010 2nd International Conference on Electronic Computer Technology. 112–116. https://doi.org/10.1109/ICECTECH.2010.5479975
13. Spanhol, F.A., Oliveira, L.S., Petitjean, C., Heutte, L.: Breast Cancer Histopathological Image Classification using Convolutional Neural Networks (2016)

Segmentation Based Preprocessing Techniques for Predicting the Cervix Type Using Neural Networks

M. B. Bijoy[1,2(✉)], A. Ansal Muhammed[1], and P. B. Jayaraj[1]

[1] CSED, National Institute of Technology, Calicut, India
[2] CDAC KP, Bangalore, India
bijoymb@cdac.in

Abstract. Cervical cancer is a disease condition which makes cells on the human organ Cervix grow out of control. Among all women, it is the second most occurring type of cancer. Human papillomavirus is the cause of this disease. Early stage detection of cancer helps in finalizing correct medication at the appropriate time. Diagnosis of cervical cancer involves human expertise to a greater extend. Advanced medical imaging enables computerized methods to identify the cancerous cells at beginning stage. Detecting the cervix type is very important as the type of treatment depends on the cervix type. We intend to propose a classification technique to predict the cervix type using Neural networks. From our earlier works, it is understood that the lack of powerful preprocessing techniques for finding Region of Interest(ROI) is missing in the literature. We have developed a Neural Network(UNet) based segmentation technique for cutting the ROI from the input image set. Finally, a CNN model is designed, developed and trained for classifying the cervix type. The applicability of transfer learning to cervix type prediction is also tried. A accuracy of 70% is obtained.

Keywords: Segmentation · Cervical cancer · Cervix type · CNN

1 Introduction

Cancer is the most dangerous disease that occurs in human beings. Statistically cervical cancer holds the second position for most occurring type of cancer in women of all age groups [12]. Cancer occurs when the cancerous cells in our body starts to replicate itself and grow beyond control. Cervical cancer starts at cervix lining cells, in lower area of uterus that contacts the upper vagina. There are few symptoms that may appear during the precancerous stage. It can be completely cured when detected at early stages. Manual tests like finding changes in cells taken from uterus of cervix region can be used for this [8].

There are different types of cervical cancer and cervix. Upon detection of cancer, the treatment plan for cervical cancer is decided based on the type of cervix. It is difficult to classify cervices just by looking as they seem very

© Springer Nature Switzerland AG 2020
S. Smys et al. (Eds.): ICCVBIC 2019, AISC 1108, pp. 717–726, 2020.
https://doi.org/10.1007/978-3-030-37218-7_81

similar to each other visually. Cervix has two separate regions called as ecto-cervix and endocervix. First region ectocervix can be seen partially from inside vagina. Endocervix is the tunnel that connects the ectocervix to the uterus. These Two areas are constituted by two types of cells and the meeting zone is called transformation zone. Cervices can be categorised into 3 basic types according to transformation zone. In Type 1 transformation region is fully vis-ible, situated completely out side cervix and can be of any size. Type 2 is fully visible from out side, has an endocervical region and may have an ectocervical component too which may be small or large. In Type 3 cervix which will have an endocervical component, may not be completely visible and may have a portion which is completely inside the cervix and may be small or large. Illustration of these types of cervix are shown in Fig. 1.

There are many methods for detecting cervical cancer which makes use of machine learning. This may use different techniques like SVM, fuzzy based tech-niques, and texture based classification. The pap smear slides are taken and the noise is removed. The features of the cell are identified and is classified as nor-mal and infected cells. But the treatment depends on the type of cervix and it is need to be found next. Methods for this require human expertise and is a time consuming task. Normally health care providers don't have expertise and infras-tucture to do this. So classification, a category of Machine Learning method can be applied for finding this. We have many machine learning methods avail-able like ANN,SVM, CNN, k-means etc for this operation. All these machine algorithm need some features using which they classify the given data points.

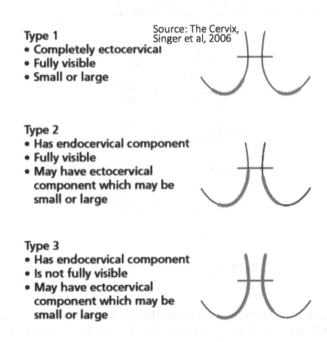

Fig. 1. Different types of cervix

Neural Networks, a supervised machine learning technique is a good option for doing this. Using the layers of Convolution Neural Networks can extract the features and then classify the cervix images. Lack of methods for segmenting the correct Region of Interest reduces the classification accuracy of existing Deep learning classifiers in this works. So developing computational methods for preprocessing and then classfiying the images is very important. In this paper, the design of a good segmentation method using UNet is explained. We have developed methods to classify the cervix image to different cervix types using Deep Neural Networks.

Our work is presented in the following sections as follows. Section 2 discusses the similar and related research works carried out in the area of cervical cancer detection and cervical type prediction. The details of proposed work and its design are depicted in Sect. 3. Important results and related discussions are provided in IV. Finally, Sect. 5 concludes the work with possible leads to the works.

2 Literature Survey

We intent to build an efficient learning algorithm to classify cervix to find out the type of treatment needed. But this is still a difficult task to be automated. Image segmentation is a crucial element in this process, as finding the proper region of interest provides to higher accuracy. Some of the relevant research works completed in the areas of cervical cancer detection, classification of cervix type, and image segmentation are explained in this section.

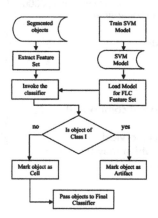

Fig. 2. SVM Classifier Algorithm

Classification of cancerous cells from the normal ones, can be done using SVM(Support Vector Machine), a type of classifier. SVM gives an optimal hyperplane which categorizes new example, when we feed it labeled training data.

Rajesh Kumar et al. [9] shows us how effectively SVM can be utilised for reducing the false positives by finding two separate support hyper-planes for the artifacts present in the cell images of similar size and shape and the cells we intend to detect. Figure 2 shows the SVM classifier algorithm used in detecting cervical cancer. Bhakti Tulpule et al. [1] showed textual difference between cancerous cells and non cancerous cells can be used for classification. He used the difference in inter capillary distance, which is high for cancerous cells.

Thanatip Chankong et al. [10] used Fuzzy C means clustering algorithm to classify the cells into four categories which helps to the selection of the treatment. In their expiriment a single cell image was successfully segmented into nucleus, cytoplasm, and background in his work.

Classification of states of cervix from microscopic images using CNN and other ML classifiers was proposed by Jonghwan Hyeon et al. [6] in his work. Figure 3 shows the basic idea of a CNN to classify cervix states.

All these work report whether the cells are cancerous or not.Therefore type of treatment to be chosen had to be done by hand. Chaitanya Asawa et al. [2] explains use of deep learning for classification of cervix types using Convolution Neural Network. Transfer learning is also explored for the benefit of this work.

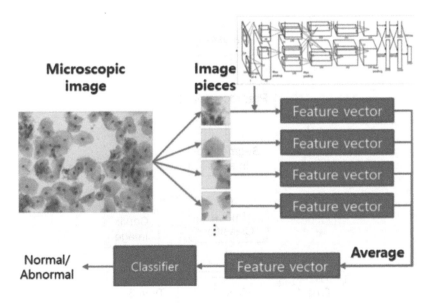

Fig. 3. Classification using CNN

We should have a good data set to build a good image classifier. Also Areas Of Interest from these images should be found and given for classification. Segmentation is used to achieve this Greenspan et al. [3] demonstrated a good way to do segmentation of cervical images into areas of interests. In this work, the strong reflections are removed in the preprocessing step and later regions are grown to find the best suited region of interest.

Lei Lei et al. [7] did some experiments with multiple choices of hyper parameters, transfer learning and was able to get good performance. He also combined manual identification of ROI and a bounding box regressor inside his to experiment.

3   Proposed Work

The design of our proposed work is explained in this section. Data is the heart of any good model. In order to get accurate training we have to pre process the data and prepare the data in standard form. After that we have to get only the relevant part of the images so that the model will only train on the relevant part and not on some extraneous objects. Once the segmentation of the images are done we have to model our classifiers which will classify the given image to different cervix type. A description of type of cervix is given in the introduction section of this paper. The work flow diagram of our design is illustrated in Fig. 4. The working of each block is explained below.

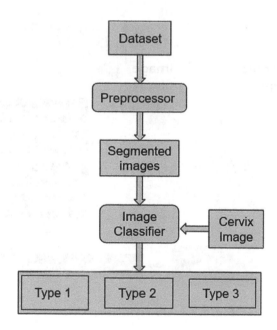

Fig. 4. Design Flow Chart

Dataset Collection. The dataset for the project is collected from Kaggle cervival cancer screening competition. Table-1 gives the details of data used along with the numbers of images of each of three types of cervices. Figure 5 shows some of the images from dataset. The classification is done depending up on the transformation zone. In many images extraneous objects are shown and some images collected are not good enough to use for ML classification.

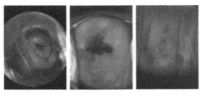

Figure: **Type 1 cervix**   Figure: **Type 2 cervix**   Figure: **Type 3 cervix**

Fig. 5. Types of cervix

Table 1. Data set

	Type 1	Type 2	Type 3
Total images	1441	2340	2337

Preprocessing of Images. For proper working of the classifier, the images used for training are processed. The region of interest (ROI) is found using a convolutional neural network, trained using manually annotated images.

Image segmentation to find the region of interest can be done using U-Net [11], a convolutional neural network designed specifically for biomedical image segmentation. This network is built based on the fully convolutional network and its architecture was reworked to do well with less training images and to get more precise segmentations. U-Net uses a 5 layered neural network involving a series of combinations of convolution and max pooling operations and combining the outputs of each level to learn and find the area of interest.

Classification of Unknown Images. This was achieved with a Neural Network with Deep Learning method. The design of these networks are explained below.

3.1   Design of Segmentation Section Using U-Net

Segmentation using UNet: UNet consists of a fully convolutional network. It's architecture constitutes of three parts, the contraction, the bottleneck and the expansion.

Contraction. This section consists of a series of convolution and max pooling operations, as shown in Fig. 6. This operation is done in different blocks. The input is given to the first block which runs two 3 × 3 convolutions (depicted by

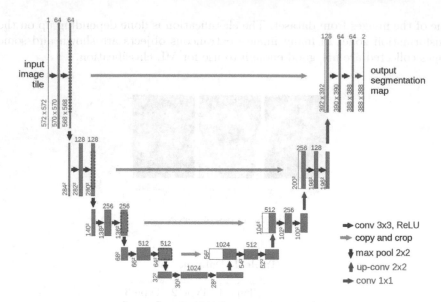

Fig. 6. U-Net architecture

the blue arrows) on it plus a ReLU operation as the activation function. Then it is followed by a 2×2 max pooling operation (depicted by the red arrows). The number of feature maps collected doubles after every block, so it learns complex structures effectively.

Expansion. Resembling the contraction path, expansion path also consists of expansion blocks. Each expansion block consists of two 3×3 convolution layers followed by 2×2 up sampling layer(depicted by the green arrows). The input is also appended by the feature maps from the corresponding layer in the contraction path. Also this time, the number of feature maps extracted halves down after each block. After that it is passed through a final 3×3 convolutional layer with the number of feature maps equal to the desired output(number of segments needed). Downward sequence of convolutions and max pooling operations results in a spacial contraction which increases the area of the expected ROI,

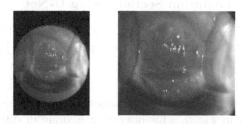

Fig. 7. Images before and after segmentation

and at the same time decreases area of the other portion. Figure 7 shows the images before and after segmentation using U-Net.

Training of U-Net. The dataset collected from kaggle is used here. The expected output is finding the required region of interest (ROI). The ROIs of the training set which were manually annotated and saved in json file was used to train the neural network. This file contains annotated image details of 250 Type I, 781 Type II and 450 Type III data sets.

3.2   Design of Classifier Using Deep Learning

A sequential model with 32 layers was created. It consists of Input Layer, Batch normalization layers, Convolutional layer, Pooling, Dropout. Batch Normalization layer will provide any layer in the neural network with input that have zero mean and unit variance. There are batch normalisation layer between other layers.

The activation function used is ReLU. There are dropout layers to reduce the over fitting of data by turning. Dropout regularization method approximates training a big number of neural net with different architectures parallally. During this process, some layer outputs are randomly "omitted out". This leads to the effect of making the layer similar and it will be treated like a layer with a connectivity to the prior layer.

2D Max pooling is done in between layers to reduce the spatial size of the representation. This is done to reduce the computations and number of parameters involved. Loss function indicates how well the network is performing and is optimization algorithm makes use of it. We have chosen to use Adam Optimiser with Categorical loss function.

The data set is divided into training and test for performance calculation. 80:20 proportion is used for this division. Using the training set the model is trained. And the test set is used for assessing the performance of the classifier. Our model is trained for Twenty epochs with 32 as the batch size. The accuracy is found using the test data set.

Another model was made by fine tuning a pre trained model. Inception v4, a model trained on huge data sets is used for this.

4   Results

The neural network for segmenting the image was trained. The best DICE coefficient was **0.78**. The segmented test images was also found to be similar to the manual annotation.

TensorFlow[5], a symbolic math library was used for the data programming with **Keras**[4], which provides a high-level neural network API. Tensorflow and Keras had GPU support and increased the speed of training significantly.

The segmented images were then used to train a CNN. The model had a accuracy of **0.6328** for **unsegmented images**. When the **segmented** images was used, the model gave an accuraccy of **0.7037**. The details are given in Table 2. This table gives the training and validation matrices like loss and accuracy. Accuracy = number of correct prediction/total.

Table 2. Model results

	Loss	Accuracy	val. acc.	val. loss.
without Segmentation	0.8740	0.6328	0.6528	0.8319
after segmentation	0.8653	0.7037	0.6828	0.8653

Table 3. Transfer learning model results

	Loss	Accuracy	val. acc.	val. loss.
Inception v4	0.9564	0.6667	0.6628	0.9648

The low accuracy of the results can be attributed to imperfections in the segmentation method. Even though the segmentation function gave a DICE coefficient of 78%, the classifier could not have an improved accuracy.

Transfer learning is a method in which we use already trained model as a base for our model. Transfer learning was attempted with fine tuning **Inception V4 model** (Table 3). This model gave low accuracy. Direct feature extraction cannot yield good results because the model was trained with images very different from cervix images. This Experiments need to be perfected to get good result and will be the future work.

5   Conclusion and Future Work

Classification of Cervix into the three types is done using a Convolutional Neural network. The segmentation of input images to find the Region Of Interest was done by training a neural network using manually annotated images. The images was segmented and the model gave a DICE coefficient of 78% for segmentation. Classifier was then trained using the segmented images. CNN model gave an accuracy of 70.37% and the transfer learning model gave an accuracy of 66.7%. The lower accuracy of transfer learning can be because of the difference in the data that was used to pre train the model and the data set that we have. The data sets used for the training of these models contains large amount of data, but our cervical type detection uses very different images which causes a slack in the accuracy. This paper is concentrated on the segmentation and its effects on the model and the details of the model design is presented as separate paper.

CNN model accuracy can be improved by using a better pre processing method. Classification of cervix type is done primarily by seeing the difference in the transformation zone in the image. Unfortunately the changes of this feature is very minute and not clearly distinguishable from images to images. The accuracy of the prediction can also be increased by specifically designing the CNN to take texture and color features by introducing color channel pre processing methods before the classification.

Acknowledgment. We express our sincere gratitude to Computer Science & Engineering Department of NIT, Calicut for providing us the facility for our experiments. We would like to thank Dr. Santhosh Kuriakose, Department of Cervical Oncology, Calicut Medical College for providing the necessary supports. We would also like to thank Mr. Vaibhav Shilimkar for setting up the initial environment for running cervical projects.

Compliance with Ethical Standards. All authors declare that there is no conflict of interest No humans/animals involved in this research work. We have used our own data.

References

1. Tulpule, B., Yang, S., Srinivasan, Y., Mitra, S., Nutter, B.: Segmentation and classification of cervix lesions by pattern and texture analysis. IEEE (2005). 0-7803-9158-6(05)
2. Chaitanya Asawa, Y.H., Sy, S.: Deep learning approaches for determining optimal cervical cancer treatment. In: International Conference on, pp. 390–393. IEEE (2017)
3. Greenspan, H., Gordon, S., Zimmerman, G., Lotenberg, S., Jeronimo, J., Antani, S., Long, R.: Automatic detection of anatomical landmarks in uterine cervix images. IEEE Trans. Med. Imaging **28**, 454–468 (2009). https://doi.org/10.1109/TMI.2008.2007823
4. Keras documentation (2018). https://keras.io
5. Tensorflow documentation (2018). https://www.tensorflow.org/
6. Hyeon, J., Choi, H.J., Lee, B.D., Lee, K.N.: Diagnosing cervical cell images using pre-trained convolutional neural network as feature extractor. In: 2017 IEEE International Conference on Big Data and Smart Computing (BigComp), pp. 390–393 (2017)
7. Lei Lei, R.X., Zhong, H.: Identifying cervix types using deep convolutional networks. Stanford university report
8. Papanicolaou, G.N.: A new procedure for staining vaginal smears. Science **95**(2469), 438–439 (1942). https://doi.org/10.1126/science.95.2469.438, http://science.sciencemag.org/content/95/2469/438
9. Kumar, R.R., Kumar, V.A., Kumar, P.S., Sudhamony, S., Ravindrakumar, R.: Detection and removal of artifacts in cervical cytology images using support vector machine. IEEE (2011). 978-1- 61284-704-7
10. Chankong, T., Theera-Umpon, N., Auephanwiriyakul, S.: Automatic cervical cell segmentation and classification in pap smears. Comput. Methods Programs Biomed. **113**, 539–556 (2014)
11. Falk, T., Mai, D., Bensch, R., Çiçek, Ö.: U-net - deep learning for cell counting, detection, and morphometry. Nat. Meth. **16**, 67–70 (2018). https://doi.org/10.1038/s41592-018-0261-2
12. WHO: Human papillomavirus (hpv) and cervical cancer. https://www.who.int/news-room/fact-sheets/detail/human-papillomavirus-(hpv)-and-cervical-cancer

Design and Implementation of Biomedical Device for Monitoring Fetal ECG

Rama Subramanian[1][(⊠)], C. Karthikeyan[3], G. Siva Nageswara Rao[3], and Ramasamy Mariappan[2]

[1] Madras Institute of Technology, Chennai 600 045, India
rama.mit2016@gmail.com
[2] Saveetha Engineering College, Chennai, India
prof.mariapan.r@gmail.com
[3] Koneru Lakshmaiah Education Foundation (K.L University),
Vijayawada, India
ckarthik2k@gmail.com, sivanags@gmail.com

Abstract. Heart disease is one of the critical diseases and World Health Organization (WHO) shows that large parentage of people is dying due to heart disease. But, the timely and accurate monitoring of heart disease still remains as a challenging problem. This paper proposes a portable biomedical device for monitoring fetal ECG signal from the abdominal signal and analysis of the fetal heart rate variability (FHR) to ensure the fetal wellbeing. The portable device for monitoring FECG was implemented using Arduino and AD8232 ECG sensor module. In this work, Fetal ECG extraction is done with a single composite maternal ECG signal. The ECG signal is processed using Wavelet transform for fetal ECG signal extraction. The experimentally observed results show the better extraction of FECG and that could lead towards the appropriate analysis of heart disease.

Keywords: Fetal ECG · Discrete wavelet packet transform · Heart rate variabaility

1 Introduction

Heart being a vital part of our body is often affected by various diseases leading to cardio vascular problems. Cardiovasular diseases are causing deaths across the world especially among the elderly.

The electrocardiogram (ECG) is an essential tool to analyse most of the heart diseases. The normal sinus rythum of human heart comprises the PQRST pattern [1] pulses as shown in Fig. 1. The fetal FECG signal provides the information about electrical activity of heart of the child and hence it identifies any abnormal heart diseases. The fetus ECG can be captured using the ECG electrodes placed on the maternal abdomen and hence this electrical signal of the fetal heart is very small inmagnitude. The recorded abdominal ECG signal consists of maternal ECG and fetal ECG. The fetal ECG can be extracted from this composite signal by applying the appropriate signal processing algorithm [2–4].

© Springer Nature Switzerland AG 2020
S. Smys et al. (Eds.): ICCVBIC 2019, AISC 1108, pp. 727–734, 2020.
https://doi.org/10.1007/978-3-030-37218-7_82

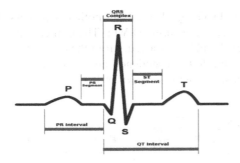

Fig. 1. Typical ECG Waveform of normal heart

Normally, the FHR esteems are going from 60 bpm to 240 bpm, while commanding physiological qualities are extending from 120 bpm to 160 bpm [5, 6]. The fetal pulse can be characterized as the time interim between continuous heart thumps in fetalECG [7, 8]. The current fetal ECG observing frameworks investigate the FHR flag that is conveyed by the bedside monitors equipped with Doppler ultrasound-based innovation [9]. This approach has poor accuracy of instantaneous FHR values and heart rate variability indices. Therefore, it is necessary to devise a new method for accurate monitoring of fetal ECG using a simple device.

The aim of this paper is to extract FECG from the abdominal composite ECG signal and to monitor the FHR variability using a portable device, which can able to sense ECG signal and to extract the features of FECG. The outline of the paper is as follows. Section 2 reviews the literature in the field of ECG and heart rate monitoring. Section 3 outlines the proposed system followed by implementation methodology in Sect. 4. The results and discussion is given in Sect. 5 and concludes in Sect. 6.

2   Related Work

Various strategies and diverse methodologies have been planned and utilized for FECG-Flag recognition: Wavelet based ECG extraction techniques [1–6], LMS versatile commotion wiping out sifting [11]. Blind Source Separation (BSS) procedure for extraction of FECG which is run of the mill daze source partition issue determined in [12, 13] which proposed a calculation utilizing Wavelet Transform to extricate fetal ECG from stomach composite signs got from mother's stomach surface and figure fetal pulse. As the amplitude of fetal ECG is always fluctuated, the existing algorithms are not able to find out all R-peaks accurately [14].

3   Proposed System

In this paper, we have proposed a portable device for sensing ECG signal and developed an algorithm to extract fetal ECG from the composite abdominal ECG signal. The proposed method for the extraction of fetal ECG is working as follows.

The captured composite ECG signal from the mother is processed and extracted features using discrete wavelet transform (DWT). Then, after applying n-point adaptive filtering, the FECG signal will be extracted. From the extracted FECG signal, R-peaks will be detected using threshold dependency and fetal heart rate was analysed. The block diagram of proposed system is shown in Fig. 2.

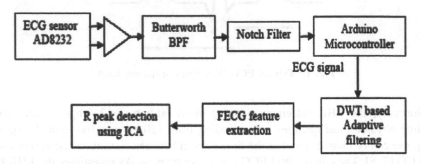

Fig. 2. System diagram for the Fetal ECG monitoring using Arduino

As shown in Fig. 2, the proposed system uses an ECG sensor AD 8232, which has three sensing electrodes respectively left arm (LA), right arm (RA) and right leg (RL). The sensed ECG signal is pre-amplified and then filtered using Butterworth band pass filter (0.5–100 Hz) and the power line interference is filtered using a notch filter with cutoff frequency 50 Hz. The filter ECG signal is fed into the Arduino Microcontroller for further processing [15]. The input ECG signal consists of mother ECG (mECG) as well as fetal ECG (fECG). Since the amplitude of mECG is predominant than fECG, it is essential to filter out the noise present in the ECG so that to distinguish fECG signal.

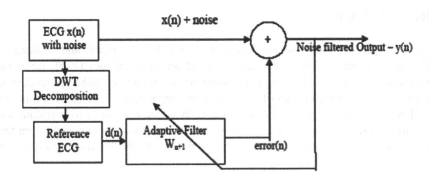

Fig. 3. Adaptive Filtering for ECG Noise cancellation

The noise filtering of ECG signal can be done by using adaptive filter with least mean square (LMS) algorithm as shown in Fig. 3. The adaptive filter is designed with discrete wavelet packet transform (DWPT) with suitable wavelet coefficients. The signal decomposition and filtering is done by DWT followed by the adaptive error correction using normalized least mean square (NLMS) method. The NLMS optimization algorithm is described below and the flow of the algorithm is shown below.

Algorithm : NLMS Optimisation
1. *start*
2. *initialise filter cosfficients w(n) =0*
3. *inputs: d(n) and u(n)*
4. *filter and buffer to gain y(n) = w(n)*u(n)*
5. *determine errors e(n)= d(n)-y(n)*
6. *computer normalised mean sq. error NLMS = μ*e(n)/|u(n)|²*
7. *if (NLMS) ==0 end*
8. *else update filter coefficients until NLMS = 0*

The ECG signal input is x(n), desired reference signal is d(n) and the error signal is e(n). The adaptive filter coefficients w(n+1) is computed using the NLMS optimization algorithm with the help of Lagrange multiplier method as follows.

$$w(n + 1) = w(n) + \frac{\mu}{\|x(n)\|^2} x(n) e(n)$$

where $e(n) = d(n) - x^T(n)w(n)$ and μ is the step-size that decides the percentage of convergence and the excess mean square error.

4 System Design

The implementation methodology used in this project is as follows. The sensing of ECG signal is done using AD8232 ECG sensor, which is interfaced with Arduino Microcontroller as shown in Fig. 4. The sensed composite ECG signal consists of maternal R wave in the range of 20 Hz to 40 Hz and fetal R wave in the range of 15 Hz to 30 Hz. The input abdominal signal is filtered by second order Butterworth filter in the frequency band of 0.5–100 Hz, followed by a notch filter with the cut-off frequency of 60 Hz to remove the power line frequency interference. Then the signal baseline is corrected by subtracting the mean value of the abdominal signal from the abdominal signal. The signal sampling rate is 1000 samples per/sec. The noise removal in ECG signal is done by using adaptive filter with wavelet decomposition method. Then, the wavelet packet transform decomposition is used to extract the fetal ECG signal. Finally, the R peak detection and fetal heart rate is analysed with thresholding method.

i.   Sensing of composite ECG signal – using AD8332 with Arduino
ii. The original signals are decomposed as the FECG and the MECG with 2 level DPT.
iii. R-peaks are been detected from the mined FECG signals.

The methodology applied to extract FECG from the abdominal signal is shown in Fig. 5. After preprocessing the composite ECG signal, it is applied with wavelet transform based adaptive filtering to remove unwanted noise frequencies. Then Discrete wavelet transform decomposition [16] method is used to extract the fetal ECG.

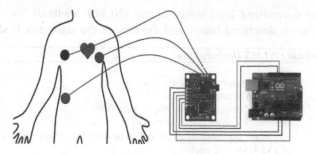

Fig. 4. Interfacing ECG sensor AD8232 with Arduiono

The wavelet packet transform decomposition of ECG signal is shown in Fig. 5. First, the input signal is decomposed into low pass frequency and high pass frequency, then each of the components further decomposed into LP and HP components and so on. This process is continued till 4 or 5 levels of decomposition so that accurate distinction between mECG and fECG signals.

  i. ***Get composite ECG signal***
 ii. ***Pre-processing & Adaptive Noise filtering***
iii. ***Wavelet transform Decomposition***
 iv. ***Fetal ECG extracted***
  v. ***Wavelet packet transform decomposition for QRS peak detection***

As shown in Fig. 5, the ECG signal sample is decomposed into low and high pass components and then downsampled by 2. In the second stage, again LP sample is decomposed into LP and HP; HP sample is decomposed into LP and HP. All the decomposed signal samples are down sampled by 2. The decomposition at every stage uses discre wavelet packet transform (DWPT) for getting accurate frequency and time resolution of the ECG signal. An R-R interval of a typical ECG signal is shown in Fig. 6 Now, from the extracted fetal ECG wave, the successive R-peaks will be detected using wavelet packet transform decomposition [17].

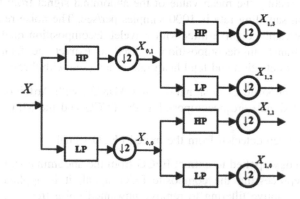

Fig. 5. ECG signal Decomposition using DWPT

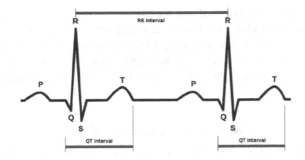

Fig. 6. R-R interval of ECG signal

It is observed from the decomposition that 4th level approximation and 4th & 5th level detail coefficients are related to QRS frequency of maternal signals [10, 18]. They are remade and included before that residual estimate and nitty gritty coefficients in alternate dimensions are focused for better recognition of maternal QRS alone. At that point, the data on maternal QRS edifices is utilized to smothen the maternal ECG in the stomach flag which makes conceivable discovery of the fetal QRS. The maternal QRS signal is eliminated by using the steps given below.

Step 1. The maternal R peak from the reconstructed signal is detected by threshold method.

Step 2. From the R peak the samples are zeroed in both sides till the baseline is reached. This is done by the following equation.

Whenever the R peak is coincident with the signal the samples are zeroed in both forward direction and reverse direction by following method.

$$\sum_{n=i-100}^{i} x(n) = 0 \quad \&\&$$

$$\sum_{n=i}^{i+100} x(n) = 0$$

Where i is the current R peak. The detected peaks are shown in Fig. 4a. Figure 4b shows the signal after maternal R wave is removed. Now the signal contains fetal component alone. The fetal signal contains the frequency of 15–30 Hz.

5   Results and Discussion

Fetal heart fluctuation is the lists, which demonstrates the primary dynamic state of the embryo. The protected scope of FHR standard (110–160BPM). tachycardia is FHR> 160 bpm for 10 min bradycardia is FHR< 110 for 10 min. These changes would be the record of fetal trouble.

Heartbeat rate per minute = 1/[pulse-period/(3 × 512 × 60)] = 92160/puls- period.

From the extracted FECG signal, the fetal R peak is detected again using threshold method. The FHR is calculated from these detected fetal R peaks. This FHR observed is shown in Table 1, which shows that the BPM values are oscillating for about 10 s, after which it becomes stable. This is due to number of levels of wavelet decompostion as time increases. Hence, this new method extracts the fetal ECG and it becomes stable within 10 s time.

Table 1. Heart rate variability of fetus

Time in Seconds	10	12	14	16	88	110	112	114	116
BPM		164	138	124	122	123	118	115	116	115

The above mentioned variation of fetal heart rate is of two types namely momentary fluctuation and long haul variation. Transient variation is the swaying of the FHR around the pattern in adequacy of 5 to 10 bpm. Long haul fluctuation is a to some degree slower wavering in pulse and has a recurrence of 3 to 10 cycles for each moment and abundancy of 10 to 25 bpm. Loss of beat-to-beat variation is more critical than loss of long haul inconstancy and might be unpropitious.

6 Conclusion

This paper proposed a new method for extracting the foetal ECG from the single maternal lead ECG and to evaluate the fetal well being by analyzing the heart rate variability of the fetal signal. The portable device for monitoring FECG was implemented using Arduino and AD8232 ECG sensor module. In this work, Fetal ECG signal was extracted from the composite maternal ECG signal using wavetlet decomposition method. The extracted fetal ECG signal is processed using Wavelet packet transform for ECG R peak detection. The experimental observed results show the better extraction of fetal ECG, accurate detection of R peaks and accurate fetal heart rate, which cn be further used to analyse the fECG for investigation of child health or any disease if any.

Compliance with Ethical Standards
   ✓ All authors declare that there is no conflict of interest.
   ✓ No humans/animals involved in this research work.
   ✓ We have used our own data.

References

1. Jezewski, J., Kupka, T., Horoba, K., Czabanski, R., Wrobel, J.: A problem of maternal and fetal QRS complexes overlapping in Fetal heart rate estimation. In: Fujita, H., Sasaki, J. (eds.) Selected Topics in Applied Computer Science. Applied Computer Science Series, pp. 122–127. WSEAS Press, Athens (2010)

2. Jezewski, J., Horoba, K., Wróbel, J., Matonia, A., Kupka, T.: Adapting new bedside instrumentation for computer-aided fetal monitoring using efficient tools for system reconfiguration. In: Mastorakis, N., Mladenov, V. (eds.) Recent Advances in Computers, Computing and Communications, pp. 80–85. WSEAS Press, Greace (2002)
3. De Haan, J., van Bemmel, J.H., Versteeg, B., Veth, A.F.L., Stolte, L.A.M., Janssens, J., Eskes, T.K.A.B.: Quantitative evaluation of fetal heart rate patterns I. processing methods. Eur. J. Obstet. Gynecol. **3**, 95–102 (1971)
4. Hon, E.H.: Instrumentation of fetal heart rate and fetal electrocardiography II a vaginal electrode. Am. J. Obstet. Gynecol. **86**, 772–778 (1963)
5. Yakut, O., Solak, S., Bolat, E.D.: Implementation of a web-based wireless ECG measuring and recording system. In: 17th International Conference on Medical Physics and Medical Sciences, Istanbul, vol. 9, no. 10, pp. 815–818 (2015)
6. Lin, B.-S., Wong, A.M., Tseng, K.C.: Community based ECG monitoring system for patients with cardiovascular diseases. J. Med. Syst. **40**(4), 1–12 (2016)
7. Spanò, E., Di Pascoli, S., Iannaccone, G.: Lowpower wearable ECG monitoring system for multiple-patient remote monitoring. IEEE Sens. J. **16**(13), 5452–5462 (2016)
8. Kupka, T., Jezewski, J., Matonia, A., Wrobel, J., Horoba, K.: Coincidence of maternal and fetal QRS complexes in view of fetal heart rate determination. J. Med. Inform. Techonol. **4**, 49–55 (2002)
9. Ibrahimy, M.I., Reaz, M.B.I., Mohd Ali, M.A., Khoon, T.II., Ismail, A.F.: Development of an efficient algorithm for fetal heart rate detection: a hardware approach. In: Proceedings of the 5th WSEAS International Conference on Instrumentation, Measurement, Circuits and Systems, China, pp. 12–17, April 16–18 (2006)
10. Ibrahimy, M.I., Reaz, M.B.I., Ali, M.A.M., Khoon, T.H., Ismail, A.F.: Hardware realization of an efficient fetal QRS complex detection algorithm. WSEAS Trans. Circ. Syst. **5**(4), 575–581 (2006)
11. Jeżewski, M., Czabański, R., Roj, D., Kupka, T.: Influence of input data modification of neural networks applied to the fetal outcome classification. In: Mastorakis, N., Mladenow, V. (eds.) Latest Trends on Computers. Recent Advantage in Computer Engineering Series, pp. 202–207. WSEAS Press, Athens (2010)
12. Yeh, S.Y., Forsythe, A., Hon, E.H.: Quantification of fetal heart rate beat-to-beat interval differences. J. Obstet. Gynecol. **41**, 355–363 (1973)
13. Yuan, L., Zhou, Z., Yuan, Y., Wu, S.: An improved FastICA method for fetal ECG extraction. Comput. Math. Meth. Med. **2018**, 7061456 (2018). https://doi.org/10.1155/2018/7061456
14. Haq, T.M.: Extraction of fetal heart rate from maternal ECG—non invasive approach for continuous monitoring during labor. MDPI Proc. **2**(13), 1009 (2018). https://doi.org/10.3390/proceedings2131009
15. Mariappan, R., Ramasubramanian, M.: Experimental investigation of impact of meditation yoga on mental and physical health parameters using EEG, Springer Briefers on Applied Sciences & Tech, January 2018 (2018)
16. Gupta, N.: Evaluation of noise cancellation using LMS And NLMS algorithm. Int. J. Sci Tech. Res. **5**(04), 69–72 (2016)
17. Mariappan, R., Ramasubramanian, M.: Design and Implementation of Cognitive Radio Sensor Network for Emergency Communication Using Discrete Wavelet Packet Transform Technique, Springer – LNCS – ISSN: 0302-9743, vol. 11319, January 2019 (2019)
18. Karthikeyan, C., Ramadoss, B.: Non linear fusion technique based on dual tree complex wavelet transform. Int. J. Appl. Eng. Res. **9**(22), 13375–13385 (2014)

A Comparative Review of Prediction Methods for Pima Indians Diabetes Dataset

P. V. Sankar Ganesh[1](✉) and P. Sripriya[2]

[1] Vistas Pallavaram, Chennai 600117, India
Sankarphd2017@gmail.com
[2] Department of Computer Applications, Vistas Pallavaram,
Chennai 600117, India
2Sripriya.Phd@gmail.com

Abstract. Diabetes Mellitus is generally considered to be a health issue that affects people and results in different types of problems like heart problems, vision problems, leg amputation and failure of kidneys if the disease diagnosis is not carried out in the perfect time. As Type II Diabetes Mellitus is a kind of abnormal disorder epidemic. The method of mining the medical information is concentrated in the recent days of research in the extraction of the significant data. This data is useful for the experts in the medical field, the enhancement of treatment and diagnosis of diabetic disorders. In this study, the review of different classification techniques and its details are implied on the dataset of benchmark Pima Indian diabetes. Neural Network (NN), Support Vector Machine (SVM), Machine learning algorithms is the different techniques of prediction that are used to predict Diabetics. This review also explains various advantages and disadvantages of the prediction methods and correlates the accuracy level of data classification. This also analyses the survey of different issues of recognizing the importance of relationship among the significant factors that lead to the growth of diabetics. The results prove that the data which is pre-processed provides a better accuracy level in classification. To validate the effectiveness of different models of prediction, we have precisely implied to Pima Indians diabetes datasets.

Keywords: Classification · Classification methods · Data mining · Diabetes mellitus · Pima indian diabetes

1 Introduction

The improvements Diabetes Mellitus (DM) is a severe defect which is featured by the increase in high blood glucose [1, 2]. Almost half of the diabetic patients are influenced by heredity factor, which is considered to be the significant feature of DM. Insufficient production of insulin due to the pancreatic failure are the pathologic causes of DM. The DM is of two types. β-cells are the pathogenesis of type 1 diabetes mellitus (T1DM), which are damaged by secretions of pancreas, that prevents from lowering blood glucose level. The Insulin resistance and insulin secretion deficiency, the pathogeneses of type 2 diabetes mellitus (T2DM) are that are considered as non-insulin

© Springer Nature Switzerland AG 2020
S. Smys et al. (Eds.): ICCVBIC 2019, AISC 1108, pp. 735–750, 2020.
https://doi.org/10.1007/978-3-030-37218-7_83

dependent DM [3]. To reduce the morbidity and minimize the influence of DM, it is significant to concentrate on a high-risk set of people affected by DM.

As there is an increasing cost in the health care industry, it is highly useful to help the victims to have control diabetics by themselves. In most of the instances, the initial information co-related with diabetics assists in avoidance of disease, curing and relevant treatment of the disease. Many computer systems were constructed by embedding the human intelligence that is helpful for the victims in the management of diabetes [4]. In the assessment of various systems like artificial intelligence, mobile phone applications and specifically designed devices for the diagnosis of diabetic disease.

The process of getting vital data from the huge set of data is called Data Mining [5, 6]. The mining of data is the technique [7, 8] has been implemented in different areas like medicine, marketing, banking, etc. In the medical industry, the technique of prediction is used to discover the epidemic at the initial phase and aids the medical experts in the preparation of treatment process.

The target is to build the prediction models by utilizing some machine learning algorithms. The machine learning algorithm is an artificial intelligence that makes the systems to analyse without embedded programming. Machine learning algorithm focuses on the construction of computer programs that allows modifying and generating when related to new information. These algorithms are generally classified as supervised or unsupervised. The technique called supervised learning helps to gain predictions on new information while unsupervised algorithms can get inferences from the set of data. The supervised learning technique is also considered as a classification. This review utilizes the classification technique to yield more approximate prediction model as it is implied as machine learning algorithm that validates the training data set and constructs an inferred function that can be utilized for the mapping operation. The target of the classification method is to predict the final class that is accurate for every set of data. The Classification Algorithms need the classes to be defined based on the values of data attributes.

Many studies on classification techniques are required to be acceptable in India to influence the people Pima Indians diabetes datasets and teaching them about the population regarding diabetes to avoid the further complex features. This is a detailed study that is related to diabetes mellitus. A thorough literature survey is performed using different titles and reports on peer reviewed search was conducted using various titles and reports on medical indexed peer reviewed journals. Literature reviews of research articles are done, that are published during 2005–2015, in English. The survey is done with the help of Pima Indians diabetes set of data.

2 Review of Machine Learning Algorithms

Karegowda et al. [9, 10] formulated the growth of a fusion model for the classification of the database namely Pima Indian diabetic database or repository (PIDD). The study contains 2 different phases. In the initial phase, the clustering technique called K-means, is utilized to recognize and remove the irrelevant examples of classification.

The discrete information is changed to groups by suspected width of the predetermined gaps depending on the suggestion of medical examiner. In the next phase, well classified technique is proposed with the help of Decision tree (DT) C4.5 by using the exactly examples of cluster from the previous stage. The cascaded K-means clustering and C4.5 has improved the level of accuracy in classification of C4.5 in the results of experiments. Also the regulations are extracted with the aid of cascaded C4.5 tree with the categorical data are minimal in numbers and seems to be easy for implementation when co-related to the extracted from C4.5 using the continuous information.

Barakat et al. [11] formulated a technique to use Support Vector Machines (SVMs) for the prediction of diabetic disease. Specifically, the utilization of extra explanation module changes the "black box" model belonging to that of SVM into a wise presentation of the SVM's prediction methods. The output of a real-life diabetic set of data prove that innovative SVMs render a challenging aid for the diagnosis of diabetic disease, where an exclusive set of rules have been extracted, with prediction level of accuracy is 94%, 93% sensitivity and 94% specificity. Moreover the generated rules are medically. Moreover, the rules extractedare medically effective and strong output of nearest medical studies.

Han et al. [11] formulated a RapidMiner technique tool to study about Pima Indians Diabetes Data Set, which gathers patient data with and without the development of diabetic disease. The concentration will be in the area of pre-processing, identification and selection of attributes, elimination of the irrelevant data, normalization of data, numerical discretization, analysis of visual data, discovery of hidden relationships and prediction of diabetic model.

Al Jarullah [12] proposed a method called a decision tree technique, utilized to diagnose diabetic developing patients. The set of data utilized is the database containing the data set such as Pima Indians Diabetes, which gathers data about the subjects who have or do not have diabetics. The survey is conducted in two areas. The initial stage is pre-processing along with selection of attributes, handling of values that were missing and discretization of numerals. The next stage contains a model for detection of diabetic disease with the aid of decision tree technique, which is a model to predict diabetes utilizing the method of decision tree. The software namely Weka is utilized for the entire stages of study.

Pradhan and Sahu [13] proposed Artificial Neural Network (ANN) based classification model, which acts as the efficient tool in the area of intelligence, used for the classification of diabetic victims into 2 broad groups. For targeting on good results, genetic algorithm (GA) is utilized for the feature selection process. It is shown from the results of simulation that the proposed model shows better performance better co-related to NN, KNN, MFS1, MFS2and FLANN model and behaves a excellent technique for several application like realtime domain, as these are effective with better outputs.

Iyer et al. [14] diabetes has infected more than 246 million people throughout the world, and the larger number is women patients. As per the WHO report, by the year 2025, this count is suspected to increase with the hike over 380 million. The disease is given this name after the fifth dreadful disease in US with the improper cure in sight. With the increase of IT sector, and medical and health care fields, the symptoms of the patients are filed. This study aims to identify solutions to predict the early diagnosis of the disease by validating the patterns that are present in the data through the classification analysis method with the implementation of Decision Tree and Naïve Bayes algorithms. The work tries to formulate fast and much more effective techniques for the prediction of the disease taking the patient to timely treatment.

Chikh et al. [15] formulated the expert systems usage and techniques of artificial intelligence techniques in the prediction of diabetics. Artificial Immune Recognition System (AIRS) is considered to be one of the popular techniques in the classification of medical illness. AIRS2 is an effective version of the AIRS algorithm. In this study, modified AIRS2 called MAIRS2 are replaced and the KNN algorithm with the fuzzy K-nearest neighbors to used to enhance the prediction level of diabetes diseases. The dataset of diabetes disease that is utilized in the study are taken from learning repository named UCI machine repository. The outputs of the AIRS2 and MAIRS2 are validated across the accuracy level of classification, values of sensitivity and values of specificity. The highest level of classification accuracy enforcing AIRS2 and MAIRS2 using 10-fold cross-validation obtain 82.69% and 89.10%.

Kavakiotis et al. [16] proposed a study for the applications of machine learning, data mining methods and diabetes research tools with regard to (a) The process of Prediction and Diagnosis, (b) Difficulties in diabetic disease (c) The study of Genetic Background and Environment, and (d) Management of the Health Care with the initial level of groups seems to be popular. A large number of algorithms involving machine learning were utilized. Generally, 85% are described by the methods of supervised learning and 15% by unsupervised approaches along with the association rules. SVM proves to be most popular and commonly used algorithm. Regarding the data type, the set of clinical data are significantly used. The applications in the identified areas prove the significance of collecting wise data which is valuable resulting to innovative ideas intense understanding and analysis of DM.

Zheng et al. [17] formulated information on framework for selection of cases with and without T2DM from EHR through engineering features and the machine learning concepts. They validate and show difference on the effectiveness of widely-used models in machine learning framework, including KNN, Naïve Bayes, DT, Random Forest(RF), SVM and Logistic Regression(LR). The formulated framework was tested on the patient samples of 300(161 cases, 60 controls and 79subjects that is not confirmed), probably identified from 23,281 diabetes correlated with cohort derived from a regional distributed EHR database with a range between 2012 and 2014.

Komi et al. [18] proposed variety traditional methods, depending upon physical and chemical tests for the purpose of prediction of diagnosing diabetes. The techniques rely on the techniques of data mining that can be efficiently implied for high blood pressure risk prediction. This study reveals the early detection of diabetes through 5 various data mining methods which include: Gaussian Mixture Model (GMM), SVM, Logistic regression, Extreme Learning Machine (ELM), and ANN. The results of study show that the renders the higher level of accuracy when compared to other techniques.

Pradeep and Naveen [19] proposed a variety of classification techniques. The target was to identify a make a better diabetic disease prediction by considering the level of blood glucose level before 2 h. At current trend, a huge number of techniques and methodologies are adapted in the diabetic disease diagnosis (DD). The reviewed technique is followed for the process of classification and prediction, on the feature identification is J48 Decision Tree algorithm. Primal Diagnosis of DD renders an effective least cost method. J48 algorithm is a popular algorithm delivering its highest accuracy.

Guo et al. [20] designed Bayes Network to detect the subjects with Type-2 diabetes. The set of data for this used was the database called Pima Indians Diabetes, which gathers the data of subjects with/without developing Type-2 diabetes. The software named Weka was utilized for the entire study. The technique rendered accurate results proving the formulated Bayes network to predict Type-2 diabetes is more powerful.

Thirumal and Nagarajan [21] formulated an application of computational intelligence with the aid of the model called fuzzy hierarchy which has the potential to conduct the early prediction of DM. To gain the success of formulated technique, cooperation is done with Laboratory in Indonesia named Eastern Jakarta hospital as a provider of the information that is required at the time of research and knowledge acquiring process is carried out by scheduling discussions interviews with 2 clinical experts at the similar clinic. The methodology is formulated depending on how a clinical expert decision correlated to information that anyone someone has the strength towards Data Mining, the technique has been modified with the information have fetched from the laboratory officials. The validation is done by co-relating the information extracted from the process with the clinical expert's decision which is adapted with the information from lab, and the output obtained is 87.46% of the 311 patients' reliable information is merely equivalent to medical examiners report. As the final statement provided with the hospitals co-operated the results proved that the method formulated met the requirements of the performance in conducting initial prediction on Data Mining and can assist the people in the awareness of DM at initial stage.

Lukmanto and Irwansyah [22] formulated the technique among the middle age and obese children and adults. To minimize the patients with diabetes mellitus, the disease must be predicted at initial phase; therefore a fast and effective disease prediction technique has to be identified. The aim of this study is to implement different techniques of data mining to diagnose diabetes mellitus and generate the patterns that are hidden from the PIMA Indian diabetes repository made available at Machine Learning Repository from University of California at Irvine (UCI).

In the area of medical industry, a huge level of information is extracted but not effectively used. Several techniques have been proposed for better usage of data. It implements data analysis tools to identify the unknown pattern which is interesting from the huge sets of data. The set of data which is utilized is Data set namely Pima Indians Diabetes, which gathers the subject's data with/without diabetes. Chattar et al. [23] proposes a variety of classification algorithms like Naive Bayesian, Random Forest (RF), and CART to identify which works better in Pima dataset.

3 Review of Hybrid Learning Algorithms

The modern trend proposes an innovative technique called machine learning (ML) that contains 2 generic algorithms relating one another to evaluate the issues which are not designed. As many machine learning algorithms are proposed for a specific task, joining many ML algorithms can enhance the final outputs by helping each other, can take over the new tasks. This section concentrates on the study of hybrid machine learning algorithms and its classification.

Ilango and Ramaraj [24] formulated a study to enhance the accuracy level of diabetic prediction by identifying the features of Database called Pima Indians Diabetes. The examples which are perfectly classified correctly identify the pattern for the prediction the disease and are used for the next stage of classification process.

Patil, et al [25] formulated a model named Hybrid Prediction Model (HPM) which utilizes an algorithm called Simple K-means clustering algorithm targeted at the validation of selected class label of the provided information and implying the algorithm designed for classification for the resulting dataset. C4.5 algorithm is utilized to construct the model which acts as a final classifier with the help of k-fold cross-validation technique. The Pima Indians diabetes dataset was retrieved from the repository named the University of California at Irvine (UCI) repository. A large variety of various methods of classification have been implied by the variety of researchers so as to identify the technique which performs better on this data set. The level of accuracy lies under the range between 59.4–84.05%. The formulated HPM fetched a classification rate of 92.38%. To validate the output of the formulated method, the sensitivity and the specificity factor values are generally utilized in the classification of the medical industry.

Giveki et al. [26] formulated an innovative technique for the prediction of Diabetic disease prediction called Feature Weighted Support Vector Machines (FW-SVMs) and Modified Cuckoo Search (MCS). The formulated work undergoes three different stages. Initially, Principal Component Analysis is implied to choose a feasible subset of features among all other features. Next, the mutual data is deployed to generate FWSVM by weighing various features depending on their significance. At last, the selection of parameters makes a significant place in the accuracy level of classification of SVMs; MCS is implied to choose the optimal values of the parameter. The formulated MI-MCS-FWSVM technique extracts an accuracy rate of 93.58% on the accuracy on UCI repository. The results of experiments prove that the formulated technique prove that the existing methods not only by providing the accurate results but also increases the performance of the classification procedure.

Choubey et al. [27] formulated a technique which has an objective of providing a better classification of the diabetic disease. There are various methods that are existing have been implemented for the purpose of diabetics' classification. In medical industry, the classification techniques extract the patients' information and make the models for the construction of prediction process. In this study, initially NBs are utilized for the classification of every attributes and GAs are utilized for the identification of attributes and NBs are utilized for the identification of attributes for the purpose of classification.

The results of experimentation prove that work on PIDD render higher classification accuracy for diabetic prediction.

Haritha et al. [27] formulated a technique called firefly algorithm and cuckoo search oriented attribute selection algorithm with the target of providing higher accuracy rate and minimizing the training overhead for PIMA Indian diabetic database from UCI repository. The experimental model has been established with the dataset from the UCI dataset utilizing KNN classifier. The level of accuracy, precision rate and recall have been evaluated as a parameter and result co-related with the technique called Cuckoo search and Firefly algorithm optimized structure, the formulated structure achieves higher accuracy rate when compared to the traditional approach.

Deepika and Poonkuzhali [29] formulated an innovative technique targets at rendering an effective hybrid classification framework for the diagnosis and monitoring the Diabetes disease. The significant target of this study main aim of this research is to recognize and create models and helps the medical experts to handle the people in an effective way to acquire healthier life. Thus the conception technique namely hybrid classifier is formulated to detect the diabetic disease prediction with the help of Feature Relevance Analysis with greater level of accuracy.

The elaborated enhancement [30] is to construct a combined kernel function rather than the individual, adding self adapting weights, and evaluate the process which containing WLS-SVM with QPSO algorithm linear equations in the training model, which could develop the effectiveness of the prediction model. By the implication of this technique in type 2 diabetes, it proves the speediness model-building is fast and the accuracy of prediction is greater with the enhanced output WLS-SVM is greater to the enhanced BP algorithm, LM algorithm neural network and the single-kernel function SVM.

Barale and Shirke [31] formulated the models for cascading for the process of classification in the repository called PIMA Indian diabetes. The technique called k-nearest neighbour is utilized to extract the missing information and the information is further processed for the process of classification. This is performed in 2 different phases. Initially, k-means clustering algorithm is utilized for gathering the patterns that are hidden among the dataset followed by the classification process using the apt classifier. The algorithm k-means is embedded with artificial neural network classifier along with logistic regression classifier that renders higher classification accuracy.

Bashir et al. [32] formulated a multi-layer classifier ensemble framework which is done depending on the feasible fusion of classifiers which are heterogeneous. The designed model called "HMV" solves the disadvantages of efficiency issues by using the seven heterogeneous ensemble classifiers. The framework is validated on 2 distinct sets of heart disease data, 2 breast cancer sets of data, 2 diabetes sets of data, 2 liver disease sets of data, 1 Parkinson's disease sets of data and 1 hepatitis sets of data are extracted from public databases. The efficiency of the formulated technique is tested by the co-relation of the outputs with many popular classifiers also the ensemble methods. The results of experimentation show that the formulated framework handles the attributes of all types with high level of classification accuracy. A case study is also represented depending on a medical set of data on real time basis, so as to prove higher performance and efficiency of the formulated model.

Kalyankar et al. [33] formulated an analysis for the diagnosis of the diabetics as an integration of different techniques of data mining, ML algorithms and statistics that utilize existing and previous sets of data to acquire the insight and to detect future problems. This study, proposed the algorithm involving machine learning technique in Hadoop MapReduce environment are enforced in data set named Pima Indian diabetes to identify the information that are missing and to find out the patterns out of it. This study predicts the diabetic type, associated complications and thepatient's complication and the type of treatment that can be given to the victims.

Ateeq and Ganapathy [34] targets in constructing an innovative hybrid classification algorithm for diabetes classification. The Modified Particle Swarm Optimization algorithm is combined with Multi Layer Perceptron Network with back propagation learning named as MPSO-NN Algorithm which is utilized to classify the diabetic patients with positive as binary 0 and with negative as binary 1. To validate the efficiency of the algorithm, the metrics like Sensitivity, Specificity, False Positive Rate, Accuracy, Mean Squared Error, Regression, and Percentage of Error is used. The technique is proven to be the formulated novel hybrid model is serves improved in total performance in the data set classification.

Gill and Mittal [35] formulated an innovative Hybrid Prediction Model (HPM) for effective prediction of diabetes. The repository called Pima diabetic database is utilized as the source of data extracted from the California University, Irvine (UCI), and the ML database. At the initial phase of the formulated HPM the feature filtration used for the recognition process of MATLAB is utilized for recognizing the predictors in distinct, showing the probability of diabetes disease existence. At the next level, a 2 layered classification is implied on the information filtration process, by merging NN and SVM, to improve the total rate of the model like recognition rate. The formulated combination model acquired 96.09% of total accuracy level. The co-relative work is done and it is proved that the formulated model acquires the notable level of accuracy in the classification. The rate of accuracy is obtained by large number of researchers in the previous years, with the similar repository that range from 59.4% to 92%. Moreover for verification and validation of the outputs, Receiver Operating Characteristics (ROC), Mean Absolute Error (MAE), Recognition rate attributes metrics are utilized. This study can assist the medical experts for the proper diagnosis of the diabetic disease at the early stage.

Ganji and Abadeh [36] formulated study proposesan Ant Colony-based classification technique to fetch a fuzzy rule sets for prediction of diabetes called FCS-ANTMINER. Certain works have been reviewed and innovative methods have been described for an effective approach that yields noticeable results for the classification of diabetes disease prediction. FCS-ANTMINER has innovative attributes that make it distinct from the current techniques that have used the algorithm named Ant Colony Optimization (ACO) for tasks of classification. The extracted rate of accuracy level of classification is 84.24% which shows that FCS-ANTMINER brings out many popular and latest techniques for the prediction of diabetes disease. During the construction of the model named classification, the information utilized to construct could include polluted information. To enhance the level of classification accuracy, the ensemble methods are proposed. Also the fusion of multiple models results in bias and variance reductions. Individual base classifier can be complied with them. Ensemble methods are acquired by the researchers.

4 Review of Ensemble Learning Algorithms

Akyol and sen [37] formulated an innovative study which variants normal level or diabetes patients consists 2 significant measures. Initially, the selection of features or weighting methods is surveyed to identify many of the efficient attributes for the prediction of the disease. Next, the efficiencies of AdaBoost, Gradient Boosted Trees and Random Forest ensemble learning Algorithms are validated. As per the validation results, the accuracy level of the prediction of the fusion of Stability Selection method and AdaBoost learning algorithm proves good when compared to certain algorithms with the level of accuracy of classification is 73.88%.

Alehegn and Joshi [39] formulated an innovative team and diagnose symptoms in clinical information; various techniques of data mining are utilized by a variety of researchers at different times. A summation of 768 instances is gathered from Pima Indian Diabetes Data Set (PIDD). In this method, the popular prediction techniques apply KNN, Naïve Bayes, Random forest, and J48. With the help of these algorithms, they made a technique by merging separate methods into one so as to maximize the efficiency and accuracy level. Bashir et al. [40] formulated the usage of multiple ensemble classification methods for diabetes sets of data. The output of experimentation and validation prove that Bagging ensemble method provides better efficiency when co-related to single and other ensemble techniques.

Nai-Arun and Sittidech [41] designed the techniques of data mining to enhance and maximize reliability in the classification ofdiabetes. The group of actual information is gathered from Sawanpracharak Regional Hospital, Thailand, was first initially verified by utilizing gain-ratio feature recognition methods. Karegowda et al. [42] demonstrates the different applications of ensemble for the enhancement of classification accuracy. The case uses the repository called Pima Indian Diabetic Dataset (PIDD). The model used for computation contains 2 phases. Initially, k-means clustering is deployed to recognize and eliminate the references that are incorrectly classified. Next, the effect of normalizing the classification is carried out. To carry out this ensemble process, The outputs of computation with the formulated method proved an enhancement of 16.14% to 22.49% as the accuracy level of classification when co-related to the review of literature. This survey concludes the importance of tiling k-means clustering with ensemble techniques in the improved level of accuracy indiabetic groups of information. Ensemble learning aids in the enhancement of machine learning output by the merging of various methods. This approach provides better performance when compared to a single model, and it can be defined as a future work. Overall representation of the review work is shown in the Fig. 1.

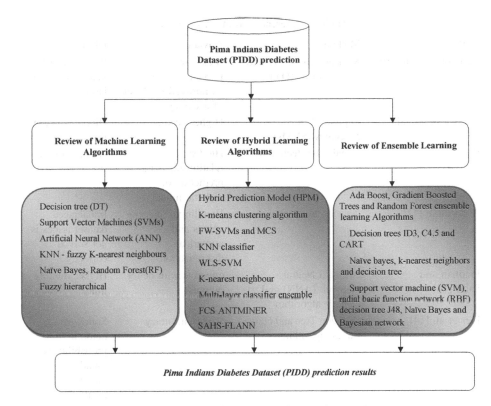

Fig. 1. Overall representation of the review work

5 Inferences from Review Work

The modern trend proposes an innovative technique called machine learning (ML) that contains 2 generic algorithms relating one another to evaluate the issues which are not designed. As many machine learning algorithms are proposed for a specific task, joining many ML algorithms can enhance the final outputs by helping each other, can take over the new tasks. This section concentrates on the study of hybrid machine learning algorithms and its classification. During the construction of the model of classification, the information utilized to build could include incorrect data. To enhance the level of classification accuracy, the ensemble methods are proposed. Also the fusion of multiple models results in bias and variance reductions. Individual base classifier can be complied with them. Ensemble methods are conducted by the researchers. The inferences from the review work are discussed in Table 1.

Table 1. Inferences from review work

Author	Method	Advantages	Disadvantages
Karegowda et al. [9]	K-means clustering and Decision tree (DT)	-Easy for implementation - Improved the level of accuracy	Selection of k values becomes issue
Barakat et al. [11]	Support Vector Machines (SVMs)	Higher accuracy	Higher running time
Han et al. [11]	RapidMiner technique	Hidden relationships and prediction of diabetic model	Assumption in rules
Al Jarullah [12]	decision tree technique	Handling of missing values and numerical discretization	Imbalanced dataset issues is not solved correctly
Pradhan and Sahu [13]	Artificial Neural Network (ANN) and Genetic algorithm	effective with better outputs	Higher running time
Iyer et al. [14]	Decision Tree and Naïve Bayes algorithms	Higher accuracy	Sometimes random probability is found so inaccurate results
Chikh et al. [15]	Artificial Immune Recognition System (AIRS)	The highest level of classification accuracy	Still more error results
Zheng et al. [16]	KNN, Naïve Bayes, DT, RF, SVM and LR	Easily applied for more samples Higher accuracy	More running time to complete task
Komi et al. [18]	data mining methods	Higher level of accuracy	Some methods need more running time
Pradeep and Naveen [19]	Decision tree algorithm	Delivering its highest accuracy	High dimensional of features
Guo et al. [20]	bayes network	Detect the patients with developing Type-2 diabetes	Probability values will not provide best results for all samples
Thirumal and Nagarajan [20]	fuzzy hierarchical model	Potential conduct of initial recognition of DM.	Still the accuracy is need to improve

(*continued*)

Table 1. (*continued*)

Author	Method	Advantages	Disadvantages
Ilango and Ramaraj [24]	Hybrid Prediction Model (HPM)	Prediction the disease and are used for the next stage of classification process	Some of the features are incorrectly removed from the dataset
Patil et al. [24]	Hybrid Prediction Model (HPM)	Classification rate of 92.38%	Still accuracy needs to improve
Giveki et al. [26]	FW-SVMs&MCS	Providing the accurate results	Still incomplete and missing value problems are not satisfied
Choubey et al. [27]	Naïve Bayes	PIDD render higher classification accuracy	Probability values will not provide best results for all samples
Haritha et al. [27]	KNN	level of accuracy, precision rate and recall is obtained	Selection of k nearest points becomes issue
Deepika and Poonkuzhali [28]	hybrid classification framework	Feature Relevance Analysis with greater level of accuracy	Filter feature selection will not give optimal results
Yue et al. [30]	WLS-SVM	Prediction is greater with the enhanced output WLS-SVM	
Bashir et al. [31]	multi-layer classifier ensemble framework	Higher performance and efficiency	All the features are used for classification
Ateeq and Ganapathy [34]	MPSO-NN	Hybrid algorithm is serves better in total performance	All the features are used and missing values are not detected
Gill and Mittal [35]	k-nearest neighbour and artificial neural network	Provides higher classification accuracy	Could include noise or incomplete dataset
Naiket al [36]	SAHS–FLANN	SAHS–FLANN is proved to be good statistically important	Higher error rate

(*continued*)

Table 1. (*continued*)

Author	Method	Advantages	Disadvantages
Ganji and Abadeh [36]	FCS-ANTMINER	Enhance the level of classification accuracy	Could include noise or imbalanced information
Minyechil Alehegn, and Rahul Joshi [38]	AdaBoost, Gradient Boosted Trees and Random Forest ensemble learning Algorithms	classification accuracy of 73.88%	Higher error rate
Bashir et al. [40]	multiple ensemble classification methods- ID3, C4.5 and CART	Shows better efficiency	Could include noise or imbalanced information
Nai-Arun and Sittidech [40]	naïve bayes, k-nearest neighbors and decision tree	decision tree algorithm (95.312%)	Higher running time
Karegowda et al. [42]	support vector machine (SVM), radial basis function network (RBF), decision tree J48, naïve Bayes and Bayesian network	Higher classification accuracy than the single models. Proved an enhancement of 16.14% to 22.49% as the classification accuracy. Ensemble learning assists in the enhancement helps improve machine learning outputs by merging variety of techniques.	

6 Conclusion and Future Work

This review work targets in the prediction of diabetic disease with Pima Indian Diabetes Data-set. It is a threat to women's health. The complication of acquiring diabetes in women is in a higher percentage due to various reasons. Therefore the idea is to diagnose the diabetic disease with the aid of certain algorithms like machine learning, hybrid methods involving in the machine learning, machine learning techniques and ensemble methods of learning. The merit of utilizing these techniques is that it supports in automation of process and makes the process like pattern recognition methods. The review work started with the introduction and significance of the worst effect of the Diabetes by demonstrating them in different methods related to it. A brief literature survey is carried out to continue a work in it. The method of Ensemble learning assists in the enhancement helps improve machine learning outputs by merging variety of models. This approach lets the production of better predictive efficiency when compared to a single model. This can be considered as scope of future work. From this review work, it is revealed that the ensemble technique provides better performance than normal hybrid classification techniques and the classifiers of the base. The observation from this study is highly useful in the guidance the selecting the classification algorithms for applications in future. Few other algorithms used for classification like stacking method might be taken as other purpose for future study.

Compliance with Ethical Standards

✓ All authors declare that there is no conflict of interest
✓ No humans/animals involved in this research work.
✓ We have used our own data.

References

1. Culver, A.L., Ockene, I.S., Balasubramanian, R., Olendzki, B.C., Sepavich, D.M., Wactawski-Wende, J., Manson, J.E., Qiao, Y., Liu, S., Merriam, P.A., Rahilly-Tierny, C.: Statin use and risk of diabetes mellitus in postmenopausal women in the women's health initiative. Arch. Intern. Med. **172**(2), 144–152 (2012)
2. Wilson, P.W., Meigs, J.B., Sullivan, L., Fox, C.S., Nathan, D.M., D'Agostino, R.B.: Prediction of incident diabetes mellitus in middle-aged adults: the Framingham Offspring Study. Arch. Intern. Med. **167**(10), 1068–1074 (2007)
3. Muller, L.M.A.J., Gorter, K.J., Hak, E., Goudzwaard, W.L., Schellevis, F.G., Hoepelman, A. I.M., Rutten, G.E.H.M.: Increased risk of common infections in patients with type 1 and type 2 diabetes mellitus. Clin. Infect. Dis. **41**(3), 281–288 (2005)
4. World Health Organization. http://www.who.int/topics/diabetes_mellitus/en/. Accessed 30 Sept 2012
5. Hand, D.J.: Principles of data mining. Drug Saf. **30**(7), 621–622 (2007)
6. Mining, W.I.D.: Data Mining: Concepts and Techniques. Morgan Kaufinann, Burlington (2006)
7. Han, J., Pei, J., Kamber, M.: Data Mining: Concepts and Techniques. Elsevier, Amsterdam (2011)
8. Witten, I.H., Frank, E., Hall, M.A., Pal, C.J.: Data Mining: Practical Machine Learning Tools and Techniques. Morgan Kaufmann, Burlington (2016)
9. Karegowda, A.G., Punya, V., Jayaram, M.A., Manjunath, A.S.: Rule based classification for diabetic patients using cascaded k-means and decision tree C4. 5. Int. J. Comput. Appl. **45** (12), 45–50 (2012)
10. Barakat, N.H., Bradley, A.P., Barakat, M.N.H.: Intelligible support vector machines for diagnosis of diabetes mellitus. IEEE Trans. Inf. Technol. Biomed. **14**, 4 (2010)
11. Han, J., Rodriguze, J.C., Beheshti, M.: Diabetes data analysis and prediction model discovery using rapid miner. In: Second International Conference on Future Generation Communication and Networking, pp. 96–99. IEEE (2008)
12. Al Jarullah, A.A.: Decision tree discovery for the diagnosis of type-2 Diabetes. In: IEEE International Conference on Innovations in Information Technology, pp. 303–307 (2011)
13. Pradhan, M., Sahu, R.K.: Predict the onset of diabetes disease using Artificial Neural Network (ANN). Int. J. Comput. Sci. Emerg. Technol. **2**(2), 303–311 (2011). (E-ISSN: 2044-6004)
14. Iyer, A., Jeyalatha, S., Sumbaly, R.: Diagnosis of diabetes using classification mining techniques. arXiv preprint arXiv:1502.03774 (2015)
15. Chikh, M.A., Saidi, M., Settouti, N.: Diagnosis of diabetes diseases using an artificial immune recognition system2 (AIRS2) with fuzzy k-nearest neighbor. J. Med. Syst. **36**(5), 2721–2729 (2012)

16. Kavakiotis, I., Tsave, O., Salifoglou, A., Maglaveras, N., Vlahavas, I., Chouvarda, I.: Machine learning and data mining methods in diabetes research. Comput. Struct. Biotechnol. J. **15**, 104–116 (2017)
17. Zheng, T., Xie, W., Xu, L., He, X., Zhang, Y., You, M., Yang, G., Chen, Y.: A machine learning-based framework to identify type 2 diabetes through electronic health records. Int. J. Med. Inform. **97**, 120–127 (2017)
18. Komi, M., Li, J., Zhai, Y., Zhang, X.: Application of data mining methods in diabetes prediction. In: 2nd International Conference on Image, Vision and Computing (ICIVC), pp. 1006–1010 (2017)
19. Pradeep, K.R., Naveen, N.C.: Predictive analysis of diabetes using J48 algorithm of classification techniques. In: 2nd International Conference on Contemporary Computing and Informatics (IC3I), pp. 347–352 (2016)
20. Guo, Y., Bai, G., Hu, Y.: Using bayes network for prediction of type-2 diabetes. In: International Conference for Internet Technology and Secured Transactions, pp. 471–472 (2012)
21. Thirumal, P.C., Nagarajan, N.: Utilization of data mining techniques for diagnosis of diabetes mellitus-a case study. ARPN J. Eng. Appl. Sci. **10**(1), 8–13 (2015)
22. Lukmanto, R.B., Irwansyah, E.: The early detection of diabetes mellitus (DM) using fuzzy hierarchical model. Procedia Comput. Sci. **59**, 312–319 (2015)
23. Chattar, S., Deshmukh, V., Khade, S., Abin, D.: Data mining techniques for prediction of type-2 diabetes. Int. J. Eng. Comput. Sci. **7**(01), 23517–23520 (2018)
24. Ilango, B.S., Ramaraj, N.: A hybrid prediction model with F-score feature selection for type II Diabetes databases. In: Proceedings of the 1st Amrita ACM-W Celebration on Women in Computing in India, p. 13 (2010)
25. Patil, B.M., Joshi, R.C., Toshniwal, D.: Hybrid prediction model for type-2 diabetic patients. Expert Syst. Appl. Sci. Direct **37**, 8102–8108 (2010)
26. Giveki, D., Salimi, H., Bahmanyar, G., Khademian, Y.: Automatic detection of diabetes diagnosis using feature weighted support vector machines based on mutual information and modified cuckoo search. arXiv preprint arXiv:1201.2173 (2012)
27. Choubey, D.K., Paul, S., Kumar, S., Kumar, S.: Classification of Pima Indian diabetes dataset using naive bayes with genetic algorithm as an attribute selection. In: Communication and Computing Systems: Proceedings of the International Conference on Communication and Computing System (ICCCS 2016), pp. 451–455 (2017)
28. Haritha, R., Babu, D.S., Sammulal, P.: A hybrid approach for prediction of type-1 and type-2 diabetes using firefly and cuckoo search algorithms. Int. J. Appl. Eng. Res. **13**(2), 896–907 (2018)
29. Deepika, N., Poonkuzhali, S.: Design of hybrid classifier for prediction of diabetes through feature relevance analysis. Int. J. Innov. Sci. Eng. Technol. **2**(10), 788–793 (2015)
30. Yue, C., Xin, L., Kewen, X., Chang, S.: An intelligent diagnosis to type 2 diabetes based on QPSO algorithm and WLS-SVM. In: International Symposium on Intelligent Information Technology Application Workshops, pp. 117–121 (2008)
31. Barale, M.S., Shirke, D.T.: Cascaded modeling for PIMA Indian diabetes data. Int. J. Comput. Appl. **139**(11), 1–4 (2016)
32. Bashir, S., Qamar, U., Khan, F.H., Naseem, L.: HMV: a medical decision support framework using multi-layer classifiers for disease prediction. J. Comput. Sci. **13**, 10–25 (2016)
33. Kalyankar, G.D., Poojara, S.R., Dharwadkar, N.V.: Predictive analysis of diabetic patient data using machine learning and Hadoop. In: International Conference on I-SMAC (IoT in Social, Mobile, Analytics and Cloud) (I-SMAC), pp. 619–624 (2017)

34. Ateeq, K., Ganapathy, G.: The novel hybrid Modified Particle Swarm Optimization-Neural Network (MPSO-NN) algorithm for classifying the diabetes. Int. J. Comput. Intell. Res. **13** (4), 595–614 (2017)
35. Gill, N.S., Mittal, P.: A computational hybrid model with two level classification using SVM and neural network for predicting the diabetes disease. J. Theor. Appl. Inf. Technol. **87**(1), 1–10 (2016)
36. Naik, B., Nayak, J., Behera, H.S., Abraham, A.: A self adaptive harmony search based functional link higher order ANN for non-linear data classification. Neurocomputing **179**, 69–87 (2016)
37. Ganji, M.F., Abadeh, M.S.: A fuzzy classification system based on Ant Colony Optimization for diabetes disease diagnosis. Expert Syst. Appl. **38**(12), 14650–14659 (2011)
38. Akyol, K., Şen, B.: Diabetes mellitus data classification by cascading of feature selection methods and ensemble learning algorithms. Int. J. Modern Educ. Comput. Sci. **6**, 10–16 (2018)
39. Alehegn, M., Joshi, R.: Analysis and prediction of diabetes diseases using machine learning algorithm: ensemble approach. Int. Res. J. Eng. Technol. (IRJET) **04**(10), 426–436 (2017)
40. Bashir, S., Qamar, U., Khan, F.H., Javed, M.Y.: An efficient rule-based classification of diabetes using ID3, C4. 5, & CART ensembles. In: 12th International Conference on Frontiers of Information Technology (FIT), pp. 226–231 (2014)
41. Nai-Arun, N., Sittidech, P.: Ensemble learning model for diabetes classification Advanced Materials Research, vol. 931, pp. 1427–1431. Trans Tech Publications, Switzerland (2014)
42. Karegowda, A.G., Jayaram, M.A., Manjunath, A.S.: Cascading k-means with ensemble learning: enhanced categorization of diabetic data. J. Intell. Syst. **21**(3), 237–253 (2012)

Instructor Performance Evaluation Through Machine Learning Algorithms

J. Sowmiya[(✉)] and K. Kalaiselvi

Vels Institute of Science, Technology and Advanced Studies (VISTAS),
Pallavaram, Chennai 600117, India
sowmiyalive@gmail.com, kalairaghu.scs@velsuniv.ac.in

Abstract. Development in the data mining approaches promotes the researches over the classification of features in the provided dataset. The applications like student performance evaluation plays important role in measuring efficiency of the instructors in the educational institution. Student evaluation feedback database, analyze the performance of the instructors based on the course Id, attendance, difficulty and repetition of selection course by the student. The student evaluation dataset collect from UCI machine learning repository. The accuracy of the feedback provided by the student's measure with deep learning algorithm. Several Instances record to improve the efficiency of the performance evaluation system. The implementation of neural network along with linear regression model, multiple regression model, feed forward network and Association rules apply for student evaluation database. The performance plots were used to compare the efficiency of the deep learning algorithm over the applied data set.

Keywords: Students feedback · Neural network · Linear regression · Multiple regression · Feed forward network · Association rules

1 Introduction

The higher education system aims to improve teaching quality by implementing student feedback procedure. The quality of the education system improve by utilising student's feedback about instructor's teaching ability to make student understand subject. The student feedback database also motivate and determine negative aspects in instructors teaching style. An overt and covert process perform to evaluate teacher performance. To improve the quality of teaching an appraisal and feedback system develop for faculties and parameters identify through which performance of different subject instructors evaluate. Feedback surveys conduct for students and data's save in repository. The collected data tabulate and apply for data set analysis. The huge data set formed through feedback system make individual assessment of student's feedback complex. Hence, the instructors performance evaluate with regression analysis tool.

Feedback of students involve rating of all modes of teaching provision provided by training institutions, which include teaching methods, library activities and information technologies. This process is represented as total student experience handling. In some other cases, the evaluation of students limit to feedback over the performance of the teachers based on the module or course level. Student evaluation of teachers refer to as

S. Smys et al. (Eds.): ICCVBIC 2019, AISC 1108, pp. 751–767, 2020.
https://doi.org/10.1007/978-3-030-37218-7_84

the student's feedback. The data from the feedback extract more carefully with data mining algorithm. The extracted data set provides information about student knowledge and progress in the course of study acquired through teaching sessions and help teachers to improve their level of teaching. Data mining techniques apply for data classification and grouping. Data classification algorithms analyse the relationship between teaching parameters and its attributes and identify the best-fit value to predict output from provided input. The classification algorithm segregates dataset and evaluates teachers performance. The accuracy of prediction is used as the measure to identify the efficiency of the algorithm. Several methods classifies student feedback dataset to generate high accuracy instances together with correct attribute space.

2 Literature Survey

Sequential patterns occur frequently from databases. Discovering these sequential patterns is a critical task. The d-dimensional sequential data yields these sequential patterns where conventional systems do not. Without detecting these sequential patterns, mining is not possible for several practical data. The sequential patterns extract by modified Apriori and Prefix Span algorithm [1].

Classification rules can be combined, if the sample data extracts these classification rules. The traditional approach combines the output of the classifier or combines the classification rule sets. The probabilistic generative classifier with multinomial distribution combines classifiers at the parameter level of classification rules. These multinomial distributions are normal-Wishart or Dirichlet distributions and these distributions refer to as second-order distributions or hyper-distributions [2].

Feature selection widely apply to reduce the processing load in data mining. Data mining on high dimensional data causes the search space to increase in size and needs more computation. The subset for the feature selection derive from this search space. This search space problem also relate with data feeds' streaming format and high dimensionality in Big data. To resolve this problem, a new lightweight feature selection apply. The design of this feature selection mainly apply for data streaming during fly through accelerated particle swarm optimization (APSO) that attains improved precision within the acceptable processing time [3].

The neural network integrate into physical model with Data assimilation algorithm. The data assimilation algorithm comprises of kalman filter. The approach increase forecasting result and assimilation process. Initially, the neural network train as each instance of observational data is applied [4].

A deep overview of different machine learning approaches such as Deep Learning is presented to support the learning and data analytics in the field of Internet of Things (IoT). The characteristics of IoT data are described. Two significant enhancements are identified for IoT data from machine learning techniques such as IoT streaming data analytics and IoT big data analytics. Various Deep Learning algorithms and architectures are presented [5].

A visual analytics technique is presented to enhance the interactivity and acknowledging ability of recurrent neural networks through visual analytics researchers, artificial intelligence scientists and medical experts. A new visual analytics tool known as RetainVis combines an interactive, enhanced recurrent neural network based model called RetainEX and users' visualization to explore EMR data during the prediction tasks [6].

Several challenges arise while performing data analytics in Big Data with machine learning. The cause-effect relation in machine learning with big data is described by categorizing the challenges related to Big Data's velocity, volume and variety. Also upcoming machine learning techniques and approaches are discussed based on the handling capability of Big Data with several challenges and aim to aid practitioners for their use cases to select relevant solutions. Then a matrix provides relation between the techniques and challenges [5].

A technique for open student models based on curriculum level is presented. The technique describes evaluation of core competencies of student through relation between core competencies and courses. Based on this technique, a visual analytics system called visual analytics of core competencies (VACC) based on curriculum-level and competency is implemented. The VACC, radar charts for performance of course work, diagnostic tools for core competency are developed that reflects the students and adjust their core competency levels [6].

Learning Management System (LMS) provides more data on student's online behavior. The data is used in several analysis to predict performance of student. This lead to several findings such as predictor variables obtained from LMS and difference in courses and becomes complex to conclude about techniques in predicting student performance [7].

A multidisciplinary research area such as the Learning Analytics acquire insights from learning the data sets accomplished through collaboration of technical experts with various disciplines. Learning Analytics collect data from the interactions between student, teacher and then translating, filtering these interactions to correct formats through different analysis techniques. The need for experiences and knowledge sharing is emphasized in the Knowledge Conference and Learning Analytics [8].

A technique called Panel for Adults Dyslexia learning Analytics (PADA) apply to visualize and inspect the difficulties in reading. PADA is a web application supporting illustrative visualizations so that the students can better understand about their structure of learning or learner models. PADA represents the knowledge in student's learner models to aid students to raise their awareness, reflect and regulate the complexities in reading. PADA provides various learning analytics on student's reading performance, so that the students can know their weaknesses and strengths in learning and correct their learning [9].

A refined Boolean Logic Analysis enhances the tractability, acknowledging capability and identifiability by recognizing a limited and ordered set of Boolean logic functions [10].

Massive Open Online Courses (MOOCs) have changed the way of learning. However, MOOCs have more dropout rates over students at different course stages. At certain stage of the course, students who are in risk and who are not satisfying the passing criteria may dropout. To resolve the issue, a model that identifies learners who have risk in studying an online course through association analysis was performed with the students' dataset who have pursued the course already. Finding these learners with risk and giving them real feedback aid the students to know about their current performance and help enhance their performance [10].

Different educational systems gather and utilize large data on faculty, staff and students. These datasets are employed to test the performance of student across their time of learning from one semester to another semester. The key demographics of student's characteristics, examination attempts in a semester and each semester's final grade provides the training dataset and creates a regression function to estimate the performance of students in upcoming semesters [11].

Virtual Learning Environments (VLE) provides additional information about student performance in terms of pedagogical challenges through graph, reports. However, statistical tools are needed to facilitate the decision of tutor, as the learning analysis is limited to frequency analysis. Hence, an analytical technique apply to process e-learning data. The aim was to find learner group through their answers and identify the answer profiles to aid a student in upcoming learning activities [12].

3   Methodology

The performance of teaching staffs in an educational institution measure by collecting feedback from student in terms of course code. Neural network based data mining algorithm use to analyse the data set generated from the student feedback system. Linear and multi regression models identify the best-fit value for the provided data set. The model evaluates teacher performance for a particular course code. The feedback evaluation comprises of parameters such as repetition of course by the student, attendance of the student and difficulty of the course. Questionnaires are framed based on the above attributes and rating for each questionnaires acquire from student evaluation feedback system. The Student evaluation data set was collected from the UCI Machine learning repository. The student evaluation data set consists of 5820 instances, which are grouped into 33 attributes. Where 28 attributes being Likert type in the form of questionnaires and remaining 5 attributes were instructor identifier, course code, course repeat, attendance of student and difficulty of the course. The information's about the attribute utilised in the student evaluation data set is listed below:

- Instr: It represents the instructor identity, which is denoted by values 1, 2 and 3.
- Class: It represents the course code, denoted by values between 1 to 13.
- Repeat: It represents total number of students repeatedly selected the same course. The value will be start from 0, 1, 2, …
- Attendance: Level of attendance generated by the students for particular course ID, denoted by values between 0 to 4.
- Difficulty: difficulty sensed by the students over the selected course.
- Q1: providing content for the course, teaching method and evaluation method at beginning of the course.
- Q2: Explaining about the objective and aim of the course at the initial stage of the periods.
- Q3: Worthiness of the course for assigned credit to it.

- Q4: Flow of the course is similar to the syllabus explained at initial stage of the period.
- Q5: Satisfactory level over assignment of homework, studies and applications.
- Q6: Status of updating the course resources and textbooks is satisfactory.
- Q7: Satisfactory level of fieldwork, discussions, laboratory, applications and other studies.
- Q8: Effectiveness of quizzes, projects, assignments and exams in improving learning.
- Q9: Interested in attending activities during the class hours.
- Q10: Fulfilment of expectation about the course at the end of the course.
- Q11: The relevancy of course and usefulness for the professional development.
- Q12: The course provided a new vision on life and the world
- Q13: Topic knowledge and relevance of instructor.
- Q14: Preparation for the class by instructor.
- Q15: Flow of the course is similar to the provided lesson plan
- Q16: Adaptation of instructor over the course duration.
- Q17: Punctuality of the trainer to the class hours.
- Q18: Simplicity in delivery and speech of the instructor.
- Q19: Effective handling of class hours by instructor.
- Q20: Willingness of instructor to explain and help the students related to course.
- Q21: Instructor provided positive approach over the course work.
- Q22: Instructors was overt over the views of student on the course work.
- Q23: Instructor promotes the interest to attend class sessions.
- Q24: Assigning of applicable project works, home works and provided guidelines to the students by the instructor.
- Q25: Response provided by the instructor related to course within or outside the course period.
- Q26: The evaluation of the instructor is based on course objective.
- Q27: Instructor shared and discussed about the answers to the exam.
- Q28: Handling of students by the instructor is even and right manner.

The above listed questionnaires are provided to the students and feedback was acquired and stored in a repository. The questionnaires from Q1 to Q28 is of Likert type for which the students assigned values between 1 to 5. The collected data set then process with linear regression and multiple regression along with the neural network techniques. Artificial Intelligence and Machine learning functionality utilises the neural network to perform decision making over the possible inputs from the data set. The inputs of the neural network were represented with weight and it is referred as linear combination. The weights represent the bias between the layers in neural network. The positive weight represents the excitant interface and negative weight represents the repressive interface. Activation function is used to control the values of generated output. The properties like adaptive learning, self-organisation, real time operation and fault tolerance through redundant information coding provides ability to extract patterns from the provided data set even it is too complex to frame. The training of neural network increases the expertization in categorising the information to be analysed. Neural network utilises techniques like linear regression, multiple regression and feed-forward network to train the input dataset.

3.1 Linear Regression

Linear regression is a fundamental and widely used method for data set analysis. The regression analysis is perform to identify whether the set of analysing variable assists to identify the outcome variable and identify the particular data which significantly predicts the outcome variable. These estimations in the regression is used to define the relationship between the dependent variable and independent variables. The simplest form of representing the dependent and independent variable is

$$Y = M + N * X. \tag{1}$$

where Y is the dependent variable, X is the independent variable, M is constant and N is the regression coefficient. The dependent variable is the outcome variable and independent variable is a predictor variable. Regression analysis perform to identify the strength of the predictor variable, forecasting the effect and trend.

There are different types of linear regression available they are

- Simple linear regression: Utilises 1 dependent and independent variable.
- Multiple Linear regression: Utilises 2 or more independent variables and 1 dependent variable (ratio/interval).
- Logistic regression: utilises 2 or more independent variables and 1 dependent variable (dichotomous).
- Ordinal regression: Utilises 1 or more independent variables and 1 dependent variable (ordinal).
- Multinominal regression: 1 or more independent variables and 1 dependent variable (nominal).

However, simple regression use in neural network for prediction.

3.2 Feed Forward Neural Network

Feed Forward Neural network allows the input independent variable to flow in uni direction format. No feedback is provided to the previous layer of the network. The output of current layer will not affect the input of the same executing layer. The feed forward neural network flows straight forward through the network related to the inputs with their outputs. Figure 1 shows the model of feed forward neural network. In feed forward neural network the input and output of the layer is mapped together however, the input of previous layer is not considered. Set of values are assigned to all input nodes of the neural network and the hidden nodes involved in neural network calculate with

$$H_n = l_1 W_1 + l_2 W_3 + B_n W_5 \tag{2}$$

Where 'I', represent the input node, 'H' represents the hidden node, 'W' represents the weight of the connection and 'B' represents the Bias node. Activation function is selected for each hidden layer.

$$S(x) = \frac{1}{1+e^{-x}} \tag{3}$$

The activation values for hidden node is given by

$$HA_1 = \frac{1}{1+e^{-H_1}}$$
$$HA_2 = \frac{1}{1+e^{-H_2}} \tag{4}$$

The values of the output node represent by

$$O_1 = HA_1 W_7 + HA_2 W_9 + B_2 W_{11}$$
$$O_2 = HA_1 W_8 + HA_2 W_{10} + B_2 W_{12} \tag{5}$$

Activation values for the output node is measured as

$$OA_1 = O_1$$
$$OA_2 = O_2 \tag{6}$$

The total error is measured using mean square error. Let's consider OAi is the output value of node 'I' then 'yi' is the desired result.

$$e = \frac{1}{n} \sum_{i-1}^{2} (y_i - OA_i)^2 \tag{7}$$

After the completion of first iteration, the error measured is significant. A back-propagation algorithm is used to adjust weights of the link to reduce error between generated output and desired output.

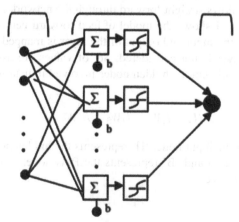

Fig. 1. Feed forward neural network

3.3   Association Rule

Association rule learning is defined as rule based machine-learning technique used to identify relationship between the parameters in a larger data set. It is utilised to discover strong rule in data set based on the scope of interestingness. New rule sets generate by rule based method by applying more parameters from the data sets. The main aim of applying more parameter to association rules is to assist the machine learning system to perform similar to feature extraction performed by human brain. In addition, it helps to improve abstract association abilities from uncategorised new data set. The definition of association rule mining is defined as follows:

Let 'I' represent the items $I = \{i1, i2, \ldots, in\}$ where 'n' be the binary attributes.

Let 'D' represents the database $D = \{t1, t2, \ldots, tm\}$ where 't' represents the set of transactions.

Unique transaction ID is specified for every transaction in database 'D' with the subset items 'I'. The rules is defined significant to the form represented as

$$X \rightarrow Y, \text{ where } X, Y \subseteq I. \tag{8}$$

Rules are framed based on two different sets namely item sets 'X' and 'Y'. Where 'X' represents the antecedent and 'Y' represents the consequent.

4   Result and Discussion

The performance of the instructors in the educational institutions are more often measured from the data set collected from the student evaluation system. The data set collected is then applied to machine learning algorithms to identify the efficiency of teacher's performance from the feedback provided by students during the course period. The machine learning algorithms like linear regression model, multiple regression model, feed forward network and association rules are applied with the neural network algorithm. The neural network classifies the provided attributes from the dataset and generates performance of each attribute as output of the system.

Linear regression with one independent variable 'X' and one dependent variable 'Y' is applied over the provided dataset. Figure 2 represents the regression output generated by applying the dataset to linear regression model. The regression value R shows 0.10527 similarity. The performance of teacher for a particular attribute based on student feedback for that particular attribute gives by

$$y = 0.12 * x + 1.3. \tag{9}$$

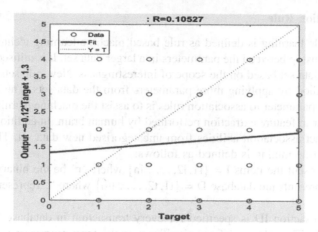

Fig. 2. Linear regression analysis

In addition, the dataset was subject to multiple regression analysis. The neural network trains the dataset in iteration process until the best performance is reached. The best performance for applied data set reach at 1000 epochs. The Fig. 3 shows the training state of the multi regression analysis and the generated gradient value of 0.046464 and no validation error occurred during the training state.

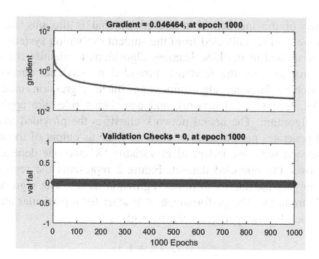

Fig. 3. Multiple regression training state

The regression values measured during training, validation, testing and overall system is shown in Fig. 4. In multiple regression, the regression value, R is achieved about 0.20802 for training set, 0.19868 for validation, 0.23243 for testing phase and 0.20826 for overall system. The regression values measured in training, validation and overall results is more or less equal which shows the stability of the network.

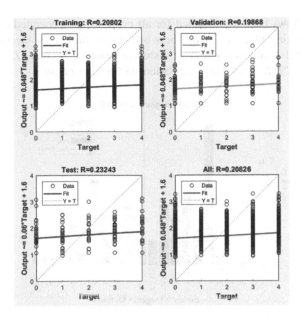

Fig. 4. Multiple regression plot.

Figure 5 shows the validation performance of the multiple regression system over the applied student evaluation dataset. The best validation point is measured about 2.1712 achieved at epoch 1000. The performance of training, testing and validation process reach closer to the best fit at 100th epoch. The performance is measured based on Mean Square Error.

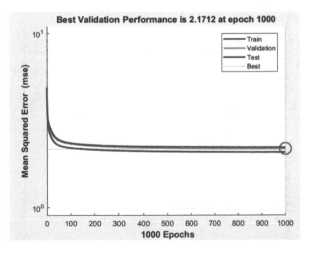

Fig. 5. Multiple regression validation performance

The Fig. 6 shows histogram plotted for all error values generated during training, validation and testing phase of neural network. Error values are measured for every instances in provided data set. The error histogram plot for 20 bins. Zero error state is achieved at 0.08229 at all training, validation and testing phase.

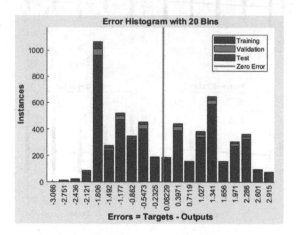

Fig. 6. Multiple regression error histogram

The dataset is then applied to the feed forward neural network model. Figure 7 shows the regression plots measured during training, validation and testing phases of neural network. The regression value R was 0.1336 for training phase, 0.12823 for validation phase, and 0.12955 for testing phase. The overall regression value R was 0.1324.

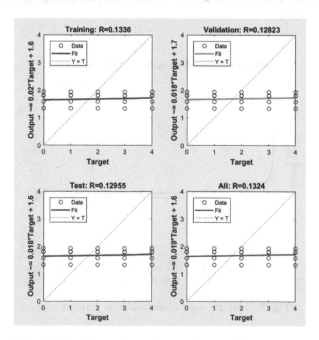

Fig. 7. Feed forward neural network regression analysis.

The best validation performance is measured about 2.1734 at 2nd epoch of the neural network. Totally 5 epochs were performed to identify the best validation performance in training, testing and validation phase. Figure 8 shows the validation performance plot of feed forward network.

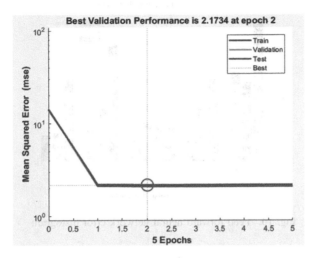

Fig. 8. Feed forward neural network validation performance

Figure 9 shows the training state plots of feed forward neural network. The gradient value is about $1.1664e^{-12}$ and mean value of $1e^{-08}$ at maximum epoch of 5. Validation check is performed at 3 instances at 3, 4 and 5th epoch.

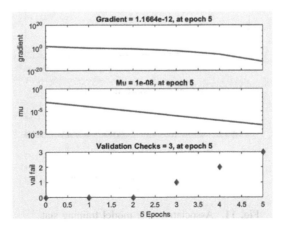

Fig. 9. Feed forward neural network training state

Error Histogram is plotted for the error value measured during the training, validation and testing phase of the system. Figure 10 represents the error histogram plot for feed forward network. Zero error point is measured at 0.02357 with 20 bins.

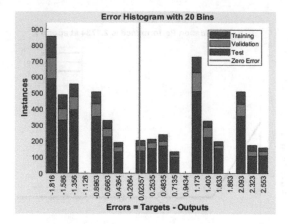

Fig. 10. Feed forward neural network error histogram

Associative rule is then applied to the student evaluation dataset for further classification. Figure 11 shows the performance of the association rule in training state of the neural network. The Gradient value is about 0.073273 with maximum epoch value of 1000 and zero validation failure occurred during the 1000 epochs.

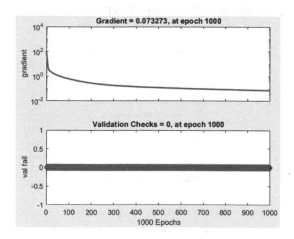

Fig. 11. Associative rule model training state

Figure 12 shows the validation performance of the association rule mining system. The best performance achieve at 1.7732 for a total epoch of 1000. The performance of train, validation and test phase reached the best-fit condition at 100th epoch.

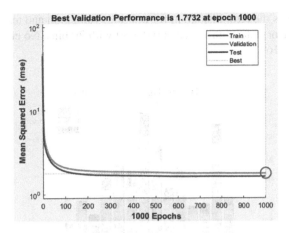

Fig. 12. Associative rule model validation performance

The regression plot generate for training, validation, testing and overall performance for the dataset. Figure 13 shows the regression plot. The regression value R measured is 0.18206 at training period, 0.12351 at validation period, 0.2089 at testing period and 0.17765 for the overall dataset.

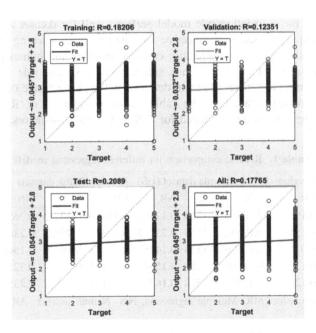

Fig. 13. Associative rule regression model

Figure 14 shows the error histogram for training, validation and testing phase of the system. the zero error state is measured at 0.02989 with 20 bins. No error histogram are generated above 3.162.

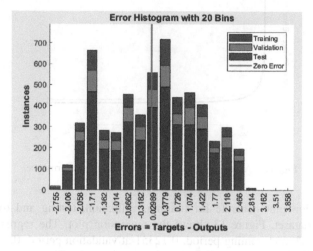

Fig. 14. Associative rule error histogram

In summary, the association rule model perform well for dataset analysis. Since, each subject require different perspective of handing and teaching methods. For example, the mathematics subject require constant practice and homework for better student performance compared to an arts subject. The associate rule selects specific parameters according to subject code and determines the performance of teacher based on feedbacks provided by students. The Table 1 show comparison of 'R' value metrics for different regression models obtained for U.S and Indian instructors.

Table 1. R-value comparison for different regression models

Instructors	R value - UCI California dataset (US)				R value (India dataset)			
	LR	MR	NN	AR	LR	MR	NN	AR
1	0.10527	0.20826	0.1324	0.17765	0.12546	0.32155	0.26548	0.36582
2	0.18159	0.19552	0.2153	0.23222	0.19325	0.21523	0.28652	0.39252
3	0.03545	0.13526	0.1326	0.14254	0.21423	0.16325	0.15469	0.29355
4	0.13654	0.24568	0.1458	0.15542	0.23155	0.26582	0.32546	0.38823
5	0.12648	0.24832	0.1578	0.16335	0.22152	0.25462	0.23594	0.35568

LR- Linear regression, MR- Multiple regression, NN- Neural network, AR- Associative rule.

The modelling shows associative rule perform well for dataset evaluation compared to other methods. The associative rule modelling of created dataset for Indian teachers has higher R value due to different metrics such as leadership, accountability, management, responsibilities, Teaching notes arrangement, teacher's nature, lecture methodology, attention to students, teachers encouragement to student, subject knowledge and teachers professional activities.

5 Conclusion

Student Evaluation dataset is considered to analyse the performance of the instructor over the specific course period. Neural network based machine learning approach is applied over the collected dataset to identify the factors affecting the performance of the instructors. The neural network approaches like linear regression, multiple regression, feed forward network and association rue mining algorithms are applied over the collected data set. The student evaluation data set is referred from UCI machine learning repository. The regression value of 0.10527, 0.20826, 0.1324 and 0.17765 was achieved for linear regression, multiple regression, feed forward network and association rule mining methods. The performance of these machine learning algorithms are compared through the regression value generated from the applied data set.

Compliance with Ethical Standards
  ✓ All authors declare that there is no conflict of interest.
  ✓ No humans/animals involved in this research work.
  ✓ We have used our own data.

References

1. Tsiakmaki, M., Pierrakeas, C., Kostopoulos, G., Kotsiantis, S., Koutsonikos, G., Ragos, O.: Predicting university students' grades based on previous academic achievements. In: 2018 9th International Conference Information, Intelligent System Application IISA 2018, no. 1, pp. 1–6 (2019)
2. Zhu, J., Hu, S., Arcucci, R., Xu, C., Zhu, J., Guo, Y.: Model error correction in data assimilation by integrating neural networks. Big Data Min. Anal. 2(2), 83–91 (2019)
3. Kwon, B.C., et al.: RetainVis: visual analytics with interpretable and interactive recurrent neural networks on electronic medical records. IEEE Trans. Vis. Comput. Graph. 25(1), 299–309 (2019)
4. Mohammadi, M., Al-Fuqaha, A., Sorour, S., Guizani, M.: Deep learning for IoT big data and streaming analytics: a survey. IEEE Commun. Surv. Tutorials 20(4), 2923–2960 (2018)
5. L'Heureux, A., Grolinger, K., Elyamany, H.F., Capretz, M.A.M.: Machine learning with big data: challenges and approaches. IEEE Access 5, 7776–7797 (2017)
6. Chou, C.Y., et al.: Open student models of core competencies at the curriculum level: using learning analytics for student reflection. IEEE Trans. Emerg. Top. Comput. 5(1), 32–44 (2017)
7. Conijn, R., Snijders, C., Kleingeld, A., Matzat, U.: Predicting student performance from LMS data: a comparison of 17 blended courses using moodle LMS. IEEE Trans. Learn. Technol. 10(1), 17–29 (2017)
8. Mejia, C., Florian, B., Vatrapu, R., Bull, S., Gomez, S., Fabregat, R.: A novel web-based approach for visualization and inspection of reading difficulties on university students. IEEE Trans. Learn. Technol. 10(1), 53–67 (2017)

9. Lan, A.S., Waters, A.E., Studer, C., Baraniuk, R.G.: BLAh: boolean logic analysis for graded student response data. IEEE J. Sel. Top. Signal Process. **11**(5), 754–764 (2017)
10. Srilekshmi, M., Sindhumol, S., Chatterjee, S., Bijlani, K.: Learning analytics to identify students at-risk in MOOCs. In: Proceedings of IEEE 8th International Conference Technology Education T4E 2016, vol. 1, no. 2, pp. 194–199 (2017)
11. Guenaga, M., Garaizar, P.: From analysis to improvement: challenges and opportunities for learning analytics. Rev. Iberoam. Tecnol. del Aprendiz. **11**(3), 146–147 (2016)
12. Fong, S., Wong, R., Vasilakos, A.V.: Accelerated PSO swarm search feature selection for data stream mining big data. IEEE Trans. Serv. Comput. **9**(1), 33–45 (2016)
13. De Morais, A.M., Araújo, J.M.F.R., Costa, E.B.: Monitoring student performance using data clustering and predictive modelling. In: Proceedings of Frontiers Education Conference FIE, vol. 201, February 2015
14. Fisch, D., Kalkowski, E., Sick, B.: Knowledge fusion for probabilistic generative classifiers with data mining applications. IEEE Trans. Knowl. Data Eng. **26**(3), 652–666 (2014)
15. Yu, C., Chen, Y.L.: Mining sequential patterns from multidimensional sequence data. IEEE Trans. Knowl. Data Eng. **17**(1), 136–140 (2005)

Advanced Driver Assistance System Using Computer Vision and IOT

M. Hemaanand[1], P. Rakesh Chowdary[1], S. Darshan[1],
S. Jagadeeswaran[1], R. Karthika[1(✉)], and Latha Parameswaran[2]

[1] Department of Electronics and Communication Engineering, Amrita School
of Engineering, Coimbatore, Amrita Vishwa Vidyapeetham, Coimbatore, India
aananddharani@gmail.com,
rakeshchowdarypolina@gmail.com, darshanslm@gmail.com,
jaggull39@gmail.com, r_karthika@cb.amrita.edu
[2] Department of Computer Science and Engineering, Amrita School
of Engineering, Coimbatore, Amrita Vishwa Vidyapeetham, Coimbatore, India
p_latha@cb.amrita.edu

Abstract. Advanced driver assistance systems (ADAS) assist the driver by providing vital information and by automating some actions. We have proposed an advanced technique based on deep learning and IOT that can monitor the driver and surroundings of the car by providing a quantifiable increase in safety and security. The sensors are embedded in the steering of the car to monitor the heartbeat of the driver. The proposed technique is to detect and classify cars, bicycles, pedestrians as well as to inform the user if there is any abnormality in the driver's heartbeat by implementing deep learning techniques and the technology of internet of things. This paper provides a well-defined structure for live data analysis using firebase which is a real time database. The proposed method is mainly supported by the real time object detection and classification which is implemented using Mobile Net and Single shot detector. The sensor that is used for pulse rate detection is INVNT_11 and the further processing is done using Arduino. CC3200 is used for sending message to medical services when there is any anomaly or abnormality in the driver's BPM. The ESP8266 is used for sending the data to firebase, where the driver's state can be monitored and analysed.

1 Introduction

Approximately 13 lakh people die in road accidents every year, on average 3,289 deaths a day, all around the world. An ancillary 2–5 crore are harmed or crippled. To overcome these, we came up with the "Advanced Driver Assistance System", which can serve as the solution for the problem. The objective of this work is to provide all necessary modules that are required for safety of driver and passengers. As semi-autonomous cars and self-driving cars are the future, the proposed solution can be implemented in cars to reduce number of accidents. Computer vision systems are employed in ADAS to identify and track any anomalies that jeopardize on the road and provide a real-time monitoring system to the driver. Computer vision system usually

S. Smys et al. (Eds.): ICCVBIC 2019, AISC 1108, pp. 768–778, 2020.
https://doi.org/10.1007/978-3-030-37218-7_85

consists of a front camera connected to a system which has trained modules. It uses deep learning techniques for analysing and extracting features from live videos captured from the on-board cameras. Automotive cameras integrated with application-specific modules for image processing technique provide an integrated computer vision system to detect pedestrians, vehicles, and even other objects in and around the vehicle while driving [5].

2 Literature Survey

The previous techniques such as hough transform [3], haar cascade classifier [4] provided knowledge on feature selection and classification for object detection. This technique captures information from multiple scales, and particularly can be computed exceptionally, but in order to increase the time constraints and to increase efficiency, improved normalization algorithm single shot detector (SSD) is introduced [1]. In the precursor techniques, there were many object detection and object classification procedures. In [2] the authors have highlighted the reduction of processing time of the detection networks by exposing the region proposal networks as an effective approach. The technologies are involved in the process of development and management but image processing techniques are already in the process of enhancement. The progressive advancements in Deep learning and Artificial intelligence techniques enhanced the possibilities of providing safety and security to people. This Single shot detector (SSD) technique is based on neural networks which has been applied to classification, feature extraction [10], clustering, forecasting, approximation and recognition problems in multifarious industries including the automobile industries. These techniques enhanced the various types of data collection like the speech recognition, object detection [12, 13] and the visual object detection. It cracks the convolved structure in a huge data set by making use of the back-propagation algorithm.

We are extremely motivated by large-scale visual identification, in which a particular type of deep learning architecture has succeeded an enormous lead on the most recent stage in the development of a product [5, 11], incorporating the latest ideas and features. These convolutional networks as stated above are supported by the back-propagation procedure through several layers of convolution filters. The next stage is the driver drowsiness system [6]. The previous systems used modules that consisted of PPG (photoplethysmographic) sensor and galvanic skin response sensor [7]. The data from sensors are sent to user device module which serves as main analysing and processing unit. Those data are analysed along with the other sensor values and processed [8, 9].

The last step is dealing with the final interface section i.e. the IOT (Internet of things) network which is the human end user interface. Internet of things [14] is the collaboration of the internet with the sensors and smart objects which includes a wide variety of embedded systems which can be used in automobile as well as in other industries. IOT (Internet-of-things) [15] can be stated as the things which belong to the Internet to supply and access the real-world information. Internet of things is the last interface through which the correspondence of the information or passage because of an aggravation to the client and their users can be made conceivable. Internet of things which makes utilization of cloud innovation gives us an added stand point of getting connected around anyplace on the planet and a quicker transmission and gathering or reception of the vital information.

3 Proposed Technique

The proposed technique has two systems for safety and security of the car and the driver. The first system has four phases. The first phase is the setup phase which is used for training the images and creating the classes using various datasets as shown in Fig. 2. The second phase is the detection phase which is used for detecting the images and classifies them. The third phase is used for displaying the output and for image processing and the fourth phase is used for real time detection of objects using trained modules. The second system has three phases collection of sensor data and uploading in the cloud storage. The second phase correlates the data between the sensor value and the threshold value stored in the cloud. The third phase is used to send alerts to users if any abnormality in BPM (Beats per minute) is observed. The alert is sent through CC3200 Launchpad as shown in Fig. 1.

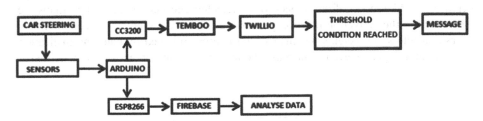

Fig. 1. Working of driver monitoring system

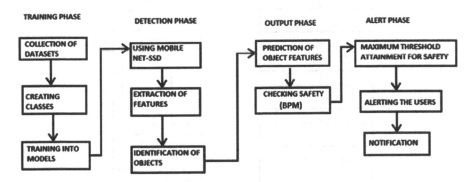

Fig. 2. Integration of different phases

4 Architecture

In this work, Mobile Net-SSD is used. SSD matches objects with default boxes of different aspects or sizes, it detects the image with the specified class or model and analyses the frame. It also draws rectangles in the image and classifies them according to the given dataset. Each element of the feature map has a number of default boxes associated which is similar to Faster-RCNN. SSD is a unified framework for object detection with a single network. The Mobile Net-SSD has many convolutional layers, which is absolutely necessary especially in real time applications like these. This Mobile Net-SSD uses depth wise separable convolutions instead of normal convolution. This makes it the fastest algorithm to work with [2]. These depth wise separable convolutions cost much lesser than normal convolutions. In architecture, the first part is a depth wise convolution layer that filters the input which is the 3 * 3 convolutional layers and then followed by batch normalization as shown in Fig. 3(a). Batch normalization increases the stability of the neural network; it normalizes the output of a previous activation layer (ReLU) by subtracting the batch mean and dividing by the batch standard deviation [9]. The activation function is the function that checks the similarity in weights and allows the detection function to continue. Feature maps are a representation of the dominant features of the image at different scales, feature maps increases the likelihood of any object to be detected and classified. Feature maps are obtained after the activation function processing. Again there are 1 * 1 convolutional layers that filters the weights obtained which is then followed by batch normalization and activation layer (ReLU) to increase the accuracy of the objects to be classified.

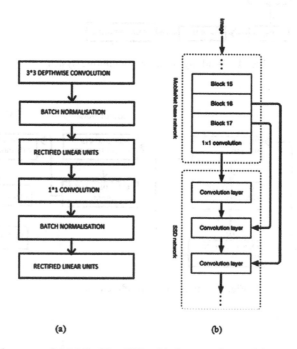

(a)   (b)

Fig. 3. (a) Architecture of Mobile Net SSD, (b) Process flow of images through Mobile Net SSD layer

5 Training Phase

The models are trained using Mobile Net-SSD architecture. The dataset is collected for each individual object that is used in detecting in this work. In car detection, dataset of Stanford University is used for training models. The Cars dataset consists of 16,185 images divided into 196 classes of cars. The data is divided into 8,144 training images and 8,041 testing images. The models for pedestrian detection were trained using GRAZ 01 dataset. The images in this dataset comprises of class that appears at different scales, backgrounds, angles, positions and different illumination conditions. Additionally, there are significant obstructions, background clutter, and lighting changes in images of this dataset which makes it a relevant dataset to use for modelling real world images. The training models for bicycle detection were trained using LIAAD, University of Porto dataset. These datasets are used as they have good collection of images at various angles, different resolution at different distance so that the detection part will be most efficient and accurate. These datasets also provide highly diverse images which in turn provide better performance. After the images have been sorted and categorized, region proposal convolutional neural network is trained and the results are obtained. These datasets are divided into training images, testing images and validation images. Higher the number of images in training set with high diversity, higher is the accuracy.

6 Detecting Heartbeat of Driver Using Pulse Rate Sensor

Pulse rate can be estimated depending on optical power variation of light as it is dispersed or consumed through its way through the blood when the heart beat changes. The pulse sensor Invento INVNT_11 is embedded on the inner part of steering to measure heart rate, which allows to detect even with single hand driving. Each unit of sensor has a pair of infrared emitter and detector diode that are placed side by side to detect clear ECG signal. The front side of the sensor is the part that touches the skin. On the foreside, a small hole can be seen, where the LED shines through from the back, and there is also a small square just under the LED. The square part is an ambient light sensor. The LED shines green light into the palm or fingertip, which the driver keeps on steering, and sensor reads the light that is reflected back from the palm or fingertip. The rear side of the sensor is where the remaining parts are mounted. This sensor has three pins. First pin is the power supply, second one is ground and the last pin is the signal pin. The technique of calculating BPM is based on the photoplethysmography (PPG) principle which uses two basic types that are transmittance and reflectance. These two basic types of PPG can be used because adult human body parts such as fingers, cheeks, hands which are semi-transparent. The Pulse Sensor is connected to Arduino for further processing. The values obtained from the sensor are not the exact values of BPM. So these values should be converted to BPM. If the raw values read by the sensor are greater than or equal to the threshold value, the number of pulses is incremented by 1. To calculate the BPM we have to count the number of pulses for 10 s and then multiply the count by 6. Based on the value of BPM, the state of the driver is monitored.

7 Working of CC3200 and Firebase

In this work, CC3200 Launchpad is used for making an alert system. When the CC3200 board is triggered, a SMS is sent. Since CC3200 uses cloud based services [8], Twilio and Temboo are used for the SMS system. Temboo is used to connect a router and the Launchpad using the authentication process. Twilio is used for the SMS alert system; by creating a dummy number which acts as the sender and the recipient maybe a real person. The board can be programmed using sublime text, Emacs, vim, Energia etc. Here Energia editor is used to program the board. Figure 4 represents the schematic working of both CC3200 and firebase.

In this work the online database management is managed by firebase which is used as a cloud management suite. The Firebase real-time Database is a cloud-facilitated database. Data is secured as json and synchronized continuously to each related client. It can be assembled with various application platforms including iOS, Android, and JavaScript SDKs. The majority of it is used for real-time database management and automatically updates itself with the latest data. The firebase store sand sync data with NoSQL cloud database and it is synced across all users in real-time. The data sent from the ESP8266 is sent to firebase and gets updated regularly.

8 Output Phase

The object is detected using bounding boxes. In today's world, Object identification is required to be fast, accurate and be able to identify a wide variety of objects [4]. With the advancements of neural networks, detection frameworks have been increasingly fast and precise. The model is trained to detect and identify the objects and classify them based on the inserted modules and trained data-set. The main objective is to detect objects which are moving at a high speed. The next objective is to alert the drivers who are impaired, abnormal or drowsy who cause most of these accidents [7]. Although many developments are done to keep drivers off the road when they are not completely attentive and focused, this work provides a different and efficient approach. The system created can reduce that type of accidents which are caused by distraction or any health abnormalities by monitoring all passive disruptive features and efficiently focusing on the solutions. This is done by using CC3200, ESP8266 and firebase. The CC3200 LaunchPad has a built-in Wi-Fi connectivity. There is an on-board emulation using FTDI which includes sensors for various purposes. This Launchpad is used for sending messages to medical services. The ESP8266 is used to transfer the data to firebase for further data analysis.

9 Experimental Results

A set of 1000 images for three different classes divided into training, testing and validation are chosen randomly from each dataset. The training dataset consists 800 images of each class and the training and validation datasets consists of 100 images each of single class. The advanced driver assistance system has been developed using the combination of Mobile net and SSD (Single Shot Detector). One advantage of this technique is a cost efficient installation and the use of futuristic technologies like Internet of Things (IOT). From the experimental results, it is inferred that there is more to do in the field of artificial neural network based automobile industries and self-driving cars. The system solves the problem by alerting the users. This technique can not only be applied in self-driving cars but also can be used in automated logistics delivery systems. The Table 1 shows the accuracy of recognition of each trained class. The Fig. 5 shows the detection of cars and the Fig. 6 shows the detection of bicycle and pedestrians. The alert messages are sent through CC3200 and the output is shown in Fig. 7.

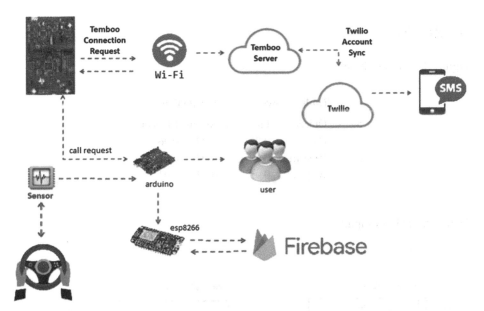

Fig. 4. Schematic workflow of driver monitoring system

9.1   Evaluation Metrics

While evaluating a standard deep learning model, the predictions are generally classified into four different classes: true positives (TP), false positives (FP), true negatives (TN), and false negatives (FN). In this model precision, recall and f1 score are used as evaluation metrics where precision is defined as the amount of positive prediction that were correct.

$$\text{Precision (P)} = \frac{TP}{TP+FP} \qquad (1)$$

Recall is defined as the percentage of positive cases caught or proportion of actual positives that were identified correctly.

$$\text{Recall (R)} = \frac{TP}{TP + FN} \tag{2}$$

Precision and recall values are used to calculate F1 Score. F1 score is the harmonic mean (HM) of precision and recall.

$$\text{F1 Score} = \frac{2(PR)}{P + R} \tag{3}$$

10 Results

Output of training phase:

Table 1. Accuracy of recognition

S. No	Object	Precision	Recall	F1 Score
1	Car	0.95	0.97	0.96
2	Pedestrian	0.98	0.94	0.96
3	Bicycle	0.95	0.93	0.94

Vehicle detection output:

Fig. 5. Detection of cars using trained model [5]

Bicycle detection and pedestrian detection output:

Fig. 6. Detection of person and bicycle using trained model [7]

Sample results from cc3200:

Fig. 7. Alert messages from CC3200

11 Conclusion and Future Work

In this work, the driver is monitored by checking the heartbeat at regular intervals and sending alert messages to medical services when required. The object detection part in our work achieved high precision and recall due to the usage of Mobilenet-SSD. As the detection is fast and accurate, this can be used in real time applications. This work can be developed in such a way that the distance can be measured and warn the driver when the vehicle is too close to the other. By including lane detection, we can make the car travel in a particular lane and also park by itself into the parking area making it semi-autonomous. Firebase can be used to monitor different drivers simultaneously which helps transportation network companies such as logistics and automobile industries to check their drivers and can be upgraded to reduce latency in the autonomous driving system and improve object detection and classification. Ensure that the system responds to a malfunction with the vehicle and alert the driver. This can be even advanced by authorizing to activate the independent automatic emergency braking (AEB) system at the time of emergency.

Compliance with Ethical Standards
  ✓ All authors declare that there is no conflict of interest.
  ✓ No humans/animals involved in this research work.
  ✓ We have used our own data.

References

1. Song, H., Choi, I.K., Ko, M.S., Bae, J., Kwak, S., Yoo, J.: Vulnerable pedestrian detection and tracking using deep learning. IEEE (2018)
2. Howard, A.G., et al.: MobileNets: Efficient Convolutional Neural Networks for Mobile Vision Applications (2017)
3. Karthika, R., Parameswaran, L.: An automated vision-based algorithm for out of context detection in images. Int. J. Signal Imaging Syst. Eng. **11**(1) (2018). https://doi.org/10.1504/ijsise.2018.090601
4. Wen, X., Shao, L., Fang, W., Xue, Y.: Efficient feature selection and classification for vehicle detection. IEEE Trans. Circuits Syst. Video Technol. **25**, 508–517 (2014)
5. Lyu, N., Duan, Z., Xie, L., Wu, C.: Driving experience on the effectiveness of advanced driving assistant systems. In: 4th International Conference on Transportation Information and Safety (ICTIS), Banff, Canada, 8–10 August 2017 (2017). 978-1-5386-0437-3/17/$31.00©2017 IEEE
6. Chellappa, Y., Joshi, N.N., Bharadwaj, V.L.: Driver fatigue detection system. In: 2016 IEEE International Conference on Signal and Image Processing (ICSIP), Beijing, pp. 655–660 (2016). https://doi.org/10.1109/siprocess.2016.7888344
7. Leng, L.B., Giin, L.B., Chung, W.: Wearable driver drowsiness detection system based on biomedical and motion sensors. In: 2015 IEEE SENSORS, Busan, pp. 1–4 (2015). https://doi.org/10.1109/ICSENS.2015.7370355
8. Bamarouf, F., Crandell, C., Tsuyuki, S., Sanchez, J., Lu, Y.: Cloud-based real-time heart monitoring and ECG signal processing. In: 2016 IEEE SENSORS, Orlando, FL, pp. 1–3 (2016). https://doi.org/10.1109/ICSENS.2016.7808911
9. Rahim, H., Dalimi, A., Jaafar, H.: Detecting drowsy driver using pulse sensor. J. Technol. **73** (2015). https://doi.org/10.11113/jt.v73.4238
10. Senthil Kumar, T., Sivanandam, S.Nb.: A modified approach for detecting car in video using feature extraction techniques. Eur. J. Sci. Res. **77**, 134–144 (2012)
11. Sandeep, A.K., Nithin, S., Ramachandran, K.I.: An image processing based pedestrian detection system for driver assistance. Int. J. Control Theory Appl. **9**, 7369–7375 (2016)

12. Senthil Kumar, T., Sivanandam, S.Nb.: An improved approach for detecting car in video using neural network model. J. Comput. Sci. **8**, 1759–1768 (2012)
13. Singh, T., Sanju, S., Vijay, B.: A new algorithm designing for detection of moving objects in video. Int. J. Comput. Appl. **96**, 4–11 (2014)
14. Tripathi, G., Singh, D., Jara, A.J.: A survey of Internet-of-Things: future vision architecture challenges and service. In: IEEE World Forum on Internet of Things (WF-IoT), pp. 287–292 (2014)
15. Xia, F., Yang, L.T., Wang, L., Vinel, A.: Internet of Things. Int. J. Commun Syst **25**(9), 1101 (2012)

Segmentation and Classification of Primary Brain Tumor Using Multilayer Perceptron

Shubhangi S. Veer (Handore)[1][✉], Anuoama Deshpande[2],
P. M. Patil[3], and Mahendra V. Handore[1]

[1] Trinity College of Engineering and Research, Pune, India
handore.shubhangi@gmail.com
[2] JJTU, Jhunjhunu, Rajasthan, India
[3] TSSM JSPM College of Engineering, Pune, India

Abstract. Brain is one of the complex organs of human nervous system where billions of neurons will form a gigantic network. Identification of infected region in such nervous system is a challenging task. The brain MRI images are mostly referred by Radiologist for identification of infected region but these images are blurred and does not contain any details about the infected region. So the identification of exact location of the tumor region, segmentation of that region and getting its exact boundaries from such a complicated nervous system is a challenge for the Radiologist. Digital image processing is used in the proposed algorithm for obtaining detail information about tumor, which will be helpful for Radiologist in the diagnosis process.

Keywords: Denoising · Enhancement · Segmentation · Clustering · KNN · SVM · GLCM · MLP

1 Introduction

The human body is made up of many cells by having their own special function. The cells in the human body grows and divides to form a new cell, these cells will replace old or damaged cells for the proper functioning of body. When the growth of cells is an uncontrollable, these extra cells often resulting in a mass of tissue called a growth or tumor. Tumor can exert pressure on different parts of brain, can increases pressure within the skull and also can damage normal cells. The brain contains about 50–100 billion neurons that forms an overlapped gigantic neural network. The tumor may be embedded in the overlapped network of a brain that may effect on the sensitive functioning of a body. The location of such infected region and vigorous spreading capacity makes treatment very complex and risky. The research of various countries has concluded that in last few decades the count of people suffers from tumor has been increased to 300 people per year [1].

This tumor can be cancerous or noncancerous. Tumor that originates in the brain itself is called Primary tumor and a tumor that originates in other part of the body such as breast cancer and kidney cancer it then spreads in the brain is called Secondary brain tumor.

© Springer Nature Switzerland AG 2020
S. Smys et al. (Eds.): ICCVBIC 2019, AISC 1108, pp. 779–787, 2020.
https://doi.org/10.1007/978-3-030-37218-7_86

So the objectives of this proposed algorithm is to identify the exact location of tumor, separate out it from gray matter and find its exact edges from Magnetic resonance imaging (MRI). This algorithm will be helpful for Radiologists to interpret the diseases quickly and accurately.

2 Related Work

The MRI images are gray images whose scale varies from 0 to 255. The identification of exact infected region is a challenging task due to variation in the gray levels. As white matter, gray matter present in brain, it is difficult to separate out white matter, gray matter and tumor region from brain MRI image. Similarly MRI contains random noise signal due to unwanted fluctuations, overheated faulty components or dust particles, variation in intensity levels or color information in an image occurs because of circuitry of a scanner or sensor during image acquisition [2].

Kalavathy and Suresh presented a paper on image denoising techniques where authors proposed an adaptive thresholding technique for noise removal. In adaptive technique, a threshold value is selected depending on the number of decomposition level. Each decomposition level represented an image by four frequency band. The low frequency again decomposed into four levels. As the number of decomposing levels increases in wavelet domain, the coefficients become more smoother. The change in magnitude of the coefficient in wavelet domain is depending on the number of decomposition level [3]. Shashikant Agrawal and Rajkumar Sahu also presented a paper on denoising MRI images using Discrete Wavelet Transform (DWT). The Discrete Wavelet Transform produced a non-redundant image. DWT provides better spectral and spatial details of an image. The DWT decomposed an image through two complementary filters that is low pass filter and high pass filter. This gives two signals, approximation and Details. The reconstruction of original signal without any loss is also possible by synthesis using DWT, it's a reverse operation. Bedi and Rati Khandelwal presented a paper on the different enhancement techniques used for medical image processing. Medical images, aerial images etc. are poor contrast and containing with noise signal. Image enhancement is one of the techniques that improve the quality of an image for removing blurriness, increase contrast and human viewing. Author tried negative transform, power log transform, piecewise slicing and histogram techniques in spatial domain to enhance the image quality [4]. There are number of image segmentations techniques like global segmentation, local segmentation, clustering etc., selection of segmentation technique is based on type of an image. The segmentation of tumor has been done by author in this paper by using clustering technique. He formed clusters by grouping set of patterns on the basis of homogeneity criteria. It is also useful to classify and to group objects based on features and minimum distance between data point and its corresponding cluster centroid [5]. The basic global thresholding and Ostu's thresholding technique tried in this paper for segmentation of brain tumor images. As the performance of these techniques yielded poor output, author used fuzzy clustering based technique for segmentation of brain MRI image. The erosion morphological operations used to extract tumor region from segmented MRI image and then the area of tumor region measured in terms of number of pixels by counting pixels having intensity 255 [6]. Madheswaran and Dhas proposed an algorithm where the features extracted from the testing region after segmentation based on their color, shape or texture properties. The extracted features are used for classification of brain tumor MRI image. Support vector machine(SVM) classifier has been used by author to classify images based on features [7]. The K-mean is a unsupervised clustering algorithm is useful for non-overlapped images. It forms clusters based on 'k' value. A cluster is formed by considering neighborhood points that belonging to a value 'k'. If the number of variables is large then K-means gives good response as it is faster than hierarchical clustering. The prediction of 'k' value is difficult and if the cluster are of different size and different density [8]. Ramaraju et al. proposed an automatic method for Brain tumor stage classification using Probabilistic neural network. In this paper author has proposed K-Means clustering algorithm for segmenting MRI image and to detect the brain tumor in its early stages [9] Medical images, aerial images etc. are poor contrast and containing noise signal. Author used spatial domain techniques for enhancement of images, they tried negative transform, power log transform, piecewise slicing etc. techniques in spatial domain to enhance the image quality. Vaishali presented a paper on feature extraction from brain tumor images using wavelet. The proposed algorithm initially removes noise from MRI images using Gaussian filter. The Gaussian filter is a smoothing filter. It is like as low pass filter. It is useful in both spatial domain and in frequency domain. The wavelet transform is used by author for feature extraction from images. Here author has used Support Vector Machine (SVM) classifier for tumor classification. SVM classifier is a supervised learning classifier, designed for a binary classification. It is useful for solving the nonlinear boundary problems [10].

3 Proposed Methodology

The objective of proposed algorithm is to enhance the quality of brain MRI image, identify tumor region, separate out tumor region and extract tumor boundaries. The steps that are followed to achieve these objectives are as follows:

- Remove noise from blur MRI image using filter.
- Enhance the quality of an image.
- Separate out the tumor region from the brain MRI.
- Find out edges of tumor region.
- Extract features of tumor region and prepare database.
- Classify primary tumor images into benign and malignant class based on features.

MRI images are noisy, poor contrast images so it is necessary to preprocess these images first. The preprocessing of these blur images is possible by filtering and enhancement techniques. The linear filters are more useful for removing the blurriness in an image and smooth an image. The linear filters assign new intensity value to each pixel by considering neighboring pixels intensities with the help of convolution technique, where a convolution of noisy image is done with a filter mask in spatial domain. Non-linear filters are also useful for removing the blurriness of an image. These filters are statistical filters and they smooth an image just by replacing center pixel intensity value by the median of the neighborhood [11]. These filters do not require any kind of mask for convolution. In the proposed algorithm nonlinear filter is used to remove noise in images (Fig. 1).

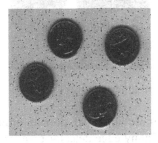

Fig. 1. Filtering of noise signal from an image [4]

The filtering gives smooth image then next task is to enhance the quality of an image. To improve the quality of an image and make it more clear enhancement technique is required. There are different image enhancement techniques to enhance an image like log transform, power law transforms, negative transform etc. Log transform is more useful for MRI images as it maps the wide range of high-level gray scale intensities into a narrow range of output intensities values (Fig. 2).

Fig. 2. Enhancement of an image [6]

Segmentation is a technique that separate out foreground and background of an image. It is basically based on similarity and discontinuities in intensity values of each pixel. It isolates the boundaries of an image in form of its segments [12]. The segmentation technique is more useful in case of medical images to identify infected region in MRI image and to study anatomical structure of human body. Global thresholding, local thresholding, regional based thresholding etc. are the various ways of segmentation. The infected region in brain MRI image is separated here by using global thresholding technique. The area of infected region is measured here from segmented image in terms of number of pixels. This segmented image is then used for extracting the boundaries of tumor region (Fig. 3).

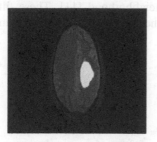

Fig. 3. Segmentation of an image [7]

There are various edge detection techniques are like sobel, prewitt, ordinary, canny etc. edge detectors to get edges of an object. Canny edge detector gives smooth edges as compare to other edge detectors (Fig. 4).

Fig. 4. Edge detection of an image [6]

4 Classification

Features are the properties that describe signal or an image. Feature extraction is a technique which represents an input image into set of features called feature vector [13]. The feature extraction technique represents large database in terms of some measuring feature these are useful for analyzing signal.

In brain tumor detection, the Gray scale co-occurrence matrix (GLCM) textural features are helpful to classify tumor into cancerous and noncancerous tumor. In a GLCM matrix, number of rows and columns are equal to number of gray levels in an image. The GLCM is useful for determining the co-occurrence matrix of an image. This determines as a pixel with intensity i occurs in relation with another pixel j, at a distance d and angle θ [14]. The textural GLCM features like contrast, variance, entropy, energy, etc. are calculated from gray scale image. These features are used for classification of tumor. There are number of classification techniques that classify images based on their features. The Fuzzy C-mean is one of the unsupervised classifier that forms cluster by assigning the membership to each pixel based on distance between every pixel and the center of cluster. Similarly K-nearest classifier is supervised classifier; it gives high accuracy in case of MRI images classification. K-nearest classifier measures distance between training point and unlabeled data points. According to Euclidean distance it classifies patters into certain classes.

The back propagation neural network is also useful for classification of images [15]. This network organizes the neurons into different layers and passes a signal in forward direction. The error occurs on the output side is propagated in opposite direction. This network adjusts weights between input and output to minimize error. It is an iterative, supervised algorithm.

The multi-layered neural network is also useful for diseases classification and recognition in medical images. In multi-layered neural network the numbers of hidden layers are more than one. It increases the complexity of network but gives good response in classification of patterns [16]. In proposed algorithm multilayer perceptron network (MLP) network is used for tumor classification based on GLCM features.

This multilayer perceptron network has been trained and tested at different combinations of training and testing data set. The database is a set of features extracted from brain tumor MRI images with the help of discrete wavelet transform. The Table 1 shows response of 60%–40%, 70%–30%, 80%–20%, 90%–10% combinations of training and testing dataset respectively of multilayer perceptron neural network. The performance of these combinations is measured in terms of accuracy, mean square error and regression factor by using multilayer perceptron neural network.

Table 1. Performance parameters of MLP

Dataset (Train–testing%)	Accuracy	MSE	Regression
60–40	80.9524	0.001099	0.99474
70–30	85.7143	0.000756	0.99745
80–20	90.4762	0.000689	0.99981
90–10	84.7143	0.000856	0.99567

By observing the performance of MLP neural network, it is observed that at 80%–20% combinations the multilayer perceptron neural network has given the best performance for the classification of brain tumor database. The Fig. 5 shows the response of regression factor during training and Figs. 6, 5 shows the response of regression factor during testing of the of multilayer perceptron neural network for a combination of 80%–20% of feature data set.

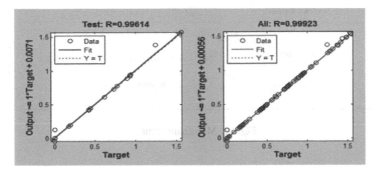

Fig. 5. Response of regression factor during training

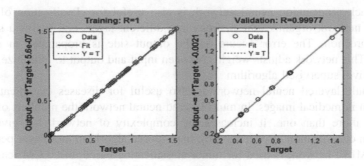

Fig. 6. Response of regression factor during testing

The classifications of primary tumor into benign and malignant class based on features by using MLP classifier is shown in Figs. 7 and 8.

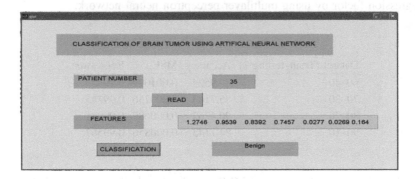

Fig. 7. Benign tumor

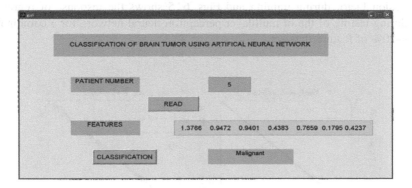

Fig. 8. Malignant tumor

5 Conclusion

Overlapped cells and complex structure of brain causes major problem in extracting accurate features from brain MRI. This complex structure affects MLP network also, where weights of neurons are not constant; they are continuously changing during Learning/Training phase. This is a main hurdle, which may cause errors in image analysis and identification of infected region in brain. The proposed algorithm overcomes the problems with the help of DWT and MLP network. The proposed algorithm has removed noise from blur brain MRI images efficiently using spatial domain filter and detected the exact location and shape of tumor region. The 'symlet' wavelet has been used for feature extraction from 2D MRI of brain tumor in proposed methodology. It is possible to fuse different wavelets for feature extraction from 3D images in future and to find out the depth of the infected region.

Compliance with Ethical Standards
✓ All authors declare that there is no conflict of interest.
✓ No humans/animals involved in this research work.
✓ We have used our own data.

References

1. Bhagwat, K., More, D., Shinde, S.: Comparative study of brain tumor detection using K-means, fuzzy C-means and hierarchical clustering algorithms. Int. J. Sci. Eng. Res. **4**(6), 626–631 (2013)
2. Verma, R., Ali, J.: A comparative study of various types of image noise and efficient noise removal techniques. Int. J. Adv. Res. Comput. Sci. Softw. Eng. **3**(10), 617–622 (2013)
3. Shen, S., Sandham, W., Granat, M., Sterr, A.: MRI fuzzy segmentation of brain tissue using neighborhood attraction with neural-network optimization. IEEE Trans. Inf Technol. Biomed. **9**(3), 454–467 (2013)
4. Bedi, S.S., Khandelwal, R.: Various image enhancement techniques- a critical review. Int. J. Adv. Res. Comput. Commun. Eng. **2**(3), 1605–1609 (2013)
5. Manikandan, R., Monolisa, G.S., Saranya, K.: A cluster based segmentation of magnetic resonance images for brain tumor detection. Middle-East J. Sci. Res. **14**(5), 669–672 (2013)
6. Tirpude, N., Welekar, R.R.: A study of brain magnetic resonance image segmentation techniques. Int. J. Adv. Res. Comput. commun. Eng. **2**(1), 958–963 (2013)
7. Madheswaran, M., Dhas, D.A.S.: Classification of brain MRI images using support vector machine with various Kernels. Biomed. Res. **26**(3), 505–513 (2015)
8. Shinde, V., Kine, P.: Brain tumor identification using MRI images. Int. J. Recent Innovation Trends Comput. Commun. **2**(10), 3050–3055 (2014)
9. Ramaraju, P.V., Baji, S.: Brain tumor classification, detection and segmentation using probability neural network techniques. Int. J. Emerging Trends Electr. Electron. **10**(3), 15–20 (2014)
10. Vaishali, R.: Wavelet based feature extraction for brain tumor diagnosis. Int. J. Res. Appl. Sci. Eng. Technol. (IJRASET) **3**(5), 776–780 (2015)
11. Dawood, A.A.-M., Saleh, M.F.: Review of different techniques for image denoising. Int. J. Innovative Res. Comput. Commun. Eng. **6**(3), 2498–2505 (2018)

12. Khan, S., Kharade, R., Lavange, V.: Segmentation in digital image processing. Int. J. Res. Eng. Sci. Manag. **2**(1), 292 (2019). ISSN 2581-5792
13. Khare, C., Nagwanshi, K.K.: Image restoration technique with non linear filter. Int. J. Adv. Sci. Technol. **39**(7), 1–8 (2012)
14. Athira, V.S., Dhas, A.J., Sreejamole, S.S.: Brain tumor detection and segmentation in MR images using GLCM and AdaBoost classifier. Int. J. Sci. Res. Sci. Eng. Technol. IJSRSET **1**(3), 180–186 (2015)
15. Amien, M.B.M., Abdelrehman, A.: An intelligent model for automatic brain diagnosis based on MRI images. Int. J. Comput. Appl. **72**(23), 21–23 (2013)
16. Mohan, M.: Digital image processing algorithms for classification, accuracy assessment and change detection of earths resources. Int. J. Adv. Sci. Res. **2**(6), 156–160 (2017). ISSN 2455-4227

Dynamic and Chromatic Analysis for Fire Detection and Alarm Raising Using Real-Time Video Analysis

P. S. Srishilesh[✉], Latha Parameswaran, R. S. Sanjay Tharagesh,
Senthil Kumar Thangavel, and P. Sridhar

Department of Computer Science and Engineering,
Amrita School of Engineering, Amrita Vishwa Vidyapeetham, Coimbatore, India
srishilesh@gmail.com,
{p_latha, t_senthilkumar}@cb.amrita.edu,
sanjay311999@gmail.com,
cb.en.d.cse17011@cb.students.amrita.edu

Abstract. Fire outbreak has become a common accident that occurs in several places such as in forests, manufacturing industries, living house and in widely crowded areas. These incidents cause severe damage to nature as well as to living creatures in the affected surroundings. Due to this, the need for efficient fire detection system has been increased rapidly. Using fire detecting sensors has proved to be an efficient solution but its effectiveness on delivering quick results depends on the affinity of fire sources. In the proposed method, we present an economical and affordable fire detection algorithm using video processing techniques which is compatible with CCTV and other stationary surveillance cameras. The algorithm uses an RGB color model with chromatic and dynamic disorder analysis to detect the fire. Fire pixels are detected by the rules of the color model which is mainly dependent on the fire pixel intensity and also the saturation of red color component in the fire pixel. The extracted fire like pixels are authorized by growth combined with the disorder of the fire regions. Furthermore, based on iterative checking the real fire is identified, if it is present then the appropriate signals will be sent. The proposed method is tested on various datasets acquired in real time environments and from the internet. This methodology can be used for fully automatic fire detection surveillance with reduced false true errors.

Keywords: Fire detection · Chromatic analysis · Dynamic analysis · Flame disorder · Image and video processing

1 Introduction

Fire accidents cause severe damage to the environment, buildings, and human lives. Thus it is considered as a great threat for humanity. The extend of fire damage depends largely on the intensity of fire. The fire caused by short circuits and gas leaks will spread very quickly, and as a result, the loss will be very severe. Though the main component of the damage is fire; smoke, fuel, heat, and soot all are responsible for the degree of destruction. [14] For a quick response to the fire accidents, the detection

© Springer Nature Switzerland AG 2020
S. Smys et al. (Eds.): ICCVBIC 2019, AISC 1108, pp. 788–797, 2020.
https://doi.org/10.1007/978-3-030-37218-7_87

methods should be real-time and very fast. Any latency in fire detection can result in a severe loss. [9, 11] Thus various state of art fire detection methods is implemented with wired/wireless sensors and vision cameras. The early methods used for detection of fire were considered to be less effective when using data from sensors. The raising of the fire alarm is delayed due to the slow response of sensors. One such method proposed earlier was a less expensive fire recognition and control system which is completely dependent on heat and smoke detection control system. It is a combination of electrical and electronic devices along with various equipment operating to find the presence of fire in any region and alert the respective officials through an audio-visual message after the detection. Then, it simultaneously initiates a system which sends SMS to the registered contact numbers and opens the water sprinkler or a fire-distinguishing pump to spray water on affected regions or fire terminating foams. The limit of this system was these sensors can detect only when the fire is within their proximity. [1] The supervision of fire alarms was under humans, which was unreliable. With Image processing, the job of fire detection was made easy by analyzing the frames in the video or image from the CCTV cameras [15]. From the available data, real-time videos are obtained from CCTV cameras which can be favorable for detecting precarious disasters such as flood [2] and fire [3, 6]. With video processing, the accuracy of detecting a fire was very low, and unreliable. [13] Thus, this led to the improvement in the algorithm for better accuracy. [4] In Gunay, Toreyin, Kose, and Cetin, an Entropy functional based online Adaptive Decision Fusion is proposed for image analysis. In their work, it is assumed that various small algorithms are fused together to get the main general algorithm for an exact application. Each and every sub-algorithm produces its own decision which denotes its confidence value. All the decision values obtained are incorporated with other weights updated online. [12] The proposed adaptive decision fusion method uses the observations from security officials in the forest regions and this is the main limitation of the system.

2 Proposed Method

The proposed architecture is shown in the flowchart (Fig. 3). The detection of fire is established through image processing and once the fire is detected, the fire emergency alarm is raised to warn the people. Widely used combustible materials are burnt only on reaction with oxygen which helps in the combustion of the material. The color of the fire is decided by the temperature of the fire. At low temperatures, the fire ranges from red to yellow, and white at higher temperatures. The saturation of fire is more in the daytime than during the night, so detection of fire during night times must consider the saturation level into consideration. Due to airflow, the fire tends to change its shape, this dynamic feature can be used to categorize fire easily. Smoke is also released when the fire is burnt, which depends on the constitution of various combustible materials. Distinguishing the fire alias objects and the real fire objects must be considered. [7] To prove that fire is real, we have to analyze certain features mentioned above. So, the aim of this experiment is to detect existing fire, and smoke and to check for the spread of fire [8].

2.1   Chromatic Analysis of Flame

The HSI (Hue/Saturation/Intensity) model can be easily perceived by humans for differentiating colors than done with the RGB model. So, the color model is mathematically transformed to HSI model. As mentioned previously, the background illuminations depend on the saturation of the flame during the day or night. In the HSI model, the hue value of the fire becomes more white when no background illumination, and less white when compared to the fire-light. Thus, during processing the video frames, the detection of fire must be done early, thus to reduce the computational complexity to a very low extent, the RGB color model conditions the value of pixels to be Red > Green > Blue for the captured fire image. Since the most important component of the fire is the R component, it must be compared with a threshold value $Red_{Threshold}$. Fire similar alias may result in the detection of fire to be false, to avoid such conditions, the saturation value of the fire must be considered. Thus based on these conditions three decisions are deduced for extracting fire pixels from the video frames. The decisions are described as follows [5]:

$$rule1 : Red > Red_{Threshold}$$
$$rule2 : Red \geq Green > Blue$$
$$rule3 : Saturation \geq \frac{(255 - Red)*Saturation_{Threshold}}{Red_{Threshold}}$$
$$If(rule1) \ AND \ (rule2) \ AND \ (rule3) = TRUE$$
$$then \ "It \ is \ a \ fire \ pixel"$$
$$else \ "It \ is \ not \ a \ fire \ pixel"$$

$Saturation_{Threshold}$ denotes the saturation value of *Red* color component of the fire pixel is $Red_{Threshold}$ for the same fire pixel. The saturation value will degrade when the *Red* component increases in the pixel.

2.2   Dynamic Analysis of Flame

Few fire-like regions from the flame image may be extracted as real fire pixel by chromatic analysis method. This happens in two particular cases: substances which are not burning but possess color similar to fire and backgrounds illuminated by solar reflections, high-intensity lights, and other similar fire like light sources. In the former case, the materials possessing a good amount of red color and fire-like color may lead to false fire pixel extraction. In the other case, when the background is illuminated with solar reflections, burning fires and artificial fire-like lights. Together, they influence fire pixel extraction, making the true fire detection complex and unreliable.

Thus, in order to confirm a burning fire is real or fake, further to chromatic analysis of fires, the dynamic (movement of fire region) disorder and movement are adopted to reduce the detection of fire aliases. Natural fire does not have a constant shape and size. The fire dynamics include sudden movement of flames due to the air flow direction, changeable shapes, and growth rate. To improve the reliability in fire identification, our proposed method uses the combination of both rate of growth of flames and the disorder characteristics of fire and to check if it is a fake fire like region or a real fire. The flame disorder value is measured from the number of fire pixels of the flame difference of two continuous video frames. The final decision rule for this segment of the algorithm is [5]:

$$If \frac{(|FireDisorder_{t+1} - FireDisorder_t|)}{FireDisorder_t} \geq FireDisorder_{Threshold}$$
$$Then \; "Real fire"$$
$$Else \; "Flame \, alias"$$

Here, $FireDisorder_t = Fire_t(x, y) - Fire_{t-1}(x, y)$

$Fire_t(x, y)$ and $Fire_{t-1}(x, y)$ denote the present and previous video frames of fire region respectively. *FireDisorder_t* and *FireDisorder_{t+1}* denote the disorder of the present and next video frames of fire region respectively. *FireDisorder_{Threshold}* is the predefined firedisorder threshold that helps to discriminate fire from fire-like objects. If all the above decisions are completely satisfied, then it can be concluded that flame is likely to be real-fire. Further, to increase the reliability of the fire detection the disorder of fire region checking should be calculated for d times. The variables, d and *FireDisorder_{Threshold}* are completely relying upon the statistical data of conducted experiments.

2.3 The Rate of Growth of the Fire Pixels

The growth of the burning fire is dominated by fuel type and air-flow. In the initial burning phase, the flame size gets towards the increasing approach. To identify this growth feature, the fire pixels in the video frame is calculated at regular time intervals and compared. Let the unknowns, k_i and k_{i+1} denotes the quantity of extracted fire pixel of the present frame and that of the next frame respectively. If $k_{i+1} > k_i$ is greater than p times at an repeated duration of t_F during the time period T, then this result concludes that there is a probable growth of the fire and this increases the authentication of a real fire. In the algorithm, the variables t_F, T and p are completely relying upon the statistical data of the conducted experiments (Fig. 2).

Fig. 1. (a) Initial burning of flame [3] (b) Extraction of Fire pixels (c) Fire spread

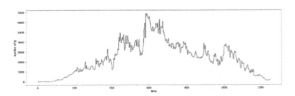

Fig. 2. Amount of fire pixels extracted during the burning of paper fuel

2.4   Smoke Detection

The burning of fire causes smoke, so this can be used as one of our feature for early detection of fire. Generally, smoke displays Grayish colors with two levels: light-gray and dark-gray. These grayish colors of smoke can be well described as the Intensity I in the HSI model. Gray levels vary from $LightGrey_1$ to $LightGrey_2$ and $DarkGrey_1$ to $DarkGrey_2$ for Light and Dark levels respectively.

$$LightGrey_1 \leq Intensity \leq LightGrey_2$$
$$DarkGrey_1 \leq Intensity \leq DarkGrey_2$$

The diffusion of the particles and the airflow causes the change of shape in smoke. Therefore, the dynamic features of the smoke can be characterized by growth and disorder of the extracted smoke pixels from the video frames. To calculate the dynamic features of the smoke, the below-mentioned rule can be used [5]:

$$If \frac{(|SmokeDisorder_{t+1} - SmokeDisorder_t|)}{SmokeDisorder_t} \geq SmokeDisorder_{Threshold}$$
$$Then \ "Real \ smoke"$$
$$Else \ "Smoke \ alias"$$

Here, SmokeDisordert+1 denotes the disorder in the smoke of the present image and the next image, SmokeDisorder$_t$ denotes the disorder in the smoke of the present image and the previous image. SmokeDisorder$_{Threshold}$ denotes the threshold value of the smoke, calculated using experimented results of the fire. The smoke growth depends mainly on fuel type and air flows. Let the unknowns, n_i signify the quantity (number) of smoke pixels derived from the video frame at any time t_i. A real smoke can be classified if $n_{i+1} \geq n_i$ for h times according to experimental results.

3 Fire Alarm Raising

The false alarm raising is discouraging to humans. The earlier algorithm states that the fire pixels are checked with an adaptive threshold value based decision function which has high sensitivity to change. [10] To overcome this problem, iterative checking of the fire pixels is done, to ensure the correctness of the raising of the alarm. Even though the spread of fire is observed from various locations through cameras, the accuracy of proving that the growth of the fire is increasing is low. To avoid the above problem,

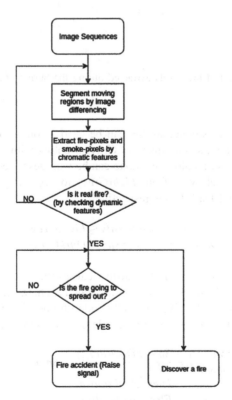

Fig. 3. Flowchart for fire alarm raising

and improve reliability, the iterative growth checking on the spread of the fire is checked. If the extracted fire pixels from the video frames increases with time, then it can be said that flame is spreading to a higher rate.

Let the variables, N denote the number of times of comparing k_i and k_{i+1} at repeated durations of time t_R during the time period T and the value R denotes $k_{i+1} \geq k_i$.

$R/N \geq 0.7$ Fire is about to spread out
$R/N \geq 0.9$ Fire spreads out
$0.7 \leq R/N \leq 0.9$ Determined by fuel and airflow

This is suitable only for the spread of general burning. For violent fuel and rapid flames, we use another strategy, where we check if $((k_{i+1}-k_i)/k_i) > S$. Here, S denotes the growth rate of flames and $S \geq 2$.

4   Result

We tested our fire detection algorithm on various video clips obtained from the internet and captured from real environments. The details about the video clips and the results are tabulated below (Fig. 4).

Video frame	Description	Resolution	No. of frames	Accuracy
	Burning of firewood during the night.	640 x 352	62	98.39%
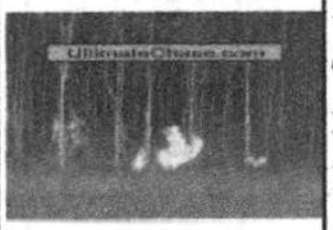	Burning of dried shrubs in forest floor in the day time	400 x 256	101	97.03%

Fig. 4. RGB model – grayscale model of fire and smoke.

	Burning of a tree in the forest during day time	400 x 256	100	100%
	Starting with no flame and gradually fire catches up in dark room.	426 x 240	1700	99.12%
	Small forest fire in the daytime.	400 x 240	101	49.50%
	Street view during the night. There is no fire in this video.	720 x 1280	573	98.78%
	Video of entry to the house in an illuminated environment.	720 x 1280	253	97.63%
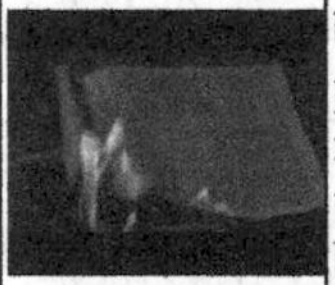	Gradual burning of paper from no-fire to total burning of paper.	176 x 144	1370	70.73%
	Burning of sofa cushion in a dark room.	426 x 172	94	75.53%
	Burning of a wooden plank in the day time.	426 x 240	500	41.4
	Fire in the corner during day-time.	640 x 360	501	99.60%

Fig. 4. (*continued*)

5   Conclusion

The proposed methodology for fire detection is based on analyzing Chromatic and Dynamic feature of the video. The aliased flame can be detected by checking the spread in fire by analyzing dynamic features. With smoke detection, the presence of fire is confirmed. Thus, using Image processing through CCTV cameras, the fire, and its spread can be detected easily, and fire alarm can be raised.

The proposed methodology for fire detection is based on analyzing chromatic and dynamic features of the video frames. The chromatic feature is adopted for fire-pixel extraction based on three decision rules applied on the RGB color model and saturation of red component. The real-flame and fire-alias are distinguished based on the dynamic features of the extracted fire-pixels. To reduce raising a false fire-alarm, the amount of fire-pixels extracted are compared over specified intervals of time. The main challenge that was faced in this research was selecting the right thresholds for the various RGB color model constraints. The main reason for this is that the color related values of flame depend on the type of fuel, amount of oxygen present and other background illuminations. The experimental results obtained from our algorithm shows that an accuracy of 94% on average can be achieved. The testing is done on various kinds of flames with different backgrounds, time and illuminations. Higher accuracy can be achieved by setting the thresholds depending upon the specific environment where the static cameras are attached. The limitation of the proposed methodology is that the dynamic features are greatly affected due to the random movements of various in the flame, but still, this cost-effective method can be implemented in forest areas, mountains where human interactions are very less.

Datasets Used

1. https://www.nist.gov/video-category/fire
2. https://www.kaggle.com/csjcsj7477/firedetectionmodelkeras-for-video/version/1
3. http://signal.ee.bilkent.edu.tr/VisiFire/Demo/FireClips/
4. https://www.youtube.com/watch?v=MnlR-J1qX68

Acknowledgment. This proposed work is a part of the project supported by DST (DST/TWF Division/AFW for EM/C/2017/121) titled A framework for event modeling and detection for Smart Buildings using Vision Systems to Amrita Vishwa Vidyapeetham, Coimbatore.

References

1. Nandi, C.S.: Automated fire detection and controlling system (2015)
2. Chi, R., Lu, Z.-M., Ji, Q.-G.: Real-time multi-feature based fire flame detection in video. IET Image Process. **11**, 31–37 (2016)
3. Lo, S.-W., Wu, J.-H., Lin, F.-P., Hsu, C.-H.: Cyber surveillance for flood disasters. Sensors **15**, 2369–2387 (2015)
4. Gunay, O., Toreyin, B.U., Kose, K., Cetin, A.E.: Entropy functional based online adaptive decision fusion framework with application to wildfire detection in video. IEEE Trans. Image Process. **21**, 344–347 (2012)

5. (Chao-Ho) Chen, T.-H., Wu, P.-H., Chiou, Y.-C.: An early fire-detection method based on image processing. In: ICIP, pp. 1708–1710 (2004)
6. Sowah, R.A., Ofoli, A.R., Krakani, S.N., Fiawoo, S.Y.: Hardware design and web-based communication modules of a real-time multisensor fire detection and notification system using fuzzy logic. IEEE Trans. Ind. Appl. **53**(1), 559–566 (2017)
7. Celik, T.: Fast and efficient method for fire detection using image processing (2014)
8. Philips III, W., Shah, M., da Vitoria Lobo, N.: Flame recognition in video. Pattern Recogn. Lett. **23**, 319–327 (2002)
9. Chen, S., Bao, H., Zeng, X., Yang, Y.: A fire detecting method based on multi-sensor data fusion. In: IEEE International Conference on Systems, Man and Cybernetics, 15–88 October 2003, vol. 14, pp. 3775- 3780 (2003)
10. Ho, C.-C., Kuo, T.-H.: Real-time video-based fire smoke detection system. In: IEEE/ASME International Conference on Advanced Intelligent Mechatronics 2009, AIM 2009, pp. 1845–1850 (2009)
11. Bondarenko, V.V., Vasyukov, V.V.: Software and hardware complex for detecting forest fire. In: 2012 11th International Conference on Actual Problems of Electronics Instrument Engineering, APEIE, pp. 138–142 (2012)
12. Rinsurongkawong, S., Ekpanyapong, M., Dailey, M.N.: Fire detection for early fire alarm based on optical flow video processing. In: 2012 9th International Conference on Electrical Engineering/Electronics Computer Telecommunications and Information Technology, ECTI-CON, pp. 1–4 (2012)
13. Chen, X.-H., Zhang, X.-Y., Zhang, Q.-X.: Fire alarm using multi-rules detection and texture features classification in video surveillance. In: 2014 7th International Conference on Intelligent Computation Technology and Automation, ICICTA, pp. 264–267 (2014)
14. Kumar, S., Saivenkateswaran, S.: Evaluation of video analytics for face detection and recognition. Int. J. Appl. Eng. Res. **10**(9), 24003–24016 (2015)
15. Athira, S., Manjusha, R., Parameswaran, L.: Scene understanding in images. In: Corchado Rodriguez, J., Mitra, S., Thampi, S., El-Alfy, E.S. (eds.) Intelligent Systems Technologies and Applications 2016. ISTA 2016. Advances in Intelligent Systems and Computing, vol. 530. Springer, Cham (2016)

Smart Healthcare Data Acquisition System

Tanvi Ghole[✉], Shruti Karande, Harshita Mondkar, and Sujata Kulkarni

Department of Electronics and Telecommunication,
Sardar Patel Institute of Technology, Mumbai, India
{tanvi.ghole,shruti.karande,harshita.mondkar,sujata_kulkarni}@spit.ac.in

Abstract. A general healthcare system monitors the parameters and simply display the readings. The records are maintained on paper. This paper aims in building an independent offline system for maintaining records in regions where network connectivity is an issue. After providing basic identification information it automatically logs vital parameters of patients acquired from the sensors. The records are maintained digitally and the system does a personalized data analysis for each patient, as "normality" is subjective to every patient. This system is trained to classify such minute details and create a cluster for each patient individually. As and when patient visit for checkup the data is added to his cluster and analysis is performed on it. Various suggestions are provided to the patient in order to avoid worsening the condition. It is a platform that provides the doctor with the patient's entire history when they visit a non-digitized area.

Keywords: Personalized healthcare · Data analysis · Machine learning · Clustering of data · Classification algorithm

1 Introduction

A general health-care systems monitors the parameters and simply display the readings. Mostly the records are maintained on paper. In recent times, there have been medical advancements leading to digitization and personalized medicines. But still there is a lack of basic health care services in the villages; causing inefficiency in checking the vital signs to identify minor disease. Even today the population in rural areas have to travel large distance even for regular checkups. Pregnant women, babies and aged people may not be able travel that far. Lack of internet connectivity, absence of basic medical facilities and non-availability of patient's medical history increases the demand for better medical solutions. Thus, the objective is to provide intelligent solution by providing a hassle-free offline health-care system.

Cluster analysis or clustering and predictive analysis can be used to provide intelligent solutions in the healthcare sector. In clustering, objects exhibiting similar properties are grouped together in a set forming a cluster.

© Springer Nature Switzerland AG 2020
S. Smys et al. (Eds.): ICCVBIC 2019, AISC 1108, pp. 798–808, 2020.
https://doi.org/10.1007/978-3-030-37218-7_88

The proposed system aims at taking it a notch higher. The records of the patients are maintained digitally and it does a personalized data analysis for each patient. As normality is subjective, it may be normal for one patient and not for the other. It also provides suggestions to the patients of what is to be done and what is to be avoided, in order to prevent worsening the condition.

2 Related Work

2.1 Literature Review

A government initiated VSTF survey was conducted on 7th June,2018 in various villages in Sudhagad taluka by students and teachers from Sardar Patel Institute of technology. While conducting the survey in Chive one of the villages, we came face to face with the problems faced by the people living over there. Some of the problems are listed below:

- Medical checkups in anganwadi are arranged every 3-months for pregnant women, primary school going children and adolescent girls.
- The caretakers appointed by government give medical suggestions based on previous experiences rather than on proper diagnosis.
- Proper medical facilities are not available in the villages (neither clinic nor hospitals).
- Villagers have to visit Pali for any medical help.

2.2 Existing Solutions on Vital Parameters

Health Care Monitoring System in Internet of Things (loT) by Using RFID [1]: This paper describes a complete IoT and RFID based monitoring system. The sensors senses the parameter and send it to mobile device using RFID technology. Limitations: The patient must be in the range of RFID to acquire the parameters [1].

Heart Rate Monitoring System Using Finger Tip and Processing Software [2]: This paper deals with measurement of heart rate from fingertip using optical methodology which involves IR transmitter and infrared detector rather than any conventional monitoring system.
Limitations: It is arduino based, so it can be used for study purpose only [2].

2.3 Existing Solutions on Zigbee Based System

A Zigbee-Based Wearable Physiological Parameter Monitoring System [3]: This system deals with blood pressure and heart rate where the sensors and circuitry was placed in wrist wrap and at finger tip.
Future Scope: It can be further improved by implementing noise reduction techniques and use of oximeter [3].

Transmitting Patient Vitals over a Reliable ZigBee Mesh Network [4]:
System is designed that converts pulses acquired from fingertips into beats per
minute. Later ZigBee is to display dynamically on a remote system [4].

2.4   Existing Solutions on Clustering Analysis

Clustering Analysis of Vital Signs Measured During Kidney Dialysis
[5]: Analysis of vital information of kidney dialysis is projected in this paper.
Vital signs such as pulse rate and respiration rate are analyzed. Analysis of
eight patient belonging to different age group was done. For this, a hierarchical
clustering method was used and to observe the similarities in vital parameters
multi-dimensional dynamic time warping distance is applied. If any parameters
deviates from the physiological rhythms then the abnormal conditions can be
determined according to the behaviour of the deviation [5].

3   Proposed System

3.1   Sensor Integration

The system comprises of various modules which are integrated together. One
module comprises of heart rate and blood pressure and the other temperature.
The band is mounted on the wrist in order to take heart rate and blood pressure
reading. Once the sensor acquire the data, it is sent to to Arm Cortex M0
Discovery Board. The data is send to a host system.

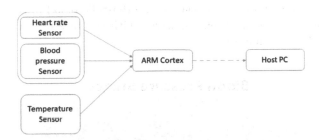

Fig. 1. Hardware block diagram for smart health-care data acquisition system

ARM Cortex. The Cortex-M0 is a processor which is highly energy efficient
with very low gate count. Its 32-bit performance is combined with high code
density by thumb instruction set. It also provides power control of system components is optimized. For low power consumption integrated sleep modes are
available. Increase in sleep mode time or slower processor clock is achieved by

fast code execution. For time-critical applications, it is deterministic and has high-performance interrupt handling. The number of pins required for debugging is reduced by serial wire debug [6].

Temperature Sensor MCP9700 At the point when the sensor interacts with the body of the patient it gains the temperature in Fahrenheit. For precise readings, temperature sensor must be held either in the palms or the armpits [7].

Features

- Analog Temperature Sensor is tiny
- Temperature Measurement Range is wide
  - Extended Temperature is $-40\,°C$ to $+125\,°C$
  - High temperature is $-40\,°C$ to $+150\,°C$
- Accuracy is $\pm2\,°C$ (max.) for $0\,°C$ to $+70\,°C$
- Operating Current is low i.e.6 µA [7]

3.2 Blood Pressure and Heart Rate Sensor

(1) Systolic Blood pressure: It is the pressure experienced by the artery walls while the blood flows when the heart beats.
(2) Diastolic Blood pressure: It is the pressure arteries experience due to blood flow between the two consecutive heart beats.
(3) Pulse Rate: It is the number of times a heart beats in a minute.

The sensor used provides us with systolic and diastolic blood pressure as well as pulse rate. These values are then compared with the threshold values given in Fig. 2 and the data is analyzed [9].

Blood Pressure Stages

Blood Pressure Category	Systolic mm Hg (upper #)		Diastolic mm Hg (lower #)
Normal	less than 120	and	less than 80
Elevated	120-129	and	less than 80
High Blood Pressure (Hypertension) Stage 1	130-139	or	80-89
High Blood Pressure (Hypertension) Stage 2	140 or higher	or	90 or higher
Hypertensive Crisis (Seek Emergency Care)	higher than 180	and/or	higher than 120

Source: American Heart Association

Fig. 2. General values for blood pressure

3.3   Test Data Generation

Figure 3 explains the block diagram of the software module of the proposed system. Java based desktop application accepts data from the patient. An excel-sheet is exported that consists of the patient data. This data-sheet is further loaded as test data. Data is sampled and the model is trained with respect to this selected data. k-mean and hierarchical clustering is used to analyze the data.

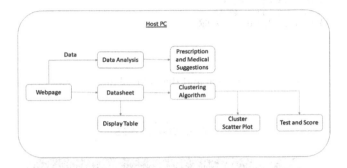

Fig. 3. Software flow

3.4   Data Analysis

Orange Data Mining Software: For data visualization, data mining and machine learning an open source software named as Orange is used. It includes a graphical front-end for scrutinized information investigation and hunch information representation, and utilized as a python library. Bio-informatics libraries are used to cluster the data of a person. This software is used for hierarchical clustering of data.

Algorithm: The data sheet is created by taking instances of vital parameters of individual patient from web-page. After every new entry the database is updated. This data is stored in .tab file which is loaded and sub-sampled. These sub-sampled instances are feed to k-mean model. [10] K-mean model uses euclidean distance between the data points and as the distance between the two points reduces the data instances are similar to each other. A dendrogram(tree diagram) represents the discovered clusters and the distance between these clusters with annotations of required parameter. Scatter plot presents the data clusters of informative projections like ECG v/s gender.

3.5   Serial Interface

Tera-Term: To read the data transmitted on the serial port of pc using ARM board we used tera-term which is a open source tool. It bolsters SSH 1 and 2, telnet and sequential port associations. It likewise has a worked in large scale scripting dialect and a couple of other helpful modules.

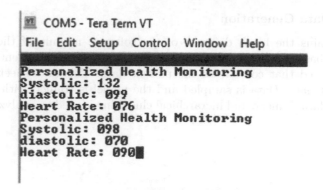

Fig. 4. Heart rate, systolic and diastolic blood pressure display

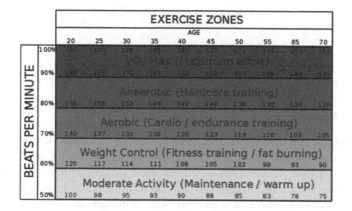

Fig. 5. Heart rate threshold

4 Experimental Results and Analysis

4.1 Hardware

The vital parameters were measured and displayed successfully. Figure 4 displays the heart rate, systolic blood pressure and diastolic blood pressure on Tera Term VT terminal. Data is serially received at the port and it's separated values are displayed respectively. For every new set of sensed data, data is displayed on new line with previously existing values. These results lie within the specified range given in Fig. 5.

Figure 6 displays the body temperature value of a person. This measured value is compared with the general threshold value for temperature given in Fig. 7.

Fig. 6. Temperature display

Method	Range (°F)		
0-2 years (Infant)			
Oral	N/A		
Rectal	97.9		100.4
Axillary	94.5		99.1
Ear	97.5		100.4
Core	97.5		100.0
3-10 years (Child)			
Oral	95.9		99.5
Rectal	97.9		100.4
Axillary	96.6	98.0	
Ear	97.0		100.0
Core	97.5		100.0
11-65 years (Adult)			
Oral	97.6		99.6
Rectal	98.6		100.6
Axillary	95.3		98.4
Ear	96.6		99.7
Core	98.2		100.2
65+ years (Senior)			
Oral	96.4		98.5
Rectal	97.1		99.2
Axillary	96.0	97.4	
Ear	96.4		99.5
Core	96.6		98.8

Normal Body Temperature Ranges

Fig. 7. Temperature threshold

4.2 Software

Data Acquisition: Fig. 8 demonstrates a desktop application that is created using java 8 to accept data from patient and keep individual records of each patient. It accepts patient's name,date of birth , gender, weight, height. Heart rate systolic and diastolic blood pressure and temperature measured directly from sensors. These details are stored in a database and xml file is generated as depicted in Fig. 9.

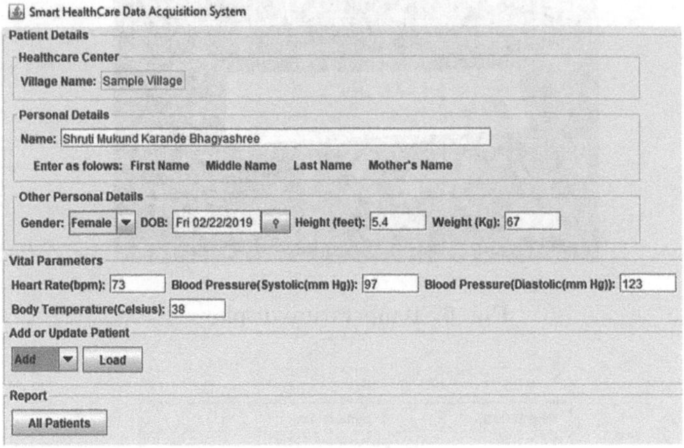

Fig. 8. Desktop application

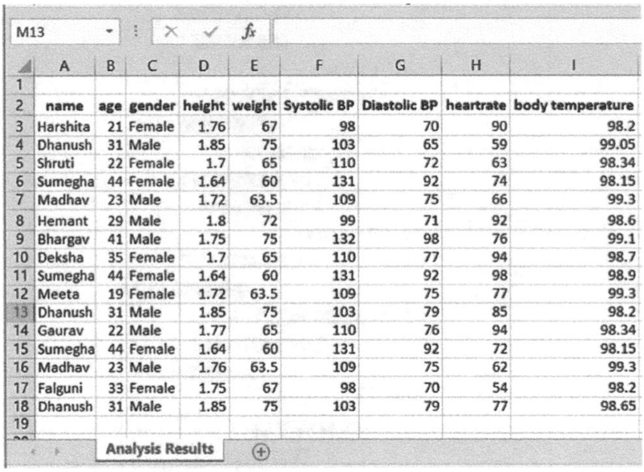

name	age	gender	height	weight	Systolic BP	Diastolic BP	heartrate	body temperature
Harshita	21	Female	1.76	67	98	70	90	98.2
Dhanush	31	Male	1.85	75	103	65	59	99.05
Shruti	22	Female	1.7	65	110	72	63	98.34
Sumegha	44	Female	1.64	60	131	92	74	98.15
Madhav	23	Male	1.72	63.5	109	75	66	99.3
Hemant	29	Male	1.8	72	99	71	92	98.6
Bhargav	41	Male	1.75	75	132	98	76	99.1
Deksha	35	Female	1.7	65	110	77	94	98.7
Sumegha	44	Female	1.64	60	131	92	98	98.9
Meeta	19	Female	1.72	63.5	109	75	77	99.3
Dhanush	31	Male	1.85	75	103	79	85	98.2
Gaurav	22	Male	1.77	65	110	76	94	98.34
Sumegha	44	Female	1.64	60	131	92	72	98.15
Madhav	23	Male	1.76	63.5	109	75	62	99.3
Falguni	33	Female	1.75	67	98	70	54	98.2
Dhanush	31	Male	1.85	75	103	79	77	98.65

Fig. 9. Exported excel-sheet

Data Analysis: Data analysis based on hierarchical clustering model of selected data is projection using visualization models i.e. scatter plot and tree diagram. Figure 10 represents a dendrogram of ECG. The annotations can be selected depending on area of focus. Here the dendrogram is generated on bases of ECG annotations and the instances in the clustered data is labeled depending on the stages it belongs i.e. normal, left vent hypertrophy and ST-T abnormal. The selected cluster in dendrogram indicates the instances belonging to same category and they are highlighted in the scatter plot given in Fig. 11.

Figure 11 is a scatter plot of ECG v/s gender. The scatter plot is used to display informative projections. The size, shape and colour of the symbols is

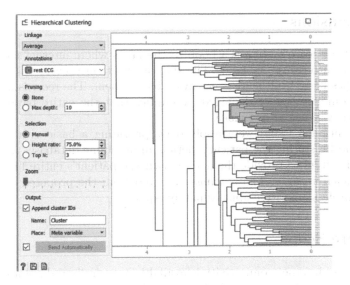

Fig. 10. Dondrogram(ECG)

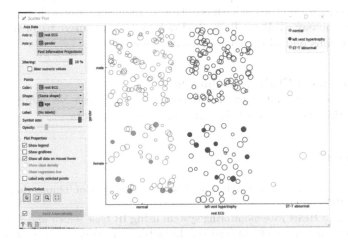

Fig. 11. Scatter plot (ECG v/s gender)

customized. There are six clusters that are created because of five categories, two on Y-axis(gender) namely male and female and three on X-axis(ECG stages) namely normal, left vent hypertrophy and ST-T abnormal. Blue coloured points are under normal category, Red points or bubbles indicate critical situation of left-vent hypertrophy and green points indicate ST-T abnormal category. The size of the bubbles indicate the age group falling under the category. The opaque bubbles indicate the selected instances of the cluster.

5   Conclusions

An offline healthcare system is a boon to the areas where internet connectivity and basic healthcare is an issue. The doctors visit once every 2 to 3 months which implies basic healthcare is not available to them. The records maintained on paper could be lost or damaged. It implies that the entire patient history may not be available at one place.

Aforementioned issues were resolved by providing a basic vital parameter checkup and by storing the patient's details in an offline database. A clustering model is also used to provide the doctor with the patient's entire history when they visit such an area. The readings observed from the proposed system and an actual professional were approximately equal.

6   Future Scope

In areas where internet connectivity is available a centralized database could be maintained so that the patient's history is readily available at a remote location. In order to make the system portable, an offline mobile application could be implemented provided the device has optimum memory and wireless connectivity or Li-Fi technology. Tele-medicine technology can be used to provide expert medical solutions and direct connectivity to doctors.

Compliance with Ethical Standards
All authors declare that there is no conflict of interest No humans/animals involved in this research work. We have used our own data.

References

1. Khan, S.F.: Health Care Monitoring System in Internet of Things (IoT) by Using RFID (2017)
2. Pawar, P.A.: Heart rate monitoring system using IR base sensor and Arduino Uno. In: 2014 Conference on IT in Business, Industry and Government (CSIBIG) (2014)
3. Malhi, K., Mukhopadhyay, S., Schnepper, J., Haefke, M., Ewald, H.: A zigbee-based wearable physiological parameters monitoring system. IEEE Sens. J. **12**(3), 423–430 (2012)
4. Filsoof, R., Bodine, A., Gill, B., Makonin, S., Nicholson, R.: Transmitting patient vitals over a reliable zigbee mesh network. In: 2014 IEEE Canada International Humanitarian Technology Conference - (IHTC) (2012)
5. Yamamoto, K., Watanobe, Y., Chen, W.: Clustering analysis of vital signs measured during kidney dialysis (2007)
6. Infocenter.arm.com (2018). http://infocenter.arm.com/help/topic/com.arm.doc.ddi0432c/DDI0432Ccor-tex-m-r0p0-trm.pdf. Accessed 10 Oct 2018
7. CVA, K.: J. Orthop. Bone Disord. **1**(7) (2017). https://www.medwinpublishers.com/JOBD/JOBD16000139.pdf

8. American Diagnostic Corporation - Core Medical Device Manufacturer. Stetho-scopes, Blood Pressure, Thermometry, and EENT. Adctoday.com (2019). https://www.adctoday.com/learning-center/about-thermometers/how-take-temperature. Accessed 28 Jan 2019

9. Blood Pressure Sensor - Serial output [1437]: Sunrom Electronics/Technologies. Sunrom.com (2018). https://www.sunrom.com/p/blood-pressure-sensor-serial-output. Accessed 9 Oct 2018

10. Kanungo, T., Mount, D.M., Netanyahu, N.S., Piatko, C.D., Silverman, R., Wu, A.Y.: An efficient k-means clustering algorithm: analysis and implementation. IEEE Trans. Pattern Anal. Mach. Intell. **24**, 881–892 (2002)

11. Disabled World. Blood Pressure Chart: Low Normal High Reading by Age Table. Disabled World, 28 Jul 2018. http://www.disabled-world.com/calculators-charts/bloodpressurechart.php. Accessed 8 Oct 2018

12. Filsoof, R., Bodine, A., Gill, B., Makonin, S., Nicholson, R.: Transmitting patient vitals over a reliable ZigBee mesh network. In: IEEE Canada International Humanitarian Technology Conference - (IHTC) (2014)

A Sparse Representation Based Learning Algorithm for Denoising in Images

Senthil Kumar Thangavel[(✉)] and Sudipta Rudra

Department of Computer Science and Engineering, Amrita School
of Engineering, Amrita Vishwa Vidyapeetham, Coimbatore, India
t_senthilkumar@cb.amrita.edu,
cb.en.p2csel7029@cb.students.amrita.edu

Abstract. In the past few decades, denoising is one of the major portions in imaging analysis and it is still an ongoing research problem. Depending upon some pursuit methods an attempt has been made to denoise an image. The work comes up with a new methodology for denoising with K-SVD algorithm. Noise information has been extracted using the proposed approach. With reference to heap sort image patches are learnt using dictionary and then it is updated. Experimentation says that introduced approach reduces noise on test. The proposed approach is tested on test datasets and the proposed approach is found to be comparatively good than the existing works.

Keywords: Image denoising · Dictionary · K-SVD · Image patches · Sparse learning · Representation

1 Introduction

In current and advanced developments for modern digital imaging analysis denoising plays a important role. Image steganography analysis is the elementary issue of noise elimination and information recovery in image processing that preserves corner and edge details of the image [1]. Image noise is also related to variation of pixels in a image. The variation is related to artifacts in the image. The variation is due to the changes in spatial and frequency domain. The noise varies from image to video. The noise can be removed by suitable algorithms.

Noise in image is unwanted and the output form it is unexpected. An image is a digital representation of pixels that means the intensity values. Noise in an image cannot be avoided but they can be minimised. Noise in image is unwanted and the output form it is unexpected. An image is a digital representation of pixels that means the intensity values. Noise in an image cannot be avoided but they can be minimised.

Here x represents original intensity value and n refers to noise. Noise can be identified based on earlier information regarding the intensity of pixels. On averaging similar pixels the probability of getting a clear patch is maximum. Noise affects information sometimes. In real time applications like face detection high noise may cause problem, also in low resolution images if SNR is high then the image is not suitable for processing. Even illumination problem results in noise, burring results in noise. The elimination of noise from an image is called denoising, there are many ways

© Springer Nature Switzerland AG 2020
S. Smys et al. (Eds.): ICCVBIC 2019, AISC 1108, pp. 809–826, 2020.
https://doi.org/10.1007/978-3-030-37218-7_89

for denoising an image. Applying filters can one of the methods for denoising and some of the filters that can be applied for denoising are mean, median and Weiner filters.

$$y = x + ny = x + n$$

Here x is intensity value and n is noise. Noise can be identified based on earlier information regarding the intensity of pixels. On averaging similar pixels, the probability of getting a clear patch is maximum. Periodic noise happens through electromagnetic inference. We can see the periodic noise as discrete spikes and the occurrence of noise reduction can be done using notch. If the raise of noise is found in borders then low pass filter is a remedy for it as it reduces the noise levels. Even the surrounding temperate could have a sufficient effect on noise. Some noise can be even caused by quantization of pixels and it can be removed by using Gaussian filter. There are some noise with random dot patterns and then can be fused with pictures and this results in snow, which is a from of electronic noise. The noise also depends on the shutter speed, even abnormal shutter speed can also result in increased noise. Noise in images can be cleared suing both spatial and frequency domain filters. Some noises are signal dependent

Image noise is also related to variation of pixels in a image. The variation is related to artifacts in the image. The variation is due to the changes in spatial and frequency domain. The noise varies from image to video. The noise can be removed by suitable algorithms. Periodic noise could be a regular changes in the image. It can be suitably found and predicted using suitable filters.

The sensors that are as part of camera also has a effect on the image characteristics. The sensors can be influenced by heat, fill factor. The sensor data can also be fused for understanding the image characteristics. The uncertainty of noise is the major challenges. It is highly challenging in video as the noise varies in location from frame to frame. Learning the parameters and model the same can help for denoising. Generated model should be able to reconstruct the error in the image set. Patch based approaches divide the image into regions. Each regions is further applied with a set of tasks for learning the noise features. They are then approximated for classifying the type of noise.

Architectures namely KrispNet are suitable for noise reduction. The noise as part of audio content in moving pictures can removed using codec. Noise can be removed by improving the preprocessing at input layers. The deep learning could be considered when there are several thousands of images. The uncertainty of image pixels contributing to noise can be considered as a motivation for using deep learning. The fully convolution layer can handle the error in image.

Image noise is a part of acquistion. It affects the brightness variation in random manner. Denoising is the basis pre-processing technique for removing the noise in an image/video. Prior to denoising, analyse the noise estimation and type of noise is an important. In most cases Gaussian noise is occurred in image during transmission. Many denoising methods are used such as Un-biased risk estimator, weighted averaging, edge preserving and segmentation for specific application. The distribution function can be identified for type of noise. The value is considered as a feature vector of the any specific noise of the image.

Gaussian noise can be handled using gaussian filter. Noise in cameras can be handled with suitable cameras. It can be used for image enhancement and reconstruction. The results vary on type of noise. Classification is a type of associating the image with a class label. The class label can be a single or multiclass. Training the data with less images will lead to overfitting.

Noise is considered to be one of the major performance affecting factor while analysing an image or video, the noise might be due changes in environment, defective lenses, motion based etc... the image can be successfully analysed only of it can be successfully removed with noise. For noise removal several filters has been proposed. The existing algorithms focusses on reducing the noise the largest noise value. That is the region or patch with noise value more than the threshold is minimized. The proposed approach aims to remove noise from image patches and recover the original image from noise-filled image [2, 3]. Researchers have come up with various image denoising formulations with an objective to safeguard the image quality [1], because the noise could have harsh effect on varying applications namely [4–8], signal processing along with video processing [9]. Michal Aharon et al. [10], proposed a technique for dictionary learning and updation of non color image regions. In the past few decades, the K-SVD algorithm has been depicted for dictionary learning for sparse image representation [11]. It has been found out that a sparse representation technique has proved extremely successful technique for Gaussian noise removal. In this paper, we proposed a novel denoising method. Our main aim is to combine the heap sorting technique [12, 13] with dictionary learning method [12, 10] with an objective to remove noise. In this section we discuss about denoising applications and the techniques that we use to denoise an image. There are state of the art methods that focuses on spatial level denoising [14–16]. Michal Aharon et al. [10] came up with an image representation using K-Means clustering technique to handle denoising issue. Dabov et al. [17], exhaustive window matching technique for noise removal. Dabov et al. [18] proposed a filter that monitors shape regions and uses PCA technique for denoising with presenting the disadvantage of very high computational complexity. Zhang et al. [19] an image-denoising process based on PCA and here neighbourhood pixels are grouped. This technique uses the block matching method to group among the pixels with similarly local structures and transformed each group of pixels through locally learned PCA. He et al. [20] a technique using SVD for learning local basis. At stage one dictionary initialization process has been done to provide data overlapping patches. In the second stage denoising phenomenon is applied and at the final stage the dictionaries get updated. L. Ganesh et al. [21] fused two images for noise removal. images are acquired by sensors and other capturing devices. Non-linear filters are used to remove noisy images. image fusion algorithm for impulse noise removal. Then on the filtered image the image fusion techniques are done and the density of noise using image ranges from 10 to 50%. J. Harikiran et al. [22] fused images to remove salt and pepper noise and it can be removed by statistic filters. Fuzzy filters can also be used for noise reduction [23] and all the above said filters reduce the impulse noise Rong Zhu et al. [24] came up with an algorithm that reduces noise and gives noise matrix. The grey value across all the pixel is replaced by median of neighbourhood. Local histogram is used to preserve the original pixel values from noise. the histogram gives the impulse noise pixel value. It is observed that the proposed approach works with better precision than existing. J. Aalavandan et al. [25] came up with a method by which the median filter is switched and noise removal is done using two stages, at stage 1 the noise regions are identified and in stage 2 the noise that surpasses above the threshold are removed. Video data has challenges due to motion. Moving pixels are hard to classify due to nature of the video. However it is possible to model the spatial motion of the pixels based on the problem [43]. PSNR can be used as a performance metric. This is part of every video codec in design. A other way is to understand the spatio temporal nature in video and model them as blocks using Laplacian. Motion characteristics can be classified as slow or fast based on the problem characteristics. The estimation error can be considered based on Peak Signal to Noise Ratio [44]. Approaches have been developed that considers the image structure and categorizes the type of noise. As part of the approach noise with PSNR between 20 and 50 db is added and is checked for its estimation in the algorithm [45]. Average PSNR is used for estimating the performance. Certain practices adopt the approach of minimization problem for handling noise. Poisson and Gaussian Noise has been considered for analysis. The approach is compared with algorithms VBM3D, VBM4D, RPCA [46]. Night vision present more challenges as the noise variations are hard to find. Approaches with Bilateral filter has also been proposed [47].

An approach has been followed by Ashish Raj et al. [30] to handle parallel image problem with gaussian noise. The problem is seen as an quadratic optimization problem that has to be minimized. Huanjing Yue et al. [31] used variance maps based on Bayer pattern. Meisam Rakhshanfar et al. [32] followed an approach to overcome noise. This approach detects the damaged edges and recovers the same using soft thresholding. Philippe Martin-Gonthier et al. [33] analysed the noise using power spectral density, they started analysing noise in frequency domain. Zhu Lin [34] used NL-means algorithm for removing noise. Singular Value Decomposition is related with matrix decomposition. The diagonal values are called as singular values. It takes a singular valued matrix by calculating the decomposition. It provides a way for achieving dimensionality reduction. Both Gaussian and impulse noise are removed using this proposed approach. Decomposing pixels into region results in image quality reduction as the pixels affected further affects the neighbouring pixels. Based on NL means algorithm an adaptive means algorithm is also proposed and it could remove mixed noise at a single step. Section 2 presents the proposed denoising approach in detail with formulation. Section 3 discusses the experimental results. Findings are presented in Sect. 4.

2 Proposed Methods

In this paper, we address the start of the art image-denoising as a radical problem for image analysis where Y is additive mean, Gaussian noise X and standard deviation σ is considered. Thus the representation becomes,

$$Y = X + \sigma$$

Motivated by K-SVD [10], we proposed the strategies for denoising with layers such as dictionary learning, patch extraction, and updating and the last layer deals with denoising methodology.

Here, we represent the basic K-SVD process with a constrain $min_{dx}\{|X - DY|_2^p\}$ representing the basic K-SVD phenomenon for image denoising process; Subject to $\forall$, where, $x_i = e_k$ for some k, $min_{dx}\{|Y - DX|_2^p\}$; Subject to $\forall$ where, $\|y\| = 1$. Again, for updating, $min_{dx}\{|X - DY|_2^p\}$: Subjected to $\forall; N_{X_i} \leq n_p$ or $min_{dx} \sum \|x_i\|$ subjects to $\forall_i$, $\|X - DY\|_2^F \leq \in \min(n_p)$. Here D is considered to be a fixed coefficient matrix.

$$\|X - DY\|_2^F = \left\| \sum X_i w_i \otimes \frac{1}{2}(X_i - E^2) \right\|$$

In this study, our proposed algorithm fills the values in the dictionary regions after the training phase [21] and updation happens continuously.

2.1 Stage-1

Among all these methods being presented in the literature section, K-SVD algorithm being presented by Michal Aharon et al. [10] has become one of the wide potential candidates. This approach aims to find the best coefficient vector x while the dictionary Φ is provided. At this stage coefficient matrix is calculated. With using the dictionary updation rule, the K-SVD extracts noisy elements. At the process of dictionary updation stage [22], the optimal function of the K-SVD algorithm is being represented as follows:

$$\|Z - DX\|_2^F = \left\| Z - \sum_M^{N-1} d_x \right\|_2^F$$
$$= \left\| \left(Z - \sum_{N \neq M} d_x \right) - d_j \right\|$$
$$= \|E_K - d_x\|_2^F$$

The kth row in X is defined as x_t^j which is not mentioned as due to the mathematical ambiguity and as well as the d_x is the kth columns in dictionary D. Matrix E_K is represented as the error matrix in where the kth atom is removed. The definition of atom d_k is as follows:

$$\delta_k = \{j | 1 \leq k, x_t^j(i) \neq 0\}.$$

With reference to the above process we could infer that after the training process the dictionary gets updated. the training set is used in an iterative manner the dictionary learning for patches gradually gets improved.

Let be σ - noisy pixels and let x is the characteristic matrix that is being defined as follows

$$x = \{1 \ if \ (i, j \in 0)0, otherwise$$

Then the above equations are being formulated as follows

$$\omega \left\| \sum E_K - d_x \right\| \otimes \left\| \left(X - \sum_{N \neq M} d_x \right) \right\|_2^F + \sum (X - E^2) + \sum_J^I \sigma_{ij}$$

In the above equation the noise matrix and characteristics matrix are multiplied. E^2 is a fidelity operator covering all the patch-based atom.

Algorithm 1: Dictionary Initialisation

Input- noisy image.
Output- unordered image regions.
Step-1: Let i, j ⊗pixels of the input data
Step-2: Set the region size (a, b), step size and loops.
Step-3: Column vectors ⊗Patches, $f(a, b) \in j$
Step-4: K- Number of iterations
Step-5: Unordered pixel extraction with the proportion
        of E^2.
Step-6: Stop

Here from algorithm depicted above, it is inferred that from an input, a window of size of $a \times b$ – reference patch from the input image and its function as $f(a, b)$. The aforementioned algorithm after continuous iteration extracts overlapped patches.

2.2   Stage-2

The overlapped and unordered image is given as input to the sequential algorithm in stage 2 and the output of the algorithm is patches for image denoising.

Algorithm-2: Linear Algorithm

Input: Crossed over image regions.
Output: Identify of regions for image noise minimization.

Step-1: Noise REGION ⊗a_0 and other regions ⊗f(a_0) = b and let the
        Loop count ⊗K=1
Step-2: region (b_{t_i}) selection and max criterion fitting $max|b_t' a_0 - 1|$
b_{t_i} Addition to the residual patches and process gets updated.
Step-3: $B_i = f(b_{t_i})f(b_{t_i})'f(b_{t_i})^{-1}$, Linear space projection
Step-4: If K < 1 ⊗→ End.
        else
        Increment K by 1 and move to step 2
Step-5: Exit the algorithm.

The noise pixels are initialized and the residual patches are selected. The iteration starts with this process. Then the selected patches are allowed to fit the criterion mentioned in step 2.

Let X $\rightarrow$ the support-vector of σ

K (X) $\rightarrow$ column vectors j of K for X.

$$M = max_{k \in T/K(X)} \left\{ \left\| K(X)'K(X)^{-1}K(X)'k \right\|^2 \right\}$$

The stopping condition satisfies when K < 1.

Given, set of atoms $x = \{x_1, x_2, x_3, \ldots \ldots \ldots \ldots x_n, x_{n+1}\}$. Here, for the sequential algorithm, that often aims at detecting a set of dictionary vector, represented as $D = \{d_1, d_2, d_3, \ldots \ldots \ldots \ldots d_n, d_{n+1}\}$. For which, each atom $n \in N$ atoms can be well represented by a combination of $\{d_J\}_{j=1,2\ldots n}^M$. Such that $y = \sum_{n=1}^m a_l b_l$ and such, coefficients a_l are zero or close to zero. The algorithm can be formulated with respect to the area and the denoising phenomenon represented as $min_{D,c_i,p_{max}} \sum_{i=1}^{P_{max}} \frac{3}{4} \|b_i - Dc_i\|^2 \otimes \|\tau c_i\|_2$ subject to $\|d_i\| = 1.0$, $1.0 \leq i \leq n$.

$$min_{dx} \left\{ |X - DY|_2^p \right\} \text{ Subject to } \forall N_{Y_l} \leq n_p$$

Here max heap property is utilized [24] and is given as is given as

$$S[Parent(i)] \leq S(i)$$

Which means the value of parent node is higher than the child. can have. Thus in the proposed heap structure the maximum value is stored at the root.

$$Z = \{1, \quad if |X - DY|_2^p \ll N_{Y_l}, 0, \quad if |X - DY|_2^p = N_{Y_l} - 1, \quad if |X - DY|_2^p \gg N_{Y_l}$$

We know that $N_{X_l} \leq n_p$ represented above. In such, the optimal performance for the dictionary D is shown below in the equation.

$$max_{E_K, z/Y} \left\| \sum_l E_K - d_y \right\| \otimes \left\| \left(Z - \sum_{N \neq M} d_Y \right) \right\|_2^F + \sum (Y - E^2) + \sum_J^I \sigma_{ij}$$

Here $E^2 > n_p$, such that the near-optimal node selection is simplified. Therefore, the time complexity for denoising is minimized. With respect to dictionary search is done with the formula given below.

$$\frac{max_{E_K, z/Y} \left\| \sum E_K - d_y \right\| \otimes \left\| \left(Z - \sum_{N \neq M} d_y \right) \right\|_2^F + \sum (X - E^2) + \sum_J^I \sigma_{ij}}{\|x_i - Dc_i\|^2} W$$

In the above Equation W represents the patch atoms provided. The max heap property is as follows.

$$
\left(1 + \frac{\dfrac{1}{max_{E_K,z/Y}\left\|\sum E_K - d_y\right\|\otimes\left\|\left(z-\sum_{N\neq M}d_y\right)\right\|_2^F + \sum(X-E^2)+\sum_J' \sigma_{ij}}}{\dfrac{\|x_i - Dc_i\|^2}{W}} \right)^N
$$

After the atoms are selected the sorting begins. To compute the sparse matrix minimization technique is followed. sparse matrix is found out and finally, with the constraint as W >> 1, the sorting process starts in ascending fashion.

$$
min_{D,c_i,p_{max}} \sum_{i=1}^{p_{max}} \frac{3}{4}\|x - Dc_i\|^2 \otimes \tau \|c_i\|_2 \times \left\|\left(Z - \sum_{M\neq N} d_y\right)\right\|_2^F + \varphi\|c_i\|_2
$$

With the induction of this process in K-SVD [10] we could get better results with respect to PSNR. The proposed approach may learn with less accuracy from the trained sets but the noise removal is inferred to be comparatively good [10]. This methodology removes impulse noise in images.

2.3   Stage-3

The next phase presents the dictionary updating process. The entire patched from the above step are given as input. So, for this phase fixing all the n_{ij} and for each atom $d_{n,}$, where n varies from 1, 2, 3….in Dictionary D. Select the region k^l which will be used by these atoms ,$k^l = (n, m)|n_{ij}(l) \neq 0$.

So for each of the patches (a, b) $\in k^l$. The residual is completed using the below said formula $\forall$ among each patch; (a, b) that belongs to k^l

$$
e_{ij}^l = d_{ij}u - Dn_{ij} + d_J
$$

and $\omega_{ij}^l = D_{ij}\omega$ represents candidate pixels among the region of small image size $\sqrt{n} \times \sqrt{n}$ from the location (a, b) of the image.

Now set $E_l = (e_{ab}^l)_{a,b \in k^l}$, i, j for the image window and is stated below in algorithm 3.

Algorithm 3: Updating process

Input → Deteriorated Image
Output→ Recovered Image

Step-1: Set $E_l = (e_{ab}^l)_{a,b \in k^l}$, a, b⊛input image segments
Step-2: Initialise U, S, V vectors forE_l;
Step-3: Across all the patches [U, S, V] = svd $(E_l) \forall$ k;
Step-4: Coefficient replacement of atom D_{ij} in image with the entries of s_1, v_1, d_{ij}.
Step-5: Stopping condition verification.
    else
       then switch to algorithm-2 (for heap sort).
Step-6: Recovered images as $E_l = \hat{E}_l$
Step-7: Stop.

---

3 Experiments and Results

We present our results with four of the grayscale test images of size $X \times Y$ such as Barbara (512 × 512), Boat (512 × 512), House (512 × 512) and Lena (512 × 512). As these grayscale test images are generally used to validate the state-of-the-art noise removal techniques. The noise deviation (σ) is varied from 10 to 50 and as the intensity of each pixel ranges from 0 to 255. the proposed approach is tested with the grayscale images as shown in Fig. 1.

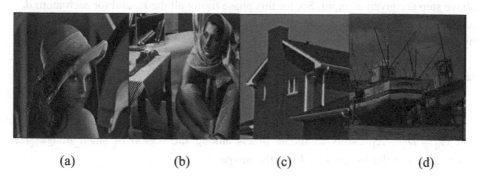

    (a)   (b)   (c)   (d)

Fig. 1. Grasycale images used to test the proposed methods (a) Lena (2) Barbara (3) House (4) Boat

(a) *Evaluation measures*

The peak signal-to-noise ratio (PSNR) and structural similarity index measure (SSIM) [26, 27] is adopted to evaluate the experimental results of denoising methods.

Peak Signal-to-Noise Ratio (PSNR): PSNR is one of the most widely used image quality measures in the imaging analysis. The PSNR for the given input is shown below.

$$PSNR = 20 * log_{10}(X) - 10 * log_{10}(X - \hat{X})/255.$$

Where, $\hat{X}$ is the restored image or the reconstructed image and X is the original image in all of the experiments with dictionary size as 64×256, whereas image patch size as 7×7. Therefore, this is enough to visualize the convergence. In this study, we have taken the window size= 10 and step size= 20.

Structural Similarity Index Measure (SSIM): Computation of similarity between two images is given based on the recovered image.

$$SSIM\left(X, \hat{X}\right) = \frac{1}{k}\sum_{J=1}^{k} SSIM_J\left(X, \hat{X}\right)$$

$$Where\ SSIM_I = \left[\frac{2\sigma_1\sigma_2}{\sigma_1^2 + \sigma_2^2}\right] \times \left[\frac{\sigma_{12}}{\sigma_1\sigma_2}\right] = \frac{2\sigma_{12}}{\sigma_1^2 + \sigma_2^2}$$

Herein, $\hat{X} -- > Recovered\ image$

X $\rightarrow$ Original Input data and the corresponding similarity values are stored in Table 1.

3.1   Results

The performance of the proposed approach is compared with methods as discussed above [21]. The programming language used to implement the proposed approach is python and visualizations are done using MATLAB. As shown in Figs. 2, 3, 4 the proposed approach is tested with various images such as Lena, Barbar and Boat respectively as shown in Table 1.

The dictionary being is shown in Fig. 5, each of size shown as 7×7 pixel. The proposed approach is executed for 1000 using the dataset with 1000 images each as shown in Fig. 5.

Figures 6, 7, 8, 9, 10 and 11 shows the variation in PSNR and SSIM for the proposed approach.

Table 1. Computing of PSNR (dB) and SSIM values of different denoising methods on test images with different noise levels

Input	Noise Levels (∫)	K-SVD		LPG-PCA		BM3D-SAPCA		SAIST		Proposed	
		PSNR	SSIM	PSNR	SSIM	PSNR	SSIM	PSNR	SSIM	PSNR	SSIM
Barbara	10	28.85	0.913	36.84	0.913	24.74	0.931	38.56	0.924	52.93	**0.932**
	20	23.20	0.896	35.58	0.896	23.72	**0.929**	36.77	0.907	46.65	0.927
	30	21.50	0.844	33.67	0.844	22.90	0.896	33.54	0.875	40.36	0.915
	40	19.39	0.819	31.89	0.819	22.29	0.873	31.56	0.843	33.37	**0.909**
	50	18.54	0.796	26.24	0.796	21.84	0.851	28.92	0.819	31.74	0.887
Boat	10	32.68	0.898	34.58	**0.927**	34.58	0.916	39.32	0.923	47.64	0.926
	20	27.48	0.856	31.56	**0.925**	31.56	0.905	37.77	0.917	45.28	0.918
	30	26.39	0.832	30.55	**0.916**	30.55	0.887	34.45	0.892	43.44	0.906
	40	25.48	0.814	29.67	0.812	29.67	0.856	30.26	**0.873**	42.67	**0.814**
	50	23.99	0.805	28.35	0.809	28.35	0.843	28.75	**0.849**	38.79	0.807
House	10	30.65	0.928	35.46	**0.936**	31.31	0.927	43.56	0.917	51.62	0.935
	20	25.70	0.916	33.25	0.929	29.67	0.917	40.75	0.879	49.77	**0.931**
	30	24.23	0.883	31.79	0.921	28.49	0.897	37.89	0.854	48.15	0.927
	40	21.53	0.854	29.32	0.810	27.23	0.876	35.14	0.837	38.55	**0.920**
	50	20.36	0.813	28.99	0.801	27.15	0.854	**31.82**	0.825	30.72	**0.917**
Lena	10	29.11	0.897	34.12	0.927	27.63	0.925	41.58	0.929	52.90	**0.931**
	20	23.95	0.786	30.24	0.861	26.70	0.897	39.24	0.917	47.90	**0.925**
	30	22.73	0.774	28.35	0.823	25.43	0.862	36.87	0.885	44.10	**0.919**
	40	20.18	0.759	26.85	0.810	25.38	0.847	31.98	0.864	**32.56**	**0.906**
	50	19.13	0.735	**25.72**	0.806	25.01	0.813	27.36	0.843	25.36	**0.899**
Avg.		24.253	0.840	31.151	0.864	27.21	0.885	35.305	0.879	**42.225**	**0.9076**

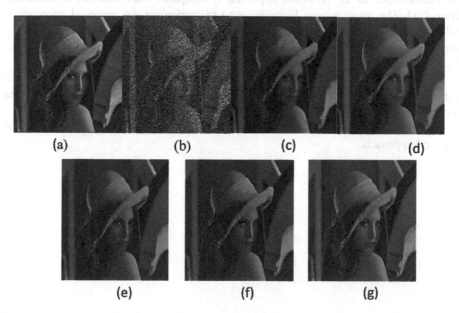

Fig. 2. (a) Original image. (b) Noisy Image (c) K- SVD (d) LPG –PCA (e) BM3D- SAPCA (f) SAIST (g) Proposed method.

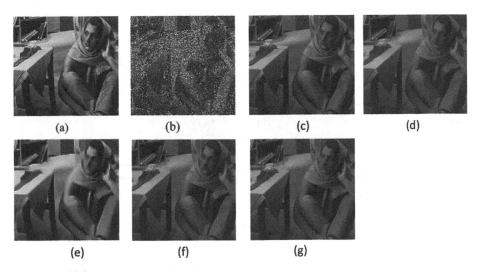

Fig. 3. Visual test results on Barbara image computed by various algorithms. (a) Original image. (b) Noisy Image (c) K- SVD (d) LPG –PCA (e) BM3D- SAPCA (f) SAIST (g) Proposed method.

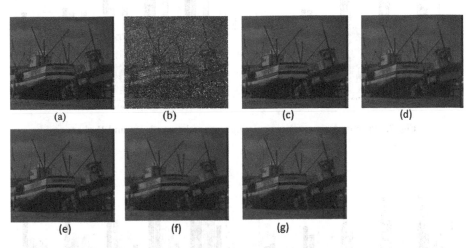

Fig. 4. Boat image computed – Existing vs Enhancement

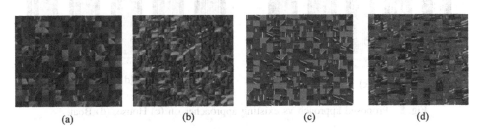

Fig. 5. Visual representation of trained dictionary images of the test images computed from proposed method. (a) Lena, (b) Barbara, (c) House, (d) Boat.

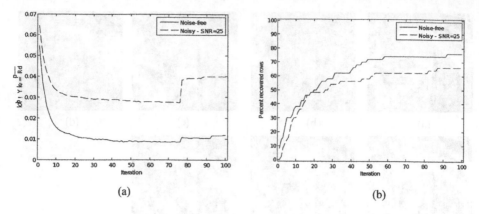

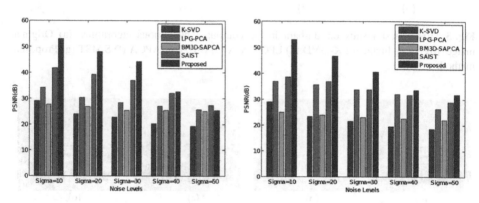

Fig. 6. .

Fig. 7. Proposed approach vs existing approaches on Lena and Barbara

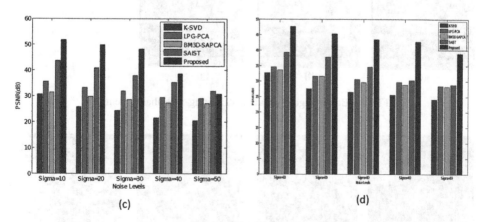

Fig. 8. Proposed approach vs existing approaches on (c) House, (d) Boat.

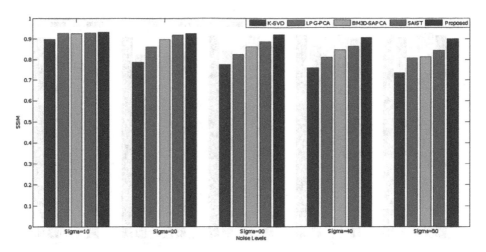

Fig. 9. Proposed approach vs existing approaches on Lena

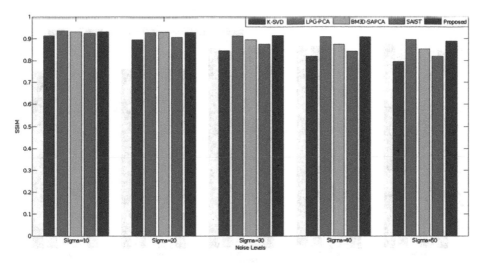

Fig. 10. Proposed approach vs existing approaches on Barbara

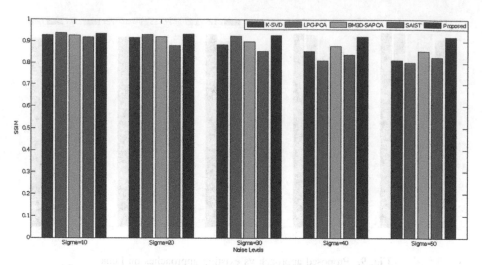

Fig. 11. Proposed approach vs existing approaches on Lena and Barbara

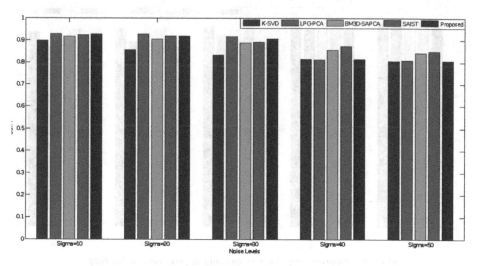

Fig. 12. Proposed approach vs existing approaches on Boat.

4 Conclusions

From this research we could infer that the formulated approach gives better PSNR, SSIM compared to other approaches due to max heap sorting strategy. The proposed approach is said to be a hybrid makeup of dictionary learning and heap sort technique. The results presented in Sect. 4 show how the proposed approach reconstructs the deteriorated images The quantitative and qualitative results on various test images show that our proposed algorithm can be used to denoise an image effectively.

Compliance with Ethical Standards
✓ All authors declare that there is no conflict of interest
✓ No humans/animals involved in this research work.
✓ We have used our own data.

References

1. Wang, G., Wang, Z., Liu, J.: A new image denoising method based on adaptive multiscale morphological edge detection. Math. Probl. Eng. **2017**, 11 (2017). https://doi.org/10.1155/2017/4065306. Article ID 4065306
2. Palhano Xavier de Fontes, F., Andrade Barroso, G., Coupé, P., et al.: Real time ultrasound image denoising. J. Real-Time Image Process. **2017**, 15–22 (2011). https://doi.org/10.1007/s11554-010-0158-5
3. Guo, Q.: An efficient SVD-based method for image denoising. IEEE Trans. Circ. Syst. Video Technol. 26(5), 868–880 (2016)
4. Han, Y., Xu, C., Baciu, G., Li, M.: Lightness biased cartoon-and-texture decomposition for textile image segmentation. Neurocomputing **168**, 575–587 (2015). https://doi.org/10.1016/j.neucom.2015.05.069
5. Liang, H., Weller, D.S.: Comparison-based image quality assessment for selecting image restoration parameters. IEEE Trans. Image Process. 25(11), 5118–5130 (2016)
6. Hao-Tian, W., et al.: A reversible data hiding method with contrast enhancement for medical images. J. Vis. Commun. Image Represent. 31, 146–153 (2015)
7. Gu, K., Tao, D., Qiao, J.F., Lin, W.: Learning a no-reference quality assessment model of enhanced images with big data. IEEE Trans. Neural Netw. Learn. Syst. **99**, 1–13 (2017)
8. Hung, K.W., Siu, W.C.: Robust soft-decision interpolation using weighted least squares. IEEE Trans. Image Process. 21(3), 1061–1069 (2012)
9. Mairal, J., Sapiro, G., Elad, M.: Learning multiscale sparse representations for image and video restoration. Multiscale Model. Simul. 7(1), 214–241 (2008)
10. Aharon, M., et al.: K-SVD: an algorithm for designing overcomplete dictionaries for sparse representation. IEEE Trans. Signal Process. **54**(11), 4311–4322 (2006)
11. Yang, J., et al.: A new approach to sparse image representation using MMV and K-SVD. In: Blanc-Talon, J., Philips, W., Popescu, D., Scheunders, P. (eds.) ACIVS 2009. LNCS, vol. 5807. Springer, Heidelberg
12. Hwang, H., Haddad, R.: Adaptive median filters: new algorithms and results. IEEE Trans. Image Process. **4**, 499–502 (1995)
13. Ko, S.J., Lee, Y.H.: Center weighted median filters and their applications to image enhancement. IEEE Trans. Circuits Syst. 38, 984–993 (1991)
14. Tomasi, C., Manduchi, R.: Bilateral filtering for gray and color images. In: IEEE International Conference on Computer Vision, ICCV - 1998, pp. 839–846 (1998)
15. Dai, T., Lu, W., et al.: Entropy-based bilateral filtering with a new range kernel. J. Signal Process. **137**, 223–234 (2017)
16. Buades, A., Coll, B., et al.: A non-local algorithm for image denoising. In: IEEE Conference on Computer Vision and Pattern Recognition, CVPR-2007, pp. 60–65 (2007)
17. Dabov, K., et al.: Image denoising by sparse 3-D transform-domain collaborative filtering. IEEE Trans. Image Process. **16**(8), 2080–2095 (2007)
18. Dabov, K., et al.: BM3D image denoising with shape-adaptive principal component analysis. In: Proceedings for Workshop of Signal Processing and Adaptive Sparse Structural Representation (SPARS), Saint-Malo, France, April 2009, HAL Id: inria-00369582

19. Zhang, L., et al.: Two-stage image denoising by principal component analysis with local pixel grouping. J. Pattern Recogn. **43**(4), 1531–1549 (2010)
20. Gan, T., Chen, W., et al.: Adaptive denoising by singular value decomposition. IEEE Signal Process. Lett. **18**(4), 215–218 (2011)
21. Ganesh, L., Chaitanya, S.P.K., Rao, J.D., Kumar, M.N.V.S.S.: Development of image fusion algorithm for impulse noise removal in digital images using the quality assessment in spatial domain. IJERA **1**(3), 786–792 (2014)
22. Harikiran, J., Saichandana, B., Divakar, B.: Impulse noise removal in digital images. IJCA **10**(8), 39–42 (2010)
23. Chauhan, J.: A comparative study of classical and fuzzy filters for impulse noise reduction. IJARCS **3**(1), 416–419 (2012)
24. Zhu, R., Wang, Y.: Application of improved median filter on image processing. J. Comput. **7** (4), 838–841 (2012)
25. Aalavandan, J., Santhosh Baboo, S.: Enhanced switching median filter for de-noising ultrasound. IJARCE **3**(2), 363–367 (2012)
26. RatnaBabu, K., Arun Rahul, L., Vineet Souri, P., Suneetha, A.: Image denoising in the presence of high level salt and pepper noise using modified median filter. IJCST **2**(1), 180–183 (2011)
27. Sreedevi, M., Vijay Kumar, G., Pavan Kumar, N.V.S.: Removing impulse noise in gray scale images using min max and mid point filters. IJARCS **2**(6), 377–379 (2011)
28. Huai, X., Lee, K,. Kim, C.: Image noise reduction by dynamic thresholding of correlated wavelet intensity and anisotropy. In: 2010 International SoC Design Conference, Seoul, pp. 28–33 (2010)
29. Chen, H.: A kind of effective method of removing compound noise in image. In: 2016 9th International Congress on Image and Signal Processing, BioMedical Engineering and Informatics (CISP-BMEI), pp. 157–161. IEEE, October 2016
30. Raj, A., Wang, Y., Zabih, R.: A maximum likelihood approach to parallel imaging with coil sensitivity noise. IEEE Trans. Med. Imaging **26**(8), 1046–1057 (2007)
31. Yue, H., Liu, J., Yang, J., Nguyen, T., Hou, C.: Image noise estimation and removal considering the Bayer pattern of noise variance. In: 2017 IEEE International Conference on Image Processing (ICIP), Beijing, pp. 2976–2980 (2017)
32. Rakhshanfar, M., Amer, M.A.: Low-frequency image noise removal using white noise filter. In: 2018 25th IEEE International Conference on Image Processing (ICIP), Athens, pp. 3948–3952 (2018)
33. Martin-Gonthier, P., Magnan, P.: CMOS image sensor noise analysis through noise power spectral density including undersampling effect due to readout sequence. IEEE Trans. on Electron Devices **61**(8), 2834–2842 (2014)
34. Lin, Z.: A nonlocal means based adaptive denoising framework for mixed image noise removal. In: 2013 IEEE International Conference on Image Processing, Melbourne, VIC, pp. 454–458 (2013)
35. Skretting, K., et al.: Recursive least squares dictionary learning algorithm. IEEE Trans. Signal Process. **58**(4), 2121–2130 (2010)
36. Michael, E., Sapiro, G., et al.: Sparse representation for color image restoration. IEEE Trans. Image Process. **17**, 53–69 (2008)
37. Carlsson, S., et al.: Heaps with bits. Theor. Comput. Sci. **164**, 1–12 (1996)
38. Gao, L., et al.: Image denoising based on edge detection and pre-thresholding Wiener filtering of multiwavelets fusion. Int. J. Wavelets Multiresolut. Inf. Process. **13**(5), 15 (2015). Article ID1550031
39. Wang, Z., et al.: Image quality assessment: from error visibility to structural similarity. IEEE Trans. Image Process. **13**, 600–612 (2014)

40. Larkin, K.G.: Structural Similarity Index Simplified. Nontrivialzeros Research, pp. 1–4, May 2015
41. Gautam, K.S., Senthil Kumar, T.: Video analytics-based intelligent surveillance system for smart buildings. Springer Soft Comput. **23**, 2813–2837 (2019)
42. Gautam, K.S., Senthil Kumar, T.: Video analytics-based facial emotion recognition system for smart buildings. Int. J. Comput. Appl. 1–10 (2019)
43. Roy, S., Sengupta, S.: An improved video encoder with in-the-loop de-noising filter for impulse noise reduction. In: 2006 International Conference on Image Processing, Atlanta, GA, pp. 2605–2608 (2006)
44. Ghazal, M., Amer, A., Ghrayeb, A.: A real-time technique for spatio–temporal video noise estimation. IEEE Trans. Circuits Syst. Video Technol. **17**(12), 1690–1699 (2007)
45. Amer, A., Dubois, E.: Fast and reliable structure-oriented video noise estimation. IEEE Trans. Circuits Syst. Video Technol. **15**(1), 113–118 (2005)
46. Li, W., Yin, X., Liu, Y., Zhang, M.: Robust video denoising for mixed Poisson, Gaussian and impule noise. In: 2017 2nd International Conference on Image, Vision and Computing (ICIVC), Chengdu, pp. 459–462 (2017)
47. Miyamae, K., Gohshi, S.: Noise level detection in general video. In: 2018 International Workshop on Advanced Image Technology (IWAIT), Chiang Mai, pp. 1–4 (2018)

A Collaborative Mobile Based Interactive Framework for Improving Tribal Empowerment

Senthilkumar Thangavel[1(✉)], T. V. Rajeevan[2], S. Rajendrakumar[3], P. Subramaniam[4], Udhaya Kumar[4], and B. Meenakshi[1]

[1] Department of Computer Science, and Engineering, Amrita School of Engineering, Amrita Vishwa Vidyapeetham, Coimbatore, India
t_senthilkumar@cb.amrita.edu,
balumeena.1992@gmail.com

[2] Department of Social Work, Amrita Vishwa Vidyapeetham, Coimbatore, India
rajeevshaarody@gmail.com

[3] Centre for Sustainable Future, Department of Chemical Engineering and Materials Science, Amrita Vishwa Vidyapeetham, Coimbatore, India
s_rajendrakumar@cb.amrita.edu

[4] Tribal Research Centre, Department of Tribal Welfare, Government of Tamil Nadu, Udhagamandalam, India
trcooty@gmail.com, udhayaanthropology@gmail.com

Abstract. Handling the user interactions and improving the responsiveness of the system through collaborative Recommendation system is the proposed framework. The application will have an interactive form that takes the user details. On valid user access the user is provided with options under a gallery as: Location wise Statistics, Tribes culture, Tribes Request and Tribal activities. The tribal users should be able to present their food products as part of the system. The tourist guest should be able to view the food products and place orders. Recommendation can be provided to the guest based on the past experience. The application will also facilitate for Geo Tagging of the users including tribes to improve their participation. The application will be developed using Android with cloud namely. The user statistics will be presented for region wise in a more visual manner. The Smart phones play a vital role in everyone's life. In smart phones we have so many applications and the applications need to be tested. The mobile applications are tested for its functionality, usability, and for its consistency. There are two types of mobile application testing. They can be of non-automatic or automatic testing. In this study automated approach is discussed. An android based mobile testing needs native test applications which is used for testing a single platform.

Keywords: Client-server model and android · Product details · Data visualization · Geo-tagging · Recommender systems · User profile · Application framework · Application testing

© Springer Nature Switzerland AG 2020
S. Smys et al. (Eds.): ICCVBIC 2019, AISC 1108, pp. 827–842, 2020.
https://doi.org/10.1007/978-3-030-37218-7_90

1 Introduction

Application provides possibilities that did not subsist before, it facilitate as the tools for native tribe to understand and compete on a world level. This gives an opportunity for economic development and the potential to be part of the mainstream society. Simple application can gain the superior of life for native tribes by introducing them to things that make their daily lives easier, in turn this can gain comforts and substantial growth of production in that social group. The important goal of the Undertaking project is to interact with users and to give the sustainable economic growth to the tribal people. The application will have a User Registration form to fill the user details. After validating the details of user it will take to the page that gives the details about Tribal community. In that page we will also have the list of Indigenous foods. The user can place the order. And also the application provide a form to hear the customer reviews. By using application We can share the link on social medias to get more users. The app can also facilitate Geo-tagging for sharing the location and sharing the photos took in tribal areas for getting more users. The app can also add a page to sale the Tribal products for improving their Economic condition.

Tribes in India have been underprivileged of possibilities due to different factors. Cardinal factor is difficulty of appropriate base for the improvement idea to scope them. It is broadly accepted that Technologies are potential to show a cardinal function for the betterment of society (Vinay Kumar, Abhishek Bansal).

A mobility today, is really a little machine in the hands of the farmer. Investing on this information, we can turn the cell phone into a device that provides fast information about various entities (*Augmented Reality in Agriculture*).

Here is a communication for an effectual user interface between the conventional and rising info systems while initiating the technologies in tribal areas. It is much than because tribal are usual for acquiring data by word of mouth from a best-known and trustworthy source. It is an ambitious undertaking to take this social group to the analogue textual based info group.

1.1 Area of Study

The Nilgiris is one of the smallest districts of Tamil Nadu (Shodh Ganga). It is also called as 'blue mountains' located on southwest of Coimbatore district. The average height of hills is around 6500 ft. bound with Kerala and Karnataka states. For administrative purpose the region is separated into six taluks viz. Udhagamandalam, Gudalur, Pandalur, Coonoor, Kotagiri and Kundah (Connie Smith-2007). Tribes like Todas, Kotas, Kurumbas, Irulas, Paniyans and Kattunayakans are found in this district; all tribes are 'Indigenous groups' of Nilgiris district (TRC and HADP- 2007). The population of tribe was 4.32% and it was grown up to 4.46% in 2011 (TRC, 2011).

Reported to 2011 census, the whole population of the Nilgiris District is 7,35,394 and recorded negative growth compared to the previous census. Scheduled caste comes around 2,35,878 and Scheduled tribe 32,813 representing 4.46% of the whole generic population. The tribal population in Nilgiri district is not equally shared to the six taluks. 3,015 of them surviving in Udhagamandalam, 944 in Coonoor, Kotagiri 1707, Gudalur 3015.

The population of Scheduled Tribes (STs) in the land, as per Count 2011 is 10.45 crore. STs constitute 8.6% of the country's total growth population and 11.3% of the total countrified population. People of ST male is recorded 5.25 crore and ST female is 5.20 crore.

2 Related Work

GoI has introduced many websites and mobile applications for uplifting tribal communities and develops their social and economic status. The mobile applications come under the fields of Agricultural, Health, Educational, Social welfare applications, e-governance and e-commerce Apps. It works both as a aid supplier and class creator for social group products. It trades social group products through the mesh of its marketing outlets 'TRIBES INDIA' in the nation. As a capability material, it also contribute education to ST Artisans and **MFP** gatherers.

TRIFED purchased tribal products worth Rs. 846.96 lakhs as on 31.12.2017. TRIFED has 1213 Individualist/Self Help Groups/organizations, etc., as its empaneled indefinite quantity which are connected to above 69247 tribal families. Data for the previous years are

Sl #	Year	Purchase tribal products (Rs. in lakhs)
1	2008–09	681.78
2	2009–10	609.34
3	2010–11	656.35
4	2011–12	719.58
5	2012–13	880.55 (as on 28.2.2013)

TRIFED has entered into statement with e-commerce platforms Like Snap deal and Amazon who will offer their customers assorted Tribal products and produce through their portals www.snapdeal.com and www.amazon.com to facilitate on-line sale. Ministry of Commerce has also made condition for sale of tribal products through TRIFED on www.gem.gov.in.

2.1 TRIFED Method of Wholesale Merchandising

TRIFED intent is to change the resource of the gathering by making up a property industry and make over concern chance for them based on their cultured knowledge and traditional skills whilst assuring fair and equitable payment. It involves designation selling possibilities for selling of social products on a property base, make overing the trade name and rendering different employment.

2.2 Source of Tribal Products

TRIFED has a mesh of 13 Location Agency crosswise the country that determines the sources of tribal products for marketing through Its wholesale selling mesh of TRIBES INDIA outlets. TRIFED has been trading source for respective handicrafts, hand-looms and natural & food products through its empaneled providers across the country. The providers consist of respective tribal artisans, tribal SHGs, Organizations/Authority/NGOs operating with tribal. The providers are empaneled with TRIFED as per the guidance for empanelment of suppliers.

2.3   Marketing of Tribal Products

TRIFED has been merchandising tribal products through its Wholesale Outlets located across country and also through expo. TRIFED has implanted a series of 35 personal showrooms and 8 consignment showrooms in association with State level Organizations encouraging their products.

2.4   Exhibitions

The exhibitions for the tribal people own products have been organized by TRIFED under national level and international level Expo in the names of AADILSHIP, AADICHITRA, and OCTAVE.

On Process, TRIFED has been dealing National Tribal Expo- Aadilship Tribal artisans/groups/organizations who are invited to take part in this Exhibition and vitrine their wealthy Tribal practice. The important accusative in keeping these consequence is to render an chance to tribal artisans to background their cultural crafts and to move directly with art lovers to learn about their taste and preferences. This aids them in varying their product designs and creations accordantly. The event also considers tribal dance performance at times. It is an try to existing tribal art and culture in sacred manner, which has been accepted well by the user.

TRIFED also organizes Aadichitra'- an presentation of tribal paintings, in which tribal paintings are solely demonstrated and sold-out. The tribal creator are also requested to demonstrate their art in this exhibitions. Pleased by the outcome, TRIFED has been dealing Aadichitra's at respective determination across the country.

TRIFED also take part in the OCTAVE – dancing celebration of North Eastern Region, arranged by GoI. TRIFED is also connected to the outcome from 2008-09 forth. TRIFED makes involvement by artisan from North Eastern Region and ply them the possibles to case and sell their products.

TRIFED participates in global exhibitions/trade fairs through Export Promotion Council for Handicrafts (EPCH) and India Trade Publicity Administration (ITPO) in different countries for displaying and merchandising the products beginning from artisans.

TRIFED is involved in the procurement and selling of tribal art and craft items through its series of 40 marketing sales outlet called TRIBES INDIA which offers a range of tribal products, which include

- Metal Craft
- Tribal Textiles
- Jewelry
- Tribal Paintings
- Cane & Bamboo

- Pottery
- Gifts and Novelties
- Organic and Natural products

The Ministry of Tribal Affairs has been highly-industrial an android based technology called Tribal Diaries for internal monitoring as also conjunctive with agency/authorized concern with implementation of schemes/programs for tribal development. The application is for licensed users. This usage caters a chance for optic activity in terms of pictures, video recording, transferring reports of official tours/inspections and sharing of best practices etc. The application is largely used to get an overview of the Ekalavya Model Residential Schools (EMRSs) funded by the Ministry of Tribal Affairs. The principals of the EMRS are being encouraged to use the application and create projects highlighting the physical infrastructure of the schools, special achievement of the students and share success stories.

In so far as grievances are concerned, the Centralized Public Grievance Amends and Monitoring System (CPGRAMS), an on line web-enabled system is the platform which initially aims to change submission of grudges by the afflicted citizens from anywhere and anytime. Besides, grievances are also physically received in the Ministry. The Ministry scrutinizes and takes action for speedy redressal of grievances besides tracking them for their disposal as well through a dedicated Division for the purpose.

(Info was rendered by Government of india Shri Jaswantsinh Bhabhor in a written reply in Lok Sabha on 19th March 2018 (PIB- GoI-Ministry of Tribal Affairs).

Non-timber forest land products (NTFPs) plays a pivotal function in improvement and living of tribal people crosswise the world. In the world more people are lodging in the forestland, being on NTFPs for subsistence financial gain and livelihood. NTFPs are considered to be important for prolonging rural support, reducing rural poverty, biodiversity conservation and rising rural economic growth.

When we see the setting from which a tribal youth operates, we see a few marks that differentiate them from the non-tribal youth. A tribal youth comes from egalitarian society bounded by traditional norms and a low technological level and with less urban impact. The tribal regions were often isolated and lacked proper means of communication which again helped them to follow their own culture and to maintain its diacritical mark (Sachindra).

In constructing the India of 21st century the capability of the tribal youth, their diligence, agility, intelligence, and perseverance will prove quite effective. With their dynamic n mental make-up and continuous efforts to make a dent on the national 4evel in all walks of life - social - political - economic - the tribal youth have the potency to rejuvenate their own group, as also the whole nation with us.

2.5 Application for the Tribal Products

The culture of any community is closely associated and assimilated with its history since time immemorial. Further the existence of culture is the base of history. Therefore, the craze to know the way of life of any community requires study and analysis of cultural history of that tribe deeply. If one thinks of cultural history of primitive tribes one must turn towards South Odissa, the hub of tribal. So, this Koraput region, the domain of tribal has become center of study and research.

GappaGoshti is a mobility based platform particularly planned for countryside India in their native language. Each of its practicality can be approached via standardized keys (left, right, up, down, select, back) and a fewer numerical keys on a phone. Writing textual matter either in English or native language using phone keypad is not an easy task for a common uneducated or semi educated person with countryside background.

GappaGoshti gives different services which are helpful for an ordinary man in country-style India. It consists of:

- social networking or micro-blogging.
- Weather forecasting in their place.
- News in the native language with respect to the user.
- Yellow Pages.
- Krushi Dyankosh- an agricultural knowledge base.

3 Literature Review

The following section gives the overview of the literature based on the improvement of tribal Welfare community using android application.

X. Qi proposes a method for user profiling in the form of vector based model. The user profiling is done based on Rocchio algorithm. The user profiling is used to store the personal information of the user and also to analyze the feedback. The process involves three steps namely data collection, profile initialization and updation.

X. Wu et al. D. Kit et al. presents a high-fidelity geo-tagging approach for identifying the geographical locations. With the use of automated geo-tagging it is easy to find the current location. Since huge amount of data is posted, it becomes challenging to handle the data. This problem is overcome by using clustering techniques. This gives us a promising result while we share large number of photos or media with the locations. Limitation of the approach is that the location in remote areas cannot be tracked.

Y. Liang, E. Kremic et al. suggest an Android based mobile application gives the user friendly environment. It helps the users to login using smart phones with android which is an open source platform. Client- server architecture is used to authenticate the user profile and to maintain the huge amount of data. The authentication and access control is also used for security purpose. Whenever the user is logging it will go client-server architecture for verification of user information.

G. Hirakawa et al, P. Pilloni et al. suggest that a Recommender system is used to filter the information from the vast amount of data using user profile. The recommender system is used for filtering the information about the users, based on their feedback about tribal food and cultural products. This customer feedback is displayed in social media platforms like Facebook twitter, etc. which helps to improve the tourist attraction in that particular location. The disadvantage of this approach is when the user's preference for the social platform changes.

Wang et al. suggests that the data cubes for aiding the decision making statement using on-line analytical processing (OLAP) systems. For Illustrating the data elements the interactive visualized system presents a qualified one dimensional cuboid hierarchical tree structure and that represents the data cubes with the usage of two dimensional graphical icons. Users can also explore interactively with the two dimensional data in hierarchical levels.

Daniel Pineo et al. executed for valuating and optimizing the visualization automatically with the use of a procedure model of person like vision. The methodology trust upon the neural network technique of early sensor activity process in retina and particular visual cerebral mantle. The neural activity resulted in the form of bionic view and then valuated to produce a measure of visualization powerfulness. Hill climbing algorithm is approached by applying the effectiveness measure and also by attaining the Visualization optimization. Use two visualization parameterizations for 2D flow methodology which can be of streaklet-based and pixel-based visualization.

Yusuke Itakura proposed that the eatable products which are shown in stores are on the foundation of worker Judgment. An probe of departmental stores discovered the leading drawbacks. The guidance given by the main office for an employer is to exhibit the eatable products on the basis of the dispersion. Then the data acquirable on the computer storage is shortly unfit for replying customer enquiries. The actual System assorts the sales data accordant to the calendar and timezone. Data is needed regarding the involvement that are mentioned by the employers on the base of eatable display. If the weather condition changes then there will be drop-off in food sale.

Anbunathan R et al. proposed an event based framework for designing the complex test cases in an android mobile. Android APK's like test cases and test scheduler has been approached. In this Framework to test the testcases they use Scheduler algorithm and android components are used to build the testing. These test are executed in an android platform and this framework is developed for an android based mobiles

Anbunathan R et al. proposed an automated test generator based on Data Driven Architecture framework. A Tool named Virtual Tool Engineer (VTE) is utilized for generating XML files. The Test cases are seizured in the kind of Sequence diagram. To reduce the test effort by separating the input case while working with the scenario like multiple test cases is an important approach.

Zhi-fang LIU et al. proposed the application used in mobility software. In this Service Oriented Architecture (SOA) framework is approached for mobile application testing. Based on SOA framework, test platform has been built. The usage of SOA framework is used to solve the problems in mobile application testing

N. Puspika et al. proposed the Sequence testing method. This testing method is used to cover all the flow of application. The Limitation of this testing method is that all the possible input flow should know to implement this method. So, the solution for this limitation is to create a model for this flow of application. The model for this problem is Colored Petri-net model (CPN). The CPN model is used to analyze the sequence of application easily and to generate the test case automatically.

S. Cha et al. proposed a middle-ware framework which provides an effective disconnected tolerant network services. The various layer of the OSI model are enforced using mobility management. To maintain mobile services effectively during the mesh separation time the mobility devices and mobile intelligent server are required

P. Chu et al. proposed a plug-in framework for the novel approach of enforcing the security policies. Existing enforcement system produces a poor result. So, the author proposes a new enforcement security policy based on an android smart phone applications

Tian Zheng et al. proposed a framework to perform safety assessment for mobile application. This security assessment is done in both android and iOS platform. In this paper the security testing platform is approached to improve their security monitoring capabilities

4   Methodology Proposed

The website and application will be a user friendly platform for the tribal communities and other users like the common public and government officials. It will work as a tribal repository which includes the details of tribal vegetation, social and demographic details of the tribes and tribal products which will be accessible for the people to buy in the platform itself. Promotion of the products is so important, in the website and app the digital promotion can be added. Sales of the tribal products would increase the economic condition of the tribes collecting the feedback from customers would help to develop further keeping aside the negatives. The devices are commonly used by rural and urban people.

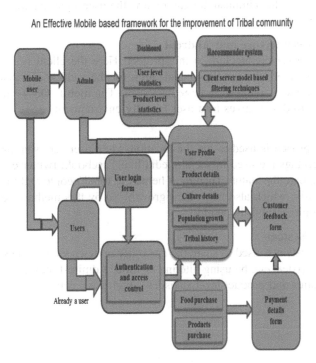

An Effective Mobile based framework for the improvement of Tribal community

We develop a mobile application for purchasing and marketing the tribal food products. The different framework are discussed below.

User Profiling

In this mobile based framework, we have two types of users. Firstly, Administrator who has the access to maintain the data and admin alone can the access the user level statistics and product level statistics. The administrator login can be accessed by groups like SHE (Self Helping Group), NGO's etc., Administrator can view the user profile and the tribal product details. Users are the one who has the registered account in this mobile application mainly Tourists who visit tribal areas. The Users who are registered already will go under authentication and access control for validating the user name and password. New users have to fill in their personal information in the login form and after validation process their account will be created. Users can view, edit and update their profile any time.

Product Details

Users can purchase food products and also cultural products like pottery, paintings, crafted materials etc. using the mobile application. The payment can be done through ready cash or through on line method by using credit or debit card etc. Then user will automatically be taken to the customer feedback form to review the products. This in turn helps us to improve the economic status of the tribal people and reach information about them globally. The privileged user can approach the administrator for a discount on the purchase and the administrator can review the user's participation to give some discount on their purchases.

Client Server Architecture and Android

We use Client server architecture in this project. Here the client is tourists and the server is administrator who run this organization for the welfare of tribal community. The client's data are maintained in large databases. The mobile OS we use is android. Android is preferred as it gives us a user friendly environment

Geo-tagging

Geo-tagging approach is used to tag the location when users are sharing their photos and videos. Geo tags are used in social media like Facebook; twitter etc. to share the media and to update the status. Thereby it helps the tribal people and their location to be visible globally, which also helps their growth. So by this method we can attract a lot of other people around the world.

Recommender System

Recommender system is used to share the customer reviews via on-line shopping sites like amazon, snap deal etc. by using filtering method. Administrator alone can view the user level statistics and product level statistics which are calculated by Recommender system.

5 Service Oriented Architecture Based Mobile Application Framework

To Develop a mobile application we need some of the features that need to fulfill the client needs. The features of a mobile based application framework are discussed below

1. Simplicity of the user: It is a helpful user interface. It make the client side to use the application without any complexity. It gives a chance to perform any activity effortlessly and this wont make any changes in an application. Like this the mobile device reaches the customer.
2. Social area integration: It helps to share the information in a clearest and a profitable way and it makes to know about us by more people. Social media is the only way to reach our thought. This integration in the mobile device becomes a tool to fulfill their needs.
3. Feedback: To develop a mobile application we need customer suggestions. So, Feedback feature in an application is very useful. This feature will fulfill the user needs.
4. Off-line Work: The Work like shopping, payment etc; needs Internet connection. Rather than this applications like Notepad, Ms Office, Memo, etc; needn't Internet connection so it can be used in off-line too. This feature is an added advantage to satisfy user needs.
5. Security: The mobile devices may have some personal information about the user. So there we need an application which built some privacy for the information. This the feature of security in a mobile application framework
6. Personalization options: This feature gives the user to make their own text formatting styles and adaptable settings option. It allows the application to look as they like. This option is also a way to reach customer needs.

6 Service Oriented Architecture (SOA) Based Testing Technique

Z. Liu et al. describes the SOA based mobile application software running on the mobile phone with the operating system like android, windows, blackberry, iOS etc needs to be tested. Testing strategy are as follows.

A. Test Designing:

- The whole field of the mobile application should be taken by the Service oriented Examiners.
- The request needs to be partitioned into separate services.
- The mobile application structure needs to be reorganized into three components that includes Information, Services and front-end request.
- The part needs to be examined, and concern scenarios should be explained clearly.

- The concern necessity needs to be categorized as common requirement and mobile application specific requirement.
- An Approachability Matrix needs to be processed, and all the test needed to be traced to concern the necessity.

B. Test Performance Approach:

All of the service component needed to be proved. Service components are divided in to three components are as follows

Integration Testing – It is a service oriented component and it should tested for authorizing the data flow through the data integrity and services.

System Testing – It is a service based component and it can be tested to authorize the data flow in between the front-end request and information.

Performance Testing – It is a service based component and it can be tested for fine tuning and performance that to be optimum.

C. SOA Testing Methods:

In SOA based Testing methods we are going to implement two methods

1. Service-Level Testing: It admits testing the components for practicality, usability, safety, action and ability. All service needs to be tested separately.
2. Regression testing: On the mobile application this method needs to be tested when there is double holding so as to enhance their stability and the accessibility of the method. The suite will be make over to cover the services which forms an essential form of an application. It can be reused for multiple holding

D. SOA based Automated Testing Tools:

There are number of SOA application testing tools. SOA Testing tools are chosen based on their execution and accurate results

SoapUI: It's a cost-free tool which is used for web testing. This tool is used for performing functional testing, performance testing and Load Testing

Apache Jmeter: It's a open source which is capable to examine the execution of soap invocation.

Jprofiler: It is a testing tool which is used to detect the memory leak or to find the bottleneck performance etc.

HP Service Test: It is a functioning tool which is also capable to support the UI Testing and for shared service testing.

7 Components and Structural Based SOA Framework

The Service oriented architecture has following components which are discussed below.

1. Adapter-It is a software module for an application or system that allows access to its capabilities through a standard compliant device interface.

2. Business Process modeling-It is a process that mapping out the business process in terms of what the human participants have to do and what the various application based on business process expected to do.
3. Enterprise Service Bus- It is the center to communicate all the services in service oriented architecture. It tends to be connectivity for various types of middle ware repository of meta data definitions, registries and interfaces of every kind.
4. Service Broker-It is a software in service oriented architecture that brings together the components associated with the rules.
5. Service Registry: The service registry is like a referential data services. The service registry contains the information about services and service definition interfaces and parameters.
6. Service Repository: A database for all Service oriented architecture software and components to keep the stuff in one place and configure the management. It is the medium for all record of polices attributes and processes.

8 Mobile Requirements

A. Operating system: Operating System are of many categories. They are Android, IOS, Windows, Symbian, blackberry etc; In this case study, Android OS is going to be implemented. Android Operating System play a vital role in day today's life. Android is an open source platform in which we develop so many mobile applications. Being Platform dependent the advantage here is off-line mobile applications can be created.
B. Mobile Devices: There are two mobile devices such as Smart Phones and mobile phones. In this Study Smart Phones are going to be discussed. Using Smart Phones with Internet connection we can do payment in on line, Shopping in on-line and foods can be ordered. These are the usage of smart phones

9 Conclusion

The world is changing into the technological era, majority of the population has either using any websites or mobile applications or using with the help of others. The tribal growth using these ICTs is the modern day changes which holistically change the living standard of the people. Tribal empowerment using technology can be set as a paradigm shift in the development of tribal lives. Uplifting of economic conditions, educational development, improving health and medical conditions and preservation of indigenous cultures. It also provides a way for holistic and sustainable development to tribal community. These developments in community, brings a democratic procedures, voluntary cooperation, self-development and participation of the members in the community. Technological advancement in our country can play crucial role in community development; same technology can be fully exploited for developing tribal community and their standard of living. The innovative ideas based on scientific principle and technological application can transform the tribal community by improving economic conditions and preserving social and cultural values of indigenous community.

Acknowledgments. We are thankful to the people who contributed their time and shared their knowledge for this research.

Compliance with Ethical Standards
  ✓ All authors declare that there is no conflict of interest.
  ✓ No humans/animals involved in this research work.
  ✓ We have used our own data.

References

1. Vinay Kumar, A.B.: Information and communication technology for improving livelihoods of tribal community in India. Int. J. Comput. Eng. Sci. **3**(5), 13–21 (2013)
2. Amber, G., Young, S.M.: Cultural Identity Restoration and Purposive Website Design: A Hermeneutic Study of the Chickasaw and Klamath Tribes. IEEE (2014)
3. Andhika, O.A.: Vege Application! Using Mobile Application to Promote Vegetarian Food. IEEE (2018)
4. Annamalai Narayanan, C.S.: apk2vec: Semi-supervised multi-view representation learning for profiling Android. In: International Conference on Data Mining. IEEE (2018)
5. Apurba Saikia, M.P.: Non-timber Forest Products (NTFPS) and their role in livelihood economy of the tribal people in upper Brahmaputra valley, Assam, India. Res. Rev. J. Bot. Sci. (2017)
6. Apurv Nigam, P.K.: Augmented Reality in Agriculture. IEEE (2011)
7. Dang, B.S.: Technology strategy for tribal development. Indian Anthropol. Assoc. **10**(2), 115–124 (1980). https://www.jstor.org/stable/41919402
8. GoI. CENSUS OF INDIA 2011 (2011)
9. Jason Henderson, F.D.: Internet and E-commerce adoption by agricultural input firms. Rev. Agric. Econ. **26**(4), 505–520 (2004). https://www.jstor.org/stable/3700794
10. Kushal Gore, S.L.: GappaGoshti™: Digital Inclusion for Rural Mass. IEEE (2012)
11. Liu Yezheng, D.F.: A novel APPs recommendation algorithm based on apps popularity and user behaviours. In: First International Conference on Data Science in Cyberspace. Changsha, China. IEEE (2016)
12. Billinghurst, M.: Augmented Reality in Education. Seattle, USA (2002). http://www.newhorizons.org
13. Manisha Bhende, M.M.: Digital Market: E-Commerce Application. IEEE (2018)
14. Murtaza Ashraf, G.A.: Personalized news recommendation based on multi-agent framework using social media preferences. In: International Conference on Smart Computing and Electronic Enterprise. IEEE (2018)
15. Nor Aniza Noor Amran, N.Z.: User Profile Based Product Recommendation on Android Platform. IEEE, Kuala Lumpur, Malaysia (2014). https://doi.org/10.1109/icias.2014.6869557
16. Radhakrishna, M.: Starvation among primitive tribal groups. Econ. Polit. Wkly **44**(18), 13–16 (2009). https://www.jstor.org/stable/40278961

17. Ramneek Kalra, K.K.: Smart market: a step towards digital India. In: International Conference on Computing and Communication Technologies for Smart Nation (IC3TSN). IEEE, Gurgaon, India (2017)
18. Sharanyaa, S., Aldo, M.S.: Explore places you travel using Android. In: International Conference on Electrical, Electronics, and Optimization Techniques (ICEEOT). IEEE (2016)
19. Lien, S.F., Wang, C.C., Su, J.P., Chen, H.M., Wu, C.H.: Android platform based smartphones for a logistical remote association repair framework. Sensors 14(7), 11278–11292 (2014)
20. Takumi Ichimura, I.T.: Affective Recommendation System for Tourists by Using Emotion Generating Calculations. IEEE, Hiroshima (2014)
21. Twyman, C.: Livelihood opportunity and diversity in kalahari wildlife management areas, Botswana: rethinking community resource management. J. South. Afr. Stud 26(4), 783–806 (2000). https://www.jstor.org/stable/2637571
22. Liang, Y.: Algorithm and implementation of education platform client/server architecture based on android system, In: 2018 International Conference on Intelligent Transportation, Big Data & Smart City (ICITBS), Xiamen, 2018, pp. 251–253. https://doi.org/10.1109/icitbs.2018.00071
23. Has, M., Kaplan, A.B., Dizdaroğlu, B.: Medical image segmentation with active contour model: smartphone application based on client-server communication. In: 2015 Medical Technologies National Conference (TIPTEKNO), Bodrum, pp. 1 4 2015.https://doi.org/10 1109/tiptekno.2015.7374546
24. Kremic, E., Subasi, A., Hajdarevic, K.: Face recognition implementation for client server mobile application using PCA. In: Proceedings of the ITI 2012 34th International Conference on Information Technology Interfaces, Cavtat, pp. 435–440 (2012). https://doi.org/10.2498/iti.2012.0455
25. Kordopatis-Zilos, G., Papadopoulos, S., Kompatsiaris, I.: Geotagging text content with language models and feature mining. Proc. IEEE 105(10), 1971–1986 (2017). https://doi.org/10.1109/JPROC.2017.2688799
26. Wu, X., Huang, Z., Peng, X., Chen, Y., Liu, Y.: Building a spatially-embedded network of tourism hotspots from geotagged social media data. IEEE Access 6, 21945–21955 (2018). https://doi.org/10.1109/ACCESS.2018.2828032
27. Kit, D., Kong, Y., Fu, Y.: Efficient image geotagging using large databases. IEEE Trans. Big Data 2(4), 325–338 (2016). https://doi.org/10.1109/tbdata.2016.2600564
28. Pineo, D., Ware, C.: Data visualization optimization via computational modeling of perception. IEEE Trans. Vis. Comput. Graph. 18(2), 309–320 (2012). https://doi.org/10.1109/TVCG.2011.52
29. Wang, X., Yi, B.: The application of data cubes in business data visualization. Comput. Sci. Eng. 14(6), 44–50 (2012). 10.1109/mcse.2012.17
30. Hirakawa, G., Satoh, G., Hisazumi, K., Shibata, Y.: Data gathering system for recommender system in tourism. In: 2015 18th International Conference on Network-Based Information Systems, Taipei, pp. 521–525 (2015). https://doi.org/10.1109/nbis.2015.78
31. Yuan, N.J., Zheng, Y., Zhang, L., Xie, X.: T-finder: a recommender system for finding passengers and vacant taxis. IEEE Trans. Knowl. Data Eng. 25(10), 2390–2403 (2013). https://doi.org/10.1109/TKDE.2012.153
32. Pilloni, P., Piras, L., Carta, S., Fenu, G., Mulas, F., Boratto, L.: Recommender system lets coaches identify and help athletes who begin losing motivation. Computer 51(3), 36–42 (2018). https://doi.org/10.1109/MC.2018.1731060
33. Hijikata, Y., Okubo, K., Nishida, S.: Displaying user profiles to elicit user awareness in recommender systems. In: 2015 IEEE/WIC/ACM International Conference on Web

Intelligence and Intelligent Agent Technology (WI-IAT), Singapore, pp. 353–356 (2015). https://doi.org/10.1109/wi-iat.2015.83

34. Leung, K.W., Lee, D.L.: Deriving concept-based user profiles from search engine logs. IEEE Trans.Knowl. Data Eng. **22**(7), 969–982 (2010). https://doi.org/10.1109/TKDE.2009.144

35. Xie, C., Cai, H., Yang, Y., Jiang, L., Yang, P.: User Profiling in elderly healthcare services in china: scalper detection. IEEE J. Biomed. Health Inf. **22**(6), 1796–1806 (2018). https://doi.org/10.1109/JBHI.2018.2852495

36. Liang, H., Mu, R.: Research on humanization design based on product details. In: 2008 9th International Conference on Computer-Aided Industrial Design and Conceptual Design, Kunming, pp. 32–34 (2008). https://doi.org/10.1109/caidcd.2008.4730513

37. Itakura, Y., Minazuki, A.: A study of the support system for displaying food products in convenience stores. In: 2010 IEEE/ACIS 9th International Conference on Computer and Information Science, Yamagata, pp. 421–426 (2010). https://doi.org/10.1109/icis.2010.15

38. Anbunathan, R., Basu, A.: Automation framework for test script generation for Android mobile. In: 2017 2nd IEEE International Conference on Recent Trends in Electronics, Information & Communication Technology (RTEICT), Bangalore, pp. 1914–1918 (2017)

39. Nidagundi, P., Novickis, L.: New method for mobile application testing using lean canvas to improving the test strategy. In: 2017 12th International Scientific and Technical Conference on Computer Sciences and Information Technologies (CSIT), Lviv, pp. 171–174 (2017)

40. Pandey, A., Khan, R., Srivastava, A.K.: Challenges in automation of test cases for mobile payment apps. In: 2018 4th International Conference on Computational Intelligence and Communication Technology (CICT), Ghaziabad, pp. 1–4 (2018)

41. Murugesan, L., Balasubramanian, P.: Cloud based mobile application testing. In: 2014 IEEE/ACIS 13th International Conference on Computer and Information Science (ICIS), Taiyuan, pp. 287–289 (2014)

42. Gao, J., Bai, X., Tsai, W., Uehara, T.: Mobile application testing: a tutorial. Computer **47**(2), 46–55 (2014)

43. Unhelkar, B., Murugesan, S.: The enterprise mobile applications development framework. IT Professional **12**(3), 33–39 (2010)

44. Cha, S., Du, W., Kurz, B.J.: Middleware framework for disconnection tolerant mobile application services. In: 2010 8th Annual Communication Networks and Services Research Conference, Montreal, QC, pp. 334–340 (2010)

45. Bernaschina, C., Fedorov, R., Frajberg, D., Fraternali, P.: A framework for regression testing of outdoor mobile applications. In: 2017 IEEE/ACM 4th International Conference on Mobile Software Engineering and Systems (MOBILESoft), Buenos Aires, pp. 179–181 (2017)

46. Chu, P., Lu, W., Lin, J., Wu, Y.: Enforcing enterprise mobile application security policy with plugin framework. In: 2018 IEEE 23rd Pacific Rim International Symposium on Dependable Computing (PRDC), Taipei, Taiwan, pp. 263–268 (2018)

47. Samuel, O.O.: MobiNET: a framework for supporting Java mobile application developers through contextual inquiry. In: 2009 2nd International Conference on Adaptive Science and Technology (ICAST), Accra, pp. 64–67 (2009)

48. Reed, J.M., Abdallah, A.S., Thompson, M.S., MacKenzie, A.B., DaSilva, L.A.: The FINS framework: design and implementation of the flexible internetwork stack (FINS) framework. IEEE Trans. Mob. Comput. **15**(2), 489–502 (2016)

49. Zheng, T., Jianwei, T., Hong, Q., Xi, L., Hongyu, Z., Wenhui, Q.: Design of automated security assessment framework for mobile applications. In: 2017 8th IEEE International Conference on Software Engineering and Service Science (ICSESS), Beijing, pp. 778–781 (2017)

50. Qin, X., Luo, Y., Tang, N., Li, G.: Deep eye: automatic big data visualization framework. Big Data Min. Anal. **1**(1), 75–82 (2018). https://doi.org/10.26599/BDMA.2018.9020007

51. Gautam, K.S., Senthil Kumar, T.: Video analytics-based intelligent surveillance system for smart buildings. In: Proceedings of the International Conference on Soft Computing Systems, pp. 89–103. Springer, New Delhi (2016). Soft Computing: 1-25.perceptron
52. Gautam, K.S., Senthil Kumar, T.: Discrimination and detection of face and non face using multi-layer feed forward perceptron. In: Proceedings of the International Conference on Soft Computing Systems, (Scopus Indexed). AISC. vol. 397, pp. 89–103. Springer, ISSN No 2194-5357

IoT Cloud Based Waste Management System

A. Sheryl Oliver[1(✉)], M. Anuradha[1], Anuradha Krishnarathinam[1],
S. Nivetha[1], and N. Maheswari[2]

[1] Department of CSE, St. Joseph's College of Engineering, Chennai, India
sherylviniba@gmail.com
[2] SCSE, VIT University, Chennai, India

Abstract. IoT based waste management system refers to the collection of waste in a smarter way using modern technologies. Whenever waste is dumped into the smart bin, the infrared sensors senses the amount of waste present inside it. This concept is achieved by using a programmable board called Arduino UNO which can be plugged in to any computer interface. The bin is partitioned into three levels where an infrared transmitter and receiver are placed at each level. Whenever waste is dumped inside the percentage of bin that is filled is tracked. The volume is determined by using Ultrasonic sensors. An infrared transmitter and receiver are positioned near the base to check whether the system is functioning properly or not. If the waste reaches level 1 then it means 30% of the bin is filled and if waste reaches level 2 then it means 60% of the bin is filled. In order to track the amount of waste collected the infrared sensors are monitored by Arduino UNO. The results are output in an IoT cloud platform called ThingSpeak. After the smart bin is 90% filled, a message is sent to the municipal corporation. This is achieved by transmitting information through web application programming interface embedded to the IoT cloud platform. The smart bins will have Ethernet module embedded with Arduino. This enables the system to access the network. The entire system is programmed in such a way that whenever the volume of waste collected reaches a certain level, the municipal corporation is notified.

Keywords: Ultrasonic sensors · Arduino · Ethernet module · IoT cloud platform · Waste management system · Sustainability · Smart bins

1 Introduction

Waste is defined as unwanted or useless materials that are unfit for our consumption or usage. Waste Management is the field which requires more concern in our society. In a country like India where the population grows by 3–3.5% every year managing waste surely possess a problem. As the population increases the amount of waste generated also rises considerably. The inevitable truth is that any kind of waste is hazardous to the society. The surroundings are affected if waste is not segregated and disposed properly. Natural elements such as air and water have higher probability of being affected by waste. Harmful gases released by waste affects the air whereas industrial wastes that are dumped into river affects the water bodies proportionately. The most peculiar problem which we notice in our daily life is lack of proper waste management. If not handled

S. Smys et al. (Eds.): ICCVBIC 2019, AISC 1108, pp. 843–862, 2020.
https://doi.org/10.1007/978-3-030-37218-7_91

properly, it may spoil the ambience of the environment. Also, lack of proper waste collection means the dumped waste tends to overflow the bins. Not only they spread diseases but also serves as breeding ground for mosquitoes. The act of collecting waste effectively have provoked thoughts in many for ages. Therefore, lessening the impact of waste disposal is the need of the hour. This project helps in maintaining the environment clean.

The concept of smart cities is in a budding stage though our Prime Minister initiated the idea a few years back. He insisted upon the idea of building smart cities all over India. With increasing number of cities, the due responsibilities also increase. Cleanliness can be achieved by disposing waste properly. This can be ensured by implementing the concept of IoT based waste collection system. The major issue faced by the current waste management system is the inability to outline a healthy status for bins. By implementing this project, the drawback can be mitigated. With the internet gaining momentum, human interaction with real life objects and machines will make the world even more interesting. As a result of high demand in computing researching focused beyond connecting computers. This sensational interest lead to the concept of Internet of Things. Here, Machine to Machine and Human to Machine interaction can take place with ease. Despite the invention of this concept years back, this field is still in the developing stage as far as technology is concerned. Industries such as home automation and transportation are rapidly growing with Internet of things.

High speed active internet connection is essential for Internet of Things. This technology can be defined as the connection between humans and other devices. Every single event and equipment in our daily life can be tracked using the concept of IoT. Most of the processes are carried out with sensor deployed systems in which the signals are converted into data and these are sent to a centralized device connected to the network.

The main advantage of the Internet of Things is that the events take place automatically by eradicating the need for manual intervention. Thus, IoT is revolutionizing the field of science and technology today. By employing this concept, the project aims at achieving its objectives such as cost savings and efficiency. Our country has a problem with disposing of wastes because of the voluminous deposits. About millions of tons of wastes are being disposed of every day. The sources for these waste includes industries, workplaces and houses. The strategies used are:

- House to House collection
- Drive in Service for collection
- Group based collection and Colony bins

Table 1. Types of waste collection along with its sustainability indicators

No	Sustainability indicators	Door to door	Curb site	Block collection	Community bins
1	Area improvement	✓	✓	✓	X
2	People convenience	✓	X	X	X
3	Staff convenience	X	✓	✓	X
4	Handling extra waste from festival	X	✓	✓	✓
5	Frequency and reliability	X	✓	X	✓

Among these, the colony bins proved to be the most common and effective way of collecting waste. But this kind of organizing waste is not very enthralling. Most often the waste tends to overwhelm the bins (Table 1).

This situation leads to problems like:

- How the waste can be disposed safely?
- How the recycling factor can be improved?

This work provides effective way of solving the problems as stated above.

- Cutting down on fuel costs.
- Automated alerts to the central municipal server.

2   Related Work

Our Cloud-based Garbage Alert System is implemented by using the concept of Internet of Things and the solution is programmed with the help of Arduino UNO. Ultrasonic sensors are utilized to track the volume of bin filled. Using this input, the conditions are framed in such a way that when the bin is almost full the municipal web server will be notified. Once the alert is sent to the municipal web server, the cleaning process takes place after verification. This is achieved by means of using Radio Frequency Identification tags which confirms the status of the cleaning process.

In order to optimize the routes that are used by the garbage collectors, Dijkstra's algorithm is used. This algorithm is implemented to optimize the routes taken and subsequently cut down on fuel costs.

Smart bin: Smart Waste Management System is a wireless sensor network-based system and is used for data monitoring and delivery. Here, the duty cycle of bins is optimized along with other operational units. This is done so as to reduce the amount of power consumed and also to maximize the operational time. Solid Waste Monitoring System is developed using the integration of technologies such as Geographical Information System (GIS), Radio Frequency Identification System (RFID), General Packet Radio Service (GPRS) and Global Positioning System (GPS). The main objective of this project is to enhance the efficiency in terms of responding to the enquiries of the customers especially in case of emergency situations.

The RFID tags are embedded in the trucks which are used for acknowledging the wastes after collection. Trucks are tracked using the GPS module which provides its position in concern with its latitudinal and longitudinal values. The bin levels are monitored continuously and once it reaches a certain threshold, a message is sent to the web server of the Municipal corporation which in turn mapped to the Geographical Information System. The Geographical Information System also contains a camera to visualize the type of waste dumped in the bins.

3   Problem Definition

Fig. 1. A municipal garbage bin turning into a wasteland

The following are some of the problems which are faced by our community (Fig. 1):

When waste is not collected, they tend to overflow the drains. This not only causes flooding but also spreads a variety of diseases.

- Generally, trash bins are open all the time. This may fascinate the cattle which are in need of food. Products like milk obtained from those cattle can have a deleterious effect on human beings.
- Insects like mosquitoes and flies tend to survive in such waste accumulated environments. This causes lots of diseases.
- Waste filled environments may provide shelter to rats. They spoil the environment by spreading diseases. Additionally, they bite electric cables causing inconvenience.
- Burning solid waste in open air leads to air pollution. It is a known fact that ozone layer is fast depleting. This reason will make the situation worse.
- Waste when left uncollected will make the environment unclean. This reduces the aesthetic value of the environment.
- Greenhouse gases like Methane and Carbon dioxide are emitted when waste is left unattended for longer time.

4   System Architecture

Smart bins use infrared sensors to detect the presence of waste that is dumped inside and thereby calculating the volume of waste filled in. Technically speaking, the infrared sensors senses each time when the waste is dumped inside the bin. Infrared sensors play the role of obstacle detection. Here the infrared sensors are placed at various levels of the bin. The sensors at the bottom ensure the system is functioning as intended. This is based upon the fact that the waste will settle at the bottom first. Depending on the percentage of waste filled and conditions specified, the information is shared with the concerned department (Fig. 2).

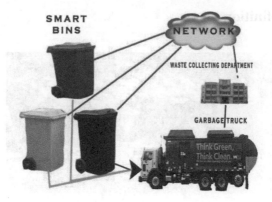

Fig. 2. The automated waste collection system.

4.1 Design of System Architecture

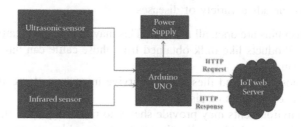

Fig. 3. Design of System Architecture

In this system, we have used both infrared and ultrasonic sensors for two reasons (Fig. 3).

Commonality: Both the sensors are used to detect the waste if they are thrown inside the bin. As a matter of fact, ultrasonic sensor measures volume of the entire bin (distance) whereas IR sensors will output binary values (true or false). Though they both serve the same purpose, ultrasonic sensors are used to address reliability issues. To make the output more accurate, we use both IR and ultrasonic sensors to cross-verify the output values. Per say one of the IR sensors is damaged, then ultrasonic sensor readings will help to understand deviation thereby stopping the system from malfunctioning. Infrared sensors are used to measure waste level by level. Typically, they are used to fix levels for bin (0%, 30%, 60%, 90%). The levels are determined depending upon the dimensions (height) of bin with 1:3 ratio for every level. For example: If height of the bin is (say) 18 cm then 18/3 = 6 cm is the difference between each level. In other words, IR sensors at level 1 would be 6 cm from base of the bin. Level 2 sensor at 12 cm from base. Level 3 sensor at 18 cm from base. The height values are for illustration purposes only. The dimensions of the bin are not part of this research. (It can be any value). The IR sensor measuring 0% is on the right side which shows true value if no waste is filled.

30%, 60% and 90% determines urgency of notification. If it's in between 60%–90% then municipal can plan accordingly to collect waste. Ultrasonic sensors measure the volume of waste inside the bin. This sensor is used to detect the presence of waste. The Arduino UNO microcontroller is programmed to interface the sensors with the ThingSpeak cloud platform. Power Supply is provided by connecting laptop to the Arduino Board.

- Arduino UNO: Programming kit that stores the program
- Power supply: Supply power for the programming kit so that it can function without interruption
- IoT web server: Send sensor (data values) via application programming interface to cloud platform. Acts as a medium that communicates between hardware and cloud platform (Fig. 4).

4.2 Design of Garbage Bin

See Fig. 4

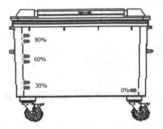

Fig. 4. Design of trash bin

5 Top View

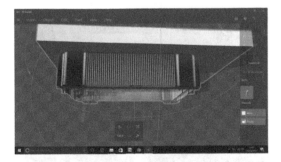

Fig. 5. Representing front view design of the bin using 3D Builder

One wall of the bin is partitioned into three levels and a pair of infrared sensors (transmitter and receiver) present at each level of the bin. These sensors are placed at fixed heights so that accurate measurements can be made. One pair of infrared sensors are placed at the base level to ensure that no malfunction takes place. The bin should be constructed in such a way that it can withstand harsh conditions like rain. In a large-scale implementation it is suggested that a bolted wall can be raised so that the electronic circuitry can be concealed (Figs. 5, 6, 7, 8).

5.1 Front View

See Fig. 6

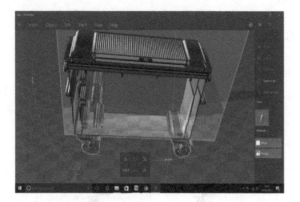

Fig. 6. Representing top view of the bin using 3D Builder

5.2 Bottom View

See Fig. 7

Fig. 7. Representing bottom view of the bin using 3D Builder

5.3   Side View

See Fig. 8

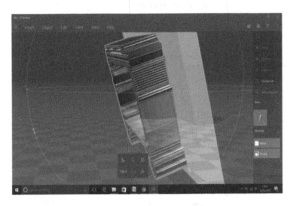

Fig. 8. Representing side view of the bin using 3D Builder.

6   Module Design

The module design is explained through a flowchart and is shown in Fig. 9.

The components used for designing the smart bin are given below. The module design consists of three phases:

Connection phase: To make the hardware interact with the cloud platform, the code programmed in Arduino kit is connected to Internet via Ethernet port that communicates via DHCP. This helps in dynamically connecting to all IoT smart bins that are available as a neighbor in the network.

Collection phase: When waste is thrown inside the thrash, the readings are recorded by IR sensors to observe if waste was thrown or not. If thrown, the different level sensor values are considered. Else IR sensor at the base level (0%) will show true value that indicates that the bin is empty

Verification phase: To address reliability issues, the values recorded by ultrasonic sensor is used for verification. This ensures that the IR sensors values sync with ultrasonic sensor values and there is no malfunctioning in the system.

For example: If IR sensor at 0% shows true value (Meaning bin is empty) but ultrasonic sensor shows 10 cm as output (Meaning bin is filled up to some level) then it means that output is contradictory. So we can predict a system malfunction and stop the program from disrupting the automated system. Once verification phase is successful, notification is sent to municipal corporation.

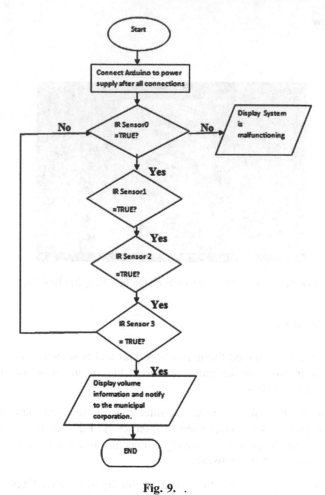

Fig. 9. .

6.1   Ethernet Module ENC28J60

ENC28J60 Stand-Alone Ethernet Controller IC features a host of features to handle most of the network protocol requirements. This Ethernet LAN module connects directly to almost all microcontrollers.

6.2 Arduino Uno

Arduino/Genuino Uno is a microcontroller board based on the ATmega328P (datasheet). It has 14 digital input/output pins (of which 6 can be used as PWM outputs), 6 analog inputs, a 16 MHz quartz crystal, a USB connection, a power jack, an ICSP header and a reset button.

6.3 Ultrasonic Sensor

The HC-SR04 ultrasonic sensor uses SONAR to determine the distance to an object like bats do. It offers excellent non-contact range detection with high accuracy and stable readings in an easy-to-use package. Its range lies between 2 cm to 400 cm.

6.4 Infrared Sensor

IR transmitter employs a LED to produce light waves of infrared wavelength. The receiver uses a specific light sensor to detect light waves of the infrared spectrum. When an object is close to the sensor, the light waves bounce off the object into the light sensor. This results in a large jump in the intensity, which we already know can be detected using a threshold.

7 Implementation

7.1 Pseudo Code

(1) Declare the pin numbers.
(2) Initialize the pin to corresponding input and output channels of the board.
(3) Initialize Ethernet connection.
(4) Get mac address and check with Ethernet buffer
(5) If mac address matches the buffer then setup DHCP connection.
(6) Initialize echo pin as input and trigger pin as output
(7) Read the pulse of Echo pin and store it in a variable "duration"
(8) Perform distance = (duration/2)/29.1 to calculate the distance in centimeters.
(9) Write the ultrasonic sensor values to the stash using web API Key.
(10) Read the digital pins as input for infrared sensor

> If IRSensor1 == TRUE then
> Print "Obstacle from sensor 1. Level 1 is full"
> If IRSensor2 == TRUE then
> Print "Obstacle from sensor 2. Level 2 is full"
> If IRSensor3 == TRUE then
> Print "Obstacle from sensor 3. Level 3 is full"

(11) Write the Infrared sensor values to the stash using Web API Key (Figs. 10, 11, 12, 13, 14, 15, 16, 17, 18, 19, 20).

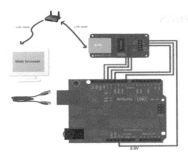

Fig. 10. Connecting Arduino pins to Ethernet module (ENC28J60)

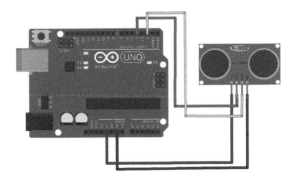

Fig. 11. Connecting ultrasonic sensor with arduino board

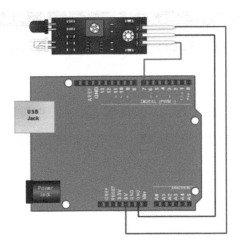

Fig. 12. Connecting IR sensor with Arduino UNO

8 Experimental Results

8.1 Ethernet Module

See Fig. 13

Fig. 13. Output displayed by serial monitor enabling internet access.

Fig. 14. Output displayed in IoT cloud platform "ThingSpeak"

Fig. 15. Distance displayed by serial monitor using ultrasonic sensor values.

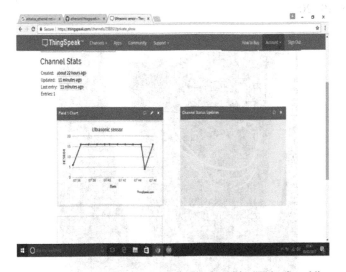

Fig. 16. Graph of sensor values displayed in "ThingSpeak"

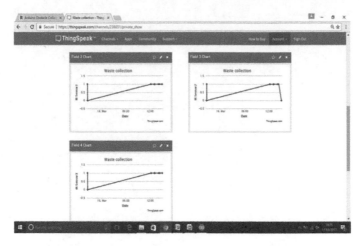

Fig. 17. Infrared sensor values in "ThingSpeak"

9 Working Model

See Figs. 18, 19, 20

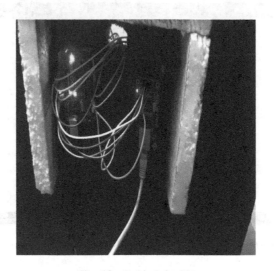

Fig. 18. Behind the bin

Fig. 19. Top view

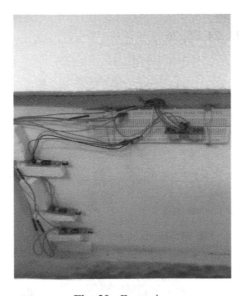

Fig. 20. Front view

10 Conclusion and Future Enhancements

This work describes a model using an automated system which reduces manual labour. Unlike the traditional smart bins manufactured by green tech life, it requires minimal effort to keep the environment clean. Though garbage bin is present a tidy environment cannot be guaranteed because of the lack of proper waste management. Issues like proper waste management and segregation are what our project aims to overcome. The smart bins located across the city can be monitored by municipal office remotely and necessary action can be taken whenever required. Hence, we conclude that employing modern technologies for collecting waste will not only benefit the public but also economical in the long run. In the current scenario, nearly 1 lakh rupees is invested for setting up a single garbage bin whereas the smart bins setup does not even sum to one-tenth of its total cost. In a large-scale implementation, many harmful gases may be emitted by the wastes in garbage's hence use of gas sensors to detect such gases and use of solar panels for power supply lies in the part of the future enhancements of this work.

Compliance with ethical standards

✓ All authors declare that there is no conflict of interest

✓ No humans/animals involved in this research work.

✓ We have used our own data (Figs. 21, 22, 23, 24, 25, 26).

Appendix

Sample Screens

See Figs. 21, 22, 23, 24, 25, 26

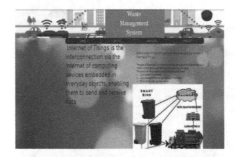

Fig. 21. Home page of the website

Fig. 22. About page of the website

Fig. 23. Check status page of the website

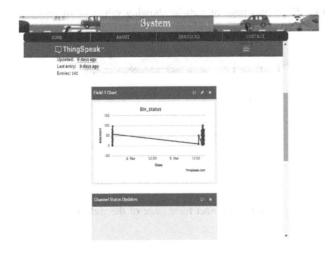

Fig. 24. Check status page showing Ethernet connection values

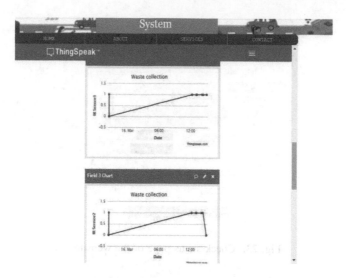

Fig. 25. Status page of website showing values of IR sensors

Fig. 26. Contact form page of the website

References

1. Kumar, N.S., Vijayalakshmi, B., Prarthana, R.J., Shankar, A.: IoT based smart garbage alert system using arduino UNO. In: 2016 IEEE Proceedings of the International Conference on Region 10 Conference (TENCON) (2016)
2. Mamun, M.A. A.I., Hannan, M.A., Hussain, A.: Real time solid waste bin monitoring system framework using wireless sensor network
3. Folianto, F., Low, Y.S., Yeow, W.L.: Smartbin: smart waste management system
4. Catania, V., Ventura. D.: An approach for monitoring and smart planning of urban solid waste management using smart-M3 platform
5. Arebey, M., Hannan, M.A. Basri, H., Begum, R.A., Abdullah, H.: Solid waste monitoring system integration based on RFID, GPS and camera
6. Ali, M.L., Alam, M., Rahaman, M.A.N.R.: RFID Based E-monitoring system for municipal solid waste management

7. Longhi, S., Marzioni, D., Alidori, E., Di Bu, G.: Solid waste management architecture using wireless sensor network technology
8. Huang, Y.F., Baetz, B.W., Huang, G.H., Liu, L.: Violation analysis for solid waste management systems: an interval fuzzy programming approach. J. Environ. Manage. **65**, 431–446 (2002)
9. Bashir, A., Banday, S.A., Khan, A.R., Shafi, M.: Concept, design and implementation of automatic waste management system
10. Hong, I., Park, S., Lee, B., Lee, J., Jeong, D., Park. S.: IoT-based smart garbage system for efficient food waste management
11. Daniel, V., Puglia, P.A., Puglia, M.: RFID: a guide to radio frequency identification, Technology Research Corporation
12. Flora, A.: Towards a clean environment: a proposal on sustainable and integrated solid waste management system for university Kebangsaan Malaysia. Report from Alam Flora (2009)
13. Sinha, T., Kumar, K.M., Saisharan, P.: Smart Dustbin. International J. Ind. Electron. Electr. Eng. **3**(5), 2347–6982 (2015)
14. Visvanathan, C., Ulrich, G.: Domestic solid waste management in south asian countries – a comparative analysis, 3 R South Asia Expert Workshop, Kathmandu, Nepal. (all reference are very old New latest reference may be added after year 2014) (2006)
15. Hannan, M.A., Arebey, M., Basri, H.: Intelligent solid waste bin monitoring and management system. Aust. J. Basic Appl. Sci. **4**(10), 5314–5319 (2010)
16. Oliver, A.S., Maheswari, N.: Identifying the gestures of toddler, pregnant woman and elderly using segmented pigeon hole feature extraction technique and IR-threshold classifier. Indian J. Sci. Technol. **9**(39) (2016). https://doi.org/10.17485/ijst/2016/v9i39/85578
17. Anuradha, M., Oliver, S.A., Justus, J.J.: IOT based monitoring system to detect the ECG of soldiers using GPS and GPRS. Biomed. Res. **29**(20) 2018

Vision Connect: A Smartphone Based Object Detection for Visually Impaired People

Devakunchari Ramalingam(✉), Swapnil Tiwari, and Harsh Seth

Department of Computer Science Engineering, SRMIST,
Chennai, Tamil Nadu, India
devakunr@ktr.srmist.edu.in,
tiwariswapnil97@gmail.com, harshseth242@gmail.com

Abstract. Visual impairments have long afflicted common people. Visual aids have also been around for a long time to help blind people to overcome the challenges and difficulties they face in everyday life. The invention of the Braille script signifies the result of incessant effort to help them lead normal, self-dependent lives. Thus, a lot of research has been done and money has been spent on various techniques to improve on these existing technologies. The proposed technique is a system for visually impaired people which not only detects obstacles for them in real-time but also helps them find various objects of their need around them. This technique is an integration of a smartphone application with an object recognition system. The object recognition module is implemented using Tensorflow. The smartphone camera is used to detect objects and the feedback is given in the form of speech output. The accuracy of the model is 85% which is higher as compared to an existing smartphone based systems using YOLO.

Keywords: Object detection · Visually impaired · Smartphone · Speech assisted

1 Introduction

The Blind People's Association, India, released a report that put the fraction of Indian people who are visually impaired at nearly 1.9% [1]. Millions of people in this world live with some form of visual affliction that hinders them from leading normal, fruitful or even independent lives. Visual challenges induce social awkwardness and low self confidence among them, and they are often alienated from their peers at a young age. Consequently, such people require constant care and help them going out and around, and often with daily routines. With time some people learn to walk with sticks but it is also just to make sure that they are not stepping on or colliding with something.

The past few years have witnessed the emergence of new technologies [2] to help visually impaired people, but most of the devices launched are expensive. The economically weak have fewer options, and fewer still are reliable.

Recent advances in Information Technology (IT), and mobile technology especially, have made it possible for people with disabilities like visual impairment to lead better, more comfortable lives using IT-based assistive technologies. Technology has,

S. Smys et al. (Eds.): ICCVBIC 2019, AISC 1108, pp. 863–870, 2020.
https://doi.org/10.1007/978-3-030-37218-7_92

in some measure, and can, significantly, help make individuals more capable of participating in social activities and leading normal, independent lives. The domain of IT-based assistive technologies, and the range of assistance they can possibly provide, is immense. Of late, ubiquitous computing, a model of human-computer interaction designed to suit the natural human environment, has seen a rising trend. With truly ubiquitous computing, a user needn't even be aware that they're using Information Technology for everyday tasks. Mobile assistive technologies are the result of numerous efforts to deliver assistance to disabled people and intelligently applied ubiquitous computing.

Mobile phones are cheaper, portable, lightweight, and easier to operate; aids delivered via phones will easily permeate into the general population, being devoid of the stigma against the more traditional aids. The proposed model is build upon this advantage with the aim of helping the visually impaired. The application will inform the user about the environment irrespective of whether it's indoor or outdoor without having to use any sensor or chip.

2   Related Work

The After the innovation of The Voice in 1974 [3], major efforts were directed towards helping visually impaired people become more and more self-dependent, so that they need not rely on a companion for seemingly simple tasks like crossing the road or finding an object. This particular problem has attracted a lot of attention ever since, both in the public and scientific spheres. The technological advancements that have been made to alleviate the effects of visual impairment have all sought to make the afflicted more capable of perceiving their immediate environment. Real time detection system [4] involved the usage of a camera to detect text written or printed on objects. They used MSER to extract the most stable features and localized using OCR. The smartphone-based system [5] used a smartphone to take the input in the form of images as well as give the output in the form of audio and a server where the processing will be done.

(a)   The Voice

The vOICe was developed at the University of Birmingham by Leslie Kay in 1974. The vOICe is an image to audio device which captures an image from a webcam. The capital letters in vOICe mean 'Oh, I see!'.

(b)   The Vibe

The Vibe [6] is a device similar to the vOICe. It is also a visuo-auditory device which employs a wide angle camera which is placed in front of the user.

(c)   Martinez et al.'s approach

Recently, an approach similar to the vOICe was developed by Martinez et al. [7]. It is also an object recognition device which converts image to speech information. It has a different image processing method as compared to the vOICe. The image is scanned from the center to the left and then to the right. The Fig. 1 is an example of an image being scanned by the Martinez et al's approach.

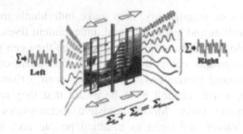

Fig. 1. Martinez et al.'s approach

(d) Prosthesis Substituting Vision by Audition [PSVA]

PSVA [8] was the result of a combination of the retina present in human eyes and the human cochlea in our ears. The combination used an inverse representation of the cochlea. This model has pixelated representations of the retina and the fovea.

(e) Features extraction

SIFT [9–11] was used to match localized key points of same objects captured in different scenes. SURF [12] is faster than SIFT in the localization process of key points. It maintains the image properties.

3 Proposed Model – Vision Connect

The proposed model uses a smartphone-based application to help the visually impaired find day to day objects as well as avoid any obstacle in real time without being dependent on someone else. The application will use the back camera of the smartphone and capture images of the objects in front of it. These images will be used by the image recognition module and the detection of the same will be performed. The final output will be given in the form of speech feedback to the user. The results indicate that people with visual impairments can gauge the direction to an object in their path, as well as know what the object is. This helps them gain a more comprehensive understanding and awareness of their immediate environment.

A visually impaired person might have a very time finding his way around or looking for some object even avoiding obstacles. Keeping this in mind, we focused on the day to day object detection in real time. If either the user is new to a particular environment or he is trying to find something, assistance from the application can be very well used.

3.1 Smartphone Application

The development of app was done for Android devices. It is one of the leading OSs for smartphones these days. Since most of the affordable smartphones have an Android operating system, we have started the development with Android for now. The mobile app was written in the Java Language using the software Android Studio. The application deals with various requirements and settings for the app to work on the smartphone. A Tensorflow trained model was used for object recognition. The processing was done using Google's Tensorflow API.

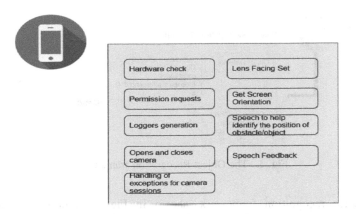

Fig. 2. Layout of Vision Connect - Speech to object Detection model

The layout of the developed android application is depicted in Fig. 2. It implements all these features when used. It will perform a hardware check and request for permissions like camera and storage. It will generate logs. It will set the back facing lens of the camera. It will also keep the screen orientation in check. When the user opens the application, it accesses the back facing lens of the mobile to capture the images. The objects in front of the lens will be continuously captured by the app and the processing will be done. Upon recognition of the object, the speech feedback is given in the form of the output. The speech output conveys the name of the object or obstacle followed by the relative position of it with respect to the camera lens. Along with the type of the object this also helps give the visually impaired person an idea of the detected object's location. Moreover, the bounding box around the detected object will also have the name of the object with the probability.

3.2 Image Recognition

The aim of the app is to detect objects and obstacles in the environment around the visually impaired and the blind people to help them in their day to day activities without being dependent on anyone else or relying on touch sensations which are neither accurate and can even turn out to be pretty dangerous for them. Object detection by image recognition is the heart of the system with the objective to assist the targeted user base. Image recognition is the process of mapping the set of pixels from an image to desired objects for which the ground truths have been provided. This pixel mapping is done by feeding the various pixel sets of the image in a neural network and then performing various convolutions on those pixels which provide us with the desired pixels for detecting the object. We perform image recognition in our model on the image data which is converted into the speech data. The application turns on the back camera of the smartphone and begins to capture images in real time via the dynamic movement of the mobile. These images are then worked upon by the system to identify the object or the obstacle in the user's environment.

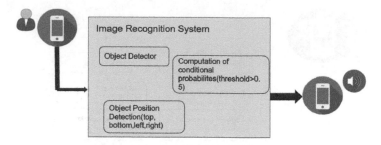

Fig. 3. The Object Recognition framework

The image recognition framework, shown in Fig. 3, comprises of an object detector, computation of conditional probabilities and object position detection module. The computation of conditional probabilities is done with a threshold of 0.5 to facilitate the decision making capability of the system in order to give the user an output when the object is recognized. This processing is done in the smartphone itself and not a server. Hence this will not put an extra burden of having an internet connection all the time and it won't fail in places where there the user may face network issues. The image recognition is done with the help of Google's API Tensorflow. It has a number of software which provides developers with the ecosystem to develop deep learning models.

3.3   Auditory Presentation

Speech is the most important pretext for communication specially for differently abled people having vision impairment. The advancements in artificial intelligence and language processing systems provide us with tools to recognize speech data and produce speech feedbacks. The final output regarding the detected image is given to the user as audio. The audio generated would have the object's name and its position. This would give the user an idea about what the object is and its relative position with respect to the mobile camera. This was implemented using TextToSpeech class which is provided by Android. This class also provides a number of useful features like setting the pitch of the output speech audio, setting the rate with which the speech is given as the output.

4 Results and Discussions

The aim was to build a recognition system which could use the live feed from the smartphone camera to carry out the object detection and recognition. The system should be insensitive of the scale and the orientation of the phone. Tensorflow was applied for this module. It is actually an open source software library used across multiple disciplines including neural networks. It is one of the most powerful open source frameworks developed by Google. It is the base on top of which Google Cloud Vision and AlphaGo are built upon. It also supports multiple languages for programming. The added advantage is that TensorFlow Lite brings model execution to a variety of devices, including IOT devices and mobile devices, providing more than a 3x boost in inference speedup over original TensorFlow. This facilitated the way to get machine learning on either a Raspberry Pi or a phone. The application was tested on various day to day objects and Fig. 4 depicts some of the recognized objects by it.

Fig. 4. Object detection by the application

The Fig. 5 is a Precision-Recall curve for Person class evaluating the prediction skill of the model. The curves are plotted for various threshold values.

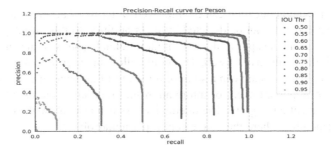

Fig. 5. Precision-Recall curve for detection of individual

One of the most widely accepted and used metric for testing accuracy of an object detection models like R-CNN, Darknet, etc is mAP (Mean Average Precision). Therefore, we have used mAP as the evaluation metric for our model with different IOU (Intersection over union) thresholds to depict the precision-recall tradeoffs over IOUs. As the recall increases, precision values become dynamic and change drastically. Therefore, it is necessary to find the mean average precision by calculating the area under the precision-recall curve for a various threshold. Our result demonstrates a mAP value of 56.38 which is well over a required 50 for a model to have true positives. Considering IOU threshold at 0.7, Precision = 0.83 and Recall = 0.95. F1 Score formulated using,

$$F1 = 2 * precision * recall/(precision + recall) = 1.03$$
$$Accuracy = (TP + TN)/(TP + TN + FP + FN) = 0.85$$

5 Conclusion

The Vision Connect model involves the implementation of an Android application which would help the visually impaired find objects around them and detect obstacles in their paths in real time. The model developed was successfully deployed in the android application. It incorporates a TensorFlow trained model which was used for image recognition in real time. The application was tested for many day to day objects that a visually impaired person may have in his/her environment. This application is actually an audio-visuo implementation and so the detected object is given as output in the form of audio feedbacks. There are many traditional ways used by the blind and the visually impaired to detect objects and obstacles in their environments. Compared with these the effectiveness of our application is good. Without putting in extra expense for a visuo-auditory device or having to rely on one's own self, this application can make the person's life so much easier. It eliminates the feeling of being dependent on someone else all the time and will give the user a sense of freedom. Tests done prove that the application meets the basic requirements and the audio output can make the user familiar to the environment around them making them feel confident about finding their way around.

In the future, the application can be improved in terms of accuracy alone with the scope of the objects that can be detected by it. In addition to this an integration of an interactive assistant in the app would make it more reliable and accurate and that would in turn make it more effective as well. This could allow the user to look for a particular object real time. The assistant can also have a feature of allowing the user to tell them to remind some important event or information like reminding the user they have an appointment or meeting to go to. The interactiveness can also allow the user to ask the assistant to remember where they have kept a particular object.

Compliance with Ethical Standards
  ✓ All authors declare that there is no conflict of interest.
  ✓ No humans/animals involved in this research work.
  ✓ We have used our own data.

References

1. Blind People's Association India, https://www.bpaindia.org/
2. Oh, Y., Kao, W.L., Min, B.C.: Indoor navigation aid system using no positioning technique for visually impaired people. In: Proceedings of 19th International Conference on Human-Computer Interaction (HCI 2017), pp. 390–397, Canada (2017)
3. Auvray, M., Hanneton, S., O'Regan, J.K.: Learning to perceive with a visuo - auditory substitution system: localisation and object recognition with 'The vOICe'. Perception **36**, 416–430 (2007)
4. Deshpande, S, Revati, S.: Real time text detection and recognition on hand held objects to assist blind people. In: Proceedings of IEEE International Conference on Automatic Control and Dynamic Optimization Techniques (ICACDOT) (2016)
5. Lin, B.S., Lee, C.C., Chiang, P.Y.: Simple smartphone-based guiding system for visually impaired people. Sensors **17**(6), 13–71 (2017). MDPI
6. Durette, B., Louveton, N., Alleysson, D., Herault, J.: Visuo-auditory sensory substitution for mobility assistance: testing the VIBE. In: Workshop on Computer Vision Applications for the Visually Impaired, Marseille, France (2008)
7. Martnez, B.D.C., Vergara-Villegas, O.O., Snchez, V.G.C., Domnguez, H.J.O., Maynez, L. O.: Visual perception substitution by the auditory sense. In: LNCS, vol. 6783, pp. 522–533, Springer (2011)
8. Renier, L., Laloyaux, C., Collignon, O., Tranduy, D., Vanlierde, A., Bruyer, R., De Volder, A.G.: The ponzo illusion with auditory substitution of vision in sighted and early-blind subjects. Perception **34**(7), 857–867 (2005)
9. Mohane, V., Gode, C.: Object recognition for blind people using portable camera. In: Proceedings of 2016 IEEE World Conference on Futuristic Trends in Research and Innovation for Social Welfare (Startup Conclave) (2016)
10. Lowe, G.D.: Distinctive image features from scale-invariant keypoints. Int. J. Comput. Vision **60**(2), 91–110 (2004)
11. Jabnoun, H., Benzarti, F., Amiri, H.: Visual substitution system for blind people based on sift description. In: Proceedings of International Conference of Soft Computing and Pattern Recognition, pp. 300–305 (2014)
12. Chincha, R., Tian, Y.: Finding objects for blind people based on SURF features. In: Proceedings of 2011 IEEE International Conference on Bioinformatics and Biomedicine Workshops, pp. 526–527, Atlanta (2011)

Two-Phase Machine Learning Approach for Extractive Single Document Summarization

A. R. Manju Priya[(✉)] and Deepa Gupta

Department of Computer Science and Engineering, Amrita School
of Engineering, Bengaluru, Amrita Vishwa Vidyapeetham, Bengaluru, India
armanjupriya@gmail.com, g_deepa@blr.amrita.edu

Abstract. Summarization is the process of condensing the information with minimal loss of information and its importance has grown with the increased information availability. This paper explores two-phase machine learning approach for Single document summarization where in Phase I sentence selection is done using Bayesian Network and in Phase II coherence summary is generated using Affinity Propagation. Phase I model has explored variety of features as incidence features, scoring features including psycholinguistic features. The proposed model is built on the benchmark dataset DUC 2001 and tested on DUC 2002 dataset using Rouge scores. The proposed two-phase machine learning approach gives significantly improved results compared to many existing baseline models.

Keywords: Extractive text summarization · Affinity Propagation · Rouge · Psycholinguistic features · Bayesian Network · Duc 2001 · Duc 2002

1 Introduction

With vast increase in the unstructured textual data that are not organized into traditional databases, summarization has become an important area of research. We need automatic text summarization to condense the text without losing the relevant information to handle the ever-increasing mass of information. Summarization is the process of condensing by losing information which are not more relevant to the concept of the document. Summaries provide sneak in to the document, without reading the entire document. Summarization requires understanding of the document, selecting the relevant sentences, reformulating the sentences and assembling the sentences in a coherent way [1]. This is a difficult cognitive task for humans and this shows the difficulty in automating summarization [2]. Benefits of summarization includes reduced time to read the document, easy document selection for researchers, improved effectiveness of document indexing and keyword extraction and is less biased than a human summarizer. Even after six decades of research the automatic summarizers are yet to reach the perfection of human summarizer. Summarization can be categorized into different types [3] based on the method of generation (Extractive, Abstractive, Sentence Compression), number of documents used to create the summary (Single Document,

© Springer Nature Switzerland AG 2020
S. Smys et al. (Eds.): ICCVBIC 2019, AISC 1108, pp. 871–881, 2020.
https://doi.org/10.1007/978-3-030-37218-7_93

Multi Document), function of the summary [4] (informative, indicative), document genre (News, Literary, Social Network), context (generic, query-guided, update) and target audience (Profile, Non-Profile).

This paper focuses on generating generic, extractive, informative single document summary with enhanced content coverage using two-phase machine learning approach. In this work, around eighty-eight features including features based on Psycholinguistics were explored to generate the summary.

Rest of paper is structured as follows: Sect. 2 describes the related work in the field, Sect. 3 describes the proposed two-phase machine learning method in detail, Sect. 4 explains the experimental setup, Sect. 5 discusses about the experimental results and Sect. 6 provides the conclusion and future work.

2   Related Work

This section summarizes the extractive summarization related papers. Kupiec et al. [5] built a Naive Bayes Classifier to identify summary sentence using fixed-phrase, sentence length cut-off, upper-case feature, paragraph feature and thematic word as features. Aone et al. built on the Kupiec approach and used the presence of signature words to exploit the linguistic dimension of the corpus. Lin modelled the problem with decision tree and concluded that for query guided summarization, certain parameters are independent of other and no isolated characteristic or individual parameter is sufficient. Other researchers proved that high performing extraction strategies is provided by Artificial Neural Networks (ANN) and Hidden Markov Model (HMM). Although machine learning approaches provide good extraction strategy, the sentences selected where not cohesive outside the context. The parameters tuned during the learning phase may not work out well for a different type of document. Methods to improve the strategy of selecting the important sentences which improves the content and context of the summary [6] has been proposed in the literature. Non-negative matrix factorization method has been used to select more meaningful sentences [7]. Identifying the sentence features that help in efficient text summarization is discussed in many papers [8, 9]. Topic heading and sentence position has come up as dominant features for text summarization. Genetic Algorithm has been applied to text summarization [10, 11]. The efficiency of various supervised machine learning algorithms on text summarization is studied in few papers [12, 13]. Yeh et al., used LSA in semantic analysis for text summarization through Text Relational Mapping [14]. Mathai et al., proposed iterative LSA for semantic based concept extraction in legal domain [15].

Graph-based algorithms were applied to extract the word relations to identify subsuming relationship between the sentences to improve the saliency of the summary [16]. Representation of the document in bipartite graph format was implemented to optimise importance of the content, coherence and non-redundancy in the summary [17]. Semi-graph concepts were explored to generate extractive summaries [18]. Chan proposed the use of textual continuity as a shallow linguistic feature for sentence extraction [19]. Discourse tree structure of the document was studied to improve the summary quality [20]. Extraction of the keywords through lexical chain and keyword extraction based on topic centrality was implemented to improve keyword based summarization [21]. Chidambaram et al., proposed combining summarization algorithms with Latent Dirichlet Allocation (LDA), a topic modeling method [22]. Remya et al., explored sparse coding techniques and dictionary learning for document summarization [23].

This paper focuses on improving the selection of summary sentences from the document such that major concepts of the document is covered through machine learning methods by exploring various features of the sentence. In this work, we compare the performance of two-phase approach with certain standard algorithms and validate the contribution of the individual phases in summary generation.

3 Proposed Methodology

Figure 1 shows the general workflow of the proposed two-phase machine learning approach for summary generation which consists of four main modules as discussed below.

3.1 Pre-processing

The raw text is subjected to pre-processing before feature extraction. First text is split into sentences and further tokenized and split into words. This is followed by POS tagging, removal of Stop words and punctuation and lemmatization. The POS tagging was done using Stanford POS Tagger.

3.2 Feature Extraction

Three groups of features are extracted as Incidence Features, Psycholinguistic Features and Scoring Features and is explained in the following.

Incidence Features: Importance of the sentence can be ascertained by presence of certain grammatical structures. Hence, the count of POS Tags present in the sentence for adjective, noun, pronouns, verbs, etc. is used for generation of Incidence features.

Psycholinguistic Features: Psycholinguistic features extracted from the MRC Psycholinguistic database [24] is shown in Table 1. In order to extract the score from Psycholinguistic database, the Stanford POS tags were mapped to syntactic category of the word in MRC database.

Scoring Features: They are divided into six groups based on source of information namely Frequency Group, Similarity Group, Wordnet Group, POS Tag Group, Position Group, and Length Group. The idea of scoring features is captured from multiple references [13, 18, 25, 26].

1. *Frequency Group*: Features under this group are predominantly dependent on the frequency of the occurrence of the word in the document. *Term Score* is average of normalized term frequency of the words in the sentence as shown in Eq. (1). Normalized term frequency of a word is computed using Eq. (2).

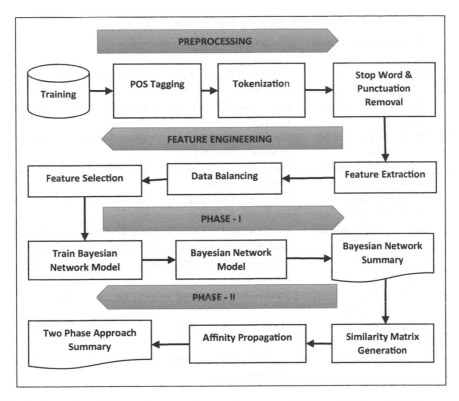

Fig. 1. General workflow of proposed two phase approach for extractive single document summarization

$$Term\ Score = \frac{\sum Normalizd\ Term\ Frequency\ of\ words\ in\ the\ sentence}{Number\ of\ words\ in\ the\ sentence} \quad (1)$$

$$Normalized\ Term\ Frequency[x] = \frac{Count\ of\ occurence\ word\ x\ in\ the\ document}{Total\ Number\ of\ words\ in\ the\ document} \quad (2)$$

where x is any word in the document.

TF-IDF of the word is calculated using the standard method. Summation of *tf-idf* score of nouns in the sentence form the *tf-idf* score of the sentence. *TF-IDF Greater than Average* is computed as shown in Eq. (3).

$$tf - idf\ greater\ than\ average = \begin{cases} 0\ tfidf_s < tfidf_a \\ 1\ otherwise \end{cases} \quad (3)$$

where $tfidf_s$ is the tf-idf score of the sentence and $tfidf_a$ is the average tf-idf score of the document.

Table 1. Psycholinguistic Features and its description

S. No	Psycholinguistic feature	Psycholinguistic feature description
1	Concreteness	Colerado, Gilhooly-Logie and Pavio norms are merged to generate concreteness
2	Kucera-Francis written frequency	Francis and Kucera norms are merged to create occurrence of word's frequency
3	Kucera-Francis number of categories	Francis and Kucera norms are merged to create occurrence of word's frequency
4	Kucera-Francis number of samples	Francis and Kucera norms are merged to create occurrence of word's frequency
5	Thorndike-Lorge written frequency	Lorge and Thorndike's L Count is used to get the occurrence frequency
6	Brown verbal frequency	Brown's London-Lund Corpus of English Conversation is used derive the occurrence frequency in verbal language
7	Familiarity	Familiarity of the word in print
8	Imageability	Imageability
9	Meaningfulness - Colorado Norms	Meaningfulness ratings from the Toglia and Battig
10	Meaningfulness - Pavio Norms	Meaningfulness from the norms of Pavio
11	Age of acquisition	Logie and Gilhooly norms are used to determine the age of acquisition

Keywords is the sum of term frequency of top k words in the document that occur in the given sentence.

2. *Similarity Group*: *Title Similarity* and *Average Cosine Similarity* are two features extracted under this group. Title Similarity is the measure of similarity of the sentence with the first sentence in the document. Average cosine similarity is the average of the similarity of the sentence with every other sentence in the document.
3. *Wordnet Group*: *Polysemy* and *Hypernymy* features are extracted from wordnet
4. *POS Tag Group*: *Proper nouns Count, Has Numeral* are computed based on number of proper nouns and presence of cardinal tag in the sentence respectively. *Centrality* is the fraction of the nouns present in the sentence is computed using Eq. (4).

$$Centrality = \frac{Count\ of\ noun\ words\ in\ the\ sentence\ that\ also\ occur\ in\ other\ sentence}{Number\ of\ unique\ noun\ words\ in\ the\ document} \quad (4)$$

5. *Position Group: Position* feature is the location of the sentence in the document and *position value* feature which represents the importance of the sentence is computed using Eq. (5).

$$position\ value = \begin{cases} 1 \ if \ i = 1 \\ 0.5^{i-1} \ if \ i \neq 1 \ and \ i \leq n/2 \\ 0.5^{n-1+1} \ if \ i > \frac{n}{2} \end{cases} \quad (5)$$

where i is the position of the sentence and n is the number of sentences in the document

6. *Length Group*: *Number of sentences* and *Normalized sentence length* are two features under this group. Normalized sentence length is the product of sentence length and average sentence length in the document.

3.3   Data Balancing and Feature Selection

As the ratio of summary sentences to non-summary sentences is close to 1:7, the data needs to be balanced before model training. Synthetic Minority Over-sampling Technique (SMOTE) [27] was used to generate synthetic data so as to balance the ratio of summary to non-summary sentences. Features were ranked based on the information gain of each attribute and top attributes that contributes significantly for the information were selected [28] Information Gain is calculated using the entropy measures as specified in Eq. (6). Entropy is computed using the Eq. (7)

$$Information\ Gain(X, t) = Entropy(t) - Entropy(t, X) \quad (6)$$

$$Entropy(S) = \sum_{i=1}^{c} -p_i \log_2 p_i \quad (7)$$

In Eq. (6), X is the attribute whose information gain is to be determined while t is the class label. *Entropy (t)* denotes the uncertainty inherent in the data. In Eq. (7) c is the total number of classes in the data, p_i denotes the fraction of data belonging to the class i and S denotes the feature for which entropy is being calculated.

3.4   Two Phase Approach

In two phase machine learning approach, summary sentence selection is done in Phase-I using Bayesian Network and coherent summary is generated in Phase-II using the Affinity Propagation (AP) clustering algorithm. Figure 2 portrays the proposed two-phase approach algorithm which explains the process of generating the final coherent summary based on the different summary length generated by Phase-I model. The Phase-I model is learnt using K2 Search Algorithm, a greedy search algorithm that recovers the underlying graphical structure based on the pre-determined order of nodes. The output of Phase-I model is used to generate sentence similarity matrix which will act as input to the Phase-II.

Similarity Matrix is computed using word similarity and sentence position as $Similarity_{ij} = TS_{ij} + P_{ij}$, where TS_{ij} is the count of common words between the two sentences and P_{ij} is the position similarity between two sentences [29]. TS_{ij} and P_{ij} is computed using Eqs. (8) and (9) respectively.

$$TS_{ij} = \begin{cases} |S_i \cap S_j|, i \neq j \\ 0, otherwise \end{cases} \tag{8}$$

where S_i and S_j are any two sentences in the document.

$$P_{ij} = \begin{cases} \frac{n}{2}, if\ abs(i-j) = 1 \\ \frac{n}{abs(i-j)}, if\ abs(i-j) > 1 \end{cases} \tag{9}$$

where n represents the number of sentences from which the summary is to be generated, i and j are position of the sentences whose similarity needs to be computed.

4 Experimental Setup

The benchmark DUC 2001 dataset which contains 292 documents is used for training the model and DUC 2002 data set that contains 562 documents is used for evaluation [30]. The standard summarization evaluation measure Rouge recall is computed using Eq. (10) and is used for performance evaluation of the proposed model. It is the ratio of number of overlapping words between the reference summary and system summary to number of words in the reference summary [31].

$$Rouge\ Recall = \frac{\sum_{S \in \{Reference\ Summaries\}} \sum_{gram_n \in S} Count_{match}(gram_n)}{\sum_{S \in \{Reference\ Summaries\}} \sum_{gram_n \in S} Count(gram_n)} \tag{10}$$

Here n is the number of grams (words) considered for comparison between the reference and system summary. Rouge-1 ($n = 1$) and Rouge-2 ($n = 2$) measures are used to evaluate the hundred-word summary generated by the proposed model. The considered dataset has only abstractive reference summary, therefore, the sentences of the document needs to labelled for training the Phase I model. The labeling is generated by calculating the cosine similarity of each of the reference summary sentence with each of the document sentence. The document sentence that has the highest cosine similarity is marked as the extractive summary sentence.

The proposed model is implemented in python using scikit-learn, nltk and pandas packages. In feature selection phase, top thirty-three features that contributes at least 0.01 information were selected. Among these thirty-three features seven are from Incidence features, eleven are from Psycholinguistic and fifteen are from Scoring features. The proposed two-phase approach is compared with three existing extractive summarization models. First two baselines are based on standard extractive text summarization algorithm and baseline 3 is a recently published work in 2018. The baselines are as follows

- *Baseline 1 (LSA):* Semantic information is used to rank the sentences based on importance and top k sentences are selected [14].

1. Select the summary sentences using the Phase-I model.
2. Pass the information from the Phase-I model to the Phase-II model.
   a. Case 1: Summary Length generated by the Phase-I Model is the desired Summary length
      i.   If the summary length is reached, do not post-process the summary using the Phase-II model
   b. Case 2: Summary Length generated by the Phase-I Model is less than the desired length
      i.   All the non-summary sentences from the Phase-I model is passed as input to the Phase-II model
   c. Case 3: Summary Length generated by the Phase-I Model is greater than the desired length
      i.   All the summary sentences selected by the Phase-I model is passed on as input to the Phase-II model
   d. Case 4: Summary Length generated by the Phase-I Model is zero
      i.   All the sentences from the document is passed on as input to the Phase-II model.
3. Once the clusters are formed by the Phase-II model, select the sentences for the summary from each cluster until the summary length is reached. The most central sentences with high tf-idf score is selected as summary sentence from each cluster.
4. The final summary is combination of results from the Phase-I model and Phase-II model based on the input passed to Phase-II model
   a. Case 1: The final summary is the one generated by the Phase-I model
   b. Case 2: The final summary is the union of sentences selected by the Phase-I and Phase-II model.
   c. Case 3: The final summary is the intersection of sentences selected by the Phase-I model and Phase-II model.
   d. Case 4: The final summary is the set of sentences selected by the Phase-II model.

Fig. 2. Proposed two phase algorithm for extractive single document summarization

- *Baseline 2 (Text Rank)*: Similarity between the sentences is calculated using count of common words and Page Rank Algorithm is applied to select the summary sentences [32].
- *Baseline 3 (ESSg)*: Extractive Summarization using Semigraph (ESSg) uses semi graphs to rank the sentences and identify summary sentences [18].

5   Experimental Results and Analysis

Table 2 lists the Rouge recall scores for the proposed model and baselines. The proposed system outperforms all the three baselines in terms of Rouge-1 score compared to Rouge-2 score. The proposed two-phase approach shows 12.45% to 15.99% improvement in terms of Rouge-1 score over the baselines. Figure 3 shows the percentage improvement of the recall score obtained by the proposed method over the state of art methods.

Table 2. Recall scores for the two-phase approach and the existing state of art models

Rouge type	Baseline-1	Baseline-2	Baseline-3	Proposed model
Rouge-1	0.371920	0.360570	0.367230	**0.418239**
Rouge-2	0.195170	0.106190	0.206470	0.189089

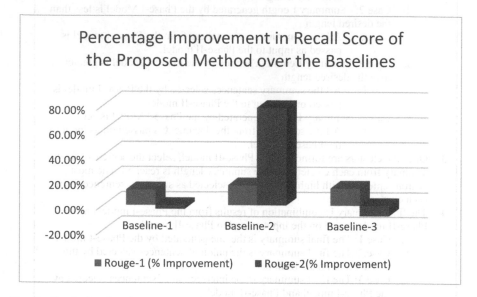

Fig. 3. Graph showing the percentage improvement of the recall score obtained by the proposed method over the state of the art methods

To prove that each of two phases have contributed significantly in generation of the final extractive text summary Single Factor ANOVA Test was conducted on the ROUGE recall score. The p-value obtained in the ANOVA test is less than 5% which implies that there is significant difference among the phases. Hence, Post-hoc analysis was done using Turkey's test at 95% confidence level and the results are shown in Table 3. In post-hoc analysis all the p-value obtained was less than 5% which proves that there is significant difference between the approaches at 0.05 significance level.

Table 3. Turkey's Post-hoc analysis test result for Rouge-1 & Rouge-2 score

Rouge type	Difference of levels	Difference of means	SE of difference	T-value	Adjusted P-value	95% Confidence interval
Rouge-1	Phase I vs Two-Phase Approach	−0.1435	0.0074	−19.3200	0.0000	(−0.1609, −0.1261)
Rouge-1	Phase II vs Two-Phase Approach	−0.0212	0.0074	−2.8600	0.0120	(−0.0386, −0.0038)
Rouge-1	Phase II vs Phase I	0.1223	0.0074	16.4600	0.0000	(0.1049, 0.1397)
Rouge-2	Phase I vs Two-Phase Approach	−0.0526	0.0059	−8.8600	0.0000	(−0.0665, −0.0387)
Rouge-2	Phase II vs Two-Phase Approach	−0.0249	0.0059	−4.2100	0.0000	(−0.0389, −0.0110)
Rouge-2	Phase II vs Phase	0.0276	0.0059	4.6500	0.0000	(0.0137, 0.0415)

6 Conclusions

Extractive single document summarization is done using two-phase approach where Bayesian Network is used in first phase to identify the summary sentences, followed by Affinity Propagation in second phase to form the final coherent summary. The position of the sentence, centrality of the sentence, TF-IDF score along with psycholinguistic features are the dominant features contributing for summary generation. The two-phase approach provides significantly improved Rouge-1 score.

The extractive summarization can be further improved by exploring the semantic space of the document using optimization algorithms. Instead of selecting the sentences until the summary length is reached, sentence fusion or compression techniques can be used to describe the desired concept with minimum words.

Acknowledgment. Dataset DUC 2001 is obtained from free web services.

Compliance with Ethical Standards
  ✓ All authors declare that there is no conflict of interest.
  ✓ No humans/animals involved in this research work.
  ✓ We have used our own data.

References

1. Torres-Moreno, J.M.: Automatic Text Summarization, pp. 4–107. Iste (2014)
2. Endres-Niggemeyer, B.: Summarizing Information. Springer, Heidelberg (1998)
3. Mani, I.: Automatic Summarization. John Benjamins, Amsterdam (2001)
4. Hovy, E.: Automated text summarization. In: Mitkov, R. (ed.) The Oxford Handbook of Computational Linguistics, pp. 583–598. Oxford University Press, Oxford (2005)
5. Kupiec, J., Pedersen, J., Chen, F.: A trainable document summarizer. In: 18th Conference ACM Special Interest Group on Information Retrieval (SIGIR 1995), pp. 68–73. ACM Press, Seattle, WA (1995)
6. Ko, Y., Seo, J.: An effective sentence-extraction technique using contextual information and statistical approaches for text summarization. Pattern Recogn. Lett. **29**(9), 1366–1371 (2008). https://doi.org/10.1016/j.patrec.2008.02.008
7. Lee, J.H., et al.: Automatic generic document summarization based on non-negative matrix factorization. Inf. Process. Manage. **45**(1), 20–34 (2009). https://doi.org/10.1016/j.ipm.2008.06.002
8. Barrera, A., Verma, R.: Automated extractive single-document summarization: beating the baselines with a new approach. In: Proceedings of the 2011 ACM Symposium on Applied Computing, pp. 268–269 (2011)
9. Barrera, A., Verma, R.: Combining syntax and semantics for automatic extractive single-document summarization. In: Gelbukh, A. (ed) Computational Linguistics and Intelligent Text Processing, LCNS, vol. 7182, pp. 366–377. Springer, Heidelberg (2012)
10. Chatterjee, N., Mittal, A., Goyal, S.: Single document extractive text summarization using genetic algorithms. In: Proceedings – 3rd International Conference on Emerging Applications of Information Technology, EAIT 2012, pp. 19–23 (2012)

11. Mendoza, M., et al.: Extractive single-document summarization based on genetic operators and guided local search. Expert Syst. Appl. **41**(9), 4158–4169 (2014). https://doi.org/10.1016/j.eswa.2013.12.042
12. Dlikman, A., Last, M.: Using machine learning methods and linguistic features in single-document extractive summarization. In: CEUR Workshop Proceedings, vol. 1646, pp. 1–8 (2016)
13. Ferreira, R., et al.: Assessing sentence scoring techniques for extractive text summarization. Expert Syst. Appl. **40**(14), 5755–5764 (2013). https://doi.org/10.1016/j.eswa.2013.04.023
14. Yeh, J.Y., et al.: Text summarization using a trainable summarizer and latent semantic analysis. Inf. Process. Manage. **41**(1), 75–95 (2005)
15. Mathai, S., Gupta, D., Radhakrishnan, G.: Iterative concept-based clustering of Indian court judgments. In: Advances in Intelligent Systems and Computing, vol. 712, pp. 91–103 (2018)
16. You, O., Li, W., Zhang, R., Li, S., Lu, Q.: A progressive sentence selection strategy for document summarization. Inf. Process. Manage. **49**, 213–221 (2013)
17. Ravinuthala, V.V.M.K., Chinnam, S.R.: A keyword extraction approach for single document extractive summarization based on topic centrality. Int. J. Intell. Eng. Syst. **10**(5), 153–161 (2017)
18. Sonawane, S., et al.: Extractive summarization using semi graph (ESSg). Evolving Syst. **10**(3), 409–424 (2018)
19. Chan, S.W.K.: Beyond keyword and cue-phrase matching: a sentence-based abstraction technique for information extraction. Decis. Support Syst. **42**(2), 759–777 (2006)
20. Kikuchi, Y., et al.: Single document summarization based on nested tree structure. In: Proceedings of the 52nd Annual Meeting of the Association for Computational Linguistics (Volume 2: Short Papers), pp. 315–320 (2014). https://doi.org/10.1006/brbi.1994.1010
21. Lynn, H.M., Choi, C., Kim, P.: An improved method of automatic text summarization for web contents using lexical chain with semantic-related terms. Soft. Comput. **22**(12), 4013–4023 (2018)
22. Subathra, P., Chidambaram, A.R.V., Ragapriya, D., Pragadeesh, C.: Document summarization using topic modeling. J. Adv. Res. Dyn. Control Syst. **10**(5), 1773–1781 (2018)
23. Remya, R., Aswathy, N.: Document summarization using dictionary learning. In: 2017 International Conference on Advances in Computing, Communications and Informatics (ICACCI), Udupi, India (2017)
24. http://websites.psychology.uwa.edu.au/school/MRCDatabase/uwa_mrc.htm
25. Subba, P., Ghosh, S., Roy, R.: Partitioned-based clustering approaches for single document extractive text summarization. In: International Conference on Mining Intelligence and Knowledge Exploration, pp. 297–307. Springer, Cham, December 2017
26. Fellbaum, C.: WordNet: An Electronic Lexical Data-Base. MIT Press, Cambridge (1998)
27. Chawla, N.V., Bowyer, K.W., Hall, L.O., Kegelmeyer, W.P.: Smote: synthetic minority over-sampling technique. J. Artif. Intell. Res. **16**(1), 321–357 (2002)
28. Azhagusundari, B., Thanamani, A.S.: Feature selection based on information gain. Int. J. Innovative Technol. Exploring Eng. (IJITEE) **2**(2), 18–21 (2013)
29. Sahoo, D., Balabantaray, R.: A novel approach to sentence clustering. In: 2016 ICCCA, pp. 1–6, Noida (2016)
30. https://duc.nist.gov/data.html
31. Lin, C.Y.: Rouge: a package for automatic evaluation of summaries. In: Text Summarization Branches Out (2004)
32. Mihalcea, R., Tarau, P.: TextRank: bringing order into texts. Association for Computational Linguistics, Barcelona (2004)

Sensor Based Non Invasive Dual Monitoring System to Measure Glucose and Potassium Level in Human Body

K. T. Sithara Surendran$^{(\boxtimes)}$ and T. Sasikala

Department of Computer Science and Engineering,
Amrita School of Engineering, Amrita Vishwa Vidyapeetham, Bengaluru, India
sitharasurendrankt@gmail.com,
t_sasikala@blr.amrita.edu

Abstract. Diabetes mellitus occurs when the pancreas is unable to produce insulin or it cannot utilize the insulin efficiently. People who have type 2 diabetes, takes insulin injection if it is not curable with medicine. Potassium drop in blood is a condition which may occur for chronic diabetic patients. Potassium is a very important mineral in blood which helps to regulate the fluid balance, muscle contractions etc. The reduction of potassium can cause many severe health problems. Diabetes patients may need to check their potassium levels also on a regular basis. Traditional methods to check the sugar and potassium level involves pricking of the finger tip and testing using the blood sample. This creates inconvenience for the patients who need to check their blood sugar and potassium level regularly. This paper proposes a dual monitoring system which can check the blood sugar and potassium level non-invasively.

Keywords: Non invasive · Glucose · Potassium · Near infra red technology · ECG analysis · T wave morphology

1 Introduction

Diabetes is the general name for Diabetes mellitus. It is one of the silent killer diseases across the globe. It is considered as a group of metabolic disorders in which blood sugar level will be high for a prolonged period. This increased level of sugar in blood can be identified by some symptoms like frequent urination, increased thirst, and increased hunger. If this disease is left untreated, it can cause severe complications. Some of the serious complications can include diabetic keto acidosis, long term complications like cardiovascular diseases, kidney problems, eye damages, ulcer, stroke and it can even lead to death at a severe stage.

Normally human body breaks down the sugar and carbohydrate into glucose. Glucose fuels the cells in the body. Insulin is produced by pancreas. Diabetes is due to either the pancreas is not producing enough insulin, or the cells are not responding to insulin properly. Continuous intake of insulin can cause the reduction of potassium in blood. Potassium is a very important mineral for the normal functioning of muscles, heart and nerves etc. The electrical rhythm of heart is maintained by the minerals in the

© Springer Nature Switzerland AG 2020
S. Smys et al. (Eds.): ICCVBIC 2019, AISC 1108, pp. 882–893, 2020.
https://doi.org/10.1007/978-3-030-37218-7_94

blood. Potassium drop in blood can result in many health issues like tiredness, breathing difficulties, heart palpitation etc. Diabetes patients have a chance of getting the potassium drop in blood.

Self-testing of blood sugar is very important for managing the treatment plan and avoiding complications in chronic diabetes patients. Nowadays varieties of electronic equipment are available in market to get the glucose measurement in a small drop of blood. Potassium measurement also plays an important role in the case of chronic diabetic patients. All available techniques for potassium detection are invasive in nature. People who take insulin to control sugar level often use continuous glucose monitoring (CGM) to know the glucose level frequently and to decide the insulin dosage to be used. It will be advantageous if a dual monitoring device is available to measure glucose and potassium Non-invasively. The currently available CGM devices can detect only glucose level by using a sensor inserted under the skin. Non-invasive potassium measurement is still under the research.

The advantage of non invasive measuring techniques over the invasive techniques is mainly the patient friendliness and the cost effectiveness. Moreover the patients will be able to do the test without any assistance. This will be very much useful for the elderly people who need to check the glucose and potassium levels many times a day. Non-invasive glucose monitoring is based on nera infra red (NIR) technology and researches on Non-invasive measuring of potassium level in blood are mainly focused on ECG based techniques. This paper tries to introduce a system which makes use of the scattering property of the infra red light to quantify the glucose level in human blood and the effect of potassium on ECG waves to detect the potassium level in blood.

The residual parts of this paper are presented in three different sections. Section 2 talks about the different research works done in the field of non invasive glucose measurement as well as potassium measurement and 3rd section explains about the design and implementation of the proposed system. The data analysis and the findings are discussed in Sect. 4, and the last section talks about the experimentation results and the analysis of the results.

2 Related Work

Non-invasive glucose monitoring devices (NGM) and continuous glucose monitoring systems (CGM) have become a major area of research [1]. Ultra wide band (UWB) microwave imaging also can be used to determine the sugar level in human body in association with the artificial intelligence technique and neural network to extract the characteristic features [2]. The ANN model can be trained in such a way that it will return the glucose value of the user.

A method which makes use of a Helium Neon gas laser to determine the sugar level of the diabetic patients is introduced by Mr. Ashok [3]. The laser is used as a monochromatic light source which operates at 632 nm wavelength. The system will be able to measure the glucose level of the patient continuously.

Radio frequency (RF) sensors are also a major type of sensor which is able to reproduce the glucose level of the patient. The RF sensors are mainly impacted by the pressure applied on the sensor module and the glucose measurement has to be taken in uniform pressure situations. Volkan Turgul and Izzet Kale have done researches related to the RF sensors and they have fabricated an RF sensor based on micro wave resonator [4]. They have also developed a pressure sensing circuit to detect the pressure applied on the sensor to find out the impact of pressure on the measurement. A portable non invasive glucose monitoring device is developed by A. Sampath reddy by using the NIR technology [5]. This device will alert the user about his sugar level and it is also designed in such a way that to suggest the insulin level for the user if it is needed.

A low current iontophoresis technology has been demonstrated by Christophcer Mc Cormic to determine both the glucose and potassium concentration in blood by passing a low voltage constant current through the patient's body with the help of a constant current source and two gel electrodes [6]. Experiments were carried out in different people and results were analyzed. The work concludes that there is a strong relation between the rise in sugar level and the potassium level.

The researches on noninvasive quantification of potassium are focused mainly on the ECG wave analysis [7]. Cristiana Corsi, Iohan Debie, Carlo Napolitano, Silvia Priori, David Mortara, Stefano Severi have done the background study and stated that potassium has a strong influence on ECG signals and the level of potassium in blood can be determined by T wave analysis of ECG signal. Based on the relationship between the T wave amplitude and slope, they have developed an estimator which can derive the value of potassium level in serum [8]. The testing has been done on different dialysis patients since there will be a remarkable variation in the potassium level during the dialysis process.

The quantification of hERG potassium channel block from the ECG waves has been studied by Johan de Bie and his team [9]. They have proved that the reduced T wave amplitude in the ECG waves is the indication of reduced plasma potassium level or Hypokalemia and, the Hyperkalemia patients will have a tall peaked T wave. They have verified the results by testing the potassium level in patients who undergoes Hemodialysis.

K A Unnikrishna Menon and his associates have done a survey on noninvasive blood glucose monitoring [10]. They have studied several related works and compared the methodologies and concluded the survey stating that the scattering property of the near infra-red-light signals while passing through glucose molecules in blood can be used as a perfect method to measure the glucose concentration non-invasively [11]. NIR technology mainly relies on the principle of change in voltage intensity at the receiver [12].

By analyzing the trend in glucose measuring techniques, it is clear that the Non-invasive Glucometer will replace the currently established invasive Glucometers in the market in nearest future. In this paper we propose a microcontroller based glucose and potassium measuring system which can predict the sugar and potassium level of the patient Non-invasively. We also justify the feasibility of our work by analyzing the data using confusion matrix. Various kinds of patients could be benefited by this real time Non-invasive method of potassium and glucose measurement.

3 Implementation

The main architecture of the proposed system consists of a glucose measuring module and a potassium measuring module. Both of the modules have the primary element as its sensor part. Glucose module uses an NIR sensor system and potassium measuring module utilizes an ECG sensor module. Both the modules use two different Arduino UNO micro controllers. The signal processing part is achieved by the filtering circuits and the output is displayed using the LCD display and the computer screen. The architectural diagram of the proposed system is shown in the figure Fig. 1.

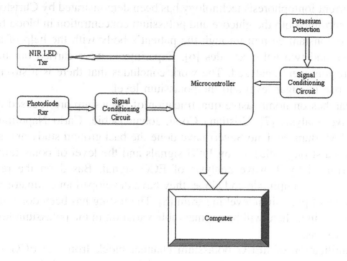

Fig. 1. Block diagram of the proposed system

3.1 Glucose Measuring Module

The glucose measuring module starts operating when near infra red (NIR) light source starts transmitting the near infra red radiation. The NIR transmitter used here is 940 nm NIR LED. The glucose absorption is very large in this particular wave length and it is transparent to the water molecules. 90–92% of human blood is composed of water. So, by using this particular wavelength the interference due to the water molecules is kept minimal. The receiver is a photodiode which is compatible with the NIR transmitter. The Tx and Rx are kept side by side inside a finger clip so that the user can insert his finger tip inside the clip. By doing so, the effect of ambient light and the pressure applied by the user can be kept uniform for every test. The reflected light will be absorbed by the photodiode and the attenuation of the signal is proportional to the glucose concentration in the blood. The attenuated signal is passed through the signal processing circuit and the unwanted frequencies will be removed by the filtering part.

The micro controller is used to convert the analog signal coming from the filtering circuit to digital voltage. The filter output voltage is fed to the analog input of the Arduino Uno and it will be converted to equivalent digital voltage by the ADC. This voltage and glucose level of the user will be printed in the LCD display connected to it. The Arduino Uno board is coded in such a way that the glucose value will be calculated from the voltage and it will be printed on the LCD display along with the equivalent voltage. The LCD display output is shown in the figure Fig. 2.

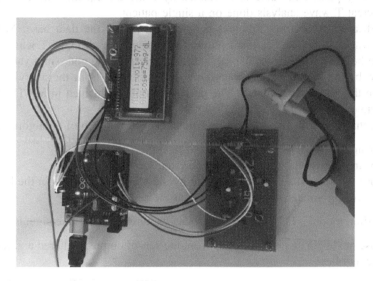

Fig. 2. LCD output for glucose measurement

3.2   Potassium Measuring Module

The hardware implementation of potassium measuring module mainly consists of the ECG electrodes along with Op-amp and the microcontroller. The single lead three electrode ECG sensors are used to get the ECG signals from the user and AD8232 Op-amp acts as a filtering and amplifying module. The small amount of bio potential signals can be filtered by the inbuilt op amp unit. The filtered signal is fed into the Arduino UNO microcontroller to convert into the digital signals. Output from the microcontroller is taken by the MATLAB program and the graphical representation of the ECG output can be viewed in the MATLAB window. The potassium concentration in user's blood will be extracted by T wave morphology. The T wave value and the potassium concentration extracted from the T wave also can be viewed as the MATLAB output. The flow diagram of potassium sensing module is shown in the figure Fig. 3.

The operational amplifier and the filters embedded in the AD8232 module will help to remove the noise and amplify the desired signal. The MATLAB program takes care of the further filtering of the signal with the help of its inbuilt filtering functions. The software is programmed in such a way that to calculate the P, Q, R, S and T values from the ECG signal acquired from the user's skin and to plot the PQRST wave on MATLAB. The PQRST wave obtained in MATLAB from the experiment is shown in the figure Fig. 4. The software program written in MATLAB is capable to find out the T value from the input signal. The T value is extracted from the signal and the corresponding potassium value is calculated by using the equation which is derived from different T wave analysis done on a single patient.

The data collected from different people is stored as the data base. MATLAB program will take this data as the input and will analyze each PQRST waves. The T value will be extracted for each T peak and the corresponding potassium value will be calculated by the program. The potassium value obtained from the analysis will be printed on the MATLAB window. The potassium output obtained in MATLAB from the experiment is shown in the figure Fig. 5.

The entire system operation can be explained with the help of an algorithm.

1. Power up the system
2. Sensor module is activated in both the systems.
3. Glucose module: NIR signal is reflected by the glucose molecules in the blood and corresponding voltage change is created at the sensor output.

   Potassium module: ECG sensor electrodes capture the heart rate signals.

4. Glucose module: Attenuated signals from the sensor output is filtered and fed to the microcontroller.

   Potassium module: unwanted signals in the ECG output is filtered, amplified and fed to microcontroller.

5. Glucose module: microcontroller converts the analog voltage into digital and the equivalent glucose level is calculated.

   Potassium module: analoge output from the filtering part is converted to digital form and fed to the MATLAB program.

6. Glucose module: Calculated glucose value and the equivalent voltage level is printed on the LCD.

   Potassium module: MATLAB program calculate the P, Q, R, S and T values from the ECG wave and the potassium value is calculated and printed on the output window along with the ECG wave plot.

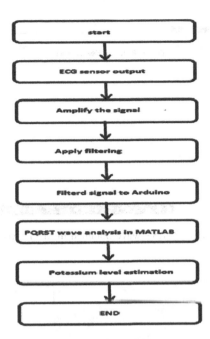

Fig. 3. Flow diagram of potassium measurement

4   Data Analysis

The Non-invasive measurement of glucose and potassium concentration in human blood is an area where the researches are being carried out for many years. Even though the researchers focuses to develop patient friendly low cost devices to make the Non-invasive measurement possible, widely accepted method is still the traditional blood sample collection and laboratory test. The effectiveness and correctness of the Non-invasive system designed in this project need to be verified and proved to propose it as a commercially accepted model and thus continue the studies to optimize and enhance it. Confusion matrix analysis has been done for both the glucose and potassium measurement modules by taking the test data and true data.

The confusion matrix is created for glucose measuring system by taking the predicted value as test data and the values taken from the Accu-check glucometer as true value. The confusion matrix can be created only by taking the data as categorical vectors. So the classification of data is done by taking the glucose values less than 80 mg/dL as false values and the glucose values greater than that as true values. The second classification is taken on the basis of the glucose values less than 90 mg/dL. The overall accuracy of the glucose measuring device is obtained as 83.8%. The confusion matrix output obtained for glucose measuring system is shown in the figure Fig. 6.

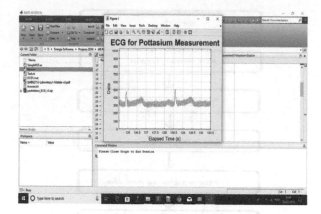

Fig. 4. PQRST wave obtained in MATLAB

Fig. 5. Potassium value obtained in MATLAB

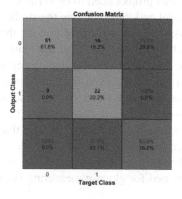

Fig. 6. Confusion matrix output for glucose measurement

The basic feature of potassium measuring technique is the T value extraction. The accuracy of potassium value obtained from the ECG analysis depends on how accurate is the T value. Once the T wave value extraction is proper, then the potassium level calculation is just putting the numerical value in the equation and the equation will return the potassium concentration in blood. So the confusion matrix is created for potassium module by taking the T wave value of the clinical ECG data as the true data and real time T value obtained from the proposed system as the test data. The classification of data is done by making them into two classes such as the values less than 8 as true values and more than that as false values. Threshold for the second set of T value data is taken as 9. The overall accuracy of T wave value extraction is obtained from the confusion matrix is 86.1%. The result obtained from the confusion matrix analysis for potassium measuring module is shown in the figure Fig. 7.

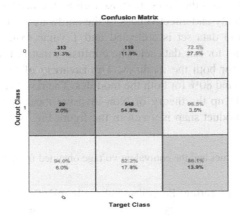

Fig. 7. Confusion matrix output for potassium measurment

The entire dual monitoring system implementation is done on a printed circuite board. Since this is an experimental set up, the hardware is not more compact and the components are visible on the board itself. The whole hardware structuer is divided in to two sections as glucose measuring part and the potassium measuring part. The confusion matrix result proves the reliability of the dual monitoring system.

5 Result Analysis

The glucose measuring module is tested with various people in the age group of 25 to 45. They have taken the test invasively and Non-invasively. The invasive measurement of sugar is done using Accu-Check instant Glucometer and immediately the test has been carried out by using the proposed system also. The glucose level and the equivalent voltage level are tabulated and compared. Sometimes the same person has shown some irregular values and those values are avoided to make the measurement uniform. More than 100 samples are collected from a group of 30 people.

The ECG signals are plotted and the data is collected from the same 30 people. ECG signal is continuous signal and it will produce the PQRST waves until it is stopped. So it was easy to get more than 100 PQRST waves from a single person. Around thousand data had been collected from 30 people and the T wave value and corresponding potassium level is calculated. The sample results obtained from glucose and potassium modules for 5 persons are given in the Tables 1 and 2 respectively.

Since the system is designed on an experiment purpose with very minimal expenditure, there are some limitations to produce the accurate and exact result. For glucose measuring system, the main challenging factors are the filtering of unwanted signals and the uniformity of testing environment. At the same time the potassium module can produce more accurate result by performing the same experiment by using more sophisticated ECG machines and 12 electrode sensing system. For both of the system the person who is going to take the test should be in a relaxed and very comfortable position to keep the blood flow as well as the heart rate as normal.

To prove the reliability and effectiveness of the hardware we used in this system, a clinically available ECG data set is collected and T value extraction and potassium calculation is carried out for this data set. The confusion matrix is also created for this true and the test data for both the modules. The accuracy of the technique obtained from the analysis is around 80% for both the modules. The results are much satisfactory and which in turn light up the theory of Non-invasive measurement of glucose and potassium. The final product snap is given in the figure Fig. 8.

Table 1. Sample sugar values and the equivalent voltage obtained from the proposed system for 5 persons.

Sl. No & age of the person	Equivalent voltage in mV	Sugar level in mg/dL
Person 1 (35)	932	99
Person 2 (30)	927	78
Person 3 (32)	929	89
Person 4 (42)	945	108
Person 5 (31)	923	75

Table 2. Sample potassium values and the T wave value obtained from the proposed system for 5 persons.

Sl. No & age of the person	T value extracted from ECG waves	Potassium level in mEq/L
Person 1 (35)	6.989	7.332
Person 2 (30)	7.170	7.653
Person 3 (32)	9.655	10.112
Person 4 (42)	9.746	10.452
Person 5 (31)	3.485	4.805

Fig. 8. Final product snap

6 Conclusion and Future Work

The number of diabetes patients is increasing exponentially due to the changed life style and unhealthy food habits. The glucose measuring device is a necessary thing for the patients to monitor their sugar level in blood frequently. The only method practiced widely is Accu-check Glucometer. In the area of potassium measurement, a Non-invasive method is still under the research. It will be a mile stone for the researches in the field of diabetes and related diseases to introduce a Non-invasive potassium measurement system. The potassium measuring unit designed here is found to be reliable and satisfactory. The confusion matrix analysis shows around 80% accuracy for both the modules. It indicates the validity of the dual monitoring system.

Further researches can be continued on the dual monitoring system by using more sophisticated sensors and the supporting systems. Glucose and potassium measurement becomes more accurate and consistent if the test set up is improved by providing uniform environment for each test. The confusion matrix analysis shows the scope and feasibility of the proposed system as an effective dual monitoring device. It will be very useful for the chronic diabetes patients who take the blood sample everyday if the proposed system is made as a commercially available device. Moreover, as compared to the existing invasive systems, this will be affordable for all kind of patients usable for a long period. The system can also be enhanced by including some other parameters like Hemoglobin, respiratory rate etc. to make it as a multy monitoring system.

Compliance with Ethical Standards. All author states that there is no conflict of interest. We used our own data. Animals/human are not involved in this work.

References

1. Aishwarya, L.K., Rashmi, R., Sadanand, S., Narayanappa, C.K., Sriram, N.: Studies on relating to monitoring blood glucose levels using non-invasive optical methods. In: 2nd International Conference on Recent Trends in Electronics Information & Communication Technology (RTEICT). IEEE, India (2017)
2. Showkath Ali, M.D., Shoumi, N.J., Khatun, S., Kamarudhin, L.M., Vijayeswari, V.: Non-invasive blood glucose measurement performance analysis through UWB imaging. In: 3rd International Conference on Electronic Design (ICED). IEEE, Thailand (2016)
3. Ashok, V., Nirmalkumar, A., Jeyashanthi, N.: A novel method for blood glucose measurement by Noninvasive technique using laser: World Academy of Science, Engineering and Technology (2011)
4. Turgul, V., Kale, I.: A novel pressure sensing circuit for non-invasive RF/Microwave blood glucose sensor, 978-1-5090-2586-2/16/$31.00 ©2016. IEEE (2016)
5. Sampath Reddy, P., Jyostna, K.: Development of smart insulin device for non-invasive blood glucose level monitoring. In: 7th International Advance Computing Conference. IEEE, India (2017)
6. Mc Cormic, C., Heath, D., Connoly, P.: Towards blood free measurement of glucose and potassium in humans using reverse iontophoresis. Sens. Actuators, B **166–167**, 593–600 (2012)
7. Corsi, C., Debie, J., Napolitano, C., Priori, S., Mortara, D., Severi, S.: Validation of novel method for non-invasive blood potassium quantification from ECG. IEEE Conference Computing in Cardiology (2012)
8. Severi, S., Corsi, C., Haigney, M., DeBie, J., Mortara, D.: Noninvasive potassium measurements from ECG analysis during hemodialysis sessions. In: Computing in cardiology. IEEE, USA (2019)
9. de Bie, J., Chiu, W.B., Mortara, D.W., Corsi, C., Severi, S.: Quantification of hERG potassium channel block from the ECG. In: Computing in cardiology. IEEE, France (2012)
10. Unnikrishna Menon, K.A., Hemachandran, D., Abhishek, T.K.: A survey on noninvasive blood glucose monitoring using NIR. In: International conference on Communication and Signal Processing. IEEE, India (2013)
11. Gayathri, B., Sruthi, K., Unnikrishna Menon, K.A.: Noninvasive blood monitoring using near infra-red spectroscopy. In: International Conference on Communication and Signal Processing. IEEE (2017)
12. Unnikrishna Menon, K.A., Hemachandran, D., Abhishek, T.K.: Voltage intensity based noninvasive blood glucose monitoring. In: 4th ICCCNT IEEE, India (2013)

A State of Art Methodology for Real-Time Surveillance Video Denoising

Fancy Joy$^{(\boxtimes)}$ and V. Vijayakumar

Sri Ramakrishna College of Arts and Science, Coimbatore, India
fancyjoyl@gmail.com, veluvijay20@gmail.com

Abstract. Video de-noising is one of the most essential tasks in video processing. Noise elimination and video reconstruction are still remaining as a challenging problem in this area. In this paper a modified two dimensional block based least mean square algorithm (TDBLMS) proposed to reduce the noise in the video, which can increasingly improve the image quality. The processing includes updating the adaptive weight matrix to filter noisy frame block by block and achieve higher peak signal to noise ratio. The performance of the modified TDBLMS approach compares with existing TDBLMS algorithm and gives better PSNR and SSIM results.

Keywords: TDBLMS · Adaptive filter · Gaussian noise · Video denoising

1 Introduction

Surveillance video analytics can be complex due to noise in video frames. Noises contaminated in video mainly during high speed capturing and transmission and it degrade the quality of video. Hence video preprocessing have a vital role in noise elimination and make ready the video in order to get better accuracy for further video processing. Clearly algorithms for image noise elimination can be used by considering video as collection of images [14]. However an ideal video noise reduction algorithm must consider the temporal information also.

In the proposed approach a modified TDBLMS algorithm is used for noise reduction in videos. TDBLMS algorithm for image de-noising is already exists, but it suffers slow convergence rate. Here first frame is processed by modified TDBLMS. Succeeding frames are estimated using block by block comparison between current and previous frame. The main contributions of this paper are follows. Firstly consider the frame and divided it into blocks and compare the similar block of current and previous frame. Secondly the surveillance videos have high similarity due to temporal redundancy the algorithm no need to evaluate similarities between adjacent frames. The result of video de-noising using proposed achieved improved peak signal to noise ratio and structural similarity results with in less computational time.

Image filtering algorithm can extend to video by considering independent video frames as an image. Generally image noise filtering categorized into two types i.e. linear and nonlinear filtering [2]. Linear filtering includes mean filter and least adaptive filter and median filtering in nonlinear method. Mean filter smoothing the image i.e.

© Springer Nature Switzerland AG 2020
S. Smys et al. (Eds.): ICCVBIC 2019, AISC 1108, pp. 894–901, 2020.
https://doi.org/10.1007/978-3-030-37218-7_95

reduce the intensity variation between pixels. It caused blur in images. Mean filter used constant weight matrix for entire image processing where as adaptive filters used adaptive weight matrix related with noise. Median filter performs similar to mean filter except it used median value instead mean value in mean filter. Median filter can efficiently use for impulsive noise removal without blurring but sometimes it also tends to alter the pixel not disturbed by noise [2].

The remaining of the paper organized as follows. Related works and video restoration concept discusses in Sects. 2 and 3 respectively. Modified TDBLMS algorithm describes in Sect. 4 and Experimental results in Sect. 5. Finally we offer conclusion in Sect. 6.

2 Related Works

Many approaches, mainly proposed for image noise elimination, have been applied to video. Some of the methods, which has been adapted successfully to videos are adaptive Least Mean Square [4, 8], Video Block-Matching and 3D joint filter [6, 7], non local means [10, 15] and spatiotemporal approaches [1, 5, 9, 11].

Keramani and Asemani [4] applied adaptive least mean square algorithm for surveillance video de-noising. Adaptive LMS approach capable of reducing the noise in non stationary frames too but the method shows slow convergence at presence of heavy noise and it requires number of iterations depends on dimensionality of input. Two Dimensional Block based LMS (TDBLMS) is an variant of adaptive LMS image de-noising algorithm can extend it for videos in an temporal filtering [3]. It shows fast convergence rate compared to LMS algorithm still it requires more time effective noise reduction. Yan and Yanfeng [8] proposed an adaptive temporal filter to track and reduce noises in video. The method it effectively reduced the noise but it produces dragging effects on moving objects [12].

Dabov et al. [7] used VBM3D approach for denoising and predictive search block matching with the motion estimation method for finding similar blocks. But improvement required for applying the same for color videos. Guo and Vaswani [6] divided into layers modeled as the sum of low rank matrix and sparse matrix. They also used VBM3D for removing noise in each layer. The method more suitable for videos contains small number of images.

Han and Chen [15] introduced dynamic non local means approach to video. They used kalman filter theory with non local means algorithm for video restoration. But computational complexity was relatively high. Ali and Hardie [10] presented recursive non local means filter for noise reduction. Authors processed the non local means approach for denoise first frame and following frames were estimated by weighted sum of pixels from current frame to previous frame. The method is able to balance noise reduction and preservation of details.

Xu et al. [9] developed very low light video denoising method. In first stage used non local means algorithm for spatiotemporal filtering followed by intermediate tone mapping. In last stage filter the YCbCr color space. The approach reduced the noise effectively but enhancement required speeding up the algorithm. Spatiotemporal gaussian mixture model for noise removal introduced by Varghese and Wang [5].

The method applied motion estimation for effective enhancement of correlation between temporal neighboring wavelet coefficients. It improved the performance of noise reduction. Yahya et al. [1] developed spatial wiener filter and temporal filter for noise removal. The approach has great capability to be adaptive in accordance with amount of noise. It preserved edge and efficient in denoising. Saluja and Boyat [11] used discrete wavelet transform and weighted high pass filtering coefficients with wiener filter spatial filtering and block based motion detector for temporal filtering. Soft thresholding required for discrete wavelet transform.

3   Video Restoration

In this section proposed TDBLMS algorithm is presented and notations used for video restoration method. Figure 1 depicts the video denoising structure. Here First frame is processed with modified TDBLMS algorithm and the following frames using block by block comparison of similar blocks in current frame to previous one.

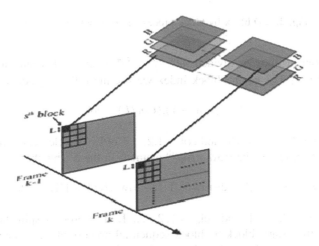

Fig. 1. Block diagram for process the video sequence used in proposed TDBLMS algorithm.

Video is collection of sequential frames. Here frame 'k' indicates the current frame and frame 'k−1' is previous frame. The proposed algorithm takes the frames, split it into its corresponding RGB (Red Green Blue) component matrices. Then divide the component matrices to blocks with block size of 'L × L' for block by block processing. Algorithm compares the RGB channels of current frame with the corresponding components of previous frame and use adaptive weight filter to remove noise in current frame.

4   Modified TDBLMS Method

Figure 2 depicts block by block processing of single component matrix, consider red component matrix. The process is same for both green and blue component matrices. Here 'Row' and 'Col' indicates height and width of the corresponding frame respectively. 'L' is the block size.

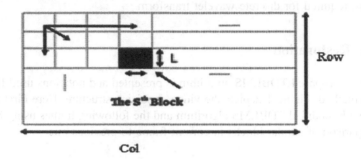

Fig. 2. 2D block by block processing of block size L × L

RGB channel of frame is divided into $\frac{Row}{L} \times \frac{Col}{L}$ blocks and S indicates the block index(row,col) [13]. For getting block index we use the following equation.

$$S = (row - 1).(Col/L) + col \tag{1}$$

Here row = 1,2,......,Row/L and col = 1,2,....Col/L. The element $d_S(row_b,col_b)$ of the frame can be taken as the $(row_b,col_b)^{th}$ element in the Sth block.

$$d_S(row_b, col_b) = d[(row - 1)L + row_b, (col - 1)L + col_b] \tag{2}$$

Here $row_b = 1, 2, ..., L$ and $col_b = 1, 2, ..., L$. The process split the frame into RGB channels then starts block by block sequential process of each channel of RGB from top to bottom and left to right, where each pixel convolved in a filter window with dimension M × N (Chen & Hsia, 2010). The modified TBLMS algorithm with respect to video as follows.

Consider the current frame as noisy frame and previous frame as reference frame, split the frames into its corresponding RGB channels. Partition these channels into blocks based on block size 'L' as discussed above. Here current frame X_S and previous frame d_S i.e. reference frame. Initialize the block size i.e. 'L', weight matrix is of size 'L × L' and step size μ. Filter the Sth block of current frame $X_S(row_b, col_b)$ using S^{th} block of reference frame $d_S(row_b, col_b)$ and weight filter $weight_S(i,j)$. Filtered output $y_S(row_b, col_b)$ is the image of Sth block after processing.

$$y_S(\text{row}_b, \text{col}_b) = \sum_{i=1}^{M} \sum_{j=1}^{N} weights_S(i,j) \times X_S(\text{row}_b, \text{col}_b)$$

$$= \sum_{i=1}^{M} \sum_{j=1}^{N} weights_S(i,j) X \begin{bmatrix} (r-1)L + r_b + (M-1) - i, \\ (c-1)L + c_b + (N-1) - j \end{bmatrix} \quad (3)$$

Current frame is adaptively filtered and subtracted from reference frame to remove noise [3]. The output will be the error 'err'. Difference between S^{th} block filtered output $y_S(\text{row}_b, \text{col}_b)$ and S^{th} block of reference frame is the error denoted by variable 'err'.

$$err_S(\text{row}_b, \text{col}_b) = d_S(\text{row}_b, \text{col}_b) - y_S(\text{row}_b, \text{col}_b) \quad (4)$$

The algorithm tries to minimize th energy of err(mean square of error). It recusively update the weight matrix 'weight' for each iteration. Value of weight matrix changes in each iteration, which depends on several parameters such as error, noisy frame, block size and step size. The weight $S + 1^{th}$ block $weights_{S+1}(i,j)$ is updated by

$$weight_{S+1}(i,j) = weight_S(i,i) + L \times \mu \sum_{r_b=1}^{L} \sum_{c_b=1}^{L} err_S(\text{row}_b, \text{col}_b)$$

$$\times X(row_b + row_{L-i}, col_b + col_{L-j}) \quad (5)$$

Where, μ- step size or learning rate.

5 Results and Discussions

To demonstrate the efficiency of modified TDBLMS compared with existing TDBLMS [13]. The proposed method used standard and publically available video dataset can be downloaded from 'https://media.xiph.org/video/derf/' for experiments. The dataset contains video sequences of various resolutions. The method select three video sequences such as tennis (352×240), flower garden (352×288) and foreman (352×288) for experiment. The video frames corrupt artificially with gaussian noise and compare the restored frames to the original frames.

The performance metrics used to test the performance were root mean square error (RMSE), peak signal to noise ratio (PSNR), and structural similarity index (SSIM). Parameters uses for finding performance measures are follows. x_{ij} is the reference frame and y_{ij} the estimated frame. μ_x and μ_y are means of x and y respectively. σ_x and σ_y standard deviations of x and y correspondingly. σ_{xy} is the sample correlation coefficient between x and y.c1 and c2 are the constants that stabilize the computations when denominators become small. For 8 bit image R is 255.

Formulae for different performance measures are:

$$MSE = \frac{1}{mn} \sum_{i=1}^{m} \sum_{j=1}^{n} (x_{ij} - y_{ij})^2 \quad (6)$$

$$PSNR = 10 \times log_{10}\left(\frac{R^2}{MSE}\right)\left(x_{ij} - y_{ij}\right)^2 \tag{7}$$

$$RMSE = \sqrt{\frac{1}{mn}\sum_{i=1}^{m}\sum_{j=1}^{n}\left(x_{ij} - y_{ij}\right)^2} \tag{8}$$

$$SSIM(x,y) = \frac{\left(2\mu_x\mu_y + c_1\right)\left(2\sigma_{xy} + c_2\right)}{\left(\mu_x^2 + \mu_y^2 + c_1\right)\left(\sigma_x^2 + \sigma_y^2 + c_2\right)} \tag{9}$$

In Table 1 PSNR, RMSE and SSIM results are presented for three image sequences each tested with eight noise standard deviations. Results were taken by TDBLMS and modified TDBLMS method with fixed learning rate and number of iterations.

Table 1. Comparison of TDBLMS and modified TDBLMS method in three video sequences

σn	Video	Tennis			Garden			Foreman		
		PSNR	RMSE	SSIM	PSNR	RMSE	SSIM	PSNR	RMSE	SSIM
5	TDBLMS	34.66	4.71	0.987	35.62	4.51	0.984	35.12	4.47	0.961
	MODIFIED	43.86	1.63	0.998	42.25	1.96	0.998	46.38	1.22	0.998
10	TDBLMS	28.78	9.26	0.952	29.26	8.77	0.942	29.41	8.62	0.866
	MODIFIED	38.98	2.86	0.995	36.73	3.71	0.994	41.12	2.23	0.993
15	TDBLMS	25.36	13.75	0.905	25.81	13.04	0.886	25.98	12.80	0.756
	MODIFIED	35.37	4.34	0.990	33.26	5.54	0.988	37.14	3.54	0.984
20	TDBLMS	22.91	18.24	0.850	23.41	17.20	0.827	23.57	16.89	0.655
	MODIFIED	32.26	6.21	0.980	30.70	7.43	0.978	33.91	5.13	0.971
25	TDBLMS	21.05	22.57	0.793	21.55	21.30	0.770	21.74	20.86	0.566
	MODIFIED	29.64	8.39	0.964	28.60	9.46	0.966	31.24	6.98	0.951
30	TDBLMS	19.49	27.04	0.734	20.06	25.29	0.716	20.20	24.89	0.492
	MODIFIED	27.22	11.09	0.942	26.95	11.45	0.952	28.92	9.12	0.927
35	TDBLMS	18.22	31.27	0.677	18.83	29.16	0.666	18.97	28.70	0.432
	PROPOSED	25.20	14.00	0.915	25.52	13.50	0.935	27.08	11.27	0.899
40	TDBLMS	17.17	35.29	0.686	17.78	32.89	0.621	17.87	32.55	0.380
	PROPOSED	23.45	17.12	0.884	24.20	15.71	0.916	25.38	13.71	0.866

In Fig. 3 comparison of denoised frame 5 from flower garden sequence with Gaussian noise 25 is presented. (a) indicates the original frame. The noisy frame shows in (b) and (c) is the denoised frame using TDBLMS algorithm. This method requires more computational time and more number of iteration to get better result. (d) Shows the result of proposed method, which shows more similar to original. The proposed one gives better result with in less computational time. It is more suitable for real time implementation than existing TDBLMS algorithm.

Fig. 3. Comparison of frame 5 from flower garden sequences (a) frame (b) frame with noise $\sigma_n = 25$ (c) TDBLMS (SSIM = 0.7704) (d) Proposed (SSIM = 0.9665)

Figure 4 shows the TDBLMS and proposed approach in tennis sequence with various standard deviations. From this figure it is clear that proposed method provides the best results.

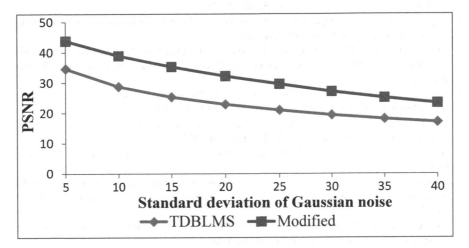

Fig. 4. PSNR versus standard deviation of Gaussian noise for TDBLMS and modified method in sequence

6   Conclusion

In this paper a robust video de-noising algorithm is used to remove the noises in surveillance video. The method improves the quality of video for further processing. The approach used partitioned the individual frames in to blocks then eliminate the noise by block matching with previous frame. Here each frame performs the block-by block operation to the original noisy image for getting suitable weight matrix for block adaptation. Adaptive weight filter update the weight matrix by considering previous weight filter, block size and error. From the experimental results it is clear that PSNR. SSIM, and RMSE proves that proposed method outperforms the existing TDBLMS algorithm in terms of convergence speed and image quality. The simplicity and low computational complexity of modified TDBLMS method shows that it is well suitable for real time implementation. Further enhancement is requires for de-noise video affected by the noise other than Gaussian.

Compliance with Ethical Standards

✓ All authors declare that there is no conflict of interest.
✓ No humans/animals involved in this research work.
✓ We have used our own data.

References

1. Yahya, A.A., Tan, J., Li, L.: Video noise reduction method using adaptive spatial temporal filtering. Discrete Dyn. Nat. Soc. **2015**, 10 (2015)
2. Gupta, B., Negi, S.S.: Image denoising with linear and non linear filters: a review. IJCSI Int. J. Comput. Sci. Issues **10**(6), 149–154 (2013)
3. Chen, C.Y., Hsia, C.W.: Adaptive filter based on TDBLMS algorithm for image noise cancellation. In: International Conference on Green Circuits and Systems, pp. 671–674. IEEE (2010)
4. Kermani, E., Asemani, D.: A robust adaptive algorithm of moving object detection for video surveillance. Eurasip J. Image Video Process. **2014**, 27 (2014)
5. Varghese, G., Wang, Z.: Video denoising based on a spatiotemporal gaussian scale mixture model. IEEE Trans. Circuits Syst. Video Technol. **20**(7), 1032–1040 (2010)
6. Guo, H., Vaswani, N.: Video denoising via dynamic video layering. IEEE Signal Process. Lett. **25**(7), 1009–1013 (2018)
7. Dabov, K., Foi, A., Katkovnik, V., Egiazarian, K.: Image denoising by sparse 3-D transform-domain collaborative filtering. IEEE Trans. Image Process. **16**(8), 2080–2095 (2007)
8. Yan, L., Yanfeng, Q.: An adaptive temporal filter based on motion compensation for video noise reduction. In: ISCIT, pp. 1031–1034 (2006)
9. Xu, Q., Jiang, H., Scopigno, R., Sbert, M.: A new approach for very dark video denoising and enhancement. In: Proceedings of 2010 IEEE 17th International Conference on Image Processing, pp. 26–29 (2010)
10. Ali, R.A., Hardie, R.C.: Recursive non-local means filter for video denoising. Eurasip J. Image Video Process. **2017**, 29 (2017)
11. Saluja, R., Boyat, A.: Spatial-temporal filtering for video denoising using weighted highpass filtering coefficients. Int. J. Adv. Res. Comput. Commun. Eng. **4**(8) (2015)
12. Hsia, S.C., Hsu, W.C., Tsai, C.L.: High-efficiency TV video noise reduction through adaptive spatial–temporal frame filtering. J. Real-Time Image Proc. **10**(3), 561–572 (2012)
13. Gupta, V., Mahle, R., Jayaswal, A.B.: Design and implementation of TDBLMS adaptive filter and comparison of PSNR values of various de-noised images. In: International Conference on Communication and Signal Processing, pp. 853–856. IEEE (2013)
14. Zhang, Y., Liu, Y., Li, X., Zhang, C.: Salt and pepper noise removal in surveillance video based on low rank matrix recovery. Comput. Vis. Media **1**(1), 59–68 (2015)
15. Han, Y., Chen, R.: Efficient video denoising based on dynamic nonlocal means. Image Vis. Comput. **30**(2), 78–85 (2012)

Disease Inference on Medical Datasets Using Machine Learning and Deep Learning Algorithms

Arunkumar Chinnaswamy[1]([⊠]), Ramakrishnan Srinivasan[2], and Desai Prutha Gaurang[1]

[1] Department of Computer Science and Engineering,
Amrita School of Engineering, Amrita Vishwa Vidyapeetham, Coimbatore, India
c_arunkumar@cb.amrita.edu, desaipruth0105@gmail.com
[2] Department of Information Technology,
Dr. Mahalingam College of Engineering and Technology, Pollachi, India
ram_f77@yahoo.com

Abstract. This paper discusses the implementation of deep learning algorithm to process the cancer dataset and identify the relevance of classification test data. Performance comparison is made with the existing classifiers. In the field of machine learning, deep learning is the emerging field that has gained a lot of interest over the past few years. It is a powerful machine learning tool that can be applied for many applications and complex problems, We have applied correlation coefficient based filter as dimensionality technique. Suitable classification algorithms like naive bayes, random forest, J48, bagging and decision stump are applied on the dimensionality reduced datasets. The same experiment is performed using deep learning algorithms. The results of classification of the proposed approaches using deep learning algorithms are compared with well-known classification algorithms. It is evident from the results that the deep learning algorithms perform better compared to other classification algorithms.

Keywords: Classification · CNN · RNN · Machine learning · Deep learning · Performance matrix · Medical dataset · Medical diseases · Ovarian · Central nervous system · Disease inference

1 Introduction

Machine Learning is the process in which a computer or any other machine is being taught to recognize a pattern of data, take help of an algorithm to understand the data and make a conclusion or inference on the supplied data. After the recognition of the pattern, the machine would learn new things by itself. Deep learning combines neural network and advances in computing power to determine complicated patterns in huge amount of data. There are many possibilities for applying machine learning in the healthcare sector but it is fully dependent on the availability of sufficient data and permissions to access and use it [3]. For healthcare dataset, MRIs (statistical report as well as images), CT scans, X-rays and tumor images are mostly used by doctors to determine or diagnose spinal injuries, heart disease and cancer [1]. One of the most

© Springer Nature Switzerland AG 2020
S. Smys et al. (Eds.): ICCVBIC 2019, AISC 1108, pp. 902–908, 2020.
https://doi.org/10.1007/978-3-030-37218-7_96

significant area of research in the medical field is disease detection. Hence artificial intelligence approaches like machine learning and deep learning have become more important and crucial for healthcare industry. Deep learning and machine learning classification approaches are used to analyze huge volumes of medical data and to make decisions [4]. Nowadays, deep learning has become an integral part of medical data analysis. To get better performance, researchers are trying to implement and analyze deep learning algorithms. Deep learning would make disease inference easier using classification models [2]. Better diagnosis could be provided to the patient after generating reliable results. Performance measures like precision, recall, accuracy etc., are used for validation of the diagnosed data. The disease inference from medical dataset is performed using deep learning and machine learning algorithms and the results are compared with existing approaches. This is a classification based problem. Deep learning assists medical professionals and researchers to discover the hidden opportunities in data and to serve the healthcare industry in a better manner. Deep learning in healthcare provides doctors the analysis of any disease accurately and helps them to treat them better, thus resulting in better medical decision [2]. Many researchers have analyzed data mining and machine learning classification techniques on these areas. Nowadays healthcare industry is providing many avenues of research in areas like fraud detection in healthcare insurance, availability of medical treatment to the patients at lower price, healthcare policies with several benefits and better relationship with patient and management [1]. The performance matrix is used to analyze the results [4]. Classification is part of the supervised machine learning in which the target attributes are divided into different classes. This research paper is divided into following 2 sections: Sect. 2 which discusses the proposed method, Sect. 3 discusses and analyzes the result and Conclusion is presented in Sect. 4.

2 Proposed Method

Our proposed method architecture for classification of medical dataset is given in below (Fig. 1):

2.1 Deep Learning Algorithms

Deep Learning is basically a neural network which has many hidden layers. It has been categorized depending upon the working of a neural network. Deep learning is a machine learning approach that allows to predict the output from a given set of input.

2.1.1 Convolution Neural Network
We can use CNN for image dataset, classification and regression prediction problems. Technically CNN is a deep learning model that can train and test each input that passes through a series of layers called the convolution layer (filters), pooling, fully connected layers and softmax functions. It produces results in terms of accuracy and loss.

Implementation of CNN: For CNN, input is two dimensional data in matrix (contains rows and columns) format. The procedure is as follows: load the required libraries, upload.csv file (which is input dataset), define a network with optimizer and split the data into training and testing set.

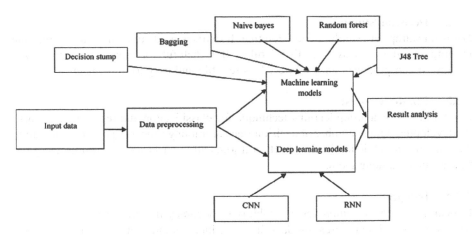

Fig. 1. Architecture of the proposed method

2.1.2   Recurrent Neural Network

Recurrent Neural Network (RNN) is a class of Artificial Neural Network (ANN). In RNN, all connections of nodes form a sequence of directed graph. The basic RNN network contains successive layers, in which each node is connected with one-way directed connection with another node in the successive layer. Each node is called a single neuron which has real valued and time varying activation. There are three types of nodes namely input nodes which receives the data from outside, hidden nodes that modifies the data and processes it and passes data to input layer to output layer and output layer that holds results given by hidden layers.

Implementation of RNN: RNN is widely used for sequence prediction. RNN is used for classification, regression prediction, text data, speech data and generative models. The steps of RNN are loading the libraries and uploading CSV file (Input dataset), Splitting data into training and testing set, defining the network and use different optimization to obtain the desired output.

2.2   Machine Learning Algorithms

2.2.1   Naive Bayes

It is a generative model. After using Bayes' theorem, it generates prediction from the evidence as observed in data. It can work on both multi and binary classification. Bayes theorem calculates the probability by combining historical data and frequency of values in historical data. This theorem is very simple to implement and to compute the accuracy and hence it is called as "Naive".

2.2.2   Decision Stump

Decision stumps are similar to decision tree. It extracts predictive information from the training set. It works as IF… ELSE rules. It is helpful for binary class prediction because it can produce one level tree. It is reliable and easy to implement.

2.2.3   Random Forest

Random forest is a group learning technique for classification and regression. It works by developing a huge number of decision trees at training time and produces the output class i.e. mean prediction in classification It joins all the decision tree results and gives best accuracy among them.

2.2.4   Bagging

Bootstrap aggregation improves the stability of accuracy in classification. It is widely used in statistical and regression data. It can reduce variance and helps to avoid over fitting problem.

2.2.5   J48 Tree

J48 tree creates the decision tree from checked dataset using information gain. To settle on a decision, the property with most surprising institutionalized information gain is used. It examines the output from the attributes of splitting data. In the wake of part the count rehashes on the humbler subsets. The part strategy stops if all cases in a subset have a spot with same class or any class. By then, the leaf node is settled on for decision tree.

3   Results and Discussion

The dataset is taken from Kentridge biomedical repository [10] and it is considered as benchmarked dataset for microarray data which are used in other standard research paper [5–9]. This dataset is binary unpaired dataset and its samples are used for training and testing of classification models which are non-overlapping and disjoint. This study contains unbalanced raw data. Both datasets are SMOTE balanced and normalized and its dimensionality is reduced by correlation coefficient. This dataset after dimensionality reduction is classified using ML algorithms like naive bayes, bagging, decision stump, random forest, J48 tree and proposed deep learning algorithms like Convolution Neural Network and Recurrent Neural Network are used for generating results with 10-fold cross validation.

Formula for FDR is given in (1) as

$$FDR = FP/(TP + FP) \tag{1}$$

The classification accuracy, one of the parameters used for evaluation in clinical medicine is computed for the datasets under study using deep learning and well known classifiers. For validating the results in a better way, kappa(k_a), an additional measure is performed that is used to compare the observed accuracy against expected accuracy.

False positive rate is computed using

$$FPR = 1 - specificity = FP/(FP + TN) \qquad (2)$$

where FP is False Positive and TN is True Negative instances. True Positive Rate (TPR) is calculated using (4) as

$$TPR = sensitivity = TP/(TP + FN) \qquad (4)$$

Accuracy is the ratio of correctly predicted observation to the total observation. Accuracy measures the percentage of accuracy of the data. Accuracy is calculated as in (5)

$$Accuracy = (TP + TN)/(TP + TN + FP + FN) \qquad (5)$$

The performance of the datasets under study are performed and tabulated in Table 1.

Table 1. Performance analysis on ovarian and CNS datasets – Conventional method Vs. CNN Vs. RNN

Measure	Ovarian dataset			CNS		
	Conventional method	CNN	RNN	Conventional method	CNN	RNN
Precision	0.941	0.953	0.977	0.667	0.804	0.765
Recall	0.981	0.951	0.977	0.810	0.860	0.765
F-measure	0.961	0.950	0.977	0.731	0.831	0.831
TPR	0.974	0.951	0.236	0.58	0.753	0.765
FPR	0.027	0.054	0.023	0.443	0.251	0.977
MCC	0.948	0.903	0.953	0.177	0.506	0.53
Kappa	0.924	0.965	0.983	0.377	0.965	0.627
TNR	0.945	0.994	0.995	0.564	0.994	0.763
FNR	0.019	0.031	0.012	0.191	0.031	0.140

The MCC, TPR, TNR and kappa values are higher for the proposed methods and FPR is lower for the proposed methods. They are represented in Table 2.

Table 2. Attribute selection model comparison

Attribute Selection method	Classification accuracy (%)
Neighbourhood approximation [12]	84.64
CFS-iBPSO [13]	84.53
ICS [14]	83.00
Scalable FS [15]	84.64

(continued)

Table 2. (*continued*)

Attribute Selection method	Classification accuracy (%)
Max dependency, relevance [16]	82.83
Conventional methods	82.68
Proposed deep learning methods	90.07

4 Conclusion and Future Work

The performance analysis was done after applying correlation based dimensionality reduction using ten-fold cross validation strategy. It is evident from the results that deep learning algorithms perform better than well-known conventional classification algorithms. The same work could be extended to analyze more medical and non-medical datasets and other classifiers like Support Vector Machine and random forest may be used in future.

Compliance with Ethical Standards

✓ All authors declare that there is no conflict of interest.
✓ No humans/animals involved in this research work.
✓ We have used our own data.

References

1. Acharya, U.R., Oh, S.L.: A deep convolutional neural network model to classify heartbeats. Comput. Biol. Med. **89**, 389–396 (2017)
2. Liu, F., Tan, H., Song, D., Shu, W., Li, W.: Deep learning and its applications in biomedicine. Genomics Proteomics Bioinform. **16**(1), 17–32 (2018)
3. Mohsen, H., El-Dahshan, E.-S.A., El-Horbaty, E.-S.M., Salem, A.-B.M.: Classification using deep learning neural networks for brain tumors. Future Comput. Inf. J. **3**(1), 68–71 (2018)
4. Kumar, G., Kalra, R.: A survey on machine learning techniques in health care industry. Int. J. Recent Res. Aspects **3**(2), 128–132 (2016)
5. Arunkumar, C., Ramakrishnan, S.: A hybrid approach to feature selection using correlation coefficient and fuzzy rough quick reduct algorithm applied to cancer microarray data. In: Proceedings of the 10th International Conference on Intelligent Systems and Control (ISCO 2016), pp. 414–419 (2016)
6. Arunkumar, C., Ramakrishnan, S.: Modified fuzzy rough quick reduct algorithm for feature selection in cancer microarray data. Asian J. Inf. Technol. **15**, 199–210 (2016)
7. Arunkumar, C., Ramakrishnan, S.: Attribute selection using fuzzy roughset based customized similarity measure for lung cancer microarray gene expression data. Future Comput. Inf. J. **3**, 131–142 (2018)
8. Apolloni, J., Leguizamon, G., Alba, E.: Two hybrid wrapper-filter feature selection algorithms applied to high-dimensional microarray experiments. Appl. Soft Comput. **38**, 922–932 (2016)

9. Pang, S., Havukkala, I., Hu, Y., Kasabov, N.: Classification consistency analysis for bootstrapping gene selection. Neural Comput. Appl. **16**(6), 527–539 (2007)

10. http://datam.i2r.a-star.edu.sg/datasets/krbd. Accessed 10 Dec 2011

11. Boughorbel, S., Jarray, F., El-Anbari, M.: Optimal classifier for imbalanced data using Matthews Correlation Coefficient metric. PLoS ONE **12**(6), 1–17 (2017)

12. Kar, S., Sharma, K.D., Maitra, M.: Gene selection from microarray gene expression data for classification of cancer subgroups employing PSO and adaptive K-nearest neighborhood technique. Expert Syst. Appl. **42**(1), 612–627 (2015)

13. Jain, I., Jain, V.K., Jain, R.: Correlation feature selection based improved-binary particle swarm optimization for gene selection and cancer classification. Appl. Soft Comput. **62**, 203–215 (2018)

14. Aziz, R., Verma, C.K., Srivastava, N.: A fuzzy based feature selection from independent component subspace for machine learning classification of microarray data. Genomics Data **8**, 4–15 (2016)

15. Jensen, R., Mac Parthalain, N.: Towards scalable fuzzy–rough feature selection. Inf. Sci. **323**, 1–15 (2015)

16. Maji, P., Garai, P.: On fuzzy-rough attribute selection: criteria of max-dependency, max-relevance, min-redundancy, and max-significance. Appl. Soft Comput. **13**, 3968–3980 (2013)

Artificial Intelligence Based System for Preliminary Rounds of Recruitment Process

Sayli Uttarwar[✉], Simran Gambani, Tej Thakkar, and Nikahat Mulla

Department of Information Technology, Sardar Patel Institute of Technology,
Mumbai 400058, India
say.uttarwar@gmail.com, gambanisimran08@gmail.com, tejdthakkar@gmail.com,
nikahat_kazi@spit.ac.in

Abstract. The process of hiring is extremely crucial for any organization. It's existing methodology, from resume screening, preliminary rounds, face to face interviews, all of it involves huge manual effort. Thus, it is crucial to implement a hiring process that reduces human intervention, screens candidates on consistent and standard criteria, makes the system flexible and trackable thus enhancing the mode of operation of the current system and consequently reducing human effort. We propose an Artificial Intelligence (AI) based recruiting system that serves as a decision support system for recruiters, especially useful for preliminary or initial screening rounds. AI for recruiting is the application of artificial intelligence to the process of recruitment, that is to automate the preliminary rounds of the recruiting system and to filter candidates from a large applicant pool with minimal human intervention. We have implemented a real-time AI based assessment system with a closed domain intelligent question answering (henceforth mentioned as QA) system in the backend and incorporated an adaptive question generation algorithm for making the process equivalent to a real interview environment.

Keywords: Deep learning · Long Short Term Memory (LSTM) · Recurrent Neural Network (RNN) · Application Programming Interface (API) · Natural Language Processing (NLP)

1 Introduction

Currently, hiring process in companies, from screening resumes to conducting phone or face to face interviews, is manual. In event of large number of applications received, it becomes tedious to filter out candidates in preliminary screening rounds and requires a huge amount of human intervention and effort. Thus, it seems crucial to implement a standard, flexible hiring process that assesses candidates with minimal human effort and involvement.

All authors have contributed equally in the making of this paper.

© Springer Nature Switzerland AG 2020
S. Smys et al. (Eds.): ICCVBIC 2019, AISC 1108, pp. 909–920, 2020.
https://doi.org/10.1007/978-3-030-37218-7_97

This paper describes a development framework of an intelligent recruiting system useful for carrying out preliminary rounds of a hiring process. The main objective of this system is to evaluate a candidate's response in real-time and adaptively generate the subsequent questions to make the interview scenario realistic. An intelligent QA system in the backend will evaluate candidate response. It will extract/retrieve answer from a closed domain subject-related dataset for a question posed in natural language. The system will take the candidate's answer as input and compare it with the answer retrieved by the QA system for evaluation of accuracy. Further questions will be generated based on how well or accurately the candidate responded to the current question or section.

The paper is organized in sections as follows - Sect. 2 related works, Sect. 3 proposed work, Sect. 4 design and implementation, Sect. 5 experimental analysis, Sect. 6 results and discussion, Sect. 7 future works and lastly references.

2   Related Works

Intelligent recruiting systems are going through a developmental phase with scope of research and innovation. There are AI and machine learning tools which analyze social media profiles of candidates and also analyze resumes received and match candidates to job profiles. Our system is built to further work in direction of automating some parts of recruitment process.

QA system forms the main component to make our proposed system intelligent and self sufficient. Significant research in the field of automated answer generation or retrieval has been witnessed in recent years. Varied forms of implementation and methodologies, from primitive to highly automated, have been developed over the years. In Ansari et al. [1], Artificial Neural Networks (ANN) were implemented to answer simple objective factoid questions based on a single document. The document was processed using Natural Language Processing (NLP) techniques like Entity Recognition (NER) and Part of Speech (POS) Tagging and pairs were formed (example: agent - main actor) called Deep cases. ANN built relations between these pairs and the knowledge base (KB) which were used for extracting and aggregating the answer. Thus, their scope was to only extract one word answers and not descriptive, one sentence answers.

Another similar model as [2] used attentive Convolutional Neural Networks (CNN) for answering questions. Their model had - entity linking and fact selection which followed as the entity identified in question and required as answer was searched for in the knowledge base and thus the relevant fact was selected and retrieved. They concluded that the attentive layer over CNN contributed to efficient performance of the model. The models in [3] used distributed vector representations of words called word embeddings. The system provided a Inverse Document Frequency (IDF) score to each word vector and cosine similarity was used to return the most matching answer as output to given query. Besides, word vectors help in identifying the context of sentences as well as dependency of different words on each other based on how near or far they are located in the obtained vector space as explained in [4]. A generative neural network model as proposed in [5], recurrent neural networks (RNN) were used to build neural architecture and to train model to understand simple factoid question frame, analyze dataset and generate answer. This led to an effort in making the process of machine comprehension even more intelligent and learning based. The improved Term Frequency - Inverse Document Frequency (TF-IDF) was proposed in [6] for weight calculation of words based on vector space model and frequency.

In order to make the process of interviewing seem realistic, questions asked must be follow-ups to previous questions and must vary in difficulty as is seen in a face to face interview. Techniques like those in [7] are used for adaptive question generation. It takes into consideration the precedence, concepts used in answer and topic relation to formulate the next questions. Thus, current state of research and past implementations have inspired the different components of our proposed system, which are described in the forthcoming sections below.

3 Proposed System

Our proposed system offers a portal for the interviewer and candidates. The interviewer posts questions on the portal, for instance Java related questions for a position requiring Java expertise. As for now, our system is closed domain and serves for a Java based interview only.

The candidate will be able to view questions posted by interviewer. He/she can attempt the questions by uploading video responses for the same. This video response serves as input to our system which is further processed, first using speech-to-text APIs to convert and retrieve candidates response as text file.

In the backend is a NLP based question answering system which also attempts to answer the same question asked to the candidate. The main steps in QA process are question and dataset analysis, question classification, answer extraction, answer ranking and selection. The QA is like an intelligent search engine that responds to questions and answers of the candidate are matched with it to calculate similarity score. Usually, in an interview questions follow from candidates response. We propose a system that will adaptively generate the next question based on performance of candidate in current question. Based on how well the candidate managed to answer current question, difficulty of next question will be decided.

Apart from response evaluation, since we have the video, cognitive analysis such as sentiment analysis can be performed to determine how confident/nervous/neutral the candidate is. The accuracy percentage and the sentiment analysis, all these results serve as a decision support system for recruiters to filter out candidates and assess them in the preliminary screening stages.

Figure 1 explains data flow between the different modules of our system. The interviewer and candidate are the two major users of the system. The diagram shows flow of input question entered by the interviewer which is processed and fed as input to the answer retrieval module. Similarly, the video interview received as input from the candidate is processed to extract candidate answer which is fed into the response evaluation model. Here, the response of candidate and answer generated by QA system are compared for evaluation and score calculation. According to score, next question is generated adaptively and presented to the candidate, while score reports are presented as output to the interviewer. This implementation is explained further in detail.

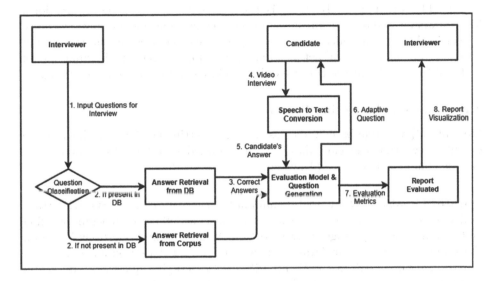

Fig. 1. System flow diagram

4   Design and Implementation

The five modules that our system can be broken down into are Video/Audio to Text, Question Answering System, Candidate Response Evaluation, Adaptive Question Generation and Report Generation, explained in detail as follows.

4.1   Video/Audio to Text

The most primitive step for the system is the interview from which candidate response features are gathered. For conducting the interview, we have used multi-threading where one thread is responsible for capturing video (OpenCV) and the other thread is responsible for audio(Pyaudio). For merging the video and audio files, FFMPEG is used. The command line instructions are executed internally using the subprocess module in python and we get 4 files as an output:

1. The video file
2. The audio file
3. The merged file
4. The text file derived from the audio file using Speech-to-Text Conversion

The video and audio quality can be set using ffmpeg but are dependent on the resolution of the webcam and proximity to the microphone. After embedding video and audio, the audio file is sent as an input to the Google Speech-to-Text conversion API and the corresponding output is received. The resultant text file is then used for finding accuracy of the candidates answers and giving a score for the same.

The file containing only video is used for detection of facial expressions, file containing audio used for speech-to-text conversion and evaluating choice of words. Apart from video response file, these can be seen by the recruiter if needed. Also these files will act as training data for the machine learning model and will thus in the future can use online training as well as offline training.

4.2 Question Answering System

We built a QA system based on a closed domain setting i.e. Java which can serve for interviews that test candidates on the knowledge of Java. Described below are the components of our QA system:

Dataset Creation: Our closed domain dataset is inspired from SQuAD (Stanford question answering dataset) [8]. SQuAD is a dataset of machine reading and comprehension passages extracted from Wikipedia and questions based on the same. We curated our dataset on similar lines but based on Java. We used a python script based on the Scrapy library for scraping questions and corresponding topic passages about Java and object oriented programming (OOP) from various online sources. All these were put together to form a consolidated dataset. Our dataset has following attributes placed in json format - passage (also referred to as context or evidence), question based on the passage, three to four answers for respective question and its difficulty level.

Data Analysis: Our aim was to build a QA system that understood the question posed in natural language. The more it understands the context of passages on which the question is based, the more well curated answer can be given as response. This requires significant analysis on the dataset. Following feature vectors of every paragraph are extracted and passed as input to our neural network model for training. The model uses bi-directional long short term memory based RNN layers to encode and embed the following features [10]. They are recurrently, sequentially fed to stacked network to capture context of passage.

1. **Word Embeddings**

   Word embedding is a way of representing words as vectors with values in the predefined vector space such that words with similar meaning, context have similar representation and are placed close together. Our text dataset is broken down into individual words which are vectorized, encoded and mapped to representative word vectors which are then fed as input to our neural network. We implemented word2vec algorithm for generating embeddings using skip-gram architecture. Because the answer retrieval part of Question Answering

System model is dependent on context, we used word vectors which are a great way of capturing words in proximity i.e. words or phrases sharing common context and meaning as proposed in [9].

2. **Token Features**

   We read the dataset file and hence for better comprehension, the json data is flattened. Passages and questions are tokenized and annotated. Major techniques used are stopword removal, removing sparse terms, stemming, dependency parsing, etc. Additional features like Named Entity Recognition(NER), Part-of-Speech Tagging (POS), Term Frequency, Lemma are calculated for each token so that it is easier to match answer and question properties. We utilized the Stanford CoreNLP Toolkit for tokenization as well as computing all the above mentioned features for every token.

Question Analysis: The question is encoded in a similar fashion as the paragraph to form embeddings. These are then passed to a RNN to extract most important keywords from the question. We used the RAKE - Rapid automatic keyword extraction algorithm which works as following components:

- Candidate word selection
- Property calculation for each word
- Scoring based on properties and selecting keywords

The algorithms main extraction function gives provision to set parameters for judging a word like length, number of words in phrase, frequency of word/phrase. This can be tweaked according to our requirements. We don't plan to restrict to using this algorithm and will explore other keyword extraction algorithms like TF-IDF based on QA system model requirement.

Answer Retrieval: Lastly, the Answer Retrieval method in a very broad sense is retrieval of most similar paragraph tokens to that of question tokens with a pre-decided context window. RNNs are the most popular implementation of neural network for instance in [11] for extractive QA systems. The input here are embeddings of word vectors of paragraph and question tokens. As stated in [12], we calculate an attention score between the paragraph features and each word of question. This score is used to predict probability of word in paragraph being the start or end of answer for the question. In our system the context-window is taken as 15. A pointer network is formed using RNN to predict the start and the end span of the most probable answer from the passage tokens.

4.3 Candidate Response Evaluation

The answer response obtained via video is evaluated by comparing it with answer generated by the QA system for the same question. Following techniques of similarity have been implemented for the same.

Answer Relevance: Similarity of answers is a very essential element in evaluation of the answers candidate gives. There are many ways to find similarity, we have compared results of Euclidean, Manhattan and Cosine Similarity and found Cosine Similarity to provide the best results. Cosine similarity is basically calculating the distance between two vectors by taking into consideration the dot product and their magnitude.

First, word vectors are computed for our Java dataset passages. Next, from the vector space model the words having most relation to the keyword used as input are returned along with the score of the extent to which they are similar. Our next step is to calculate semantic similarities between sentence vectors to evaluate the answers of candidates.

Sentiment Analysis: An addition to the evaluation criteria of answer relevance, using cognitive analysis like sentiment of candidate answers to know how positive he/she is on the basis of words chosen to answer the question can also enhance evaluation. There are two text-based ways of calculating Sentiment Analysis: Lexicon based and Deep Learning based.

The Lexicon based approach often fails in cases where sarcasm and other indirect positive/negative sentiments are included. That is why we preferred to use the Deep Learning Method. Here LSTM is trained by a dataset already classified into positive and negative sentiment. We used the Twitter dataset to train our model. We achieved an accuracy of 60% for positive sentiment and around 80% for negative sentiment classification which can be improved further by increasing the training dataset.

4.4 Adaptive Question Generation

The interviewer can input the difficulty level along with the question in order to make the process adaptive. The difficulty of next question to be asked will depend on the performance of candidate in current question. The first question will be a seed question of low difficulty. Suppose the candidate manages to score over an expected threshold in current question, the next question posed will be of same or higher difficulty. If candidate fails to score over expected threshold, the difficulty of next question is lowered. Judging the response's accuracy and difficulty level of questions answered and attitude of the candidate can be useful for the interviewer as a decision support.

4.5 Report Generation/Decision Support System

Our system acts as a decision support system for the interviewer, the decision being whether to consider the candidate for further interview rounds. It will not give a perfect yes/no answer whether to hire a candidate. But, it will provide the interviewer with a detailed analysis of the performance of every candidate that gave the video interview. The report will consist of mainly two parts: Questions answered correctly along with their difficulty and candidates attitude based on their choice of words. The interviewer may choose to view every individual report to select which candidates will be called for further interview rounds, or predefine a measure on the basis of which filtered candidates will be presented. Also if the interviewer wishes, he/she can self-monitor the video responses of candidates.

5 Experimental Analysis

To verify the components of our design, we compared different techniques and implemented the ones best suited for efficient performance of our system.

5.1 Embedding Layer

In our embedding layer, skip-gram model which comes under the word2vec method of constructing embeddings was implemented with some changes in dimensions to suit our model. Further on we used GloVe i.e. global vectors for representing words that are pre-trained globally on a word-word co-occurrence matrix for a given corpus. Using pre-trained GloVe is not as efficient or reliable as simple word2vec embeddings. Also we used spaCy, a open source NLP library to extract features like NER, POS-Tagging, text classification, etc.

5.2 Similarity Techniques

The answer generated by QA system and candidate answers are compared for evaluation using similarity techniques. Table 1 summarizes results obtained by applying various algorithms like Manhattan, Euclidean and Cosine to test similarity between two statements like
Statement 1: 'I am good at sports.' Statement 2: 'I love playing sports.'

Table 1. Similarity techniques

Technique used	Similarity score
Manhattan	43.56%
Euclidean	59.1%
Cosine	79.8%

5.3 Question - Answering System

The QA system requires a passage or evidence as input. In our case, the passages from dataset are passed to the QA system pipeline as input. Being a machine comprehension model, questions related to input passage are answered by the QA system. Figure 2 depicts a sample working of the QA System with input evidence as passage about Methods in Java and a question. The response generated by the QA system as well as the time taken are as shown in the figure.

6   Results and Discussion

For system to work justifiably and for it to be able to correctly evaluate the candidate, it is necessary that the QA component works to best of it's ability. The original QA model that was released by Facebook [9] was modified by us as described in the paper above. The main model was open domain, had more number of modules like document retriever unlike our model which only caters to a closed domain dataset of Java, as a proof of concept.

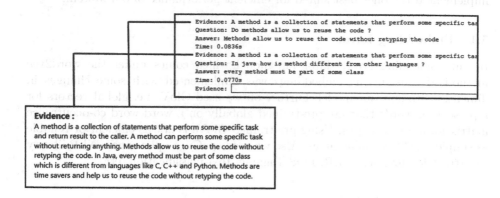

Fig. 2. Output of QA system

For the answers retrieved in both models, they are judged on the basis of two scores EM and F1. EM is the exact match score metric for predictions as per [8]. Figure 3 shows EM scores that we obtained over the training period. It peaked around epoch number 25 and remained around it more or less further on. Figure 4 shows the F1 scores obtained by our model. F1 score is the measure of accuracy for any test data, and it takes into consideration both precision and recall. Our model was able to achieve the best EM and F1 score as 69.76% and 78.6% respectively over 40 epochs as depicted from the graph of epochs vs EM and F1 scores in Figs. 3 and 4 respectively.

Table 2 provides a comparison between scores of our model and others using the SQuAD question answering set as their training and testing dataset.

Thus, in terms of relevance and correctness of answers generated by our QA component, we observed that our modified neural network based model gave comparable performance to that of other existing QA model implementations (as stated in Table 2), though not better. Statistically, EM score of our model is better than other QA models while F1 score is almost the same. With further modifications to our LSTM-based neural network architecture, there is scope of improvement.

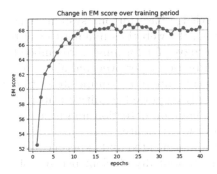

Fig. 3. EM score over 40 epochs

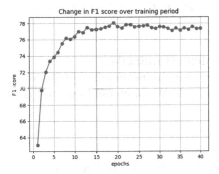

Fig. 4. F1 score over 40 epochs

Table 2. Performance comparison with existing QA models

Model	EM score	F1 score
Co-attention networks [13]	69.5%	78.8%
BiDAF [14]	68.0%	77.3%
R-Net	71.3%	79.7%
DrQA	69.5%	78.8%
Our model	**69.76**%	**78.6** %

7   Future Works

This paper presented an AI based approach to conduct interviews for preliminary filtering of candidates using QA systems.

The statistics and results mentioned above can be further improved by changing and scaling different components of our system. The implemented system has its scope restricted to Java, an OOP language. The scope of corpus can be extended to encompass other technical topics to scale the system. Currently, the questions for the interview along with their difficulties are part of a json file which serves as corpus or knowledge base for our system, it can be further replaced by fetching questions from the web thus making the system open domain. The system currently uses word2vec for forming word embeddings. An improvement could be implementing attention layer over and above word2vec using Convolution Neural Networks to compute more composite word embeddings.

Along with the currently employed cognitive feature of sentimental analysis, other features like facial, emotion detection and tone analysis can be employed to enhance evaluation of the candidate.

Usage of GRU can be implemented and experimented with, instead of LSTM as it might be more computationally efficient because of its less complex structure. Currently the QA system works on the principle of machine reading and comprehension. Research is being carried on in field of conversational QA systems [15] which are appropriate for application in automated interviewing systems.

Compliance with Ethical Standards
All authors declare that there is no conflict of interest. No humans/animals involved in this research work. We have used our own data.

References

1. Ansari, A., Maknojia, M., Shaikh, A.: Intelligent question answering system based on Artificial Neural Network. In: 2016 IEEE International Conference on Engineering and Technology (ICETECH) (2016)
2. Yin, W., Yu, M., Xiang, B., Zhou, B., Schütze, H.: Simple question answering by attentive convolutional neural network. In: COLING 2016, 26th International Conference on Computational Linguistics, Proceedings of the Conference: Technical Papers, 11–16 December 2016, Osaka, Japan, ACL 2016 (2016)
3. Feng, M., Xiang, B., Glass, M., Wang, L., Zhou, B.: Applying deep learning to answer selection: a study and an open task. In: IEEE Automatic Speech Recognition and Understanding Workshop (ASRU) (2015)
4. Bendersky, M., Metzler, D., Croft, W.B.: Learning concept importance using a weighted dependence model. In: Proceedings of the Third ACM International Conference on Web Search and Data Mining, WSDM 2010, New York, NY, USA. ACM (2010)
5. Yin, J., Jiang, X., Lu, Z., Shang, L., Li, H., Li, X.: Neural generative question answering. In: Proceedings of the Twenty-Fifth International Joint Conference on Artificial Intelligence (IJCAI 2016) (2016)
6. Su, B.-H., Kuan, T.-W., Tseng, S.-P., Wang, J.-F., Su, P.-H.: Improved TF-IDF weight method based on sentence similarity for spoken dialogue system, pp. 36–39 (2016). https://doi.org/10.1109/ICOT.2016.8278973
7. Lmati, I., Achtaich, N.: Towards adaptive generation of mathematical exercises. Int. J. Comput. Appl. **147**(6), 0975–8887 (2016)
8. Rajpurkar, P., Zhang, J., Lopyrev, K., Liang, P.: Squad: 100,000+ questions for machine comprehension of text. arXiv preprint arXiv:1606.05250 (2016)
9. Chen, D., Fisch, A., Weston, J., Bordes, A.: Reading wikipedia to answer open-domain questions. CoRR Vol. abs/1704.00051 (2017). (arXiv:1704.00051)

10. Wang, Z., Mi, H., Hamza, W., Florian, R.: Multi-perspective context matching for machine comprehension. arXiv preprint arXiv:1612.04211 (2016)
11. Lee, K., Kwiatkowski, T., Parikh, A., Das, D.: Learning recurrent span representations for extractive question answering.arXivpreprint arXiv:1611.01436 (2016)
12. Luong, M.-T., Pham, H., Manning, C.D.: Effective approaches to attention-based neural machine translation. arXiv preprint arXiv:1508.04025 (2015)
13. Xiong, C., Zhong, V., Socher, R.: Dynamic coattention networks for question answering. arXiv preprint arXiv:1611.01604 (2016)
14. Seo, M., Kembhavi, A., Farhadi, A., Hajishirzi, H.: Bidirectional attention flow for machine comprehension. arXiv preprint arXiv:1611.01603 (2016)
15. Reddy, S., Chen, D., Manning, C.D.: CoQA: a conversational question answering challenge. arXiv preprint arXiv:1808.07042 (2018)

Eye Movement Event Detection with Deep Neural Networks

K. Anusree and J. Amudha$^{(\boxtimes)}$

Department of Computer Science and Engineering, Amrita School
of Engineering, Bengaluru, Amrita Vishwa Vidyapeetham, Bengaluru, India
anusreegoplan@gmail.com, j_amudha@blr.amrita.edu

Abstract. This paper presents a comparison of event detection task in eye movement with the exact events recorded from eye tracking device. The primary goal of this research work is to build a general approach for eye-movement based event detection, which will work with all eye tracking data collected using different eye tracking devices. It utilizes an end to end method based on deep learning, which can efficiently utilize eye tracking raw particulars that is further grouped into Saccades, post-saccadic oscillations and Fixations. The drawback of deep learning method is that it requires a lot of preprocessing data. At first, we have to build up a strategy to enlarge handcoded information, with the goal that we can unequivocally augment the informational index utilized for preparing, limiting the run through time on coding by a human. Utilizing this all-encompassing hand-coded information, we instruct neural networks model to process eye-development fixation grouping from eye-movement information in the absence of any previously defined extraction or post-preparing steps.

Keywords: Event detection · Deep learning · Fixations · Saccades · Eye tracking

1 Introduction

Event detection in eye-following information is related with numerous difficulties. One of those is that a wide range of sorts of clamor and aggravations may happen in the recorded signs which start both from the eye-tracker and from individual contrasts among the clients. This fluctuation among estimations and people may make flags that are hard to break down. The challenge is thusly to create calculations that are sufficiently adaptable to be utilized for signs that contain different sorts of occasions and aggravations, and that can deal with both distinctive people what's more, extraordinary kinds of eye-trackers. An extra test in Event detection of eye-following signs is the means by which to assess and look at changed calculations. In eye development area, the objective of event detection is to mighty separate events, like fixations and saccades, from the surge of gaze information tests from an eye tracker. Quite a while, wide classes of calculations were utilized: Initial one is velocity-based algorithms that recognize saccades and expect the rest of the events fixations. Another one algorithm that commonly used is dispersion based algorithms rather recognize fixations and accept the

S. Smys et al. (Eds.): ICCVBIC 2019, AISC 1108, pp. 921–930, 2020.
https://doi.org/10.1007/978-3-030-37218-7_98

rest to be saccades. Both the velocity based and the dispersion based algorithms accompany edges that the client needs to set.

Deep Learning models have reformed many research handle as of now. Nevertheless, the raw eye development information is still normally prepared into discrete occasions by means of edge based calculations or manual naming. Change of information in to many features. Features are unmistakable properties of information designs that assistance in separating between the classes of information designs. The ongoing increment of dimensionality of information represents a serious test to many existing element choice and highlight extraction strategies as for proficiency and viability. The most critical undertaking of an AI application is include extraction. Eye following need to focus on organizing what eye improvement estimations are used, how they are implied, and how they should be deciphered concerning interface plan. Issue of removing the eye features and its multifaceted nature was examined by made an examination on features utilized in various eye following applications. In feature determination process a subset from accessible features information are chosen for the way toward learning calculation. The upside of feature determination is that essential data identified with a solitary component is not lost yet in the event that a little arrangement of features is required and unique features are extremely assorted, there is possibility of data lost as a portion of the element must be discarded. Then again, with dimensionality decrease otherwise called feature extraction, the span of features space can be regularly diminished without losing data of the first component space.

High dimensional information is tricky for order calculations because of high computational expense and memory utilization. We examined the methods utilized for highlight extraction and different methodologies received for critical thinking. The effect of feature extraction utilized in a profound learning engineering and how might we receive that benefits in our application is searching for. Deep learning techniques have successfully used to give incredible improvement in different research fields, for example, AI, Image processing and computer vision. High dimensional information is hazardous for arrangement calculations because of high computational expense and memory use. We broke down the systems utilized for feature extraction and different methodologies embraced for critical thinking. The current methods can accomplish high exactness anyway the precision relies upon features they use and features are removed by a different model for feature extraction. The hand created feature become improper for the application and it might prompt a misguided course of your exploration. The current strategies can accomplish high exactness anyway the precision relies upon feature they use and features are removed by a different model for feature extraction. The handmade highlights become unseemly for the application and it might prompt a misguided course of your exploration.

2   Literature Review

Various researchers worked to develop an assistive technology for reading disabled people. Different approaches and methodologies have proposed in this area. The investigation centers on the adequacy of assistive innovation systems and programming in helping understudies with perusing handicaps. Assistive innovation portrayed as any device or things that can utilized to increment keep up or enhance the capacities of people with in capabilities. Customary methodologies used to prepare poor users have been viable to some degree. Nevertheless, for particular assignments computer based systems perform better. The acquaintance of innovation with this preparation field has rearranged the preparation procedure however not the sum total of what highlights have been executed completely. The system-based methodologies have demonstrated powerful in actualizing the techniques utilized a computer. Human-Computer collaboration for reading in abled has turned into a noteworthy zone of research. Event detection plays an important role in eye movement researches told by Zemblys [1]. Event detection solely done utilizing PCs by applying a gaze calculation to the crude eye tracking information. Thusly the outcomes created by these calculations dramatically affect more elevated amount examinations also, measures utilized in eye development examine. A noteworthy downside of recent development location techniques is that client left with number of parameters, which must be balanced dependent on eye-development information quality. He compared 10 machine-learning algorithms used to find out the event detection-using eye tracking information. Human Computer communication is especially engaged for individuals with psychological and learning incapacities [2].

Recognizing Eye Tracking Traits for Source Code Review also giving the approach towards the applicability of Eye tracking in Research areas. Deep learning approaches have accomplished leap forward execution in different areas. In any case, the division of raw eye-development information into discrete occasions yet done prevalently either by hand or by calculations that utilization handpicked parameters and edges. Mikhail Starsev proposed and make freely accessible a little 1D-CNN related to a bidirectional long momentary memory arrange that characterizes look tests as fixations, saccades, at the same time allocating names in windows of up to 1 s [4].

Human Computer Interaction in assistive innovation, the fundamental objective of this is to assist impaired individuals with the utilization of human computer association. The people who have the inability require interfaces that suite their aptitudes and help them in defeating physical and intellectual obstructions. The paper written by Patrik Hlavac and Marian Simko proposed a strategy for recognizing truly perused pieces of reports dependent on gaze information from eye tracker [5]. This work manages the conceivable outcomes of distinguishing client connection with (online) archives. Their calculation considers client's eye obsession data and maps their directions onto word-level components. These are handled regarding their relative word remove. Christiantia Wirawan and colleagues proposed an android application for perusing utilizing gaze estimation technique [6]. Face and eye detection, eye gaze estimation, tracking pupil, are the strategies used. The fundamental criteria of work is utilizing entirely of pixels sensitivity. Utilizations the appearance based strategy for look estimation and following of eye tracking bearings for looking of the screen is the approach. Visual observation in dyslexic individuals has principally been analyzed for perusing assignments or in different sorts of reading and non-reading errands in the investigations of visual consideration hindrances [7]. Human-PC association thinks about that utilization eye movement with individuals with dyslexia have typically centered in finding the most available contents. For the statistical model to group readers with and without dyslexia, they utilized a Support Vector Machine classifier. Individuals with dyslexia make longer fixations, more fixations, shorter saccades and a greater number of relapses then readers without dyslexia [8]. The perception which is getting and the highlights in dataset, for example, look pints, time length and so forth can be utilized to dissect where an individual is concentrating which will be explained in the work Devoloping an Application Using Eye Tracker. This can be utilized in different applications [9].

The authors Rello and Miguel tried to build up the primary factual model to anticipate readers with or without dyslexia utilizing eye-tracking measures [10]. The investigation demonstrates that the age of the individuals indicates clearer contrasts in their reading execution. To build the prototype with the objective of over unpredictability by diminishing the quantity of highlights exhibited to the individual at any one time, and by upgrading essential parameters [11], Fryia, Renata Wachowiak-Smolikova proposed one approach for this. The paper on Assistive technology analyzing the contribution of HCI in assistive technology. Introduce assistive medium from the perspective of HCI is the area of concentration of the author [12]. They were trying to define how HCI was helping the disabled people in different modalities as an assistive technology. Silbert and Gokturk together proposed a framework for therapeutic reading guidance that utilizations outwardly controlled sound-related inciting to assist the individual with acknowledgement and elocution of words in 2000. The technique included with eye tracking information is utilized for discover the directions where eye settling. The word choice algorithm was utilized to trigger articulation and featuring of words. They have executed a framework, which utilizes a reader's visual checking example of the content to distinguish, and articulate, words that the individual is experiencing issues perceiving [13].

3 Proposed Method

Existing event detection algorithms for eye-development information solely depend on thresholding at least one hand-made features includes, each registered from the group of eye gaze inforrmation. Additionally, this thresholding is generally left for the end client. Another advancement is the presence of event detection strategies dependent on AI procedures. Up till presently, these finders have still utilized the equivalent high quality includes same as the previous calculations, however the scaling, arrangements are found out as information which performed naturally with the event detector. Zemblys relatively examined machine learning algorithms and in that Random forest classifier produced high performance and accurate result within ten machine learning algorithm. In this approach to utilize a completely start to finish deep learning methodology, uses crude eye-following information as info and orders it into fixations, saccades and PSOs. As opposed as previously mentioned algorithms based on Artificial Intelligence, the deep learning method that adopted in this paper consequently adapts all characteristics and suitable edges as includes in the information. It moreover figures out how the different haul setting of every crude information test influences what the present example can be delegated, in this way guaranteeing labeled events are delivered in reasonable successions out of any requirement for postprocessing, as a mortal master would create them.

Approach Overview

The architecture of this work is an end to end event detection model, like collect data from eye tracker with respect to the stimulus as well as the method of data collection and use appropriate neural network model to predict the exact event that in eye movement during a specific timestamp. In this work we take a different approach, using a I2MC algorithm as well as RF classifier for the event detection. These all methods wrapped in to a pretrained model using neural network. The architecture and training protocol is a succession to-arrangement LSTM with a Combined Density Network as a layer of yield (Fig. 1).

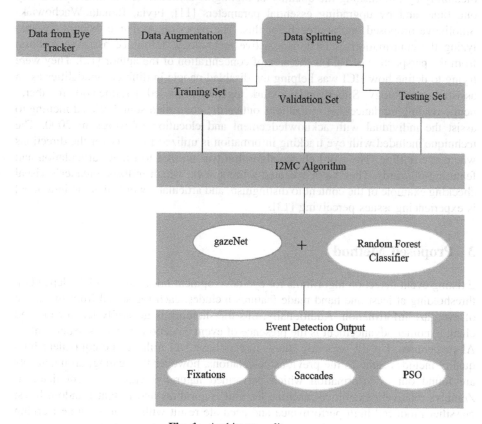

Fig. 1. Architecture diagram

A Combined System a Fully Connected layer, however rather than straightforwardly utilizing its yields to make forecasts about the following look test, the system yields a parameterized, comprising of features which used for probabilistic dissemination of the area for the following look test. Naturally, this Mixture of Gaussians could be translated as the quantity of decisions the system. Two convolutional layers are used in this model pursued by bi-directional repetitive layers with a completely associated layer on top. The convolutional layers utilize 2Dimensional channels and are intended to separate profound highlights from gaze info information, while the intermittent layers model occasion arrangements and are in charge of distinguishing onsets and counterbalances of fixations, saccades and PSOs.

Two means clustering, is to accomplish consistent naming of fixations when there might be substantial contrasts in quality of data among members and other preliminaries, as is of-ten experienced in. This means that the I2MC calculation ought to accomplish fixation labeling across a scope of short periods of information misfortune might be present. The I2MC calculation is made out of three separate steps: interpolation of missing information, two-implies grouping, and finally fixation naming. It is significant to mention that the fact that model qualities will be given the algorithm for all parameters, alongside an inspiration for the specific esteem, these qualities may should be adjusted to well matched a particular information set. To augment the measure of eye-following information that which will be used for fixation location, ascription of brief times of missing data is done using interpolation. Following interpolation, a two-means clustering technique is done for every window. A variation of k-means clustering is Two-means clustering, a strategy in which various perceptions are iteratively clustered into k clusters. The perceptions having a place with each group are which that are more closer to the mean of that group's perceptions than to the mean of some other group. A cutoff is utilized to decide fixation applicants from the clustering weight flag. A cutoff of the mean clustering weight here we used in addition to two standard deviations. Be that as it may, various shorts might be need for various datasets. All times of grouping weight flag which is below this cutoff are named as fixation applicants, and there after back to back fixation hopefuls are consolidated. At long last, short fixation competitors are avoided from the yield. The settings for consolidating obsession applicants may rely upon the stimuli used in the investigation, the commotion level in the eye-movement information, or the extent of the saccades of premium.

We utilize a Random forest classifier to play out the underlying grouping of every crude information test. We employ heuristics to the yield of the classifier, for example, the converging of adjacent fixations, expulsion of saccades and fixations that are excessively small, to create the last occasions. Settings will result from these heuristics with no client versatile. Preference of utilizing an arbitrary random forest classifier rather than other AI calculations is that we can utilize the features as they seem to be. There is really no compelling reason to scale, focus, or change them in any capacity. A random forest classifier works by creating numerous choice trees. Every tree, from its root to every one of its leaves, comprises of a progression of choices, made per test in the information, in view of the parameters that we furnish the classifier. Each tree hub—rising to a particular sensible suggestion—is a state on a solitary component, compelled to different hubs in a tree with on the off chance that provisions, which conveys the calculation closer to choosing whether the example has a place with example a saccade or fixation. These choices are like how conventional hand-made event detection calculations work. These additionally take various highlights, as info and, by methods for principles and limits set on these highlights by the calculation's fashioner, determine which occasion the example likely has a place with.

It comprises of monocular eye-development information of members seeing pictures. Using SMI tracker we were able to identify fixations, saccades from raw data itself. After detecting events using the neural network we were make a comparison of the event detection output with the raw data. This was a kind of test case to validate the output from the event detection model (Fig. 2).

	t	x	y	status	evt	
2	0	0	-0.03773	0.611697	FALSE	0
3	1	0.001	-0.06467	0.592833	FALSE	0
4	2	0.002	-0.05389	0.606307	FALSE	0
5	3	0.003	-0.07276	0.611697	FALSE	0
6	4	0.004	-0.08892	0.622474	FALSE	0
7	5	0.005	-0.07815	0.557803	FALSE	0
8	6	0.006	-0.0485	0.603613	FALSE	0
9	7	0.007	-0.09162	0.565886	FALSE	0
10	8	0.008	-0.08084	0.568581	FALSE	0
11	9	0.009	-0.10779	0.536245	FALSE	0
12	10	0.01	-0.08623	0.514687	FALSE	0
13	11	0.011	-0.07545	0.501213	FALSE	0
14	12	0.012	-0.09701	0.511993	FALSE	0
15	13	0.013	-0.07276	0.557803	FALSE	0
16	14	0.014	-0.08353	0.611697	FALSE	0
17	15	0.015	-0.11587	0.57936	FALSE	0
18	16	0.016	-0.061198	0.609001	FALSE	0
19	17	0.017	-0.02425	0.57397	FALSE	0
20	18	0.018	-0.0997	0.57936	FALSE	0
21	19	0.019	-0.1024	0.549719	FALSE	0
22	20	0.02	-0.10509	0.576666	FALSE	0
23	21	0.021	-0.06737	0.622474	FALSE	0
24	22	0.022	-0.0997	0.576666	FALSE	0
25	23	0.023	-0.07815	0.598223	FALSE	0
26	24	0.024	-0.03234	0.590139	FALSE	0
27	25	0.025	-0.1024	0.592833	FALSE	0
28	26	0.026	-0.07545	0.590139	FALSE	0
29	27	0.027	-0.14551	0.560497	FALSE	0
30	28	0.028	-0.07006	0.611697	FALSE	0
31	29	0.029	-0.05928	0.555107	FALSE	0
32	30	0.03	-0.11587	0.590139	FALSE	0

lookAtPoint_EL_S2

t	time in seconds
x	horizontal gaze direction in degrees
y	vertical gaze direction in degrees
status	status flag. False means track loss
evt	#0 :undefined
	#1 :fixation
	#2 :saccade
	#3 :post saccadic oscillation

Fig. 2. Event detection output

4 Results

The fundamental reason for the project was to construct the initial completely start to finish event-identifying model for eye-development information. Our system uses raw eye-following information as info and groups it into post-saccadic motions, fixations, saccades and out of any post-handling. It comprises of three bi-directional repetitive layers pursued by two convolutional layers and a completely associated layer on top. The convolutional layers are implied to extricate profound highlights from the raw eye development data information, while the repetitive layers model occasion arrangements and are capable for distinguishing features and balances of saccades, PSOs and fixations. At last, the completely associated layer yields probabilities of each example being a saccades, PSOs or fixations. RF was created utilizing Python 2.7 programming language and number of bundles for information control and preparing AI calculations.

A noteworthy test with deep and machine learning approach as a rule is the uneven event of the test groups. Algorithm significantly demonstrates that hand-creating features is not fundamental for event identification. It is completely conceivable to construct an event identifier with astounding execution without presenting any unpredictable features or on the other hand unmistakable thresholding on the first signal We present here is named eye movement information which is the just thing needed for the featureless event detection. The gaze net event detection model which we developed using random forest classifier resulted in event detection data with reduced error rate. To evaluate the output of random forest gazeNet event detection model, we further created a MATLAB script with the change in event detection metrics, and the plot which representing X-axis time period in seconds and Y-axis corresponding to gaze coordinate position. The proposed evaluation script resulted in same error rate of the gazeNet event detection model which proves the accuracy of the system. To compensate the frequency mismatch problem in SMI eye tracking data used the gaze net model data, which is synthesized in the proposed system is, used which further explains the accuracy (Fig. 3).

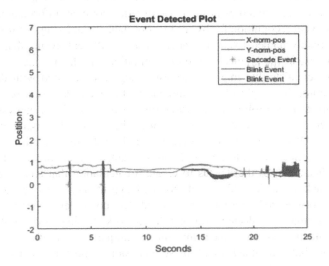

Fig. 3. Detected events on data collected using SMI

The existing event detection algorithm producing the events in eye movement for only saccades and fixations. Using the model created as a combination of gaze Net and Random forest classifier will extend the way towards the new features other than saccades and fixations like PSO, blink. When prepared appropriately utilizing a wide scope of various information, an event detector based on artificial intelligence can sum up well to information of various inspecting frequency scales still out of any need of pre-process and post-process information or physically assigned limits. The main obstacle in accomplishing which is the accessibility of preparing data.

5 Conclusions

In this work, we have demonstrated about the event identifier, which is developed through profound training that remains dependent on just a couple of little portions of hand created information and functions self-sufficiently without requiring any previously defined extraction or post processing method to arrange any limits that can execute and mortal master coders in creating PSO, saccades, and fixation. This demonstrates and builds event indicators by using artificial intelligence is considered as an encouraging methodology for the forthcoming research areas, and it can ideally empower the formation of event detectors that can manage more kinds of eye developments all the while than a hand-created calculation ever could.

There is anyway an exchange off made while picking the AI method rather than conventional methodology of hand-creating an event indicator. The features Hand crafting gives the analyst with knowledge in what way the subsequent events characterized as far as flag properties of the basic eye-development information. The point of performing event recognition for by far most of analysts utilizing eye following is to obtain appropriately named fixations, saccades also, different events that they would then be able to use as the premise of their examinations. In this manner, outside of the little horde of scientists that might be to some degree unsatisfied with the Artificial Intelligence approach in light of their enthusiasm for event detection procedures and the flag resources of eye-following information, for most by far of analysts.

AI approach is an important and, as we appear here, a good working technique for event detection. Other than fixations, saccades and PSOs, there are a lot of different events in eye-development information hanging tight to be distinguished algorithmically. A significant test to the field is to create a calculation that creates events from different eye tacking devices like head-mounted eye trackers, or eye trackers incorporated in head-mounted presentation utilized for computer generated originally, includes the information collected with such frameworks having a loaded blend of head and eye movements. We can expect that the technique to seek after for creating an Event detector that can manage all these occasions immediately using Artificial Intelligence approach.

Compliance with Ethical Standards
✓ All authors declare that there is no conflict of interest.
✓ No humans/animals involved in this research work.
✓ We have used our own data.

References

1. Zemblys, R., Niehorster, D.C., Holmqvist, K.: gazeNet: end-to-end eye-movement event detection with deep neural networks. Behav. Res. Methods (2018). https://doi.org/10.3758/s13428-018-1133-5
2. Zemblys, R.: Eye-movement event detection meets machine learning. In: 20th International Conference Biomedical Engineering (2018)

3. Behroozi, M., Lui, A., Moore, I., Ford, D., Parnin, C.: Dazed: measuring the cognitive load of solving technical interview problems at the whiteboard (2018). https://doi.org/10.1145/3183399.3183415
4. Startsev, M., Agtzidis, I., Dorr, M.: 1D CNN with BLSTM for automated classification of fixations, saccades, and smooth pursuits. Behav. Res. Methods 51 (2018). https://doi.org/10.3758/s13428-018-1144-2
5. Hlavac, P., Simko, M.: Detecting genuinely read parts of web documents. 6–11 (2017). https://doi.org/10.1109/smap.2017.8022658
6. Wirawan, C., Qingyao, H., Yi, L., Yean, S., Lee, B.S., Ran, F.: Pholder: an eye-gaze assisted reading application on android. 350–353 (2017). https://doi.org/10.1109/sitis.2017.64
7. Meland, Z.: Eye movements during audiovisual speech perception with dyslexia, Master Thesis in psychology. Norwegian University of Science and Technology (2017)
8. Chandrika, K.R., Amudha, J., Sudarsan, S.D.: Recognizing eye tracking traits for source code review. In: 22nd IEEE International Conference on Emerging Technologies and Factory Automation, 12–15 September 2017
9. Venugopal, D., Amudha, J., Jyotsna, C.: Developing an application using eye tracker. In: IEEE International Conference On Recent Trends In Electronics Information Communication Technology, 20–21 May 2016
10. Rello, L., Ballesteros, M.: Detecting readers with dyslexia using machine learning with eye tracking measures. 1–8 (2015). https://doi.org/10.1145/2745555.2746644
11. Fryia, G.D., Wachowiak-Smolikova, R., Wachowiak, M.P.: Human-computer interface design in an e-Learning system for individuals with cognitive and learning disabilities. In: 4th International Conference on Digital Information Management, ICDIM 2009, pp. 146–151 (2009). https://doi.org/10.1109/icdim.2009.5356784
12. Liffick, B.: Assistive technology as an HCI topic. J. Comput. Sci. Coll. 19, 142–144 (2003)
13. Sibert, J., Gokturk, M., Lavine, R.A.: The reading assistant: eye gaze triggered auditory prompting for reading remediation, pp. 101–107 (2000). https://doi.org/10.1145/354401.354418

A Review on Various Biometric Techniques, Its Features, Methods, Security Issues and Application Areas

M. Gayathri[✉], C. Malathy, and M. Prabhakaran

SRM Institute of Science and Technology, Kattankulathur,
Chennai 603 203, India
gayathrm2@srmist.edu.in

Abstract. Biometrics is the emerging technology in the era of internet and mobile communication. The IoT revolution has enabled the things around us to communicate as it emerges as a smart system, hence security should be considered as primary issue as everything around us is going to be connected. Biometric Technology is considered to be the future of all electronic security which provides authentication and security management. This paper provides a detailed survey on existing biometric technology, different types of biometric traits, techniques adopted for feature extraction of various biometric traits and application areas of different biometric traits. The importance of biometric technology related to the various fields of security is also discussed through this paper. The biometric technology connected with smart systems helps in monitoring the human activities all over the world, thus providing a good security level. Biometrics acts as a major support to various fields of automobile security, Internet of Things (IOT) Security, health care security, workforce management of organization, government security, banking, and retail industry.

Keywords: Authentication · Unimodal · Multimodal · Security · Feature extraction · Verification

1 Introduction

Biometric recognition and verification are considered to be important aspect in the current scenario of research. Nowadays in smart world, passwords are replaced by biometric traits in many places including companies, educational organization, defense, public sector, private sector or research labs. Hence the focus on biometric data security has gained much importance. Biometric systems use individual physical characteristics and behavioral characteristics as input. Multimodal systems combine more than two or three unimodal systems individually into a single system which overcomes the drawback of Unimodal system in terms of security issues, verification and recognition. Multimodal technique adopted will provide security enhancement and the hackers may find it difficult to crack the data. The biometric data which is obtained from different sensors are combined together by different fusion techniques. (i.e. feature level, score level, rank, sensor, and decision level). Biometric systems involve verification (one to one process) and identification (one to many process). The input given to a system is

© Springer Nature Switzerland AG 2020
S. Smys et al. (Eds.): ICCVBIC 2019, AISC 1108, pp. 931–941, 2020.
https://doi.org/10.1007/978-3-030-37218-7_99

pre-processed, its feature is extracted and it is stored in the database. (For identification and verification).

1.1 Security Problems

Security is one of the primary concerns of mankind in today's world. We give great priority to our privacy and security. Getting our data in others hand is almost a life threat these days. This made us delve in the various problems with respect to both the authentication and data storage security problems in biometric systems. Different, efficient biometric parameters have to be chosen as the core for authentication.

1.2 Need for Biometric Security

The world is marching towards the biometric era in terms of security and access control. The hacking rate has increased worldwide which leads to loss of data and more cybercrime issues. Many countries have started to focus on research areas of biometric security.

General Process of Security
The general process is capturing a biometric trait of an individual as an image, pre-process the image, apply feature extraction techniques, apply hash function and cryptographic techniques on the template obtained after feature extraction. The transformed template will be stored in database for further processing of recognition and verification. The general process of the security is illustrated in Fig. 1.

2 Literature Survey About Attacks

There are many attacks carried on biometric data namely replay attack, spoofing attack. etc. Few attacks are discussed in the following points. Presentation of an dishonest data to get authorization (e.g.: a fake fingerprint). Replay attack: Biometric data can be digitally stored. This stored data can be resubmitted to the system during the matching phase to obtain authentication. Thus, a previously obtained data is "replayed" into the biometric system without the usage of the sensor during the matching phase. An example would be the "presentation of a digital copy of fingerprint image or recorded speech" [20]. An attack with "a trojan horse program on the feature extractor to produce predetermined extracted features" [20]. An attack with a trojan horse program on the matcher to always produce desired result. Replacement of legitimate extracted feature sets that act as an input, with synthetically produced feature sets. An example would be the transmittance of the minute of fingerprints to a remote matcher (let's say via internet). Modification or removal of enrolled templates from the database; introduction of new templates to database leading to false negatives and positives respectively. Sending of enrolled templates to matcher via a communications channel which itself is attacked for the purpose of making changes to the templates database [20]. Overriding of the final decision of the biometric system by an attacker with its desired choice of output. "Even if the Biometric system had an excellent efficiency, it has been rendered useless due to the exercising of the overridden result" [20]. This attack is on the entire work of the biometric system is to degrade the performance. It involves "the submission of a stolen, copied or synthetically replicated biometric trait to the sensor to defeat the biometric system security in order to gain unauthorized access" [20]. It can be carried out against a wide spectrum of biometric traits including finger- prints, palm print, iris, face etc. Having being carried out directly on the sensor, it is also called "direct attack".

3   Biometric Systems Categories

Biometric systems are divided into two categories.

Cancellable Biometric
The original image is intentionally distorted by suitable technique and stored in the database. If the data is stolen by the hacker, we can provide a new template by changing parameters and producing a distortion in image. This biometric is very helpful in preserving the privacy. The revocable and noninvertible transformations are used to produce cancellable biometric templates.

Soft Biometrics
Soft biometrics provides information about the personal attributes like height color, age, weight which can be derived from physical and behavioral characteristics of the individual.

1. Physical biometrics
2. Behavioral biometrics

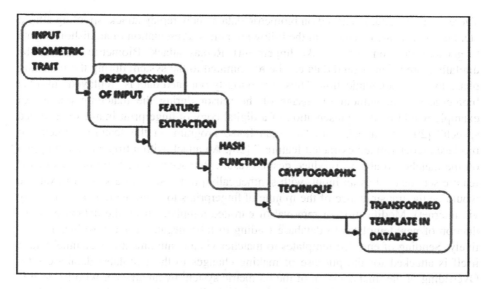

Fig. 1. General process of security

3.1 Physical Biometrics

Biometrics uses the physiological traits. Table 1 illustrates the different physical biometric traits.

Fingerprint

Fingerprint biometric technique is widely accepted as cost is low and gives high efficiency. A fingerprint sensor captures the image of the fingerprint of the person. Image enhancement is done followed by the feature extraction and template protection. Ridges (minutiae points), valley, island, lake are the features extracted from the fingerprint image. Feature vectors are acquired from the acquired image. Many sensors like capacitive sensors, thermal sensors, and ultrasound sensor are used to acquire the fingerprint image [1].

Iris

Iris biometric system is considered to be the most accurate system. Iris image is collected and feature extraction is done by filters to extract the features. From the image, features like eyelids, eyelashes are extracted. Image acquisition is done by infrared illuminations, the steps in iris image processing is Image segmentation, image normalization, feature extraction and finally classification [2].

Face

Face recognition is a commonly used system for many commercial applications related to public security. The image captured from sensors will have noise, blur and the quality of the image may be weak. To enhance the quality of image, enhancement techniques like pre-processing is done. Fisher discriminant analysis is done for the feature extraction process and feature vectors are generated for recognition process. The features extracted from face are eyes, nose, and mouth [3].

Palm Print

Palm print recognition uses a scanner to capture the features of palm in human. Thermal or optical scanners may be used. Ridges, wrinkles are the features extracted. The methods used for feature extractions are line based, subspace based (PCA, LDA) and statistical (mean, standard deviation). Neural network is been deployed for classification of final decision [4].

DNA

DNA biometric systems involves Identification of person by using his DNA as a trait. DNA code is comprised of four bases: adenine (A), guanine (G), cytosine (C), and thymine (T). DNA can be extracted from saliva, blood and hair. The investment cost for setting up laboratory for DNA biometrics is high. DNA biometric system helps in crime cases to find the victim and in medical diagnosis [5].

Periocular

Periphery of ocular (periocular) is biometric system designed as an alternative to iris system. It contains the features in visible spectrum and infrared spectrum. It has reduced feature set and hence the matching accuracy is fast and suitable for mobile based applications. The key point is detected using reduced phase intensive local pattern method [6].

Palm Vein

The vein patterns of the palm are captured by the infrared light sensor. The features extracted are black lines. The veins within the hands are internal to the body and it is difficult to forge. It is mainly used in banking sector for ATM, remote banking [7].

Finger Vein

There is a unique finger vein pattern inside a finger and it can be captured by IR light transmission. Pre-processing is done on the vein image and noise is removed by applying various filtering techniques and normalization is carried out. Parameters like mean, skewness and response time is calculated and it is used during the matching process [8].

Retina

The retinal image is captured from the camera and its blood vessel pattern is recorded for the further processing. The Vascular features and non-vascular features can be extracted from the image. Non-vascular features (mean) does not require blood vessel pattern. Retinal verification is considered to be the fastest processing as its feature vector size is small [9].

Hand Geometry

Infrared sensors are used to capture the vein patterns inside the person hand. Vascular patterns are captured and principal component analysis is applied on the extracted feature to achieve a very good matching and verification rate. Edge detection methods and texture-based methods are used to extract the features.

Ear

The 2D image or 3D ear image are captured by the camera and features are extracted. The features extracted are grouped under holistic, local, geometric and hybrid. Many feature extraction methods are employed like SIFT, SURF, LBP convolutional neural network and deep learning. Recently deep learning has gained much importance for recognition where the sample dataset is large in size. In few cases, the features like mean, centroid, and distance between the points is calculated as feature vector [10].

EEG

In EEG brain oscillations are recorded as EEG patterns with the help of electrodes tied to the head of the person.co variance matrix is calculated as feature vector. EEG signals suffer from more noise and hence ensemble is adopted and the wavelet transform is mostly used to extract the features. Features extracted are given as input to artificial neural network for classification [11, 12].

3.2 Behavioral Biometrics

Behavioral biometrics identifies the unique pattern in human behavioral aspects. Table 2 illustrates the behavioral biometric traits.

Gait

The emergence of computer vision has paved the way for smart surveillance system. Gait recognition is mostly used for the surveillance applications. Gait biometrics records the way the person walks, cyclic motion. The background subtraction method is used to find the object in motion. As huge amount of data is involved and hence machine learning techniques are adopted for training the data [13].

Voice

Voice print is the unique voice of individual to identify and verify the system. A spectral feature from the voice is measured. It is a pattern recognition approach which uses the speech signal to extract the features. Mel-frequency cepstral coefficients, Linear Prediction cepstral coefficients are the few techniques adopted for feature extraction. Modelling the voice is done by neural network [14].

Handwriting

Handwriting biometrics helps in identifying the person by his written text. It is extraction of textural features of an individual handwriting and involves various parameters like segment counts, number of pen ups events, number of pens down events, duration of writing, pressure, spatial coordinate etc. The data acquisition is done by digital tablet [15].

Signature

The signature is captured by Pen tablets. Signature verification can be done in two ways namely online signature and offline signature. The image is acquired, pre processed, feature is extracted and selected template is generated and stored in database. Features like X, Y coordinates, Pressure, Event type is extracted and signature embedding is done in barcode. Precision, F measure and recall are calculated. Signature verification is mainly used in many applications like patent office, E procurement, etc. [16].

Keystroke Dynamics

It is process of analyzing the typing pattern of the individual. The timing features are extracted; overall typing speed frequency of errors while typing is recorded. Classification and training the patterns is done by artificial neural network and support vector machine. Online courses use this technique of keystroke dynamics to check the verification of the individual during the tests conducted [17].

Touch Dynamics

All modern devices are touch enabled, hence touch dynamics is considered to be trending biometric. The touch actions like scrolling, swiping, alphabet input, numerical input is monitored and captured and recorded. Probabilistic modelling and cluster analysis are employed for pattern recognition approach of touch patterns. Main application is on the Android smart phones which use the touch logger app [18].

Mouse Dynamics

Mouse dynamics is about the capture of mouse movements of the individual. The various mouse movements are drag, drop, move and click. Only few datasets are available for mouse movements. Segmentation is done on the mouse movements and statistical features like mean, standard deviation, min and max is calculated between the mouse movements from one position to other position. Statistical and probabilistic models are used for the feature extraction and classification of mouse movements' patterns [19].

4 Feature Extraction

In image processing application we can extract and manipulate the particular needed portion of the image, and we can represent it as a reduced feature vector, this process is termed as feature extraction. Many automatic feature extraction algorithms are available in deep learning models. Few of them are Local binary patterns, speeded up robust features and histograms of oriented gradients. Feature is considered to be the quantifiable measurement obtained from the image. Features obtained are classified into low level and high-level features. Low level features can be derived directly from the image and high-level features can be derived by locating shapes in images and it relies on low level features. Feature extraction process gives the set of feature vectors which are then applied to the classification process for obtaining the target result and completing the recognition process.

5 Feature Selection

It is the process of selecting the important features from the available set of features using filtering methods. It is a learning algorithm to select an optimal input from the set of available data. Based on the important features selected the model can be built. From Original features set few are selected for feature selection and in feature extraction method new set of features are formed from old set of feature sets. Various Filter methods and wrapper methods are adopted for feature selection. Few techniques of feature selection are correlation co efficient, genetic algorithm approach and recursive elimination method. It helps in dimensionality reduction. The different features extracted from the image are Statistical features, Structural features, Textural features, spatial features, Histogram features, Edge and boundary features.

Table 1. Physical biometric trait.

Biometric trait	Features extracted	Methods adopted for feature extraction	Application of biometric trait
Fingerprint	Ridge flow pattern, Minutiae points Pores	Convolutional neural network	Time and management, home security
Iris	Isolates eyelids, eyelashes	Hough transform, wavelet transform	Banking security
Face	Extracting face component features like eyes, nose, mouth	LBP	Surveillance
Palm print	Principal lines, wrinkles and ridges	Gabor wavelet	Criminal applications
DNA	Sequence related to codons, amino acids, and proteins	Hidden Markov chains genetic programming	Forensic cases, Genetic genealogy

(continued)

Table 1. (*continued*)

Biometric trait	Features extracted	Methods adopted for feature extraction	Application of biometric trait
Periocular	Region of interest (ROI), key points	Local Binary Patterns (LBP) and, to a lesser extent, Histogram of Oriented Gradients (HOG) and Scale-Invariant Feature Transform (SIFT) key points	Smartphone authentication
Palm vein	Geometry, the principal line, and the delta point	Principal component analysis and self-organizing maps	Palm vein authentication device at the ATM
Finger vein	Point, line and plane	Wavelet and Fourier transforms	Retail industry application pay by finger
Retina	Binarization, circle fixing and blood vessel extraction	Fuzzy C-Means (FCM) clustering, SVM	Medical diagnostic application (hypertension)
Hand geometry	Boundary of hand (2D/3D), Length/width of fingers or palm, finger perimeter, aspect ratio of finger or palm, finger area, angles between inter fingers	Gaussian filtering	Access control
Ear	Shape, mean, centroid and Euclidean distance between pixels	Gradient descent and k nearest neighbor, CNN	ERGO ear biometric app
EEG	Mean, amplitude, statistical feature	Principal component analysis, independent component analysis and local discriminant bases	Psychiatry brain computer interface
ECG	Amplitudes and intervals	Support vector machine, discrete wavelet transform	Body area sensor networks (fitness tracker)

Table 2. Behavioral biometric trait.

Biometric Trait	Features extracted	Methods adopted for feature extraction	Application of biometric trait
Gait	Height, length of limbs, stride length and speed	Independent component analysis, linear discriminant analysis	Pediatric orthopedics
Voice	Mean, median, variance values of energy plateau, energy contour	Linear predictive analysis (LPC)/ Linear predictive cepstral coefficients (LPCC)/perceptual linear predictive coefficients (PLP)	Call centre authentication (Speech pro voice key)
Handwriting	Spatial coordinate, pressure, azimuth, inclination	Backpropagation neural network technique	Document recognition, forensic application
Signature	Angle between the strokes, tangential acceleration	Hidden Markova model, neural network	**Indian Patent Office, E- Procurement**

(*continued*)

Table 2. (*continued*)

Biometric Trait	Features extracted	Methods adopted for feature extraction	Application of biometric trait
Keystroke dynamics	Latency between consecutive keystrokes. Duration of the keystroke, hold-time. Overall Typing speed. Frequency of errors	Particle swarm optimization techniques and neural network	Keytrac, API, ID control
Touch dynamics	Timing, finger pressure, area of finger pressure	Probabilistic modelling, cluster analysis	Electronic musical instruments
Mouse dynamics	Average click time, percentage of mouse action per mouse movement direction, Average distance per mouse movement direction	Statistical models	User identity verification for login system, Intrusion detection

6   Challenges and Countermeasures on Security Attacks

Enrolled data in Biometric system is mostly stored in the form of encrypted images. As long as the data is not stored in a distributed storage system, there is always a huge risk of data breach involved. To secure these vulnerable data there are various methods in practice such as traditional cryptography, steganography, water marking and visual cryptography. Encryption can be done by strong cryptographic algorithms, secure hashing techniques and message digest. Digital enveloping is also one of the methods to share the data secretly. Of all the methods Visual cryptography gains more importance in recent days to share the biometric information secretly. This is one of the primary ways of protecting any data that is visually understand to human eyes. Using this, the data of an individual can be broken up into parts of codes during enrolment; each part stored in different storages some on the servers or some on the user's end. To gain access to all the parts, the particular decryption key is necessary which can be accessed only by the authorized person. Thus, since the data is not useful if not all of it is present, it is difficult to break have access to the fully usable data.

7   Inference from the Survey

The physical biometrics is stronger biometric technology and widely adopted on variety of platforms for authentication process. The behavioral biometrics is slowly gaining importance in the verification and identification of the person by recording his pattern of individual activities in day today life. Through this survey we analyze that the security of biometric data is primarily important as it contains much sensitive information of the human.

8 Future Enhancement

Template protection of biometric data is a primary concern as we are relying on the biometric security on all aspects. There is more scope for the research in template protection of the biometric data. Biometric data template protection should be focused to avoid problems related to forgery and hacking. Block chain combined with biometrics could be good solution to security issues.

9 Conclusion

The study has inferred that the biometric technology has gained importance in daily activities of an individual due to the various issues relating to the privacy and security in many fields. Future scope of this survey can be related to multimodal security issues. Multimodal security is gaining popularity as it is much complex for the hacker to identify the original source. All over the world Immigration process, wearable devices have started incorporating the biometrics into their process to strengthen their field security and monitoring. Biometric technology has reduced the problems of theft level but not fully eradicated, even in biometric technology the hackers find a way to breach your data and hence the biometric template protection must be strong. Few examples of current trend biometrics include certain application like cars with ECG biometrics control for monitoring the driver and passenger's health, palm recognition in smart panda bus (autonomous vehicle).

Compliance with Ethical Standards
✓ All authors declare that there is no conflict of interest.
✓ No humans/animals involved in this research work.
✓ We have used our own data.

References

1. Joshi, M., Mazumdar, B., Dey, S.: Security Vulnerabilities Against Fingerprint Biometric System, Cryptography and Security (cs.CR), May 2018. https://arxiv.org/abs/1805.07116
2. Ammour, B., Bouden, T., Boubchir, L.: Faceiris multimodal biometric system using multi-resolution Log-Gabor filter with spectral regression kernel discriminant analysis. IET Biometrics (2018). https://doi.org/10.1049/iet-bmt.2017.0251
3. Liu, X., Pedersen, M., Charrier, C., Bours, P.: Can image quality enhancement methods improve the performance of biometric systems for degraded face images? In: 2018 Colour and Visual Computing Symposium (CVCS). IEEE (2018). 978-1-5386-5645-7/18/31.0
4. Kong, A., Zhang, D., Kamel, M.: A survey on palm print. Pattern Recogn. **42**(7), 1408–1418 (2009)
5. Hashiyada, M.: DNA Biometrics Tohoku University Graduate School of Medicine, Japan. www.intechopen.com
6. Bakshi, S., Sa, P.K., Wang, H., Barpanda, S.S., Majhi, B.: Fast periocular authentication in handheld devices with reduced phase intensive local pattern. Multimedia Tools Appl. **77**, 17595–17623 (2018)

7. Sarkar, I., Alisherov, F., Kim, T.H., Bhattacharyya, D.: Palm vein authentication system: a review. Int. J. Control Autom. **3**(1), 27–34 (2010)
8. Mulyono, D., Jinn, H.S.: A study of finger vein biometric for personal identification, May 2008. https://doi.org/10.1109/isbast.2008.4547655. Source: IEEE Xplore
9. Mazumdar, J.B., Nirmala, S.R.: Int. J. Adv. Res, Comput. Sci. **9**(1) (2018). ISSN No. 0976-5697
10. Eyiokur, F.I., Yaman, D., Ekenel, H.K.: Domain adaptation for ear recognition using deep convolutional neural networks. IET Biometrics. Special Issue: Unconstrained Ear Recognition, ISSN 2047–4938, E-First on 13th February 2018
11. Gui, Q., Jin, Z., Xu, W.: Exploring EEG-based biometrics for user identification and authentication. This research was supported by NSF grants SaTC-1422417 and SaTC-1423061, and Binghamton University Interdisciplinary Collaboration Grant (2014)
12. Silva, H., Lourenco, A., Canento, F., Fred, A.L., Raposo, N.: ECG biometrics: principles and applications, January 2013. https://doi.org/10.5220/0004243202150220
13. Boyd, J.E., Little, J.J.: Biometric gait recognition. In: Tistarelli, M., Bigun, J., Grosso, E. (eds.) Biometrics School 2003, LNCS, vol. 3161, p. 1942. Springer, Heidelberg (2005)
14. Singh, N., Agrawal, A., Khan, R.A.: Voice biometric: a technology for voice based authentication. Adv. Sci. Eng. Med. **10**, 16 (2018). www.aspbs.com/asem
15. Zhu, Y., Tan, T., Wang, Y.: Biometric personal identification based on handwriting. This work is funded by research grants from the NSFC, the 863 Program and the Chinese Academy of Sciences (2000)
16. Querini, M., Gattelli, M., Gentile, V.M., Giuseppe, F.: A new system for secure handwritten signing of documents. Int. J. Comput. Sci. Appl. **12**(2), 37–56 (2015)
17. Singh, B., Sonawane, S., Shah, Y., Singh, V.: Literature survey on keystroke dynamics for user authentication. Int. J. Recent Innov. Trends Comput. Commun. **5**(5), 280–282 (2017)
18. Ellavarason, E., Guest, R., Deravi, F.: A framework for assessing factors influencing user interaction for touch-based biometrics. In: 2018 26th European Signal Processing Conference (EUSIPCO) (2018)
19. Kasprowski, P., Harezlak, K.: Fusion of eye movement and mouse dynamics for reliable behavioral biometrics. Pattern Anal. Appl. **21**, 91103 (2018). https://doi.org/10.1007/s10044-016-0568-5
20. Bhable, S.G.: A survey of security of multimodal biometric systems. Int. J. Eng. Res. Appl. **5**(12 (Part - 4)), 67–72 (2015). ISSN: 2248-9622

Blind Guidance System

A. S. Nandagopal[✉] and Ani Sunny

Department of Computer Science and Engineering,
Mar Athanasius College of Engineering, Ernakulam, India
iamnandagopalas@gmail.com

Abstract. Dazzle is a condition of helpless to see, which unfavorably impacts to mindfulness and judgment. WHO state that 36 million people far and wide exist in a visually impaired condition. Still India does not have much cutting-edge innovation and supplies to help the visually impaired for route and every day prerequisites. So as to take care of the issue, this research work has built up a visually impaired direction framework that comprises of a few modules for helping the visually impaired individuals proficiently. The structure includes hindrance location framework, object recognition framework and transport distinguishing proof framework which make vision far outwardly tested. The shrewd unit contains a deterrent location shoe with ultrasonic sensors, which help for discovering hindrance in way. Another module is RFID based transport distinguishing proof framework and item discovery is completed by utilizing an eye glass furnished with camera, an ear telephone. The microcontroller UDDO x86 is utilized in the task where processing is done. The camera present at the eye glass catches the planned picture of the client as depictions and exchange to the UDDO x86 where it gets handled and produce the predetermined sound portrayals of the object found in the previews. The set of ultrasonic sensors place in the shoe will give signs about the nearness of snags around and RFID innovation is utilized for transport recognizable proof framework to furnish the insights concerning the transport regarding its goal.

Keywords: Object recognition · Obstacle detection · Ultrasonic sensors · RFID

1 Introduction

We know in India innovation is continuously expanding today as a pattern. Yet at the same time India haven't no ongoing innovation for helping blind individuals. Out of the absolute populace of the world, 35 million individuals are visually impaired. India comprises of 14 million visually impaired individuals. Dazzle individuals think that its hard to meet their every day necessities particularly in voyaging, acquiring and so on. These will unfavorably impacts their freedom. They are dependably attempt to change their existence as for circumstance. By utilizing white sticks and guided mutts dazzle individuals get help up as far as possible, But it haves a few disadvantages. Insight lackness is the fundamental issue so it is hard to give bearings particularly in unvisited condition. A human go about as guide for visually impaired individuals and gives the insight yet it makes high conditions, so a sentiment of weight is arised [1].

© Springer Nature Switzerland AG 2020
S. Smys et al. (Eds.): ICCVBIC 2019, AISC 1108, pp. 942–949, 2020.
https://doi.org/10.1007/978-3-030-37218-7_100

People have visual hindrances that give trouble for route. They have to utilize a wide scope of types of gear to for portability. Yet, that types of gear does not give a total answer for route. Another technique utilized by the visually impaired individuals is direct mutts. It has major disservice, preparing has mind-boggling expense and hard to oversee. Daze individuals faces bunches of trouble in their day by day life. Present day society ensures all opportunity for individuals. These are given by cutting edge correspondence innovation. Route reason for visually impaired individuals offer two parts for safe way. The principal procedure incorporates quick condition identification and discovering objects present on the way. The route decides the present position status as for wanted position status alongside course of development. There are numerous routes for route. Outside signs are not utilized in route dependent on quickening [2]. The straight and rotational quickening regarding twofold combination for decide the advancement and course.

Through this, I will build up a visually impaired direction framework which comprises of a few modules for helping the visually impaired individuals effectively. The plan includes snag location framework, object recognition and transport recognizable proof framework which make vision for outwardly tested. The savvy pack contains an impediment recognition shoe furnished with ultrasonic sensors then RFID based transport ID framework and an eye glass furnished with camera, an ear telephone and UDDO x86 where handling is done. The camera present at the eye glass catches the proposed picture of the client as previews and exchange to the UDDO x86 where it gets handled and produce the predefined sound depictions of the article found in the previews. The set of ultrasonic sensors place in the shoe will give signs about the nearness of obstructions around and RFID innovation is utilized for transport recognizable proof framework to furnish the insights regarding the transport as for its goal.

2 Literature Survey

There are numerous gadgets present for giving frameworks benefits that assistance dazzle individuals in their day by day lives. The primary assistive innovation was built up in 1961. It makes answer for day by day issues faces by the visually impaired. It intends to give route orientated guides to individual consideration.

2.1 ETA-Electronic Travel Aids

Estimated time of arrival gadgets works by tunes in to the encompassing improvement and accumulate insights regarding the relative condition. The subtleties given to the client by utilizing computerized camera, Various sensors, sonar, scanners and so on [3, 4]. The working of ETA is fill in as pursues:

- Obstacles assurance as for the earth
- Given guidance about the development to client
- Determine object around the obstructions
- Calculation of separation of hindrance from client
- Details accommodate self-introduction

2.2   EOA-Electronic Orientation Aids

EOA offer bearings to clients on obscure spots [5]. The EOA functioning as pursues:

- Shortest course to way is characterized
- User area is determined
- Displaying directions for portability

2.3   PLD-Position Locator Devices

PLD capacity to discover the exact position of its holder. The gadget utilized GPS innovation for finding exact position. It gives basic administrations for visually impaired individuals. The basic administrations are identification of deterrents and decide most brief way by utilizing GPS [6].

3   Methodology

3.1   The Components Used in Hardware and Software

- UDDO X86
- Arduino board
- Camera
- Ultrasonic Sensors
- Earphone
- Membrane keypad
- Battery
- RFID

4   Proposed System

To introduce a wearable gadget to help outwardly disabled clients for achieve goal safely. The model is ultrasonic sensors implanted on shoe by different points and inserting a high goals camera on specs, Using RFID will assist us with finding the goal transport accessibility. All the handling is finished by Arduino and UDOO X86 chip. The proposed framework engineering comprises of predominantly six segments. They are camera, UDDO X86 and Arduino board for doing over all preparing, speaker for delivering sound flag and ultrasonic sensor are settled on shoe on different plots for deterrent discovery and utilizing RFID assist us with finding goal place.

4.1   Obstacle Detection

A gadget is to create for outwardly debilitated individuals for safe route. It points is keep risk from snag. It comprises of ultrasonic sensors, Arduino board, battery and headphone. The sensor ultrasonic is utilized to discover the hindrances in the way. Ultrasonic sensor is send a flag to the way then when any hindrance found on way, the flag is reflected back. Along these lines, nearness of deterrent present on way is found and it are recognized by utilizing ear telephones. The deterrents present from floor to face level impediment are distinguished very effectively by setting ultrasonic sensors at 60° edges. It produce the sound to demonstrate the individual that diversion will be in their direction. The extent of item is to make simplicity of route for the visually impaired individuals and outwardly impeded who can't discover their way without utilization of an express device or some different people help. The thought is to utilize every day use attire that is, a shoes to direct the client to his/her goal with portrayal of deterrents in his/her way [7] (Fig. 1).

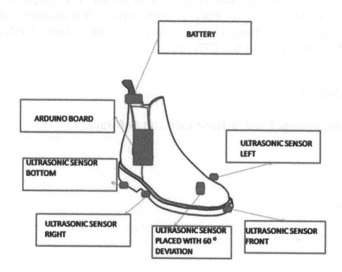

Fig. 1. Proposed obstacle detection architecture

The deterrent location is given by shoe without utilizing some other outside gear. This shoe is an incredible advancement for helping the visually impaired individuals for route. A sound gadget is available for giving guidance for safe route. The application planned by utilizing Google Maps interface for continuous route. This item starts and performs following real functionalities (Fig. 2):

- For route inside call the Google API
- Voice directions for perusing and coordinating the ideal ways
- Detection and arrangement of the items close to the client's condition.
- Intuit the client through voice directions for the current position of the client and the hindrance which is in its region alongside its sort.

Continuous preparing might be the most basic execution of the proposed framework. The product ought to have a capacity to do constant handling of the hindrances distinguished by the sensors. The important identification, preparing and exchange of these subtleties to the device by means of Bluetooth in least required time, is basic. For compelling utilization of battery life of the batteries present in the shoes, it is essential that the calculations be kept to a base. The content to discourse change, to advise the

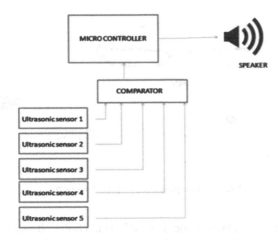

Fig. 2. Obstacle detection sensors arrangement

client of any conceivable impediments, must be done in least conceivable time. No deferral can be endured in this as the whole handling time spared will be rendered futile if the content to discourse is done insufficiently. The accompanying advances:

- Read users destination input
- Pass the input to Google Maps API
- Start the navigation
- If Ultrasonic sensor returns true then
  - Calculate the distance of obstacle
  - Return the obstacle detection with direction as voice
- Else move according to the Google Map API
- Repeat till running
- Stop

4.2 Object Detection

Recognizable proof of item can without much of a stretch identify by people. The human can perform complex undertaking in quick and precise way. In the event that they are such a large number of information are available, little gadget like raspberry pi have not capacities for preparing the information. So instead of raspberry pi, UDDO x86 is utilized. UDDO x86 has productive preparing necessity are available. In article recognition, You just look once (YOLO) is the calculation is utilized. In article discovery, any item picture is caught then it look at the article picture in database and state the word through headphone. The proposed article recognition is appeared in Fig. 3.

YOLO [9] is exceptionally straightforward. A system which is convolutionally in nature is utilized for predicts jumping encloses which is various number and boxes class probabilities. Preparing is finished utilizing entire pictures and discovery execution is improved specifically [10]. Complex pipeline isn't required here. Edge

Fig. 3. The proposed object detection

discovery is bargain as a relapse issue. For anticipate choice, It essentially run neural system on picture. The 46 outlines for every second is handled without group preparing. The most recent quick form keeps running at in excess of 150 fps [11].

4.3 Bus Identification System

Transport distinguishing proof framework will helps the visually impaired for transportation. It increment the personal satisfaction of visually impaired individuals. The procedure utilized for transport ID framework is by remote sensors system and Radio Frequency recognizable proof (RFID) [12]. By utilizing sound gadget and RFID Reader help dazzle in the territory of transportation. RFID tag is set in transport and send flag to RFID reader present in the users visually impaired gadget. The handling of flag occur in inward memory where it contain every one of the subtleties of transport data [13].

4.4 Docker Image and Quantization

A model is quantized it implies it tends to be fit into any little gadget by making littler. Quantized Model methods make it littler to fit on a little gadget. Rasberry PI like little gadget has constrained intensity of calculation and little memory is existed. Neural system preparing is happen by giving numerous little pushes to the loads, and utilizing these drifting point exactness to work in little additions. Running interface of pre-prepared model is distinctive in nature. The commotion present in system can without much of a stretch adapt by utilizing profound neural system. For preparing neural system utilized and it heavily expend plate space. For instance, 300 MB drifting point group is utilized by AlexNet. The neural associations in neural system has take distinctive size. In 32 bit coasting point, hubs and loads in edges of neural system is put away. By busing idea of quantization, can diminish the record measure by contracting and putting away it in min and max for each layer, at that point pressure occur by believer glide an incentive to 8 bit whole number. The extent of record is decreased up to 80%. Docker image produce a simple way of training. process of training a model is unnecessarily difficult to simplify the process, created a docker image would make it easy to train (Fig. 4).

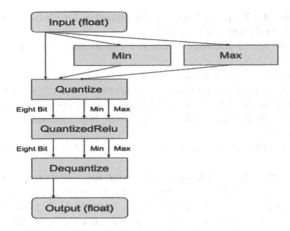

Fig. 4. Quantization model

5 Conclusion

Proposed framework is effectively planned and manufactured. In India, number of visually impaired individuals are more than anyplace on the planet still India does not have much trend setting innovation and types of gear to help the visually impaired for route and day by day necessities. So as to take care of the issue I will build up a visually impaired direction framework which comprises of a few modules for helping the visually impaired individuals effectively. The structure includes obstruction discovery framework, object location framework and transport recognizable proof framework which make vision for outwardly tested. Utilizing hindrance discovery, deterrent present in way is effectively discovered, object recognition utilized for discovering object lastly transport distinguishing proof framework utilized for discovering transports in precise time for visually impaired.

Compliance with Ethical Standards

✓ All authors declare that there is no conflict of interest.
✓ No humans/animals involved in this research work.
✓ We have used our own data.

References

1. Utaminingrum, F., et al.: A laser-vision based obstacle detection and distance estimation for smart wheelchair navigation. In: 2016 IEEE International Conference on Signal and Image Processing, ICSIP, pp. 123–127 (2016)
2. Nguyen, V.D., Nguyen, H.V., Tran, D.T., et al.: Learning framework for robust obstacle detection recognition and tracking. IEEE Trans. Intell. Transp. Syst. (99), 1–14
3. Kun (Linda), L.: Electronic Travel Aids for Blind Guidance An Industry Landscape Study (2015)

4. Yang, Y., Wang, W., Zeng, L., Chen, J.: An algorithm for obstacle detection based on YOLO and light filed camera. IEEE (2018)
5. Whitney, G.: The use of electronic orientation and mobility aids by blind and partially sighted people
6. Vehicle tracking system. www.Wikipedia.com
7. Nguyen, T.-H.: Obstacle detection and warning system for visually impaired people (2016)
8. Asif, U., Bennamoun, M., Sohel, F.A.: RGB-D object recognition and grasp detection using cascaded forests. IEEE Trans. Robot. (99), 1–18. https://ieeexplore.ieee.org/document/8603600/Hierarchical
9. Zhang, R.: An algorithm of obstacle detection based on YOLO and light filed camera. IEEE (2018)
10. Corovic, A.: Real time detection of traffic participants using YOLO algorithm. IEEE (2018)
11. Bishop, T.E., Favaro, P.: The light field camera: extended depth of field aliasing and superresolution. IEEE Trans. Pattern Anal. Mach. Intell. **34**(5), 972–986 (2011)
12. Hannan, M.A.: UKM campus bus identification and monitoring using RFID and GIS. IEEE (2009)
13. Fu, Y.: A new RFID-USB key. IEEE (2007)

A Novel Multimodal Biometrics System with Fingerprint and Gait Recognition Traits Using Contourlet Derivative Weighted Rank Fusion

R. Vinothkanna[1(✉)] and P. K. Sasikumar[2]

[1] Department of ECE, Vivevakanandha College of Engineering for Women,
Tiruchengode, India
rvinothkannaphd@gmail.com
[2] Department of CSE, Vivekanandha College of Technology for Women,
Namakkal, India

Abstract. We propose a novel multimodal biometric system that uses fingerprint and gait recognition traits. Fingerprint images are gathered with the help of a touchless optical sensor with total internal reflection and a capacitive line sensor. Gait samples are obtained as continuous images in 3 frames from a camera in burst mode. Features from these images are extracted with the help of Minutia Cylinder-Code (MCC) minutia descriptor. Further, the images are compared with the gathered sample database of 80 sample images. The comparison score is normalized using geometric mean. Fusion of the normalized images is done using contourlet derivative weighted rank fusion. A combination of these techniques provides an improved performance for authentication using biometrics.

Keywords: Biometric system · Minutia descriptor · Geometric mean · Rank fusion

1 Introduction

Human behaviour can be characterized as biometric that satisfies criteria such as uniqueness, performance, permanence, circumvention, acceptability, collectability and universality [3]. Uniqueness implies that the characteristic must be distinct for every individual. Performance refers to the accuracy of identification of the trait in varied environmental condition. Permanence refers to the invariance of the quality with time. Circumvention is the reliability of the system when exposed to counterfeit methods. Acceptability signifies the level of acceptance of the system by individuals. Collectability indicates the obtainability of the characteristic and quantitative measurement. Universality denotes that the characteristic should be present in every individual.

Biometric systems used for authorization and authentication provides multiple benefits such as robustness, distinctiveness, availability and accessibility. A uni-modal biometric system [6] is based on a single characteristic such as face recognition, fingerprint, hand geometry, keystroke dynamics, hand veins, gait, iris, retina, signature, voice, facial recognition and DNA. But, uni-modal biometric system have several limitations such as lack of individuality, noisy sensor data, and lack of invariant

© Springer Nature Switzerland AG 2020
S. Smys et al. (Eds.): ICCVBIC 2019, AISC 1108, pp. 950–963, 2020.
https://doi.org/10.1007/978-3-030-37218-7_101

Table 1. Assessment of biometric characteristics

Characteristics	Fingerprint	GAIT
Uniqueness	High	Low
Performance	High	Low
Permanence	High	Low
Circumvention	High	Medium
Acceptability	Medium	Medium
Collectability	Medium	High
Universality	Medium	Medium

representation, non-universality and susceptibility to circumvention. These drawbacks can be resolved by using multimodal biometric systems. A multimodal biometric system [7] relies on multiple characteristics to provide access. It combines the modalities of two biometrics and wrap it into a single system to make unified decision for authentication. Fingerprint and GAIT characteristics and its comparison is formulated in Table 1.

Unlike passwords, patterns, tokens, PINs and so on, biometrics can never be forgotten or stolen [10]. An explicit link is established between the identity of the person and biometric authentication as characteristics used are based on human physiology and behaviour. Various active research is being performed on fingerprint recognition to improve quality and performance [11]. It is also used widely in several commercial applications. Gait recognition is often used in surveillance system and accelerometer based systems [12]. Gait is distinct for every individual. However, performance of gait recognition is not as accurate as that of fingerprint recognition.

Physical biometric signals are represented as gait. Several researches are made in improving the gait recognition systems. Various environmental factors, outdoor settings, illumination, occlusion, posture, steep and so on affect the output of the gait recognition systems. Multimodal fusion [16] is an optimal solution that helps in overcoming the vulnerabilities of the limitations offered by gait recognition system. We present a multimodal biometric system in this paper that works on fingerprint and gait recognition for authentication and authorization. The main focus lies on reducing the user effort and error. Both the modalities are dealt with in different settings and finally merged together as the images are processed for identification of the individual. The multimodal system depends on the availability of multiple confirmations that can match the trait to the person's identity.

2   Literature Review

An extensive study of the biometric attributes, their fusion levels, approaches and implementation techniques is done and the data is analyzed based on the number of test samples and error rate. It is evident that merging at the feature vector level improves the authenticity of the results. But the implementation comes with certain difficulties due to the incompatibility of the attributes. Uni-modal system [14] can overcome certain drawbacks of multimodal biometrics like universality, intra class variations, noise and lack of uniqueness and so on, reduce spoofing and failure to capture over failure to enrol rates. Palm print, hand geometry and fingerprint traits are analysed. Fusion is done at feature level in order to combine the data from the obtained traits. Cancellable biometrics concept improves the system performance by providing diversity, reusability and non-invertibility.

The biometric trait studied in this paper [15] is the geometry and structure of footprints. Among several commonly used techniques such as Binarization, Principal Component Analysis (PCA), Hidden Markov Model (HMM), Centre of Pressure (COP), SOM, ICA, ROI and so on, PCA is reviewed and found to provide optimal results and better recognition rate. Footprint biometric is advantageous for use in infants and people with congenital disability. Future scope suggests that this trait can reveal information such as height, weight, gender, age, and individuality trait and health status of the individual (Fig. 1).

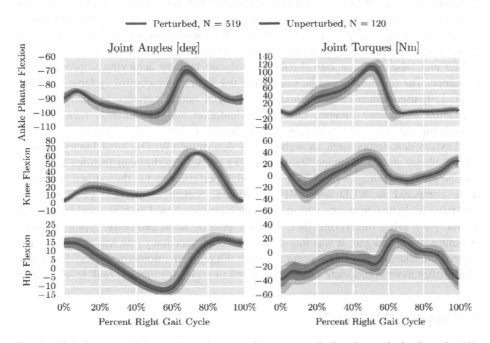

Fig. 1. Right leg mean, joint angles and torques from unperturbed and perturbed gait cycles [4]

Numerous research publications have conversed over the employment of gait recognition system. Gait proves to be advantageous while dealing with real time surveillance with large number of video frames [17]. Classification of the 2D spatiotemporal template of gait is done using neural network. The rich perceptual information from the video is used to detect the human body, segment the parts, identify the viewing angle, align, register, extract the feature and classify the image. The paper [18] proposes Gait Energy Image (GEI) for spatiotemporal representation of gait in order to characterize the walking properties of individuals. The silhouette sequence is trained to compute the original templates and silhouette distortion is simulated by generating synthetic templates from the original training sequences.

A solution to overcome the issue of gait recognition during night time with the assistance of thermal infrared imagery is presented in [19]. It uses a combination of Gaussian mixture modelling and extraction of human silhouettes along with head-torso-thigh (HTI) characterisation. This enables ease of use of biometrics while dealing with low-contrast infrared images. Vera et al. [20] proposed the linking of gait and footstep biometrics extracted from video sequence and floor sensors respectively. The human body parts are distinctly identified as individual components thereby improving the recognition rate. He et al. [31] used adaptive Gabor filter for every modality individually and further projected to a vector regression model that is supported and trained in order to make the final decision.

A secure and efficient fingerprint biometric system for staff attendance in a tertiary institution is developed in [21] that can help eradicating fake records and improve data integration. A new algorithm for a biometric system that works on recognition of finger vein pattern is developed by [22]. Singular Value Decomposition (SVD) and random transform are used for feature extraction and classification. The false acceptance rate (FAR) and false rejection rate (FRR) are reduced thereby achieving improved performance. The technology can be put to best use in entrance control systems. Ibrahim et al. [30] used Eigen-faces method and artificial neural network to recognize the individual using face images. Distance, orientation and illumination factors were considered and the system was proposed for access control of office door.

Texture descriptors such as weber local descriptor (WLD), local binary pattern (LBP), binary statistical image features (BSIF) extract discriminant features from the ear and palm-print images [23]. BSIF offers high performance but the extraction time is more compared to WLD and LBP descriptors. Sarhan et al. [24] performed an extensive study by comparing various biometric systems based on biological traits such as iris, palm print, fingerprint, face, vascular pattern, hand geometry, typing pattern, signature, voice and gait. It was found that the highest accuracy rate was achieved with fingerprint analysis and least accuracy with finger vein biometrics. Decision level fusion offer more appropriate results compared to score, feature and rank level fusion.

3   Proposed Method

Figure 2 shows the block diagram of the proposed multimodal biometric system that uses two distinct hardware for capturing fingerprint as well as gait based images. The features are extracted from the images individually using MCC minutia descriptor and compared with the pre-collected sample database of 80 images. Normalization of the images are done using geometric mean method and fusion by weighted rank fusion technique. Further, the decision is made based on authentication and accessibility the success and error is analysed based on this output (Fig. 2).

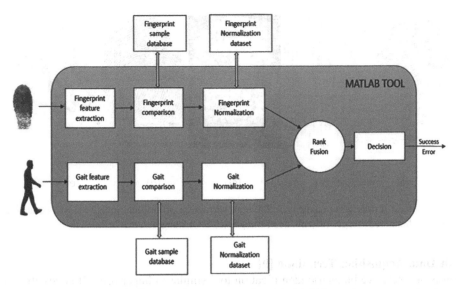

Fig. 2. Block diagram of the proposed multimodal biometric system

Fingerprint Data Acquisition Technique [8]

One of the most commonly used and reliable personal authentication tool is biometric fingerprint. This technique has a very less likely possibility of being attacked ore replicated by intruders. This biometric trait is authentic as it is unique for every individual. It is more safe and reliable than using other verification tools like PIN, token or password. In order to improve the matching score, the images are enhanced using image processing techniques. The most suitable image is used for identification and feature extraction for authentication.

The fingerprint data is captured with the help of a commercial sensor IDEX SmartFinger IX 10-4 with a resolution of 500 DPI and 8-bit grey scale. It offers image acquisition size of 10 × 4 mm and can operate at temperature ranges between −40 to 85 °C. The device dimensions are 10 × 4 × 0.8. It offers highly proficient fingerprint verification with a compact, portable and cost efficient touch sensor. It has optical sensor with total internal reflection along with capacitive line sensor [29].

The optical sensors consists of LED light source and a charge coupled device that is positioned on the side of the glass on which fingerprint image s acquired. The CCD captures the fingerprint when LED is illuminated. This helps in easy capture of the ridges and valleys in the fingerprint. The capacitive sensor consists of capacitor plates that form a 2D array at the sensing surface. Based on the distance between the fingerprint and the plates, the capacitance is formed and images are captured. Factors such as core, cross over, ridge ending, bifurcation, island, pore and delta in the fingerprint are recognized. The captured fingerprint images are converted into digital format for processing (Fig. 3).

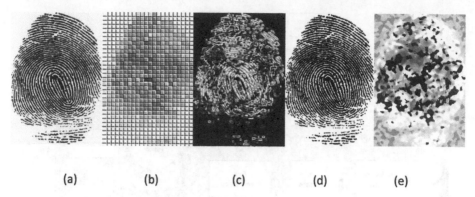

(a)   (b)   (c)   (d)   (e)

Fig. 3. Stages of fingerprint analysis and feature extraction (a) original image [4], (b) equalized image, (c) binarized image, (d) enhanced image, (e) ROI image

Gait Data Acquisition Technique [9]

Gait is an exclusive biometric identification tool similar to finger print. It is possible to identify individuals and detect abnormalities with this information. Gait can be termed as a signature of human walking as it enables identification of individuals based on their walking characteristics. Gait is movement with the active engagement of almost every body part. It involves the coordination of the motor as well as sensory system of the human body. Activities such as walking, jumping, jogging and running can be also be briefed as gait.

The walking manner of the 20 participants is recorded with the help of a camera in burst mode. Three frames of images are captured per person. With the assistance of the gait, it is possible to identify the person from a distance of about 165 feet [5]. This feature makes it effective and significant in terms of security in video surveillance and crime scenes. Usually, the gait frequency is less than 8 Hz. Hence, a low-pass filter can be used to access the gait information. Information such as height, step size, step regularity, acceleration and time can be approximated from these images. These data vary from person to person and also based on varied environment. Smartphones can also be used to assess gait in both indoor as well as outdoor environment.

In real time implementation, we propose the use of a surveillance camera with a standard resolution. By selecting efficient characteristics that can be computationally analysed based on the silhouette of the subject and time variation, features are extracted from the video frames. Markov thresholding can be used in iterative mode to eliminate the background noise and binarize them for further cropping and scaling (Fig. 4).

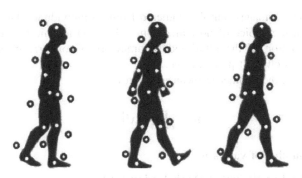

Fig. 4. Gait recognition system [4]

Minutia Cylinder-Code (MCC) Minutia Descriptors [1]

In this technique, we use a predefined step of fixed radius to rotate the main minutia. With each rotation, the cell-discretized cylinder is fused with a slice offering a predefined number of cells. The numerical value per cell is calculated by gathering the neighbouring minutiae contributions. The number of cells along the cylinder diameter and slice are given by N_S and N_D respectively. The length of the vector of each MCC descriptor is given by $N_S \times N_S \times N_D$. MCC has been engaged successfully in several dormant fingerprint identification research. It is found that MCC offers better identification rate when compared to other minutia descriptors such as m-triplets and neighbouring minutia-based descriptors.

MCC [25] depends on an invariant coding of fixed length for every minutia. Hence it enables easy local structure computation. The coding and matching allows managing issues with borders. Smoothened functions provides error tolerance in occurrence of local distortions. An XOR operation is performed individually on the extracted minutia feature blocks. MCC descriptor improves the speed and efficiency of the matching process. Few of the several advantages of the MCC descriptor are invariance to rotation and translation and robustness to small distortions in the image. The individual local structures can be stored in binary format as they have fixed feature size. MCC does not have a discriminative representation as it is redundant and high dimensional.

Normalization Using Geometric Mean [13]

The uneven intensity across the images are adjusted to produce a uniform image by means of normalization. It allows coverage of grey level and adjusting the intensity within an expected range without affecting the clarity. The brightness and contrast of the image are also adjusted with the help of histogram equalization. Geometric mean is used for normalization by removing Gaussian noise [28] and preserving the edge features. It helps in a smooth transition between the distributions in a framework of simulated annealing. In the recent days, geometric mean is emerging in multiple neural networks as the distribution of prediction average.

The matching scores that are produced from gait and fingerprint images are not homogenous. Hence, the need for data normalization arises. The numerical scores are then transformed into a common range and domain. The weighted sum and simple sum rules are used to perform the fusion of the fingerprint and gait modalities. We conduct the experiment on 80 images samples gathered from 20 individuals. The experimental results shows that the choice of normalization and fusion technique decides the successful implementation of the multimodal framework. Geometric mean based normalization provides efficient performance.

For a sample X, the geometric mean is given by

$$m = \left[\prod_{i=1}^{n} xi\right]^{\frac{1}{n}} \tag{1}$$

Where n is the number of values in X.

Contourlet Derivative Weighted Rank Fusion [26]

Rank level fusion consolidates the results of identification and allows reliable personal identification. The biometric sample character is determined from the highest ranks that are reimbursed by the distinct biometric in the contourlet derivative weighted rank fusion. The fusion template matching is performed by the normalized images that are obtained by the geometric mean. This improves the recognition rate of the multimodal biometric recognition system. This system improves the fingerprint as well as gait recognition rate. In case of gait imitation, the traditional fusion methods lag in performance. The proposed technique obtains maximum simplification and improved performance. Fusion can be performed at feature level as well as score level. Feature level fusion requires more processing time. Rank level fusion performs better than score level fusion and consumes lesser processing time compared to feature level fusion.

The weighted rank level fusion [27] works on extracting features from both fingerprint and gait images and fusing them. The fused data is stored in a database and matched with the images already available. Based on the type of image fed to the system, the fused rank is described. There is inconsistency while interpreting the multimodal biometric samples. For instance, the interpretation output is more reliable for fingerprint images rather than gait based images. Based on the expected outputs, weights are allocated to the corresponding images and the individual biometric sample ranks are produced and analysed.

The fused rank scores in weighted can be computed using the following formula

$$m_k = \sum_{i=1}^{M} w_i r_i(k) \tag{2}$$

Where M is the count of biometric matchers (subsystem), w_i is the corresponding weighs that are assigned to the ith matcher and $r_i(k)$ is the rank assigned to user k. A matcher generates a marching score by comparing the templates containing biometric samples that are stored during enrolment with the samples extracted during operation.

Decision Making

Since the sample dataset is limited, simple matching algorithm is used for decision making regarding authentication. Along with identification, data such as age, gender, height, subject details are also provided in a user friendly interface. In case of large samples or real time application, deep learning algorithm can be used due to its vast scope. Several training samples can be included and the database can be maintained that would allow easy matching and access of information. The modalities and their features such as data depth, RGB and so on are analysed and fused.

4 Experimental Results

We conducted an analysis on 60 samples of gait (3 per person) and 20 fingerprint samples (1 per person) collected from 20 subjects to create a test database of 80 images. MATLAB simulation environment is used for processing the images. The height and age based analysis of the subjects are represented graphically as shown in Fig. 5(a) and (b).

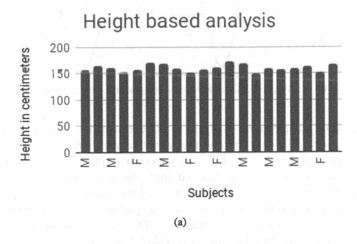

(a)

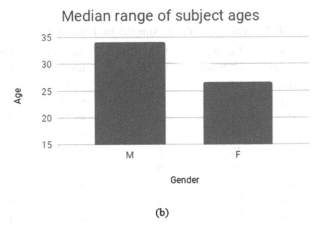

(b)

Fig. 5. Analysis of subject data based on (a) Height and (b) Age

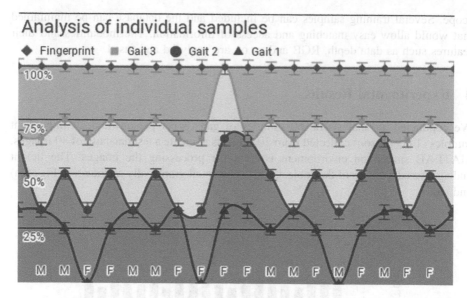

Fig. 6. Analysis of individual sample images and their authenticity

The individual sample result analysis is as represented in Fig. 6 where 3 gait samples from every individual and one fingerprint sample is represented. The 3 gait samples are represented as Δ, O and □. The fingerprint samples are represented as ◊. The level of success and error is deduced from the graphical representation. The combination of the weights gives a greatly reduced percentage of error.

Weights are applied to the fingerprint as well as gait images in a range between 0.1 –0.9 such that the sum of weights is equal to 1. The success rate and error from the samples is tabulated and plotted as shown in Table 2 and Fig. 7.

Table 2. Trial results the multimodal biometric system

Fingerprint weight	Gait weight	Samples identified	Unidentified samples	Success rate in %	Error rate in %
0.9	0.1	78	2	97.5	2.5
0.8	0.2	77	3	96.25	3.75
0.7	0.3	77	3	96.25	3.75
0.6	0.4	76	4	95	5
0.5	0.5	75	5	93.75	6.25
0.4	0.6	75	5	93.75	6.25
0.3	0.7	74	6	92.5	7.5
0.2	0.8	73	7	91.25	8.75
0.1	0.9	70	10	87.5	12.5

Samples identified and Success Rate from a database of 80 samples

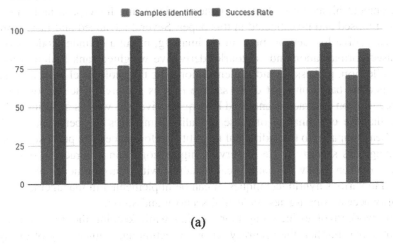

(a)

Unidentified Samples and Error Rate from a database of 80 samples

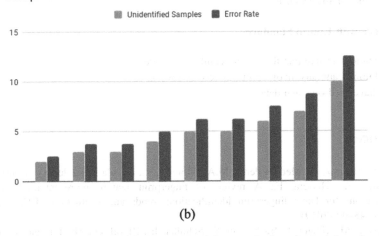

(b)

Fig. 7. (a) Success rate (b) Error rate of the samples based on trial result

From this result, we conclude that when there is less weight on gait recognition, we get an acceptable rate of success, whereas, while increasing the gait recognition weight, the success rate decreases. Hence to obtain optimum output from this system, proper matching of the modalities is to be done.

5   Conclusion and Future Work

The multimodal biometric system for authentication using fingerprint and gait recognition is proposed and investigated in this paper. Several advanced and fast techniques such as capacitive line sensor, burst mode imaging, minutia cylinder code, geometric mean based normalization and contourlet derivative weighted rank fusion is used for sample collection, processing and authentication. The threshold level is set and security analysis is done based on a set of 80 sample images gathered. The proposed method and its recognition accuracy is analysed both qualitatively as well as quantitatively. It is deduced that the combination of these modalities improves the performance significantly when compared to the individual modality performance. The proposed biometric system heightens security while preserving high recognition accuracy. It offers better performance when used with a small-scale database with minor influence of noise when compared to various hybrid techniques. It can be implemented in management systems for identity access, smart-gates, industrial security and so on.

Future work involves including more modals while keeping the dimensions of the system minimal. Further, the accuracy, error correction and authenticity of the system are to be improved significantly. Speed of recognition is also to be improved despite the increase in the number of images in the database and the number of modals used for comparison. High resolution datasets can be stored in a cloud database to further reduce the bulkiness of the system.

Compliance with Ethical Standards

✓ All authors declare that there is no conflict of interest.
✓ No humans/animals involved in this research work.
✓ We have used our own data.

References

1. Valdes-Ramirez, D., Medina-Pérez, M.A., Monroy, R., Loyola-González, O., Rodríguez, J., Morales, A., Herrera, F.: A review of fingerprint feature representations and their applications for latent fingerprint identification: trends and evaluation. IEEE Access **7**, 48484–48499 (2019)
2. Hofmann, M., Geiger, J., Bachmann, S., Schuller, B., Rigoll, G.: The tum gait from audio, image and depth (gaid) database: multimodal recognition of subjects and traits. J. Vis. Commun. Image Represent. **25**(1), 195–206 (2014)
3. Soltane, M., Bakhti, M.: Multi-modal biometric authentications: concept issues and applications strategies. Int. J. Adv. Sci. Technol. **48** (2012)
4. Moore, J.K., Hnat, S.K., van den Bogert, A.J.: An elaborate data set on human gait and the effect of mechanical perturbations. PeerJ **3**, e918 (2015)
5. Little, J., Boyd, J.: Recognizing people by their gait: the shape of motion. Videre J. Comput. Vis. Res. **1**(2), 1–32 (1998)
6. Oloyede, M.O., Hancke, G.P.: Unimodal and multimodal biometric sensing systems: a review. IEEE Access **4**, 7532–7555 (2016)

7. Ghayoumi, M.: A review of multimodal biometric systems: fusion methods and their applications. In: 2015 IEEE/ACIS 14th International Conference on Computer and Information Science (ICIS), pp. 131–136. IEEE (2015)
8. Hassan, H., Kim, H.-W.: CMOS capacitive fingerprint sensor based on differential sensing circuit with noise cancellation. Sensors 18(7), 2200 (2018)
9. Chen, X., Yang, T., Xu, J.: Multi-gait identification based on multilinear analysis and multi-target tracking. Multimedia Tools Appl. 75(11), 6505–6532 (2016)
10. Sadikoglu, F., Uzelaltinbulat, S.: Biometric retina identification based on neural network. Procedia Comput. Sci. 102, 26–33 (2016)
11. Marasco, E., Ross, A.: A survey on antispoofing schemes for fingerprint recognition systems. ACM Comput. Surv. (CSUR) 47(2), 28 (2015)
12. Chellappa, R., Veeraraghavan, A., Ramanathan, N.: Gait biometrics, overview. In: Encyclopedia of Biometrics, pp. 783–789 (2015)
13. Mansoura, L., Noureddine, A., Assas, O., Yassine, A.: Multimodal face and iris recognition with adaptive score normalization using several comparative methods. Indian J. Sci. Technol. 12, 7 (2019)
14. Kaur, G., Yadav, A.K., Chaudhary, S.: An improved approach to multibiometrics security. Int. J. Comput. Sci. Commun. 5(1), 181–187 (2014)
15. Gupta, A., Raj, D.: Multimodal biometrics for personal identification using human foot prints: a review (2019). SSRN 3350259
16. Jagadiswary, D., Saraswady, D.: Biometric authentication using fused multimodal biometric. Procedia Comput. Sci. 85, 109–116 (2016)
17. Alotaibi, M., Mahmood, A.: Automatic real time gait recognition based on spatiotemporal templates. In: 2015 Long Island Systems, Applications and Technology, pp. 1–5. IEEE (2015)
18. Han, J., Bhanu, B.: Individual recognition using gait energy image. IEEE Trans. Pattern Anal. Mach. Intell. 28(2), 316–322 (2005)
19. Tan, D., Huang, K., Yu, S., Tan, T.: Efficient night gait recognition based on template matching. In: 18th International Conference on Pattern Recognition (ICPR 2006), vol. 3, pp. 1000–1003. IEEE (2006)
20. Vera-Rodriguez, R., Fierrez, J., Mason, J.S.D., Orteua-Garcia, J.: A novel approach of gait recognition through fusion with footstep information. In: 2013 International Conference on Biometrics (ICB), pp. 1–6. IEEE (2013)
21. Adewole, K.S., Abdulsalam, S.O., Babatunde, R.S., Shittu, T.M., Oloyede, M.O.: Development of fingerprint biometric attendance system for non-academic staff in a tertiary institution. Development 5(2), 62–70 (2014)
22. Wang, D., Li, J., Memik, G.: User identification based on finger-vein patterns for consumer electronics devices. IEEE Trans. Consum. Electron. 56(2), 799–804 (2010)
23. Hezil, N., Boukrouche, A.: Multimodal biometric recognition using human ear and palmprint. IET Biometrics 6(5), 351–359 (2017)
24. Sarhan, S., Alhassan, S., Elmougy, S.: Multimodal biometric systems: a comparative study. Arab. J. Sci. Eng. 42(2), 443–457 (2017)
25. Cappelli, R., Ferrara, M., Maltoni, D.: Minutia cylinder-code: a new representation and matching technique for fingerprint recognition. IEEE Trans. Pattern Anal. Mach. Intell. 32 (12), 2128–2141 (2010)
26. Gunasekaran, K., Raja, J., Pitchai, R.: Deep multimodal biometric recognition using contourlet derivative weighted rank fusion with human face, fingerprint and iris images. Automatika 1–13 (2019)
27. Monwar, M.M., Gavrilova, M.L.: Multimodal biometric system using rank-level fusion approach. IEEE Trans. Syst. Man Cybern. Part B (Cybern.) 39(4), 867–878 (2009)

28. Zhang, T., Li, X., Tao, D., Yang, J.: Multimodal biometrics using geometry preserving projections. Pattern Recogn. **41**(3), 805–813 (2008)
29. Marcialis, G.L., Roli, F.: Fingerprint verification by fusion of optical and capacitive sensors. Pattern Recogn. Lett. **25**(11), 1315–1322 (2004)
30. Ibrahim, R., Zin, Z.M.: Study of automated face recognition system for office door access control application. In: 2011 IEEE 3rd International Conference on Communication Software and Networks, pp. 132–136. IEEE (2011)
31. He, F., Liu, Y., Zhu, X., Huang, C., Han, Y., Chen, Y.: Score level fusion scheme based on adaptive local Gabor features for face-iris-fingerprint multimodal biometric. J. Electron. Imaging **23**(3), 033019 (2014)

Implementation of Real-Time Skin Segmentation Based on K-Means Clustering Method

Souranil De$^{(\boxtimes)}$, Soumik Rakshit, Abhik Biswas, Srinjoy Saha,
and Sujoy Datta

School of Computer Engineering, KIIT University,
Bhubaneswar 751024, Odisha, India
{1605237,1605235,1605168,1605238,
sdattafcs}@kiit.ac.in

Abstract. Localization and detection of body parts (in this case, hands) is an exigent issue in image processing, since it is a prerequisite for applications like hand gesture recognition. Through this paper, a proposal for implementing an efficient way for segmenting the skin tone is developed by applying appropriate Computer Vision techniques and K-means Clustering on each frame captured by the camera under different illumination conditions as well as in the complex backgrounds, is developed. In this method, detection of skin portions using color identification, and segmentation from the image is done for its implementation in real-time systems.

While the paper focuses on using skin segmentation for the detection of hand movements, the application of this approach can easily be implemented in various applications involving Human Computer Interaction (HCI). Examples of such include mouse cursor movement, media player application, writing text on electrical documents, controlling robot, detects the pointing location, sign language, hand posture and face recognition.

Keywords: Computer Vision · Skin segmentation · K-means clustering · Hand detection · Binary mask

1 Introduction

A process for creating an efficient means for detecting human skin has always been a prerequisite for certain image processing practices, such as face detection and tracking, gesture recognition and interpretation, de-identification, privacy-protection and other human computer interaction domains.

Detection of human skin [10] has been a challenging task because many factors affect skin appearance in images. These factors include illuminating conditions, camera capabilities, ethnicity, background characteristics, etc. There are three main problems when designing a skin color based method. These are:

(1) *What color space to choose?*
(2) *How to model skin and non-skin pixel distribution?*

© Springer Nature Switzerland AG 2020
S. Smys et al. (Eds.): ICCVBIC 2019, AISC 1108, pp. 964–973, 2020.
https://doi.org/10.1007/978-3-030-37218-7_102

(3) *How to classify the modeled distribution?*

These problems are the primary factors of motivation to develop a more robust technique for skin segmentation. In this paper, an attempt is made to a region based method for skin detection based on skin color information. The primary steps for detection of skin in this method include−

(1) *Converting the image to grayscale by separation of Red color from Red Blue Green (RGB) colorspace*
(2) *Using K-Means clustering for unsupervised classification with new data set for pre-processing.*
(3) *Using binarization technique to define the threshold value of the skin and detect the edges.*

2   Related Works

While the skin segmentation methods [6, 8, 9] vary from methods based on manipulation of color-space channels to more sophisticated statistical modeling and machine learning methods [1], the former have been the most common methods in literature, and they are in general considered as computationally effective.

With the help of K-Means clustering [3, 13], clustering of similar pixels is possible [5], with an equal cardinality constraint, hence refining the image to get a better output. The proposed method is fast and efficient when compared to the existing state-of-art segmentation methods for skin detection. As there is no requirement of pre-processing a large training dataset, the method has a low computational cost. Upon implementation, the results have shown that changes in illumination conditions and viewing environments do not affect the quality of skin detection. The method does not have any limitation on choice of color or ethnicity and works modestly on complex backgrounds.

3   Proposed Technique

In order to obtain the gestures on taking images from each frame of the video captured from the webcam, some measures have been taken to improvise the technique of capturing these images and feeding it to the model for analysis.

The proposed method (shown in Fig. 1) is used to preprocess the images and produce consistent results across the dataset. This method consists of following steps -

(1) *Read the image in consideration*
(2) *Extracting the different layers by slicing.*
(3) *Choosing the most prominent color channel (i.e. Red) and isolating it from the other color channels.*
(4) *Apply K-Means clustering to achieve intensity based clusters.*
(5) *Apply a binary mask for efficient skin detection.*
(6) *Use Bitwise AND operation to combine the RGB/BGR image with the binarized image and derive the segmentation result.*

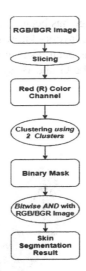

Fig. 1. Flowchart of proposed method

3.1 Image Segmentation and Skin Detection

It is the first step for image analysis, and recognition of pattern. There are several methods to perform image segmentation [13] including thresholding, clustering, transform and texture methods.

The first step was to convert the pixel values in the image from RGB to only the red channel. The red channel has been considered to work upon, as in the skin, red color is the most prominent color, hence making it the right color space to work upon.

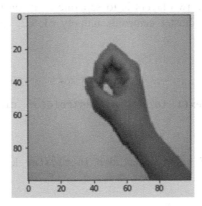

Fig. 2. Initial image represented in RGB format.

As observed from Fig. 2, the image has been represented in RGB [2]. The image in reference can be segregated into three layers - known by *R(red), G(green) and B(blue)*. In this context, only the red layer has been taken into consideration, that is present in the image. So, it is separated from the G and B layers and obtain the rest, which has been represented below in Fig. 3.

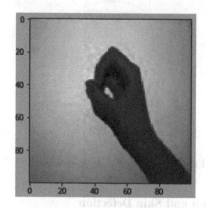

Fig. 3. Image after Red channel is isolated from Green & Blue Channels.

3.2   Binarization of Image Using K-Means Clustering

As observed from Fig. 3, the image consists of two major portions:
1. **The hand gesture** 2. **The background**
To specify these aspects to the model that is going to be trained, there needs to be an algorithm that separates the actual gesture from the background. The possible method for the separation of these two layers is to separate them on the *basis of clusters* by randomly initializing 'N' cluster centroids namely c1, c2, c3 ... ck. The algorithm follows as

```
Repeat
{    for j <- 1 to n

        cent(j)= index(1 to N) of centroid(of clusters) closest to
a(j)

    for k <- 1 to N

        c(k)= mean of points included in cluster k

}
```

This Algorithm is Known as K-Means Clustering. [3] This is a simple method which uses distance measure for grouping data into N pre-defined number of groups (clusters). Applying the algorithm to the image, it is binarized into 0's and 1's by separating the image into two clusters as per our requirement [5]. Further, the

S. De et al.

background is set to all zeroes at first, which sets the R, G, B values to (0, 0, 0) respectively, thus, imparting the cluster containing the *background of the image*, the color *black*. Similarly, setting the values of R, G, B to maximum, i.e. (1, 1, 1) for the cluster containing the gesture, the color *white* is imparted to the *gesture*. The image after this processing is represented in Fig. 4:

Fig. 4. Binarization of image using K-Means clustering

The step that follows is the inversion of this image i.e. the background is set to all ones (1, 1, 1) and the gesture to all zeroes (0, 0, 0). In this particular manner, the following image is obtained (Fig. 5):

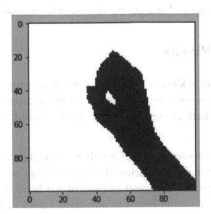

Fig. 5. Inversion of the binarized image.

Furthermore, these two images are taken into consideration and then element-wise multiplication is applied to obtain the actual gesture that will be used for a more accurate skin detection [1].

4 Implementation of Element-Wise Multiplication

In order to obtain the result, element-wise multiplication was applied, to the results obtained on filtering the red portion of the actual image and the mask.

```
plt.imshow(element_wise_multiply(red, mask), cmap = 'gray')
plt.show()
```

The result hence obtained from this has been displayed in Fig. 6.

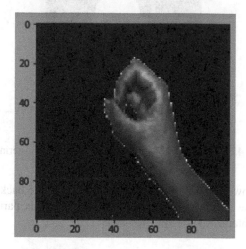

Fig. 6. Image after applying element wise multiplication.

5 Comparison of Results

We propose the use of a new kind of pipeline for preprocessing the dataset. We need to perform *background subtraction* on the images in the dataset to remove unnecessary pixels from the image. There are several conventional methods to achieve it. Some of these are:

- **Contour Detection:** This method [12] involves converting the image into grayscale and detecting the contours(boundaries of abrupt change in brightness). In an ideal situation, it should produce a perfect outline of the hand, but in reality, it is very difficult to achieve because the intensity threshold values that define the contours is dependent on how brightly the scene is lit. This method has been tested on the dataset and it yielded inconsistent results throughout the dataset.
- **Colorspace Thresholding:** This method [11] involves the following steps

    1. *It involves converting the image to a certain Colorspace in which the features of the desired object might be more prominent [4].*
    2. *Deciding a Lower Threshold Coordinate and an Upper Threshold Coordinate.*

3. *Eliminate all the pixels that do not lie in between the Lower Threshold Coordinate and the Upper Threshold Coordinate.*
4. *Combine the resultant mask with the original image to get the desired output.*

The main problem with this method is that like the Contour Detection Method mentioned earlier, it is also lighting dependent which makes it very difficult to decide the threshold coordinates [7], thus yielding inconsistent results throughout the dataset. Below is a sample result of the *binary mask* generated using the Colorspace Thresholding Method (Fig. 7) –

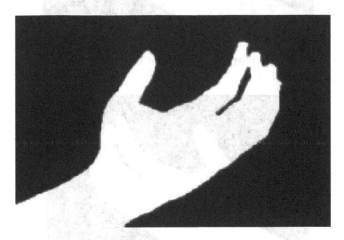

Fig. 7. Binarized image generated using colorspace thresholding method.

- **Proposed Pipeline** - We propose a new kind of pipeline to preprocess the images and produce consistent results across the dataset. The pipeline consists of two steps - Firstly, a single channel was taken from the image colorspace such that the selected channel represents the desired feature most prominently. For this purpose, it is the Red channel in the RGB colorspace that most prominently represents the skin color. A single channel, in essence, gives a type of representation of the image that is similar to its grayscale representation [5]. Secondly, now that we have a grayscale representation, the image is converted into a binary mask using K-Means Clustering where K = 2. This produces a better binary mask compared to all the other previously mentioned methods.

Below is a sample result of the binary mask generated using the Proposed Pipeline (Figs. 8, 9 and Table 1):

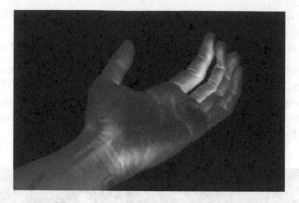

Fig. 8. Image after Red Color has been isolated

Fig. 9. Image post application of Binary Mask using K-Means Clustering.

Table 1. Comparison of the techniques involved

	Contour detection	Colorspace thresholding	Proposed pipeline
Slicing	On basis of contours	On basis of threshold set	On basis of hand position
Lighting	Highly dependent	Less dependent	Least dependent
Performance	Works well in some cases	Works well in very few instances	Works well in most cases
Drawbacks	Disturbing backgrounds leads to undesired output	Improper lighting or threshold selection will yield inconsistent results	Hardware intensive for real-time implementation

6   Conclusion

A real-time hand recognition method based on pre-identification and K-Means clustering was depicted in this methodology. Using the means of clustering, we separated the image into two clusters. Post clustering, the image was binarized and the colors were inverted, upon which element wise multiplication was used to detect the gesture. In order to remove the background pixels, a color channel was searched for, that is prominent in the subject. This was then converted into a binary mask, using K-Means clustering, which resulted in a cleaner binary mask. It was found out that the proposed pipelining technique performs better than the existing ones in terms of effectiveness and efficiency. One of the applications of this methodology was to control the simple Google's T-Rex Run Game (shown in Fig. 10), with the use of hand gestures.

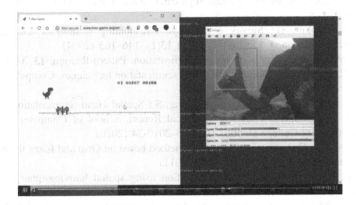

Fig. 10. Real-time application to control Google's T-Rex Run Game

6.1   Future Scope

By proper mapping, many similar Input/Output (I/O) operations can be done. More input-intensive games/applications might find usage of this framework. Hand segmentation can not only simulate keyboard operations, but also mouse operations. Further complex applications include, complete I/O using both hands or sign language detection.

Compliance with Ethical Standards

All author states that there is no conflict of interest. We used our own data. Animals/human are not involved in this work. Author's hand used in this work and given consent.

References

1. Dwina, N., Arnia, F., Munadi, K.: Skin segmentation based on improved thresholding method. In: 2018 International ECTI Northern Section Conference on Electrical, Electronics, Computer and Telecommunication Engineering (ECTI-NCON), pp. 95–99 (2018)
2. Poynton, C.A.: Digital Video and HDTV: Algorithms and Interfaces. Morgan Kaufmann (2003). ISBN 1-55860-792-7
3. Kanungo, T., Mount, D.M., Netanyahu, N.S., Piatko, C.D., Silverman, R., Wu, A.Y.: An efficient k-means clustering algorithm: analysis and implementation. IEEE Trans. Pattern Anal. Mach. Intell. **24**, 881–892 (2002)
4. Otsu, N.: A thresholding selection method from gray-scale histogram. IEEE Trans. Syst. Man Cybern. **9**, 62–66 (1979)
5. Ray, S., Turi, R.H.: Determination of Number of Clusters in K-Means Clustering and Application in Colour Image Segmentation
6. Pal, N.R., Pal, S.K.: A review on image segmentation techniques. Pattern Recogn. **26**, 1277–1294 (1993)
7. Sezgin, M., Sankur, B.: Survey over image thresholding techniques and quantitative performance evaluation. J. Electron. Imaging **13**(1), 146–165 (2004)
8. Fu, K.S., Mui, J.K.: A survey on image segmentation. Pattern Recogn. **13**, 3–16 (1981)
9. Haralick, R.M., Shapiro, L.G.: Survey image segmentation techniques. Comput. Vis. Graph. Image Process. **29**, 100–132 (1985)
10. Alvarez, S., Llorca, D.F., Lacey, G., Ameling, S.: Spatial Hand Segmentation using Skin Color and Background Subtraction, Technical Report, School of Computer Science and Statistics of Trinity College Dublin, TCD-CS-2010-34 (2010)
11. Tang, Z., Wu, Y.: One image segmentation method based on Otsu and fuzzy theory seeking image segment threshold, pp. 2170–2173 (2011)
12. Zheng, Y., Zheng, P.: Hand contour detection using spatial homomorphic filtering and variational level set, pp. 172–176 (2015)
13. Trivedi, S., Khunteta, D., Narayan, S.: Hand Segmentation using Modified K-Means Clustering with Depth Information and Adaptive Thresholding by Histogram Analysis, ICPCSI-2017 (2017)

An Image Processing Based Approach for Monitoring Changes in Webpages

Raj Kothari[✉] and Gargi Vyas

Department of Information Technology Engineering, Dwarkadas J. Sanghvi
College of Engineering, Mumbai 400056, Maharashtra, India
rajk3770@gmail.com, gargi.vyas@gmail.com

Abstract. As the website content is changing dynamically and many users rely on
change detection and notification-based systems. Prominent ones are crawling-
based change detection and notification systems. But this system has drawback,
such as requirement of huge computational resources, which make them a poor
choice for less frequently changing websites. In this paper, we propose, an alternative approach which is based on Image Processing and Image Comparison. The
proposed system includes functionalities for fetching webpage from the Internet
and allowing the users to select zone in webpages to monitor. Using Image
Comparison Algorithm, two different versions of webpage captured at different
time, are compared and the modified portion of the webpage is highlighted. This
approach is validated with the help of prototype implementation.

Keywords: Web page change detection · Change Detection · Web Crawling ·
Image processing based web page change detection · Scheduling · Change
notification · Web page segmentation · Change monitoring · Zone selection of
web page

1 Introduction

Keeping track of changes in the World Wide Web (WWW) is becoming increasingly
difficult. Users are interested in knowing the content that has changed since their last
visit. The content changes are made regularly due to the dynamic nature of modern
websites. For example, website users are interested in monitoring, E-commerce web
sites, Wikipedia pages, blogs, etc. Multinational Organizations need to monitor their
competitor's website for gaining a competitive edge.

At present, the most widely used techniques is Crawling-based Change detection
and Notification. Web crawlers or spiders crawl through different webpages, compare
the content with the previous version of the webpage and notify in case there is a
change in the content of the webpage. This serves as a good solution if the webpages
needed to monitor change very frequently (say, every hour or every day). But, if the
webpages to be monitored change less frequently (say, every 15 days or a month), an
organization might not want to utilize much resources for monitoring change on
webpages.

For such cases and in general, the authors propose an alternative approach to
monitor webpages. Our proposed approach is based on detecting change in screenshots

© Springer Nature Switzerland AG 2020
S. Smys et al. (Eds.): ICCVBIC 2019, AISC 1108, pp. 974–982, 2020.
https://doi.org/10.1007/978-3-030-37218-7_103

of webpages which are taken at different time instances and accordingly sending a Notification. This approach saves much resources required for monitoring less frequently updated webpages and has additional benefit of detecting User Interface (UI) changes in webpages.

The rest of the paper is organized as follows, Sect. 2 provides Literature Review in which existing and similar systems are discussed along with their drawbacks. In Sect. 3, our Image Processing based approach which has various merits over existing system is presented. Experimental results of our proposed approach are also provided. The paper ends with Conclusion and future scope.

2 Literature Review

This section describes existing Web page change detection and notification systems along with similar products available in the market.

2.1 Crawling Based Change Detection System

This system uses a Web Crawler or a Web Spider, which is a software or programmed script that browses the WWW in a systematic, automated manner and downloads numerous webpages from a seed.

2.1.1 Working Mechanism

Crawling Based Change Detection System consists of three main processes: Scheduling, Web Crawling, and Change Detection and Notification.

The process of change detection starts with fetching a Uniform Resource Locator (URL) from browsing list. Browsing list contains list of desired URLs in which the user wants to detect changes. The browsing list also contains other important details such as URL Id, timestamp of URL last checked and frequency for change detection of URL. The scheduler checks which URLs need to be crawled considering timestamp of URL when it was last checked as well as the frequency to detect change.

The URLs added by the scheduler are then crawled one by one or in parallel fashion. The Internet can be considered as a "directed graph" where nodes and edges represent webpages and hyperlinks respectively. The nodes or webpages are connected by edges or hyperlinks. This graph is traversed by the Web Crawlers. They visit webpages and the content is downloaded for indexing. A seed URL (here, URLs fetched by scheduler) is used to begin crawling and visit each page. Pages are downloaded and the URLs in it are retrieved. Newly obtained URLs are kept in a queue and this process is repeated as the crawler travels from page to page. Figure 1 explains the process of Web Crawling [2].

Each page visited by crawler is parsed and the content is then compared with the previous version of the webpage. Different tools use different methods to parse and compare webpages. The browsing list database is also updated, specifically timestamp of the URL last changed. This field will be checked by the crawler, when it starts the crawling process next time. Some products, algorithms and techniques used for comparison of webpages are given below [1].

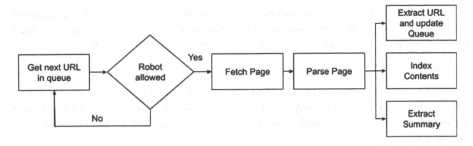

Fig. 1. Web crawling process

Structural and content changes are detected by Document Tree based Approach [2]. Two trees are formed, one for the modified web page and another for the older version of the same webpage. Nodes of these trees are then compared to detect changes. It gives relevancy to the web pages and notifies the user about the detected changes.

The facilities provided by WebCQ [5] are personalized delivery for change notifications, prioritization and summarization of changes made in the Webpage. Raw text changes can be detected between any two tags and changes can be notified to the user via e-mail.

Copernic Tracker [6] can be used to monitor a particular text as well as changes in images or text of a webpage can be detected. One drawback of Copernic Tracker is, it does not provide facility of selecting a particular region to monitor. Another drawback is, data related to performance such as accuracy and speed is not revealed.

It is based on Image-based change detection and notification system. Limited freedom is offered for selecting an area of webpage to monitor and lacks an intuitive graphical user interface where changes can be shown.

WYSIGOT [4] is also used to detect changes between HTML pages. It is a commercial application and can only be used if installed on the local machine.

AT & T Internet Difference Engine (AIDE) [7, 11] detects changes in webpages using tokens. Changes between immediate previous version and current version are detected by AIDE.

CX-DIFF algorithm [8, 10] is used for detecting changes in XML Documents. It is also based on LCS Algorithm and consists of steps such as signature computation, object extraction, filtering of unique inserts/deletes and finding the common order subsequence between leaf nodes of the given trees. Both the above systems are based on LCS algorithm, which requires extensive computation and can overload server.

2.1.2 Observations on Crawling Based Change Detection System

There are certain limitations to this approach. They are as follows:

1. Blocking of IP address

Web Crawlers can cause high load on the website's server and website administrator might have a suspicion of his/her website getting crawled. The crawler's Internet Protocol (IP) address might also get blocked due to the same reason. Most organization keep a pool of IP's for their crawler so that when one IP gets blocked by some website, another IP can be used. For this mechanism, IP addresses are required to be changed dynamically. This incurs huge cost in terms of resources for the company as they need to invest in buying many public IP addresses to overcome the issue [9, 12].

2. Load on Server

Running a crawler with changing IP addresses and in parallel fashion (crawling multiple websites simultaneously on different threads) requires huge computational power. Server with proper load handling capacity is required to facilitate crawling process efficiently.

3. Less frequent changes

Since there is a huge computational load on server for implementing crawling-based change detection system, an organization might not want to invest their resources in crawling less frequently changing websites. For example, if a new article is added to a particular website every 15–20 days, there is no need to run the crawler everyday and check if the content of the webpage has been updated or not.

4. Detects only content changes and not UI changes

A web crawler can only crawl the content of a webpage. It is useful in detecting changes in content of website such as title, body or anchor text links. If there is a change in UI and hence change in structure of website, a crawler might not give expected results.

5. Code for each webpage

The structure of Hypertext Markup Language (HTML) Document Object Model (DOM) for each website is different and the useful content required might be present at different locations in different websites in the DOM tree. Hence, for each webpage, one has to write different code. A slight change in structure of website to be crawled could lead to failure of crawling process or collection of wrong data from the website.

Considering the above issues in existing system, we decided to address them through our approach. Now let us see our proposed approach and system.

3   Proposed System

We propose a novel approach to detect changes in webpages by comparing the screenshot of the webpage at different time instances. Using image processing the task of detecting changes can be quick. This method consumes less resource and can be used to monitor any website without depending on the structure of the website.

The user marks the area to monitor on the website using an image cropper and sets the frequency to monitor the website. For example, every minute or every hour. The coordinates of the cropped image of the website are stored, to be used later. A cron job is scheduled that would take screenshot and crop the image to the required dimensions.

The latest and the recent images are then compared using image processing techniques. In case of no change, the recent image is simply replaced with the current image. Otherwise, the modified content on the website is highlighted by a rectangular box. Highlighted image is compressed and stored in a queue. The queue stores images of different website that were updated, identified by a unique name and timestamp.

3.1   System Architecture

The architecture of the proposed system is shown below. The major components of the architecture are The Scheduler, Crawling using Image Processing Module and Change Detection Module.

Scheduler: Scheduler generates a list of URLs according to the frequency set for monitoring each webpage. This list is then passed to Crawling Module.

Crawling Using Image Processing Module: The function of crawling module is to capture the webpages and crop them to the user-specified dimensions.

Change Detection Module: Version repository contains several screenshots for each URL captured at different instants of time. Similarity between crawled image and most recent, image obtained from Version Repository of the corresponding URL is calculated using mean squared error (MSE). Changes are then detected using image comparison algorithm mentioned in Sect. 3.2.3.

The latest image showing the change along with its respective URL is appended in queue. Changes in the content of the webpage are notified through interfaces such as SMS alert, email or push notifications on a webpage (Fig. 2).

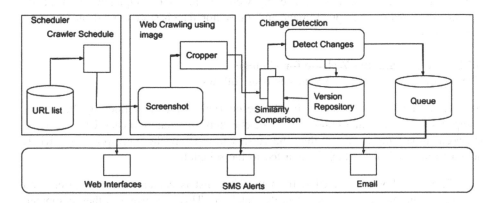

Fig. 2. System architecture

3.1.1   Database Schemas

1. Url (url VARCHAR)
   url – path of the webpage to be monitored.
2. Main (url VARCHAR, x1 INT, y1 INT, x2 INT, y2 INT, freq INT, last_checked DATETIME, to_be_checked DATETIME)
   url – path of the webpage to be monitored.

x1, y1, x2, y2 – selection of area from the entire webpage which has to be monitored.

freq – number of hours/days after which the URL has to be visited again to detect changes.

last_checked – timestamp denoting when the URL was last checked.

to_be_checked – timestamp denoting when the URL has to be checked again

3. Queue (queue_url VARCHAR, queue_img_path VARCHAR)

   queue_url – URL of the webpage in which the algorithm has detected changes.

   queue_img_path – path of the image in which changes are highlighted.

3.2 Working

The work flow of the system is broadly divided into four sections: Setting the initial dimensions, monitoring website for changes, Image Comparison algorithm and Image Compression. Setting the initial dimensions is a one-time task and it is done for each URL to be monitored.

3.2.1 Setting the Initial Dimensions

The user can enter the URL or URLs can be retrieved from the list. Each website is displayed in browser in the form of a screenshot, and user selects the area to be cropped using mouse clicks and sets the frequency to schedule the monitoring of the intended website. PyQt4, a python software is used to develop the Graphical User Interface (GUI) and it takes screenshots. For every website, the URL, frequency to check the website, dimensions of cropped image and timestamp data is stored in a database, to be retrieved later.

3.2.2 Monitoring the Website for Change

The algorithm for monitoring websites is as follows:

Step 1: A cron job executes after fixed time intervals, which is defined by the user. It compares the current timestamp with 'to be checked' timestamp of each URL. If 'to be checked' timestamp is greater than the current timestamp, the corresponding URL is added to the URL list, which is then fed to the scheduler.

Step 2: Screenshot of each URL from the URL list is captured by the system and cropped to the dimensions defined in the database. This screenshot is then compared with the latest stored copy.

Step 3: Mean Squared Error is used to find dissimilarity between two images. If the value of mean squared error for the images is zero, then there is no change and the current image replaces the recent image. If the mean squared error value is not zero, then the two images are different. These images are then compared using Image Comparison Algorithm. Changes in the image are marked with blue color and enclosed within a rectangular box.

Step 4: The highlighted image is then compressed and added in the Queue. The queue contains images depicting changes in the webpages and their corresponding URLs.

3.2.3   Image Comparison Algorithm

The basic element of an image is referred to as picture elements, image element, pels or pixels. Two images are compared for difference, by calculating the difference between their pixel values. Colored images are first converted to gray scale for calculating the difference. This maps Red Green Blue (RGB) value to values between the ranges 0 to 255 by using the formula:

$$Gray = 0.2989 * r + 0.5870 * g + 0.1140 * b$$

Next, the mean squared error (MSE) is calculated to measure the dissimilarity between two images and the pseudo code for it is given below:

```
sum = 0.0
for(x = 0; x < width;++x){
    for(y = 0; y < height; ++y){
        difference = (A[x,y] - B[x,y])
        sum = sum + difference*difference
    }
]
mse = sum /(width*height)
```

If the value of MSE obtained is approximately equal to zero, then the two images are similar, otherwise they are different. After calculating the MSE, if the value is greater than zero, the algorithm tries to find the changed portion of the image. Unequal parts between two images are highlighted in blue color (0, 0, 255). The pseudo code is as follows:

```
if ((pix[x, y][0] - pix1[k, 1][0]) > n and (pix[x, y][1]
- pix1[k, 1][1]) > n and (pix[x, y][2] - pix1[k, 1][2]) >
n)
        while  ((pix[x,  y][0]  -  pix1[k,  1][0])  >  n
        and(pix[x,  y][1]  -  pix1[k,  1][1])  >  n  and
        (pix[x,y][2] - pix1[k, 1][2]) > n):

            if (1 < im.size[1] - 1):
                1 += 1

            pixels[k, 1] = (0, 0, 255)
```

3.2.4   Image Compression

Joint Photographic Experts Group (JPEG) lossy compression is used for reducing the size of image as the motive is just to indicate where the change has occurred. The image is compressed to 5% quality of original, which is sufficient for viewing the changed areas. This is required, as many images for a website will be stored, which would consume a lot of space, if stored in original high quality. Therefore, for utilizing the space efficiently, image is compressed and stored in a Queue.

3.3 Experimental Results

Figure 3 shows the screenshot taken of a news webpage at a particular time instant, say 10:00 am. After predefined time interval, say 2 h, cron job is executed again and next screenshot (Fig. 4) is saved with its timestamp. The algorithm compares this version (12:00 am) with the previous version (10:00 am) of the webpage to check if there's change in content. If change in content is detected, it is highlighted in blue color and a rectangular box is formed enclosing the modified portion as can be seen in Fig. 5. The URL of the webpage along with the image highlighting the modified content is appended to the queue.

Fig. 3. Screenshot taken at particular time instant [3]

Fig. 4. Screenshot taken predefined time interval [3]

Fig. 5. Changes highlighted in blue color and enclosed in green rectangular box by Image-Comparison algorithm

4   Conclusion and Future Scope

Thus, in this paper we describe a novel approach to monitor website changes and notify the user about it. The approaches using image comparison techniques are available in resource efficient ways to monitor changes. It can be used to monitor time crucial events, such as discounts, offers or arrival of new products and for competitor intelligence. Future scope for this approach includes predicting the frequency of checking the webpages or detect areas to monitor. Text could be extracted from the screenshot using optical character recognition in order to provide the user with the changed content.

Compliance with Ethical Standards
All author states that there is no conflict of interest. We used our own data. Animals/human are not involved in this work.

References

1. Shobhna, M.C.: A survey on web page change detection system using different approaches. Int. J. Comput. Sci. Mobile Comput. **2**(6), 294–299 (2013)
2. Mallawaarachchi, V., Meegahapola, L., Alwis, R., Nimalarathna, E., Meedeniya, D., Jayarathna, S.: Change Detection and Notification of Webpages: A Survey
3. Aignesberger, M.: Watcher Product (2006). http://www.aignes.com
4. Wysigot. http://www.wysigot.com
5. WebCQProduct (2006). http://www.cc.gatech.edu/projects/disl/WebCQ
6. Copernic Technologies: Copernic Tracker Product (2006). http://www.copernic.com/en/products/tracker/tracker-features.html
7. Douglis, F., Ball, T.: Tracking and viewing changes on the web. In: 1996 USENIX Technical Conference
8. Chakravarthy, S., Subramanian, C.: Automating change detection and notification of web pages. In: Proceedings of 17th International Conference on DEXA. IEEE (2006)
9. Hosting Facts. https://hostingfacts.com/website-monitoring-services/
10. Jain, S.D., Khandagale, H.P.: A web page change detection system for selected zone using tree comparison technique. Int. J. Comput. Appl. Technol. Res. **3**(4), 254–262 (2014). ISSN: 2319–8656
11. Kuppusamy, K.S., Aghila, G.: CaSePer: an efficient model for personalized web page change detection based on segmentation. J. King Saud Univ. Comput. Inf. Sci. **26**, 19–27 (2014)
12. Woorank. https://www.woorank.com/en/blog/how-will-changing-ip-address-impact-seo

Vehicle Detection Using Point Cloud and 3D LIDAR Sensor to Draw 3D Bounding Box

H. S. Gagana[1(✉)], N. R. Sunitha[1], and K. N. Nishanth[2]

[1] Department of CSE, SIT, Tumkur, India
gaganahs2016@gmail.com, nrsunithasit@gmail.com
[2] Solution Architect, KPIT Technologies, Bengaluru, India
nishanth.jois@gmail.com

Abstract. In recent times, Autonomous driving functionalities is being developed by car manufacturers and is revolutionizing the automotive industries. Hybrid cars are prepared with a wide range of sensors such as ultrasound, LiDAR, camera, and radar. The results of these sensors are integrated in order to avoid collisions. To achieve accurate results a high structured point cloud surroundings can be used to estimate the scale and position. A point cloud is a set of Data points used to represent the 3D dimension in X, Y, Z direction. Point cloud divides data points into clusters that are processed in a pipeline. These clusters are collected to create a training set for object detection. In this paper, the cluster of vehicle objects and other objects are extracted and a supervised neural network is trained with the extracted objects for the binary classification of the objects representing vehicles or other objects. By learning global features and local features the vehicle objects represented in the point cloud are detected. These detected objects are fitted with a 3D bounding box to represent as a car object.

Keywords: Autonomous drive · 3D LiDAR · Vehicle detection · 3D bounding box · Point cloud

1 Introduction

Vehicle detection and classification are critical for traffic management. Various sensing techniques are used for vehicle detection, including loop detectors and video detectors. Almost all of the vehicle detection and classification tools in conventional practice are based on fixed sensors mounted in or along the right side of the vehicle [13, 14]. This research examines the possibility of considering vehicles mounted with sensors instead of the fixed detectors, thereby enabling alternative monitoring techniques such as the moving observer [4, 5] that were previously too labor intensive to be viable.

In autonomous driving, 3D LiDAR scanners play a vital role and create depth data of the environment. Though generating huge Point cloud datasets with point cloud labels involves a substantial quantity of physical annotation, this exposes the effective growth of supervised deep learning algorithms which is frequently data-hungry. The data that we have collected from this framework supports both auto driving scenes as well as user-configured scenes.

© Springer Nature Switzerland AG 2020
S. Smys et al. (Eds.): ICCVBIC 2019, AISC 1108, pp. 983–992, 2020.
https://doi.org/10.1007/978-3-030-37218-7_104

LiDAR is one of the critical technologies in the development and deployment of autonomous vehicles as shown in Fig. 1.

Fig. 1. LiDAR [3]

It is also a part of the development mix in other hot topics like border security, drone mapping and much more. In autonomous driving LiDAR is comprised of a number of spinning laser beams and is usually mounted on the roof of a car, and spins 360° around its axis, measures light intensity and distance, frequency of LiDAR is 50 k to 200 k pulses per second(HZ). We can also collect the LiDAR data in different forms like Airborne, Ground, and Satellite.

In this paper, LiDAR (Light detection and Ranging) is used to monitor the ambient traffic around a probe vehicle. The LiDAR does not distinguish between vehicles and non-vehicles, as such, much of the work is focused on segmenting the vehicles from the background. A vehicle classification method based on the length and height of detected vehicles is developed to classify the vehicles into six categories. The classification accuracy is over 91% for the ground truth data set.

2   Related Work

In the recent years, Vehicle detection is extensively examined by many researchers.

In [1, 3] the usage of convolution neural network to train the machine to identify objects and draw bounding boxes around them is explained. Also, in [3] authors have made use of RGB depth images and made 3D bounding boxes on camera image as well as on LiDAR image. In [2] a methodology is shown on how to identify the moving objects.

In [7] authors have made mobile mapping system showing camera and LiDAR together and have provided some features such as 3D images based modeling and visualization of street view along with synchronization between camera and LiDAR images. They have also mentioned about the automated tools to extract 3D road objects. In [10], the working of the LiDAR system and creation of the Digital Elevation Models is discussed. In [8], authors have done a case study on object recognition in public places.

In [11], authors have made use of the KITTY Datasets and compared the camera images with LiDAR images. In [6] 3D detection of pedestrians is carried out. Also, similar work is shown in [3]. In [6], authors have split the pedestrians diagram into two parts namely trunk and legs. This is done to ensure that other objects are not considered as pedestrians and obtain a very accurate human profile.

In [9], authors have done the work on buildings. They have visualized LiDAR and 3D RGBD images together. Also, in [9, 10] they have established a Region of Interest where all the detection should take place which removes the time taken for unwanted detections. In [3], detection using 2D bounding information is processed and converted to 3D projection of point clouds into images by quantization and later convolution neural network applied. But this would affect the natural 3D variation and accuracy of the point cloud.

3 Approach and Model Architecture

With the help of feature descriptor, 3D features are extracted and techniques of machine learning applied on point cloud data of input image to observe dissimilar forms of object classes. Then, the same procedure is applied to categorize different images captured by LiDAR data.

Figure 2 is the proposed architecture. Here, LiDAR uses more point cloud datasets to filter unwanted points like noise, ground removal, etc. This process is known as segmentation. These segmented and filtered images are sent as input to the neural network to get the output of point cloud mapping. Then we detect the objects like car, pedestrian, road using two data sets used namely CITYSCAPES DATASET [12] shown in Fig. 4 and KITTY DATASET [11] shown in Fig. 5. These two datasets are used for training purpose.

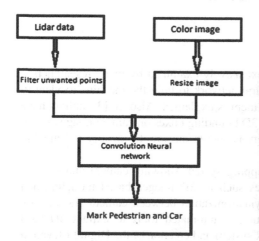

Fig. 2. System architecture

Fig. 3. Image of point cloud [4]

This paper deals with the displaying of LiDAR datasets. The data sets are read from point clouds and displayed as a 3D mesh. This is achieved in two parts here the first part being the conversion of data to a suitable format. Then the second is data is to display a data as a 3D mesh using applications like VTK/Mayavi.

A point cloud is a set of data points in space and are usually taken from 3D scanners, signifies a group of multi-dimensional points (usually in X, Y, and Z) which represents 3D information about the world. Each 3D point can contain more information like RGB colour, distance, intensity. Point clouds are integrated from hardware sensors like Time –of –flight cameras, 3D Scanners, etc. Each 3D point can contain more information like RGB colour, distance, intensity. Point clouds are integrated from hardware sensors like Time–of –flight cameras, 3D Scanners, etc.

The LiDAR measures the distance of the objects surrounding the test vehicles using the point cloud as shown in the Fig. 3. Convolution neural network is a class of deep learning used to analyze the visual imagery in computer vision applications. Here we use 3D convolution neural network to achieve for Visualization, animation, rendering, and mass customization applications.

The proposed approach is implemented in the following steps and explained clearly in Experiments and Results (Fig 6):

- Visualizing the point cloud data
- Identifying and classifying objects of interest: vehicle and other objects
- Storing the identified data in efficient manner.
- Detecting the object.
- Drawing the 3D bounding box

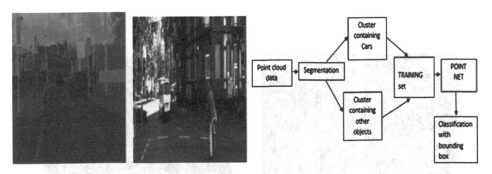

Fig. 4. Cityscape dataset image [9]   **Fig. 5.** Kitty dataset image [5]   **Fig. 6.** Workflow of proposed approach

Implementation Details

We are using CITYSCAPE DATASET and KITTY DATASET as a training dataset. The input point cloud is passed to segmentation process in order to extract clusters. Then the clusters are extracted by Noise filtering and Ground removal. The Noise filtering process is done by statistical outlier removal process using PCL library which removes the noisy data represented by outlier. Ground removal process makes a significant role in segmentation and extraction of objects. To extract the ground from point cloud before segmentation we focus on the interest of object segmentation but not ground segments, for this ground removal model fitting method RANSAC was used. For this method PCL includes functionality and is used in implementation, these point clouds passed to the Point Net network model as training set for binary classification. Then this trained data set is evaluated with some unseen data and then we draw bounding box to represent the identified objects.

4 Experiments and Results

To analyze the vehicle detection, the Input point cloud data set is segmented into many clusters to create a training set for the point net classification which involves the following steps like Noise filtering, Ground Removal, and clustering.

4.1 Noise Filtering

LiDAR scans the surrounding environment and creates point cloud datasets. The densities of the point cloud depend on the nature of the surroundings. The measured errors in outliers tend to be corrupt since it is away from the LiDAR source. It is very difficult to estimate exact representative like curvature or surface normal which is solved by statistical analysis. Each point on outlier is used as input and can be defined by global distance and standard deviation considered and this errors can be trimmed from the set of data. The Fig. 7a and b shows the result of Point cloud Image Before and after filtering. The Fig. 7c shows the filtered noisy data.

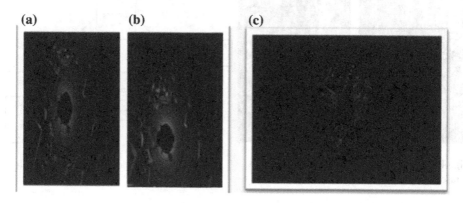

(a)   **(b)**   **(c)**

Fig. 7. a: Before filtering image [10] **b**: After filtering image **c**: Noisy data filtered

4.2 Ground Removal

To extract the ground from point cloud before segmentation we focus on the interest of object segmentation but not ground segments. For this ground removal model fitting method RANSAC, was used. PCL used in the implementation.

RANSAC ALGORITHM

For the given fitting problem, estimate the parameters.
Assumption:

1. From N data items parameters can be projected.
2. Assume M data items totally.

Then the algorithm is:

1. Select N data items randomly.
2. Estimate X parameter.
3. With parameter X find in what way data items [M] appropriate the model in the interior of a given image and considered this K.

Here K has to depend on a number of structures in the image and percentage of the data that belongs to the structure being fit. In the case of multiple structures, remove the fit data and redo RANSAC after a successful fit.

Figure 8a and b shows the results of before ground removal and after ground removal.

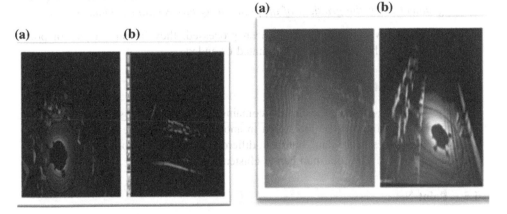

Fig. 8. a: Before ground removal [10] b: After ground removal

Fig. 9. a: Outlier image b: Inlier image

It is repeated for the exact number of times, every time model is rejected because a small number of points categorized as Inlier or model is redefined as a measurement error. Then to know whether this point is Inlier or outlier, the threshold is set to ground which specifies how long the point is Inlier as shown in the Fig. 9a and b. In our instance, 30 cm threshold is most effective. It will help to cut some part of clusters like pedestrians.

4.3 Clustering

Figure 10a and c shows the results of clustered car objects and Fig. 10b shows the result of a clustered tree object.

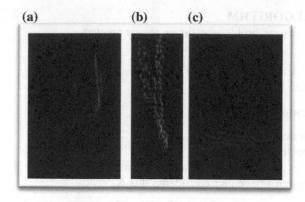

Fig. 10. a: Cluster 1 **b**: Cluster 2 **c**: Cluster 3

Assume structure of *kd*-tree to find nearest neighbors then algorithm steps would be:

1. *We create a **kd**-tree Pi to represent a Input point cloud;*

   - *Enhance Q to the gradient of collections C, and rearrange Q to an unfilled gradient when the gradient of all points in Q has remained administered.*

2. *When all points* of the set P have been processed, they become a part of point clusters C and the algorithm is terminated or ended.

4.4   Training Data

After the clustering process, the clusters containing car objects are stored in one folder as cars and remaining clusters are stored in another as others. The same object could occur multiple times in the data, but from different viewpoints all objects are collected as different objects. The total number of clusters extracted is over 2000 car objects

4.5   Point Net

It is a significant kind of symmetrical data configuration. Due to its asymmetrical arrangement many researchers transmit this information to 3D voxel networks or Image groups, even though rendering of data is challenging. Point net provides combined architecture for application, ranges from segmentation part, semantic process scene and classification object.

4.6   Training and Validation

For implementing machine learning in real-time applications, one needs to evaluate all the necessary parameters so that we can obtain an optimized application. Due to the over fitting problem, the efficiency of the training set will not be considered efficient for unseen objects. So it should be taken care that all unseen data is evaluated before the validation of the detected data.

4.7   3D Bounding Box

The 3D bounding boxes are a method for feature extraction. Firstly the covariance matrix of the point clouds is calculated and its Eigen values and vectors are extracted. We consider the resultant Eigenvectors as normalized and the right-hand coordinate system (major Eigenvectors) represents x-axis and minor vector represents z-axis (minor Eigenvectors) as shown in Fig. 11. On each iteration a major Eigenvector is rotated, rotation order is always the same and it's performing around other Eigenvectors provides the invariance to the rotation of the point cloud. Now, this rotated major axis vector is treated as a current axis vector. For every current axis, the moment of inertia is calculated. Also, the current axis is used for calculating eccentricity. The current vector is treated as a normal vector of the plane and the input point cloud is projected on to it. This axis is used as a reference to obtain eccentricity and projection.

Fig. 11. 3D Bounding box

4.8   Object Detection

For training and validation the data is split into 2000 and 1000 respectively. The model is trained with training set containing cars and other objects. The remaining gives the efficiency of the testing process. The Fig. 12a and b shows the result of visualized detected object in front and top view.

(a)   **(b)**

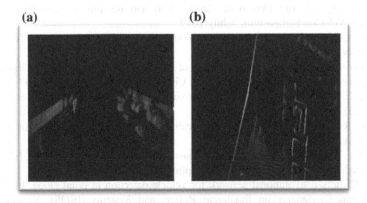

Fig. 12. a: Detected object (Front view) **b**: Detected object (Top view)

The performance on the training set is not a respectable display on the performance of hidden data due to the problem of over-fitting, consequently the trained model is usually evaluated with some unobserved data and obtained 91% accuracy.

5 Conclusion and Future Scope

In autonomous driving, object detection plays a major role in the collision avoidance. LiDAR actively creates hypotheses and detect objects, although a vision based classifier is used for object validation or classification. Processing the point cloud for the extraction of objects makes a prominent role in classification. Point Net which takes input as raw point cloud without voxeliztion that absorbs both global and local point features, as long as effective, simple and efficient method, which classifies the point cloud with high accuracy and helps in detecting the object of interest. With this proposed framework the objects are detected accurately which helps to prevent the collision with the obstacle. The results formed are capable of showing the observer a 3D simulated environment which lets the observer see all the directions of the vehicle. Also, it helps the observer to identify the obstacles like car and plotting a bounding box around it. It cannot identify other obstacles present like pedestrian, cyclist. So further we can make a data set that can store all the possible obstacles and provide a name and classify all the obstacles by providing a 3D bounding box.

Compliance with Ethical Standards

✓ All authors declare that there is no conflict of interest.
✓ No humans/animals involved in this research work.
✓ We have used our own data.

References

1. Asvadi, A., Garrote, L., et al.: DepthCN: vehicle detection using 3D-LIDAR. In: IEEE 20th International Conference on Intelligent Transportation Systems (ITSC) (2017)
2. Azim, A., Aycard, O.: Detection, classification and tracking of moving objects. In: Intelligent Vehicles Symposium, 3 July 2012
3. Qi, C.R., Liu, W., Wu, C., et al.: Frustum PointNets for 3D Object Detection from RGB-D Data, arXiv, 13 April 2018
4. Charlesworth, G., Wardrop, J.: A method of estimating speed and flow of traffic from a moving vehicle. In: Proceedings Institute of Civil Engineers, Part II, vol. 3, 158–171 (1954)
5. Wright, C.: A theoretical analysis of the moving observer method. Transp. Res. 7(3), 293–311 (1973)
6. Wolcott, R.W., Eustice, R.M.: Visual Localization within LIDAR Maps for Automated Urban Driving
7. Wang, R.: 3D building modeling using images and LiDAR: a review, Canada, 05 July 2013
8. MMS Inc, A White Paper on LIDAR Mapping (2001)
9. Li, B.: 3D fully convolutional network for vehicle detection in point cloud. In: IEEE/RSJ International Conference on Intelligent Robots and Systems (IROS), Vancouver, BC, Canada, 24–28 September 2017

10. Choi, M., Ulbrich, S., Lichte, B., Maurer, M.: Multi-target tracking using a 3D-Lidar Sensor. In: 16th International IEEE Annual Conference on Intelligent Transportation Systems, The Hague, Netherlands, 6 October 2013
11. Geiger, A., Lenz, P., Stiller, C., Urtasun, R.: Vision meets robotics: the KITTI dataset. Int. J. Robot. Res. **32**(11) (2013)
12. Cordts, M., Omran, M., Ramos, S., Rehfeld, T., Enzweiler, M., Benenson, R., Franke, U., Stefan, R., Schiele, B.: The Cityscapes dataset for semantic urban scene understanding. In: IEEE Conference on Computer Vision and Pattern Recognition (2016)
13. Harlow, S., Peng, C.: Automatic vehicle classification system with range sensors. Transp. Res. Part C, 231–247 (2001)
14. Gelenbe, E., Abdelbaki, H.M., Hussain, K.: A laser intensity image based automatic vehicle classification system. In: Proceedings of ITSC 2001, vol. 2, pp. 277–287, August 2001

An Experiment with Random Walks and GrabCut in One Cut Interactive Image Segmentation Techniques on MRI Images

Anuja Deshpande[1]([⊠]), Pradeep Dahikar[1], and Pankaj Agrawal[2]

[1] Department of Electronics, Kamla Nehru Mahavidyalaya,
Nagpur, Maharashtra, India
anuja_1978@yahoo.com, pbdahikarns@rediffmail.com
[2] Department of Electronics and Communication Engineering, G.H. Raisoni
Academy of Engineering and Technology, Nagpur, Maharashtra, India
pankaz_agr@rediffmail.com

Abstract. This research work proposes the Random Walks and GrabCut in One Cut interactive image segmentation techniques using MRI images, particularly those posing segmentation challenges in terms of complexity in texture, indistinct and/or noisy object boundaries, lower contrast, etc. We have computed accuracy measures such as Jaccard Index (JI), Dice Coefficient (DC) and Hausdorff Distance (HD) besides Visual assessment to understand and assess segmentation accuracy of these techniques. Comparison of the ground truth with segmented image reveals that Random Walks can detect edges/boundaries quite well, especially when those are noisy, however, has tendency to latch onto stronger edges nearby the desired object boundary. GrabCut in One Cut on the other hand sometimes needs more scribbles to achieve acceptable segmentation.

Keywords: Accuracy · Graph cuts · Hybrid segmentation · Random Walks · GrabCut in One Cut

1 Introduction

Segmenting an image has been a stiff challenge for many decades now. The objectives of image segmentation being different for different needs, the techniques developed to solve those were oriented towards solving the specific problem, and did not have generality towards all applications. Complexity of the images further made it more challenging and researchers had to take different approaches to solve specific segmentation problems. This has resulted in multitude of image segmentation techniques [1], falling under manual, automatic and semi-automatic image segmentation categories. Over past several decades, hundreds of image segmentation techniques belonging to various categories [2] have come into existence and each one solved a specific problem. A general-purpose segmentation technique that can segment all types of images is still in development. Automatic segmentation is still a challenging task that is not yet perfected although there have been many attempts towards the same over last few decades. In image segmentation, accuracy of the segmentation process decides if

© Springer Nature Switzerland AG 2020
S. Smys et al. (Eds.): ICCVBIC 2019, AISC 1108, pp. 993–1008, 2020.
https://doi.org/10.1007/978-3-030-37218-7_105

the extracted object is good enough to be fed to next application, if not, pre and/or post processing becomes critical to ensure that the segmentation results are acceptable.

2   Literature Survey

As put forth in [3], Intensity images can be segmented using mainly four approaches [4, 5]; these approaches are based on threshold, boundary, region and hybrid (these employ boundary and region criteria both).

Thresholding based methods [6] discard spatial information and does not perform effectively in the presence of noise or if the foreground object has blurred boundary or textural complexities. Thresholding methods assume that adjacent pixels belong to either foreground or background. Assignment of foreground or background is based on grey levels for greyscale images or color value for RGB images, belonging to specific range. In boundary based methods [3], Sobel or Robert filter [3, 7] is employed as gradient operator and pixel value changes swiftly at the boundary between two regions [8]. Complementary to boundary based methods [4], is regions based methods [9]. In this method, pixels belonging to one region demonstrate attributes which are similar. This category has given rise to a new family i.e. region growing. Split and Merge technique [3, 10] is one of the best-known technique in the category. In Region Growing method, homogeneity is evaluated by comparing one pixel with its neighbor based on its connectivity (4 or 8), however, these methods are influenced by noise and even moderate success is critically dependent on homogeneity criterion [3, 11–13]. The Hybrid methods [3] combine boundary and region criterion and employ morphological watershed segmentation [14] and variable-order surface fitting [4]. With the watershed algorithm it is certain that closed boundaries will get formed along the ridges. Region boundaries prevent water from getting mixed up when the topology gets flooded and yield segmentation [15]. Random Walks falls under this category of hybrid method.

The Random Walks interactive segmentation technique as described in [16], formulates the problem on graph. In this method, the user identifies few pixels as foreground and background, called seeds. The remaining pixels in the image are imagined to release a random walker and probability is computed that it reaches the seed point.

In Graph partitioning methods, hard constraints are imposed by marking pixels belonging to foreground (object to be extracted) and background that is discarded. In [17] Hard and soft constraints are imposed to segment the images. Through min-cut/max-flow the global optimum of soft constraints is achieved that satisfies hard constraints. Cost function incorporating region and boundary properties, as a soft constraint is defined. For the graph cuts [17], algorithm [18] has been developed to facilitate interactive segmentation based on user input and generally requires more seeds to be placed to overcome small-cut problem. There have been quite a few extensions to overcome shortcomings of graph cuts like, multi-level banded graph cuts [19], lazy snapping [20], GMMRF [21] and GrabCut [22] however the later needs more computational power and the box interface is not always feasible. Random walks has earlier been proposed by [23] for texture discrimination and different aspects of random walk have been explored in [24, 25]. Automatic clustering using Random Walks has also been proposed in [26, 27].

In the GrabCut in One Cut method [28] the authors propose a new energy term, that through one graph cut can be globally maximized and demonstrate that the NP-hard [29] segmentation functional can be avoided using this approach. The technique further replaces the iterative optimization techniques based on block coordinate descent quite effectively.

This paper discusses experiments conducted using Random Walks [16] and GrabCut in One Cut method [28] image segmentation techniques on MRI images. We have organized this paper as follows. We have discussed accuracy measures in Sect. 3. The Sect. 4 illustrates our approach towards this experiment and the experiment details in Sect. 5, the results and findings are described in Sect. 6 followed by observations of this experiment in Sect. 7 and supported by critical discussion in Sect. 8. The concluding comments are categorized in Sect. 9.

3 Accuracy Measures

Without a scientific evaluation method to assess correctness and completeness of image segmentation technique, both from coverage as well as spatial overlap perspective, accuracy of the image segmentation technique cannot be ascertained. Further, without establishing accuracy, we cannot ascertain effectiveness of the image segmentation technique either. Various evaluation methods have evolved over time to ascertain sensitivity and specificity of image segmentation techniques, viz. F1 score, SIR (Similarity Ratio), TPR (True Positive Ratio), FPR (False Positive Ratio), etc. Worst-case needs to be additionally worked out for these methods towards sensitivity and specificity. For the same methods like AHE (Average Hausdorff Error), AME (Average Mean Error), etc. are used. Similarly, owing to squaring function, MSE (Mean Squared Error), weighs outliers heavily. ROC Curve [30, 31] is another method but it also requires computation of TPR and FPR. While these methods have significance in certain context, however, require additional methods for conclusive results.

Hence, it is necessary that we chose such evaluation methods which are conclusive and to ensure the same, in this experiment, we have employed accuracy measures - Jaccard Index (JI), Dice Coefficient (DC) and Hausdorff Distance (HD) for the assessment of segmentation accuracy of these techniques. These methods are computationally inexpensive, faster, and conclusive and are convincing as against other evaluation methods mentioned above.

3.1 Jaccard Index (JI)

The Jaccard Index compares the similarity as well as diversity between two data sets A and B and is expressed in (1) below.

$$J(A, B) = \frac{|A \cap B|}{|A \cup B|} = \frac{|A \cap B|}{|A| + |B| - |A \cap B|} \tag{1}$$

By subtracting Jaccard Index from 1 we get the Jaccard Distance and is expressed as

$$d_J(A, B) = 1 - J(A, B) = \frac{|A \cup B| - |A \cap B|}{|A \cup B|} \tag{2}$$

3.2 Dice Coefficient (DC)

The Dice Coefficient, is used for determining similarity (presence or absence) of two data sets A and B and is expressed in (3) below.

$$QS = \frac{2|A \cap B|}{|A| + |B|} \tag{3}$$

Where QS stands for quotient of similarity.

3.3 Hausdorff Distance (HD)

The Hausdorff distance (spatial distance) indicates how different two data sets are from each other and expressed in (4) as –

$$d_H(X, Y) = \inf\{\epsilon \geq 0;\ X \subseteq Y\epsilon\ \text{ and }\ Y \subseteq X\epsilon\} \tag{4}$$

Where

$$X\epsilon := \bigcup_{x \in X}\{z \in M\ ;\ d(z, x) \leq \epsilon\} \tag{5}$$

4 Our Approach

In this experiment, instead of executing the algorithm once as usually practiced, we have implemented Random Walks technique iteratively, particularly two iterations. The first iteration filtered most of the background and in the second, we have achieved final segmentation. We have performed multiple runs of the segmentation process with varying parametric values (Beta, Threshold, etc.), marked different foreground and background seeds on the image during each run and have selected only the visually best results for further assessment.

For the GrabCut in One Cut technique, we have performed multiple iterations for different parametric values (Number of Colour bins per channel and Colour Slope) to get the best result. Additional scribbles were marked to get the best results.

We have studied Random Walks and GrabCut in One Cut image segmentation technique and conducted this experiment using implementation code [32] and [33] respectively. We have ascertained accuracy of these segmentation technique through –

1. Visual Assessment
2. Accuracy Measures
   a. Jaccard Index
   b. Dice Coefficient
   c. Hausdorff Distance

5 Experiment Setup

The image dataset [34] we have used for this experiment consisting of MRI images is the Left Atrial Segmentation Challenge 2013 (LASC'13). The dataset included original images in RAW format as well as ground truth constructed by human subjects. We have used an open source Insight Segmentation and Registration Toolkit (ITK) [35] to extract Image snapshot from the RAW MRI image for the purpose of this experiment. We have performed below steps towards image acquisition and pre-processing for this experiment –

1. Download LASC'13 Dataset consisting 10 MRI images.
2. Improve Contrast using ITK Toolkit on each data set.
3. Using ITK Toolkit, export RAW image slice to PNG format.
4. Perform Lucy-Richardson deconvolution on exported images to enhance contrast and image edges.

In the Lucy-Richardson deconvolution filtering step, we have implemented disc based Point Spread Function with radius of two. This has enhanced the image contrast and edges significantly better.

During the segmentation process, we have visually compared segmented image and the ground truth, and if found acceptable, we have completed the segmentation. Post completing all the image segmentations we have then computed the accuracy measures to establish the accuracy, statistically.

We have accepted the segmentation output if the algorithm is unable to segment the image correctly even after varying requisite parameters – Beta and Threshold for Random Walks and Number of Colour Bins and Colour Slope for GrabCut in One Cut and performing numerous segmentation runs for each of these changed values.

For all segmented images, we have assessed accuracy by computing accuracy measures - Jaccard Index, Dice Coefficient and Hausdorff Distance. We have performed segmentation process as shown below (Fig. 1).

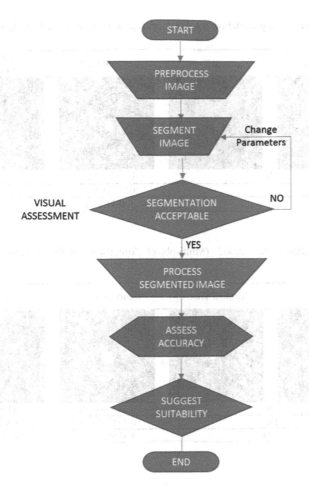

Fig. 1. Segmentation process followed in this experiment

6 Experiment Results

Below are the segmentation results of our experiment on MRI images using Random Walks and GrabCut in One Cut (Tables 1, 2, 3, 4, 5, 6, 7, 8, 9 and 10).

Table 1. Segmentation set A001 (Foreground with noisy boundary)

Original Image	Ground Truth	Random Walks	GrabCut in One Cut
JI		0.857340720	0.870056497
DC		0.923191648	0.930513595
HD		3.464101615	2.236067977

Table 2. Segmentation Set A002 (Multi-object foreground with overlap and indistinct boundary)

Original Image	Ground Truth	Random Walks	GrabCut in One Cut
JI		0.380698638	0.819523269
DC		0.551457976	0.900810979
HD		9.165151390	2.645751311

Table 3. Segmentation Set A003 (Foreground with overlap but sharper boundary)

Original Image	Ground Truth	Random Walks	GrabCut in One Cut
JI		0.802176697	0.933717579
DC		0.890230906	0.965722802
HD		5.744562647	2.0

Table 4. Segmentation Set A004 (foreground with overlap and indistinct boundary)

Original Image	Ground Truth	Random Walks	GrabCut in One Cut
JI		0.679456907	0.900738310
DC		0.809138840	0.947777298
HD		5.744562647	2.828427125

Table 5. Segmentation Set A005 (foreground with overlap and distinct boundary)

Original Image	Ground Truth	Random Walks	GrabCut in One Cut
JI		0.790398918	0.878469617
DC		0.882930514	0.935303514
HD		5.477225575	2.449489743

Table 6. Segmentation Set A006 (Multi-object foreground with partial overlap and mostly distinct boundary)

Original Image	Ground Truth	Random Walks	GrabCut in One Cut
JI		0.667215815	0.852422907
DC		0.800395257	0.920332937
HD		6.0	3.605551275

Table 7. Segmentation Set A007 (Blurry foreground with indistinct boundary)

Original Image	Ground Truth	Random Walks	GrabCut in One Cut
JI		0.875173370	0.873622337
DC		0.933431953	0.932549020
HD		4.582575695	2.828427125

Table 8. Segmentation Set A008 (Foreground with distinct sharp object boundary)

Original Image	Ground Truth	Random Walks	GrabCut in One Cut
JI		0.946229913	0.949658173
DC		0.972372182	0.974179152
HD		3.872983346	1.732050808

Table 9. Segmentation Set A009 (foreground with blurry but mostly distinct boundary)

Original Image	Ground Truth	Random Walks	GrabCut in One Cut
JI		0.879618594	0.958133971
DC		0.935954344	0.978619426
HD		3.872983346	2.00

Table 10. Segmentation Set A010 (foreground with blurry but mostly distinct boundary)

Original Image	Ground Truth	Random Walks	GrabCut in One Cut
JI		0.895893028	0.926530612
DC		0.945088161	0.961864407
HD		4.242640687	2.236067977

7 Observations

During this experiment, we have observed that the Random Walks algorithm, although capable, however, in its current implementation form, does a remarkable job of segmenting the image by marking just one foreground and one background seed. Most other interactive image segmentation techniques including GrabCut in One Cut require various scribbles to be marked as foreground and background to get desired results, at times requiring manual tracing along the noisy or blurred boundary. Random Walks has performed very well to detect noisy object boundaries in most of the cases. Since GrabCut in One Cut allows more interaction to mark additional pixels as foreground and the background, we could achieve desired segmentation in most of the cases.

For segmentation sets A001, both the techniques could achieve only around 90% accuracy, GrabCut in One Cut performed slightly better. The blurry edge at the bottom of the image, while it was detected by both successfully, it is not part of the foreground as can be ascertained using the ground truth.

Segmentation sets A002 and A006 are classic cases of multi object segmentation. Random Walks although capable, however, in its current implementation form is not suitable for such demands and could not detect as the findings suggest. GrabCut in One Cut, owing to the ability of marking additional scribbles could detect those and achieve better output, however as the accuracy measures suggest, the output from both techniques is unacceptable for most applications.

GrabCut in One Cut could segment the dataset A003 much better with accuracy ranging around 95% using additional scribbles. The edge is practically non-existent along the left side of the object leading to the failure by Random Walks.

Segmentation set A004 and A005 is another case where object boundary is not distinct, particularly towards lower left bottom edge. GrabCut in One Cut could achieve about 90% accuracy owing to more scribbles; however, for Random Walks it is lower.

Segmentation set A007 not only has blurry edges but also does not have a distinct boundary separating the foreground and background. Both techniques could achieve only around 90% accuracy.

Table 11. Comparison of Random Walks and GrabCut in One Cut

Feature	Random Walks	GrabCut in One Cut
Orientation	Medical images	Natural images
Relative speed of execution	Slow	Fast
Noisy edge detection	Excellent	Weak
Multi-object segmentation capability	Capable, but not in current implementation form	Yes
Foreground and background markers	One pixel each	Many scribbles

Both the techniques have been successful and yielded around 95% accuracy on the segmentation set A008, here the object boundary is quite distinct, although bit blurry.

On segmentation set A009, GrabCut in One Cut has been quite successful resulting in around 95% accuracy. We needed more scribbles though at the top of the foreground to handle the blurry area. Random Walks on the other hand could only achieve around 90% accuracy. While most of the foreground has distinct boundary, the blurry area on the top has caused such an outcome.

Segmentation set A010, while the foreground object has distinct boundary the entire boundary is blurred. GrabCut in One Cut has performed better with about 93% accuracy as compared to Random Walks that was barely touching 90%.

8 Discussion

Image segmentation being a very vast subject, we will restrict this discussion to supervised and/or graph based methods. Other techniques such as Intelligent Scissors [36] also treats the image as graph problem and imposes connectivity structure. In this method, user marks points along the boundary and using Dijkstra's [37] algorithm computes shortest path. This Intelligent Scissors algorithm is quite fast and implementation is also quite simple however, it requires intuitive handling when the foreground object has noisy or low contrast boundaries. In Active Contour [38] or Level Sets [39], which has numerous implementations, expects user to mark initial seeds near object boundary and the algorithms then evolve the object boundary to local energy minimum by employing different energy functional and domain knowledge. In the graph theory category, one of the first approaches are by [40, 41], however, normalized cuts introduced by [42] drew lot of interest and is an excellent work to follow. Subsequently, many algorithms, which came later, had focused on spectral properties of the graph, as explained in [43, 44]. The isoperimetric algorithm proposed by [45] and Swendsen-Wang algorithm [46] are quite an exceptions.

Various approaches to improving results using Graph Cut based methods [47, 48] have been proposed in last one decade. These largely include work on shape priors, additional constraints, etc., for varied applications ranging from biomedical images [49–53], industrial images [54], natural images [55–58], etc. to motion or video segmentation [59, 60]; each one of them trying to overcome some or the other limitation.

We however continue to study the vast field of image segmentation pursuing newer techniques with the aim to develop a general algorithm that can segment all images.

9   Conclusion

During the course of this experiment, both techniques did not yield acceptable segmentation at first and we had to introduce a pre-processing step in our experiment, which significantly improved the segmentation results. We introduced contrast enhancement and de-blur using Lucy-Richardson algorithm as pre-processing step and it then had significant impact on the overall segmentation outcome.

Random Walks could detect weak or noisy boundaries very well in most of the cases, while GrabCut required more scribbles for the same. As we witnessed during this experiment, Random Walks has shortcomings particularly when a strong edge is in proximity of desired object boundary then the algorithm tends to latch onto it instead of picking up desired boundary.

As the findings of this experiment suggests, multi-object segmentation could not be achieved using Random Walks in its current implementation form, although capable. GrabCut in One Cut on the other hand has successfully achieved it, although with low accuracy.

GrabCut in One Cut has orientation towards segmenting Natural images, however, this experiment suggest that it can also be used on MRI images, with few more scribbles or effort. Below table summarizes the comparative findings for quick reference (Table 11).

Overall, GrabCut in One cut has segmented the images comparatively better than Random Walks, largely due to additional scribbles applied to delineate the foreground, sometimes involving manual tracing along the object boundary.

While strength of algorithms is certainly critical for the segmentation success, equally important is the subject knowledge. Overlapping organs do pose a very difficult segmentation challenge and ascertaining object of interest itself could be challenging then. We have found that pre-processing of the images is required to achieve acceptable segmentation using both the techniques.

We believe this paper will be helpful for fellow researchers to understand these techniques and their applications with facts and scientifically assessed accuracy.

We intend to continue studying various image segmentation algorithms and techniques by conducting similar experiments, to understand effectiveness and accuracy on various natural as well as MRI images using the same accuracy measures to maintain consistency and provide mathematical foundation to the assessment. During study and experimentation, we will observe and assess, which algorithms are more suitable or effective for specific segmentation needs. We expect this research shall add value and assist fellow researchers since the recommendations are based on scientifically conducted experiments.

Acknowledgement. Anuja Deshpande thanks Leo Grady, the author of Random Walks interactive image segmentation technique, for his excellent work on this algorithm and making publicly available Matlab code, which enabled us to perform this experiment. His comments and suggestions shared with us during the experiment were pivotal to the successful completion of the experiment.

Compliance with Ethical Standards
✓ All authors declare that there is no conflict of interest.
✓ No humans/animals involved in this research work.
✓ We have used our own data.

References

1. Kaur, D., Kaur, Y.: Various image segmentation techniques: a review. Int. J. Comput. Sci. Mobile Comput. **3**, 809–814 (2014)
2. Adams, R., Bischof, L.: Seeded region growing. IEEE Trans. Pattern Anal. Mach. Intell. **16**(6), 641–647 (1994)
3. Besl, P.J., Jain, R.C.: Segmentation through variable-order surface fitting. IEEE Trans. Pattern Anal. Mach. Intell. **10**(2), 167–192 (1988)
4. Haralick, R., Shapiro, L.: Image segmentation techniques. Comput. Vis. Graph. Image Process. **29**(1), 100–132 (1985)
5. Sahoo, P., Soltani, S., Wong, A.: A survey of thresholding techniques. Comput. Vis. Graph. Image Process. **41**, 233–260 (1988)
6. Ballard, D., Brown, C.: Computer Vision. Prentice-Hall, Englewood Cliffs (1982)
7. Davis, L.: A survey of edge detection techniques. Comput. Vis. Graph. Image Process. **4**, 248–270 (1975)
8. Zucker, S.: Region growing: childhood and adolescence. Comput. Vis. Graph. Image Process. **5**, 382–399 (1976)
9. Horowitz, S., Pavlidis, T.: Picture segmentation by a directed split-and-merge procedure. In: Proceedings of 2nd International Joint Conference on Pattern Recognition, Copenhagen, Denmark, pp. 424–433 (1974)
10. Cheevasuvit, F., Maitre, H., Vidal-Madjar, D.: A robust method for picture segmentation based on split-and-merge procedure. Comput. Vis. Graph. Image Process. **34**(3), 268–281 (1986)
11. Chen, S., Lin, W., Chen, C.: Split-and-merge image segmentation based on localized feature analysis and statistical tests. CVGIP: Graph. Models Image Process. **53**(5), 457–475 (1991)
12. Pavlidis, T., Liow, Y.: Integrating region growing and edge detection. IEEE Trans. Pattern Anal. Mach. Intell. **12**(3), 225–233 (1990)
13. Meyer, F., Beucher, S.: Morphological segmentation. J. Vis. Commun. Image Represent. **1**, 21–46 (1990)
14. Vincent, L., Soille, P.: Watersheds in digital spaces: an efficient algorithm based on immersion simulations. IEEE Trans. Pattern Anal. Mach. Intell. **13**(6), 583–598 (1991)
15. Grady, L.: Random walks for image segmentation. IEEE Trans. Pattern Anal. Mach. Intell. **28**(11), 1768–1783 (2006)
16. Wallace, R., Ong, P.-W., Schwartz, E.: Space variant image processing. Int. J. Comput. Vis. **13**(1), 71–90 (1994)

17. Boykov, Y.Y., Jolly, M.P.: Interactive graph cuts for optimal boundary & amp; region segmentation of objects in N-D images. In: Proceedings of Eighth IEEE International Conference on Computer Vision, Vancouver, BC, Canada, vol. 1, pp. 105–112 (2001)
18. Boykov, Y., Kolmogorov, V.: An experimental comparison of min-cut/max- flow algorithms for energy minimization in vision. IEEE Trans. Pattern Anal. Mach. Intell. **26**(9), 1124–1137 (2004)
19. Lombaert, H., Sun, Y., Grady, L., Xu, C.: A multilevel banded graph cuts method for fast image segmentation. In: Proceedings Tenth IEEE International Conference on Computer Vision (ICCV 2005), Beijing, vol. 1, pp. 259–265 (2005)
20. Li, Y., Sun, J., Tang, C., Shum, H.: Lazy snapping. In: Proceedings of ACM SIGGRAPH, pp. 303–308 (2004)
21. Blake, A., Rother, C., Brown, M., Perez, P., Torr, P.: Interactive image segmentation using an adaptive GMMRF model. In: Proceedings of ECCV, pp. 428–441 (2004)
22. Kolmogorov, V., Blake, A.: GrabCut—Interactive foreground extraction using iterated graph cuts. In: ACM Transactions on Graphics, Proceedings of ACM SIGGRAPH 2004, vol. 23 (3), pp. 309–314 (2004)
23. Wechsler, H., Kidode, M.: A random walk procedure for texture discrimination. IEEE Trans. Pattern Anal. Mach. Intell. PAMI **3**, 272–280 (1979)
24. Gorelick, L., Galun, M., Sharon, E., Basri, R., Brandt, A.: Shape representation and classification using the poisson equation. IEEE Trans. Pattern Anal. Mach. Intell. **28**(12), 1991–2005 (2006)
25. Grady, L., Schwartz, E.L.: Isoperimetric partitioning: a new algorithm for graph partitioning. SIAM J. Sci. Comput. **27**(6), 1844–1866 (2006)
26. Harel, D., Koren, Y.: On clustering using random walks. In: Proceedings of 21st Conference on Foundations of Software Technology and Theoretical Computer Science, London, U.K, vol. 2245, pp. 18–41 (2001)
27. Yen, L., Vanvyve, D., Wouters, F., et.al.: Clustering using a random-walk based distance measure. In: Proceedings of the 13th Symposium on Artificial Neural Networks, pp. 317–324 (2005)
28. Tang, M., Gorelick, L., Veksler, O., Boykov, Y.: GrabCut in One Cut. In: 2013 IEEE International Conference on Computer Vision, Sydney, NSW, pp. 1769–1776 (2013)
29. Vicente, S., Kolmogorov, V., Rother, C.: Joint optimization of segmentation and appearance models. In: 2009 IEEE 12th International Conference on Computer Vision, Kyoto, pp. 755–762 (2009)
30. Powers, D.: Evaluation: from precision, recall and F-measure to ROC, informedness, markedness & correlation. J. Mach. Learn. Technol. **2**(1), 37–63 (2011)
31. Fawcett, T.: An introduction to ROC analysis. Pattern Recogn. Lett. **27**(8), 861–874 (2006)
32. Grady, L.: Random Walker image segmentation algorithm. http://leogrady.net/wp-content/uploads/2017/01/random_walker_matlab_code.zip. Accessed 20 July 2018
33. Gorelick, L.: OneCut with user seeds. http://www.csd.uwo.ca/~ygorelic/downloads.html. Accessed 06 Apr 2018
34. Tobon-Gomez, C.: Left Atrial Segmentation Challenge 2013: MRI training, 05 August 2015. https://figshare.com/articles/Left_Atrial_Segmentation_Challenge_2013_MRI_training/1492978/1. Accessed 20 July 2018
35. The Insight Segmentation and Registration Toolkit (ITK) toolkit. https://itk.org/ITK/resources/software.html. Accessed 20 July 2018
36. Mortensen, E., Barrett, W.: Interactive segmentation with intelligent scissors. Graph. Models Image Process. **60**(5), 349–384 (1998)
37. Dijkstra, E.: A note on two problems in connexion with graphs. Numer. Math. **1**, 269–271 (1959)

38. Kass, M., Witkin, A., Terzopoulos, D.: Snakes: active contour models. Int. J. Comput. Vis. **1**(4), 321–331 (1987)
39. Sethian, J.: Level set methods and fast marching methods. J. Comput. Inf. Technol. **11**(1), 1–2 (1999)
40. Zahn, C.T.: Graph-theoretical methods for detecting and describing gestalt clusters. IEEE Trans. Comput. **20**(1), 68–86 (1971)
41. Wu, Z., Leahy, R.: An optimal graph theoretic approach to data clustering: theory and its application to image segmentation. IEEE Trans. Pattern Anal. Mach. Intell. **15**(11), 1101–1113 (1993)
42. Shi, J., Malik, J.: Normalized cuts and image segmentation. IEEE Trans. Pattern Anal. Mach. Intell. **22**(8), 888–905 (2000)
43. Sarkar, S., Soundararajan, P.: Supervised learning of large perceptual organization: graph spectral partitioning and learning automata. IEEE Trans. Pattern Anal. Mach. Intell. **22**(5), 504–525 (2000)
44. Perona, P., Freeman, W.: A factorization approach to grouping. In: Proceedings of 5th European Conference on Computer Vision, ECCV 1998, vol. 1, pp. 655–670. Freiburg, Germany (1998)
45. Grady, L., Schwartz, E.L.: Isoperimetric graph partitioning for image segmentation. IEEE Trans. Pattern Anal. Mach. Intell. **28**(3), 469–475 (2006)
46. Barbu, A., Zhu, S.: Graph partition by Swendsen-Wang cuts. In: Proceedings Ninth IEEE International Conference on Computer Vision, Nice, France, vol. 1, pp. 320–327 (2003)
47. Vu, N., Manjunath, B.S.: Shape prior segmentation of multiple objects with graph cuts. In: Proceedings of IEEE Conference on Computer Vision and Pattern Recognition 2008, Anchorage, AK, pp. 1–8 (2008)
48. Wang, H., Zhang, H: Adaptive shape prior in graph cut segmentation. In: Proceedings of 2010 IEEE International Conference on Image Processing, Hong Kong, pp. 3029–3032 (2010)
49. Yu, Z., Xu, M., Gao, Z.: Biomedical image segmentation via constrained graph cuts and pre-segmentation. In: Proceedings of 2011 Annual International Conference of the IEEE Engineering in Medicine and Biology Society, Boston, MA, pp. 5714–5717 (2011)
50. Wu, X., Xu, W., Li, L., Shao, G., Zhang, J.: An interactive segmentation method using graph cuts for mammographic masses. In: Proceedings of 2011 5th International Conference on Bioinformatics and Biomedical Engineering, Wuhan, pp. 1–4 (2011)
51. Freedman, D., Zhang, T.: Interactive graph cut based segmentation with shape priors. In: Proceedings of IEEE Computer Society Conference on Computer Vision and Pattern Recognition (CVPR 2005), San Diego, CA, USA, vol. 1, pp. 755–762 (2005)
52. Furnstahl, P., Fuchs, T., Schweizer, A., Nagy, L., Szekely, G., Harders, M.: Automatic and robust forearm segmentation using graph cuts. In: 5th IEEE International Symposium on Biomedical Imaging: From Nano to Macro, Paris, pp. 77–80 (2008)
53. Ali, M., Farag, A.A., El-Baz, A.S.: Graph cuts framework for kidney segmentation with prior shape constraints. In: MICCAI, vol. 4791, pp. 384–392 (2007)
54. Zhou, J., Ye, M., Zhang, X.: Graph cut segmentation with automatic editing for Industrial images. In: Proceedings of International Conference on Intelligent Control and Information Processing, Dalian, pp. 633–637 (2010)
55. Wu, X., Wang, Y.: Interactive foreground/background segmentation based on graph cut. In: Proceedings of Congress on Image and Signal Processing, Sanya, Hainan, pp. 692–696 (2008)
56. Lempitsky, V., Kohli, P., Rother, C., Sharp, T.: Image segmentation with a bounding box prior. In: Proceedings of IEEE 12th International Conference on Computer Vision, Kyoto, pp. 277–284 (2009)

57. Lang, X., Zhu, F., Hao, Y., Wu, Q.: Automatic image segmentation incorporating shape priors via graph cuts. In: Proceedings of International Conference on Information and Automation, Zhuhai, Macau, pp. 192–195 (2009)
58. Jeleň, V., Janáček, J., Tomori, Z.: Mobility tracking by interactive graph-cut segmentation with Bi-elliptical shape prior. In: Proceedings of IEEE 8th International Symposium on Applied Machine Intelligence and Informatics (SAMI), Herlany, pp. 225–230 (2010)
59. Hou, Y., Guo, B., Pan, J.: The application and study of graph cut in motion segmentation. In: Proceedings of Fifth International Conference on Information Assurance and Security, Xi'an, pp. 265–268 (2009)
60. Chang, L., Hsu, W.H.: Foreground segmentation for static video via multi-core and multi-modal graph cut. In: Proceedings of IEEE International Conference on Multimedia and Expo, New York, NY, pp. 1362–1365 (2009)

A Comprehensive Survey on Web Recommendations Systems with Special Focus on Filtering Techniques and Usage of Machine Learning

K. N. Asha[✉] and R. Rajkumar

School of Computing Science and Engineering, Vellore Institute of Technology,
Vellore 632014, Tamil Nadu, India
{cuashin, vitrajkumar}@gmail.com

Abstract. In present scenario, to improve the consumers buying/purchasing experience, use of technologicalinnovations like Recommender system is very prominent. This small yet powerful utility analyzes the buying pattern of a consumer and suggests which items to buy or use. Recommender systems can be applied to varied fields of consumer's interest like online shopping, ticket booking and other online contents. Most of the e-commerce sites are attracting huge number of potential customers by providing useful suggestions regarding buying a product or service. This technique is mainly based on the use of Machine Learning, which enables the system to make decisions efficiently. Earlier, these recommendations were mainly relying on filtering process such as collaborative filtering, content based, knowledge based, demographic and hybrid filtering. These filtering techniques, which constitute the recommender system, are discussed in detail in this study. The survey is conducted to describe the various means and methods of recommendation to consumer in real time. This survey presents a comparative study among different types of recommender system based on various parameters and filtering schemes. Moreover, it shows a significant improvement in recommender system by using machine learning based approaches.

Keywords: Web recommendation · E-commerce · Filtering techniques · Machine learning

1 Introduction

Due to large scale data digitization and economic means of data processing, various companies and establishments have opted for customized production/service based on individuals' needs rather than the past trend of mass production [1]. With the deep penetration of internet over the time has filled the World Wide Web with overflowing information and web contents. E-commerce platforms have become the need of the hour as it is beneficial for companies which can offer customized products and services for consumers, who can choose services or products from multiple available options. This process required collection of large amount of data regarding various aspects of consumers' choice, need, buying pattern etc. Later this bulk data is processed and

© Springer Nature Switzerland AG 2020
S. Smys et al. (Eds.): ICCVBIC 2019, AISC 1108, pp. 1009–1022, 2020.
https://doi.org/10.1007/978-3-030-37218-7_106

analyzed to provide fruitful output which enables different companies in decision making regarding the customization and improvement in their offerings to consumers.

The major function of a recommender system is to create a list of products or service and along with this it offers suggestions to an individual consumer that the recommended products or services is best suited as per his/her liking or requirements [2]. Hence, the recommender system can also be termed as a Personalized Agent which caters to the requirements of individual consumer by suggesting accurate recommendations, which are most likely to be picked by the consumer. The response of a recommender system is customized and personalized for each individual.

As it is a general point that the recommender systems are based on different filtering techniques viz. content based, collaborative, knowledge based and demographic. These techniques, its merits and demerits are studied in detail in this paper. Using the combination of these techniques in harmony so as to obtain optimal results, a recommender system is called as the hybrid recommender system.

Recommender system's ability to provide unique suggestions to every individual depends heavily on the information received by the knowledge source. The knowledge source can have different origin which is extracted by processing of large amount of data.

1.1   Classification of Recommender System

The figure below [3], depicts the working of four types of filtering techniques, these techniques are discussed briefly (Fig. 1):

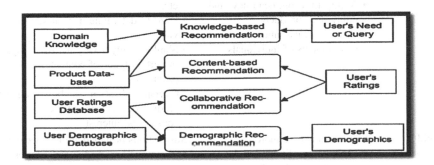

Fig. 1. Recommendation techniques and their knowledge sources

- **Collaborative:** As the name suggests this technique sorts the similar response and choices of different consumers in order to provide suggestions accordingly [4].
- **Content based:** This technique operates on the basis of product's specifications and qualities and rating of that product by an individual. This content based info system recommends the consumer based on individual's preferences [5].

- **Demographic:** This technique uses the demographic profile of an individual or group of individuals having similar demographic profile and provides suggestions to consumers belonging to similar demographic profile [6].
- **Knowledge based:** This recommender technique is based on knowledge sourced from an individual's requirements and choices. Based on this knowledge, the recommender system finds the product that suits the most as per an individual's choice [7].
- **Hybrid:** This technique combines useful technique to obtain optimal results. Majority of e-commerce giants rely on this method. Collaborative technique and content based technique are popular combinations under the hybrid recommendation [11]. Hybrid technique is more efficient in eliminating the limitations of traditional techniques (Table 1).

Table 1. Comparative analysis of filtering techniques: merits and demerits

Filtering technique	Merits	Demerits
Collaborative	• Least requirement of domain knowledge • Produces more accurate results even without user and product features	• Presence of cold start problem i.e. absence of product rating • Standardized products are required
Content based	• Usage data not required. Cold start problem is absent • Ability to recommend new and rarely used products • Catering to the unique requirements of individuals	• Incapable of observing consumers choice. e.g., Pictures, Music, Movies etc. • Hard to determine item or service value • Shallow: only a very narrow analysis of certain types of content can be provided • Overspecialization: Unable to suggest product or services outside consumers' interest
Hybrid technique	Combines two or more techniques by eliminating the limitation of each technique	Suggestions provided are less accurate and cover a broad spectrum of users
Demographic	Independent of the ratings provided by consumer	Chance of inaccurate classification can't be ruled out
Knowledge-based	It is based on explicit information without the need of consumer ratings	Probability of data overload is evident as it is challenging to store consumers profile on large scale

1.2   Relevance of Recommender Method

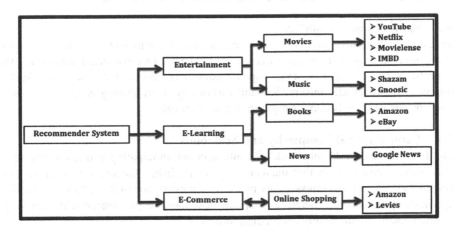

Fig. 2. Application of recommender system

The Recommender system is more relevant in the process and services which require instant response just like in online shopping platforms. Through this, one gets to know about the uses of recommender system. The Fig. 2 shows a categorization of different applications of recommender system that has been introduced which are being utilized by the entertainment world along with e-learning and e-business etc.

1.3   Barriers that are Being Faced by the Web-Based Recommender System

Automatic recommender systems aims at offering recommendations of the items and services that can be appreciated by the user, by considering their past interest and choices depending on the ratings, and history of purchase. These shared filtering techniques fail when they are transplanted into decentralized developments. Arranging recommender systems on Semantic Web involves miscellaneous and complicated issues. The major issues that can be seen in semantic web recommender systems are:

1.3.1   Ontological Commitment
The Semantic Web recommender system is featured with machine-decipherable content that has been dispersed on the Web. For assuring that agents can get the knowledge and understand about latter information, semantic interoperability via ontologies should be prepared [8].

1.3.2   Interaction Facilities
In the initiating time, decentralized recommender systems were being subject to multi-agent research projects. It enables people to communicate with their peers through making synchronized communication exchange feasible. The Semantic Web comprises

an intrinsically data- centric environment replica. Thus it can be evaluated that message exchanges have become restricted to asynchronous communication.

1.3.3 Security and Credibility

Congested communities normally own proficient and competent mediums for controlling user identity and castigate malicious manners. Decentralized frameworks, in the midst of those shared systems, open business centers and the Semantic Web, moreover, cannot evade misdirection and dishonesty. Caricaturing & character producing in this manner become simple to accomplish [9].

1.3.4 Computational Complexity and Scalability

Centralized method and framework take into account anticipating and restricting network estimate and may in this manner adapt their sifting frameworks to guarantee adaptability. Registering comparability measures for every one of these "people" in this way winds up infeasible. Therefore, adaptableness must be guaranteed while limiting last calculations to sufficiently limit neighborhoods.

1.3.5 Low Profile Overlap

Concentrated profiles are normally characterized by vectors demonstrating the customer's area of interest and their choices for different products. In order to reduce the dimensionality and ensure profile overlap, Recommenders, [10] function in the sphere where product sets are moderately small. There it requires incorporating few new efficient methods for ensuring the solution of profile overlap issue along with making profile similarity measures meaningful.

2 Related Work

2.1 Content Based Filtering Techniques

According to Gemmis et al. [16] discussed that Content-based recommender systems (CBRSs) are based on product and customer depictions for making product representations and user profiles which will be efficiently utilized to recommend products and services similar to those items which people liked the most in the previous times or similar like those which hold a great clientele base in the past decades. This chapter is trying to introduce such techniques that can be efficient and competent in offering better suggestions to the CBR methods which will help business to become more prosperous and fruitful (Fig. 3).

Bocanegra et al. [17] have stated that on channel like Health and fitness related videos are quite famous, but the matter of issue is their quality, they are generally of low quality but still they are being followed by people. One of the best methods for improving the quality videos is generating impressive and educational health content, to offer information about how to gain fitness and maintain health. This paper put emphasize on investigating the possibility of framing a content-based method for recommendation which links consumers to repute health informative website like Medline Plus for a given health video from YouTube.

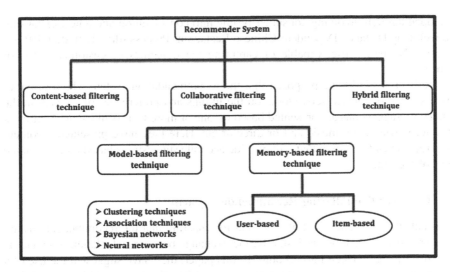

Fig. 3. Application of recommender system through implementing different methods

Bianchini et al. [20] have suggested that the idea of food recommender system is apt and suggestive in providing customers- modified, personalized as well as healthy menus along with considering consumer's short/long-term likes and medical prescriptions. Actually the chief novel contribution of the projected scheme is the usage of recommendation types linked with customer's profile that offers the idea of their medical condition and thus help in selecting food according to their behaviors too (Table 2).

Table 2. Comparative analysis of content-based filtering techniques

Article	Involvement	Performance evaluation	Dataset	Application region	Algorithm name
Bocanegra et al. [17]	Video recommendation	Mean Accuracy	Videos on Youtube	Videos related to health	HealthRec-Sys
Bianchini et al. [20]	Recommendation the basis of food stuff selection	Average response time and excellence of recommendation	Method dataset acquired through the BigOven.com	Food Prescription and formation of the menu	PREFer
Bagher et al. [21]	Consumer behavior	Accuracy, remind and F1- Evaluation	Data derived from twitter's page, New York Times and BBC articles	News recommendation	Evolutionary recommender system
Reddy et al. [26]	Provides movie recommendation based on the movie's genres	Euclidean distances	Movie Lens dataset	Publication recommender	Content-based filtering using genre correlation

In the words of Bagher et al. [21], this study portrays the theory of trend for calculating the interests of user while choosing items in the midst of different groups of that are offering similar items or services. Trend based user (TBU) approach is developed through involving consumer profile into consideration and a new expansion in developing Distance Dependent Chinese Restaurant Process (dd-CRP). dd-CRP is a Bayesian Nonparametric, capable of capturing the dynamics of customer's area of interest.

Reddy et al. [26] have proposed that the recommendation system which provides movie suggestions to viewers, works on the likings and preferences of viewers in the past. These suggestions are presented based on one or more factors like that of actors in a movie, movie genre, music and or director etc. Here they have presented a content based recommender system that will provide accurate suggestions based entirely on the genre of the film.

2.2 Collaborative Filtering Recommender System

It was calculated by Nilashi et al. [24] that for ensuring that the recommendations are being provided efficiently and instantly, the exchange between calculation time and its correctness must be taken into account. In this paper, they have suggested the use of a fresh hybrid recommendation technique using Collaborative filtering (CF) method. Use of fresh technique employs dimensionality reduction technique (DRT) and ontology method, which shall provide effective remedy of two major limitations i.e. Shortage and Expansion (Table 3).

Table 3. Comparative analysis of collaborative filtering techniques

Article	Contribution	Performance measurement	Dataset	Application area	Algorithm name
Nilashi et al. [24]	Dimension reduction, SVD, clustering and Ontology	MAE, Precision, recall, F1-score	MovieLens	Online Movie recommendation	User-and-item-based + SVD + EM + Ontology
Parvin et al. [29]	Rank, rating, trust and ACO optimization	MAE, RMSE, Rate coverage	FilmTrust, Epinions and Ciao	Online trust and opinion mining	TCFACO
Linda et al. [33]	Fuzzy logic and context aware model	MAE, coverage	LDOS-CoMoDa, Restaurant-Customer	Movie and restaurant recommendation	Context-aware RSs (CARSs)
Xiong et al. [34]	Combination of collaborative filtering and textual content along with deep learning	Precision, Recall, Normalized discounted cumulative gain, F1-score	Private real-world dataset	Web service recommendation	Deep hybrid collaborative filtering approach for service recommendation (DHSR)

Parvin et al. [29] have proposed a fresh approach based on CF technique, which forecasts about the absent ratings, correctly. This approach is termed as TCFACO - Trust-aware collaborative filtering method based on ant colony optimization. It performs three functions; firstly, it segregates the consumers based on the ratings and social trust. Next, values are allocated to the consumer in order to find resemblance with the potential consumers. Later, a cluster of users with similar preferences are selected to determine the unidentified ratings.

Linda et al. [33] suggested that despite being extensive use of CF, it faces the issue of insufficiency. To address the issue they present the addition of fuzzy trust into CARSs, this will solve the existing problem of insufficiency without compromising on the quality if output. This is a two-step process; they first introduce fuzzy trust and establish this to context-aware CF technique for enhanced suggestions. Later introduce fuzzy trust propagation to mitigate insufficiency problem and provide accurate suggestions.

Xiong et al. [34] stated that machine learning finds extensive use in matrix factorization based CF recommender system for selecting a web service. It is difficult to confine the composite relations between web applications and web components, thus resulting in inferior suggestions. To solve this problem a fresh deep learning hybrid model is developed by the union of CF text/print content and to identify the composite relation between web applications and web components, deep neural network is utilized.

2.3   Hybrid Filtering Techniques

Kouki et al. [18] presented a system termed as HyPER (HYbrid Probabilistic Extensible Recommender), establishes and questions over a large spread of data origins. Such data sources/origins comprise of various consumer-consumer, product- product identical qualities, web content and social details. HyPER is designed to harmonize various information signals to intelligently make recommendations. This ensures enhanced and better recommendations by employing graphical representation termed as hinge-loss Markov random fields.

According to Tarus et al. [19], various recommender systems have proved to be successful yet suffer from insufficient correctness, particularly in case of online education because of the dissimilarity in students' understanding level, grasping power, and learning sequence. To address this issue, they suggest a hybrid recommender system based on knowledge that uses ontology and Sequential Pattern Mining (SPM) to suggest online education resource to students. This method explores learners' chronological learning model. Also, the hybrid technique can tackle cold start and data shortage issues.

Vashisth et al. [23] discussed the use of fuzzy logic to house multiplicity and vagueness in consumers' preferences and concerns. This assures in providing more accurate suggestions with dissimilar preferences. They have devised a hybrid recommender system based on fuzzy logic with the fresh use of type 2 fuzzy sets to generate consumer models competent enough to identify the natural vagueness of human actions in terms of different choices among different consumers. This system enhances the correctness of the recommendations by minimizing the room for error (Table 4).

Table 4. Comparative analysis of hybrid filtering techniques

Article	Contribution	Performance measurement	Dataset	Application area	Algorithm name
Kouki et al. [18]	Probabilistic soft logic approach for prediction	RMSE and MAE	Yelp academic dataset and the Last.fm dataset	Online Movie recommendation	HyPER
Tarus et al. [19]	Ontology creation, similarity computation for prediction, generating top learning items and Final recommendation	MAE, Precision, recall	Learning management system (LMS) of a public university	e-learning	(CF + Onto + SPM)
Vashisth et al. [23]	Interval Type-2 fuzzy rule set	Recall, Confusion matrix, RMSE	Book purchase and shopping women apparels	Product recommendation system	Fuzzy feature combination hybridization
Neidhardt et al. [35]	Seven factors are considered based on user profile	Distance similarity	Private data	Tourism recommendation	PixMeAway GUI

In their research, Wan et al. [30] have presented a hybrid filtering(HF) recommendation system integrated with Learner Influence Model (LIM)-to collect interpersonal data, Self-Organizing based (SOB) recommendation approach-to suggest best possible student set for active students and strategy and sequential pattern mining (SPM) to suggest learning objects (LOs) to students.

It was proposed by Neidhardt et al. [35] that travel and tourism services pose complex test to a recommender system. In this study, they present a novel way by using pictorial depiction to help completely extract visitors' choice for tourism services/products. The model is created using 7 attributes, later the traveler selects the images of his/her choice and an attribute is scored on its basis.

2.4   Application Based Recommender Systems

Chau et al. [22] presented the description of Content Wizard; it is also a concept-based RS for proposing learning and reading materials that can match up with the requirement and instructor's educational objectives while designing an online programming course. In this technique the instructors are required to offer a set of code instances that in cooperation represent the learning objectives which are supposed to be associated with every course unit. This Content Wizard tool help the instructor to minimize the time he needs to spend in task management thus this tool is time efficient.

Porcel et al. [25] offered a scheme called Sharing Notes, which is an educational and academic societal network which helps people in developing personalized recommendations for enhancing the quality of teaching and learning practices. To accomplish this thought and idea it requires integrating a hybrid RS which utilizes ontology to describe the maintenance of trust amongst its users in the network. Along with this it adopts the furry linguistic modeling for improving the illustration of information (Table 5).

Table 5. Comparative analysis of application based recommender system

Article	Application	Algorithm	Dataset	Performance
Chau et al. [22]	Programming course instruction	Content Wizard	Java Course data from University of Pittsburgh	Precision, Recall and F1- Score
Porcel et al. [25]	Teaching & learning process	SharingNotes	Epinions dataset	Precision
Sridevi et al. [28]	Hospital selection	Finding right doctors and hospitals	Private dataset	Weightage of votes and ratings

Sridevi et al. [28], presented a personalized data concerning the healthcare suppliers and hospitals in an easy, modified, and well thought-out way for the users and thus assisting them in managing their health issues. Under this approach user is being offered by the opportunity of creating his ownprofile by considering input rating, and after that they are being provided by a list of popular hospitals and doctors using mutual filtering with the help of the recommender system. A combined module is projected by considering the available demographic data, alike user's rating forecast, total number of reviews and ratings supported on average ratings will help to solve cold start newuser problem.

2.5 Machine Learning Based Approaches

Meza et al. [27] concluded that the earning sources of different nationwide and local jobs that come under government are all based on taxation. Escalating the earning with tax payers has always been a major challenge for government institutions. In the present study Maze et al. have established a fuzzy-based RS approach and its preliminary outcomes that are being implemented to a dataset from the Municipality of Quito, offer their suggestions to the local peoples about their payments behavior. This projected approach enhanced people's awareness towards the tax payments and thus it helps in increasing governmental institutions' financial profits.

According to Sattar et al. [31], one of the main challenges which this theory is facing is of making useful recommendation from the available sources and set of millions of items with sparse ratings. Different and various types of approaches have been projected for enhancing the quality and accurateness, but somewhere they all have overlooked possible barriers that can hinder the process of RS. Few of the issues are sparsity and cold start problems. Sattar et al. [31] tried to suggest a new hybrid recommendation approach which unites content-based filtering with collaborative filtering that overcome above mentioned issues.

Subramaniyaswamy et al. [32] have pointed out the all the issues associated with recommender system approach, like scalability, sparsity, and cold-start. They have designed knowledge-based sphere particular ontology for developing individualized recommendations. They have also established two new and dissimilar approaches, first is ontology-based prognostic models and prominent representation model. This prediction model is persuaded by data mining algorithms by correlating the interests of the user. They have introduced a novel alternate of KNN algorithm forthe collaborative filtering based recommender system known as Adaptive KNN.

Table 6. Comparative analysis of machine learning based recommender system

Article	Application	Algorithm	Dataset	Performance
Meza et al. [27]	RS for tax payments	Fuzzy logic	Municipality of Quito (payment behavior)	NA (new discount methods suggested)
Sattar et al. [31]	Movie recommendation	KNN, SVM, Naïve Bayes, decision tree	IMDB movie dataset	MAE (Mean Absolute Error)
Subramaniya swamy et al. [32]	Movie recommendation	Adaptive KNN	Movie lens	Precision, recall and accuracy

2.6 Merits and Demerits of Content-Based Filtering Methods

Problems faced in collaborative filtering method are solved in the content based (CB) filtering technique. These are independent of ratings of products/services; as a result one can get suggestions without compromising the correctness of the recommendations. In case the consumer changes his choice, the CB filtering method easily adapts to such changes and quickly provides fresh suggestions. It provides suggestions without revealing the identity of consumers; this maintains confidentiality [12]. CB filtering method does face many issues as mentioned in the paper [13]. Metadata is of great importance for CB filtering method as they need detailed information of products and well maintained profile of consumers. This is termed as Limited Content Analysis. Overspecialized content [14] also challenges this technique as the products are suggested to consumers is similar to the products already mentioned in their profiles.

2.7 Merits and Demerits of Collaborative Filtering Methods

Collaborative filtering methods benefits over CBF as the former easily executes in areas where there is limited information about the products and is tough to get it analyzed by a computer. Although, CF techniques are widely used, it suffers some probable challenges, discussed below.

2.7.1 Cold-Start Problem

This means that in absence of relevant information about consumers as well as products, recommender system can't produce desired results [2]. This affects the correctness of the suggestions. As the new consumers have not provided any personal details nor rated any product, so their profile remains incomplete making it impossible for the recommender system to know about his preferences.

2.7.2 Data Sparsity Problem
This issue refers to shortage of required data, i.e., when a consumer has reviewed only a few products in comparison to other consumers who have rated in large numbers [15]. This results in thin consumer-product template, i.e., incapability find perfect neighbors thus ultimately production of inaccurate suggestions.

2.7.3 Scalability
This refers to the linear growth in the consumers and products count, which poses challenges in providing suggestions [15]. It is therefore vital to develop recommender methods that are able to scale up with the growth of items and users data. Dimensionality reduction techniques, like (Singular Value Decomposition) SVD method, are incorporated to address the issue and generate more accurate suggestions.

2.7.4 Synonymy
When similar products have multiple names, it creates a situation of Synonymy. In order to handle this problem several approaches, like automatic term expansion thesaurus construction, and SVD, particularly Latent Semantic Indexing are used. These approaches also suffer from a limitation where additional terms may have different meanings from what is essential, thus lowering the quality of suggestions.

3 Conclusion

In this paper, we have presented a detailed survey report on various recommender systems that have been studied how data analytics and multiple filtering techniques are applied in real time applications to generate most of the desirable suggestions. It was found while discussing various filtering techniques like collaborative, content based and hybrid filtering methods, that hybrid filtering is a proven to be the most suited technique for producing the most accurate suggestions to consumers.

Present era of social networking sites and e-commerce/e-learning platforms have penetrated the web worldwide and is growing exponentially. As a result, there is a huge data available, regarding social network information, on the web. This vital information can be utilized by the recommendation system to generate the most accurate recommendations to its users. In order to maintain the secrecy of users' information it is suggested to develop a novel recommender system to address privacy issues in the near future.

References

1. Aggarwal, C.C.: Recommender Systems, pp. 1–28. Springer, Cham (2016)
2. Burke, R.: Hybrid Web Recommender Systems, pp. 377–408. Springer, Heidelberg
3. Burke, R.: Hybrid recommender systems: survey and experiments. UMUAI **12**(4), 331–370 (2002)
4. Ekstrand, M.D., Riedl, J.T., Konstan, J.A.: Collaborative filtering recommender systems. Found. Trends® Hum. Comput. Interact. **4**(2), 81–173 (2011)

5. Wang, D., Liang, Y., Xu, D., Feng, X., Guan, R.: A content-based recommender system for computer science publications. Knowl.-Based Syst. **157**, 1–9 (2018)
6. Al-Shamri, M.Y.H.: User profiling approaches for demographic recommender systems. Knowl.-Based Syst. **100**, 175–187 (2016)
7. Carrer-Neto, W., Hernández-Alcaraz, M.L., Valencia-García, R., García-Sánchez, F.: Social knowledge-based recommender system. application to the movies domain. Expert Syst. Appl. **39**(12), 10990–11000 (2012)
8. Golbeck, J., Parsia, B., Hendler, J.: Trust networks on the semantic web. In: Proceedings of Cooperative Intelligent Agents, Helsinki (2003)
9. Ziegler, C.N., Lausen, G.: Analyzing correlation between trust and user similarity in online communities. In: Jensen, C., Poslad, S., Dimitrakos, T. (eds.) Proceedings of the 2nd International Conference on Trust Management. Volume 2995 of LNCS, pp. 251–265. Springer, Oxford (2004)
10. Miller, B., Albert, I., Lam, S., Konstan, J., Riedl, J.: MovieLens unplugged: experiences with an occasionally connected recommender system. In: Proceedings of the ACM 2003 Conference on Intelligent User Interfaces. ACM, Chapel Hill, NC, USA (2003)
11. Smyth, B.: Case-based recommendation. In: Brusilovsky, P., Kobsa, A., Nejdl, W. (eds.) The Adaptive Web, pp. 342–376. Springer, Heidelberg (2007)
12. Shyong, K., Frankowski, D., Riedl, J.: Do you trust your recommendations? an exploration of security and privacy issues in recommender systems. In: Emerging Trends in Information and Communication Security, pp. 14–29. Springer, Heidelberg (2006)
13. Adomavicius, G., Tuzhilin, A.: Toward the next generation of recommender system. a survey of the state-of- the-art and possible extensions. IEEE Trans. Knowl. Data Eng. **17**(6), 734–749 (2005)
14. Zhang, T., Vijay, S.I.: Recommender systems using linear classifiers. J. Mach. Learn. Res. **2**, 313–334 (2002)
15. Park, D.H., Kim, H.K., Choi, I.Y., Kim, J.K.: A literature review and classification of recommender systems research. Expert Syst. Appl. **39**(11), 10059–10072 (2012)
16. De Gemmis, M., Lops, P., Musto, C., Narducci, F., Semeraro, G.: Semantics-aware content-based recommender systems. In: Recommender Systems Handbook, pp. 119–159. Springer, Boston (2015)
17. Bocanegra, C.L.S., Ramos, J.L.S., Rizo, C., Civit, A., Fernandez-Luque, L.: HealthRecSys: a semantic content-based recommender system to complement health videos. BMC Med. Inform. Decis. Making **17**(1), 63 (2017)
18. Kouki, P., Fakhraei, S., Foulds, J., Eirinaki, M., Getoor, L.: Hyper: a flexible and extensible probabilistic framework for hybrid recommender systems. In: Proceedings of the 9th ACM Conference on Recommender Systems, pp. 99–106. ACM, September 2015
19. Tarus, J.K., Niu, Z., Yousif, A.: A hybrid knowledge-based recommender system for e-learning based on ontology and sequential pattern mining. Future Gener. Comput. Syst. **72**, 37–48 (2017)
20. Bianchini, D., De Antonellis, V., De Franceschi, N., Melchiori, M.: PREFer: a prescription-based food recommender system. Comput. Stand. Interfaces **54**, 64–75 (2017)
21. Bagher, R.C., Hassanpour, H., Mashayekhi, H.: User trends modeling for a content-based recommender system. Expert Syst. Appl. **87**, 209–219 (2017)
22. Chau, H., Barria-Pineda, J., Brusilovsky, P.: Content wizard: concept-based recommender system for instructors of programming courses. In: Adjunct Publication of the 25th Conference on User Modeling, Adaptation and Personalization, pp. 135–140. ACM, July 2017
23. Vashisth, P., Khurana, P., Bedi, P.: A fuzzy hybrid recommender system. J. Intell. Fuzzy Syst. **32**(6), 3945–3960 (2017)

24. Nilashi, M., Ibrahim, O., Bagherifard, K.: A recommender system based on collaborative filtering using ontology and dimensionality reduction techniques. Expert Syst. Appl. **92**, 507–520 (2018)
25. Porcel, C., Ching-López, A., Lefranc, G., Loia, V., Herrera-Viedma, E.: Sharing notes: an academic social network based on a personalized fuzzy linguistic recommender system. Eng. Appl. Artif. Intell. **75**, 1–10 (2018)
26. Reddy, S.R.S., Nalluri, S., Kunisetti, S., Ashok, S., Venkatesh, B.: Content-based movie recommendation system using genre correlation. In: Smart Intelligent Computing and Applications, pp. 391–397. Springer, Singapore (2019)
27. Meza, J., Terán, L., Tomalá, M.: A fuzzy-based discounts recommender system for public tax payment. In: Applying Fuzzy Logic for the Digital Economy and Society, pp. 47–72. Springer, Cham (2019)
28. Sridevi, M., Rao, R.R.: Finding right doctors and hospitals: a personalized health recommender. In: Information and Communication Technology for Competitive Strategies, pp. 709–719. Springer, Singapore (2019)
29. Parvin, H., Moradi, P., Esmaeili, S.: TCFACO: trust-aware collaborative filtering method based on ant colony optimization. Expert Syst. Appl. **118**, 152–168 (2019)
30. Wan, S., Niu, Z.: A hybrid e-learning recommendation approach based on learners' influence propagation. IEEE Trans. Knowl. Data Eng. (2019)
31. Sattar, A., Ghazanfar, M.A., Iqbal, M.: Building accurate and practical recommender system algorithms using machine learning classifier and collaborative filtering. Arab. J. Sci. Eng. **42**(8), 3229–3247 (2017)
32. Subramaniyaswamy, V., Logesh, R.: Adaptive KNN based recommender system through mining of user preferences. Wirel. Pers. Commun. **97**(2), 2229–2247 (2017)
33. Linda, S., Bharadwaj, K.K.: A fuzzy trust enhanced collaborative filtering for effective context- aware recommender systems. In: Proceedings of First International Conference on Information and Communication Technology for Intelligent Systems, vol. 2, pp. 227–237. Springer, Cham (2016)
34. Xiong, R., Wang, J., Zhang, N., Ma, Y.: Deep hybrid collaborative filtering for web service recommendation. Expert Syst. Appl. **110**, 191–205 (2018)
35. Neidhardt, J., Seyfang, L., Schuster, R., Werthner, H.: A picture-based approach to recommender systems. Inf. Technol. Tourism **15**(1), 49–69 (2015)

A Review on Business Intelligence Systems Using Artificial Intelligence

Aastha Jain$^{(\boxtimes)}$, Dhruvesh Shah, and Prathamesh Churi

Mukesh Patel School of Technology and Management and Engineering,
Vile Parle (West), Mumbai 400056, India
aasthajainl.nmims@gmail.com,
dhruveshshah.nmims@gmail.com,
Prathamesh.churi@nmims.edu

Abstract. Artificial Intelligence [AI] is paving its way into the corporate world focusing at the fundamentals of the business systems which are related to manufacturing, entertainment, medicine, marketing, engineering, finance and other services. Businesses today have a lot of data which can be used in the field of artificial intelligence in the form of data training, learning live data and historic data. Dealing with complex and perplex situations can be done immediately with these automated systems thus saving a lot of time. This paper puts forward a framework with different methods that makes use of artificial intelligence in business systems today. These methods include Swarm intelligence and Port intelligence which can even exceed human intelligence and be of great help to the corporate world. These help in constructing frameworks and algorithms that have helped the business systems in achieving the optimum results.

Keywords: Artificial intelligence · Business intelligence · Data mining · Swarm intelligence · Port intelligence

1 Introduction

Artificial Intelligence has a fundamental purpose of creating intelligent machines and computer programs that are capable of understanding and learning and sometimes exceeding the power of human intelligence [1, 9]. Presently AI is used in many equipments, games, automations software etc. [8, 11]. It heavily depends on the data training, historic data and the live data to learn (Fig. 1).

Emergence of AI in Business:
After the arrival of web enabled infrastructure and quick steps made by the AI development community, AI has been increasingly used in business applications [11]. Performing mechanical competition using program rules is very easily performed by the computers and these simple monotonous tasks are efficiently and robustly solved better than the humans. A form of the computers even better than the humans. But unlike humans' computers may have difficulty in understanding some specific scenarios and situations which is tackled by artificial intelligence. Thus, most of the year the search is based on understanding the intelligent human behaviour which involves

© Springer Nature Switzerland AG 2020
S. Smys et al. (Eds.): ICCVBIC 2019, AISC 1108, pp. 1023–1030, 2020.
https://doi.org/10.1007/978-3-030-37218-7_107

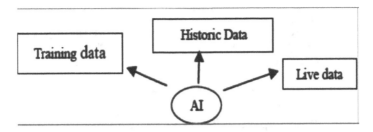

Fig. 1. Learning journey of AI [2]

problem solving skills abstract thoughts, high-level reasoning and recognition of patterns [3]. Artificial intelligence helps us comprehend this procedure by creating scenarios and recreating it potentially helping us boost beyond our abilities. It includes the following:

- Attributes of intelligent behaviour
- Reasoning and thinking
- Reasoning to solve problems
- Repeating and understanding from experience
- Applying the acquired knowledge
- Exhibiting creativity and imagination
- Dealing with complex and perplex situations
- Responding rapidly and efficiently to new situations
- Handling ambiguous and random information

1.1 Components of Business Intelligence

Businesses today have admittance to a lot of data. Business intelligence puts the data of these series of methods and makes more profit. The major components of business intelligence are (Fig. 2):

1. OLAP (Online Analytical Processing)

This component of BI allows administrators to categorize and choose combinations of data for their monitoring. The foremost purpose of OLAP is to recapitulate information from the large database for the decision-making process [7]. The report is then generated using OLAP and can be accessible in a format as per the user's requirements [10]. It has many advantages including quick response to queries, flexible working with data and helps user create views using spreadsheet.

2. Data mining

Data mining is the process of examining data from various sources and giving a brief information about it which can be used to increase profits and decrease the costs. It's chief purpose is to find associations or patterns among loads of fields in huge databases.

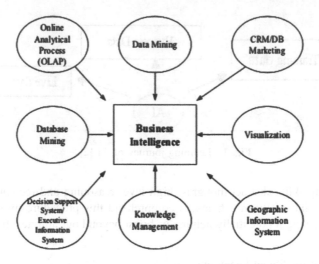

Fig. 2. Components of business intelligence [5]

3. CRM marketing

To successfully manage customer relationships, the organization must serve their customers in a particular manner and to understand their buying behavior and keeping in mind their needs. Customer satisfaction for various markets, products and services should be kept in mind too.

4. GIS

The collective use of Geographical Information Systems and Business Intelligence today is helping businesses function and increase productivity and decrease the costs. Applications allow companies to track employees, machinery, assets and also collect and deduce data in real time to make knowledgeable decisions.

5. Knowledge management

Knowledge management helps the business to gain awareness and understanding from its own experience. Detailed knowledge management activities help emphasis on obtaining, stowing and exploiting knowledge for active learning, problem solving, tactical planning and decision making.

2 Related Works

2.1 Swarm Intelligence

Swarm Intelligence is an algorithm which is a recent and a growing example in intelligent computing which is inspired by biology. It is used in the real-world applications for building adaptive systems. Swarm intelligence used tracking algorithms which are encouraged by the shared behavior of animals which display regionalized, self-organized patterns in the scavenging procedure [12].

1. Genetic Algorithm (GA)

The Genetic Algorithm is an algorithm which discovers improved explanations with the help of nature enthused by evolutionary biology like reproduction, selection, mutation, and also recombination. We find estimated solutions using this search technique. A generation is every next iteration. Iterations from 50 to 500 or more generations are natural for any GA. Run is a complete set of generations. A single or more significantly appropriate chromosomes in the population after the final run ends. Report statistics are frequently averaged over multiple runs of this algorithm on the exact same problem by the researchers.

2. Honeybee Foraging Mechanism

Bees adopt the well-organized communication of food source and their foraging activities are inspiring. Employed bees collect the food in the hive. Bees go over different flowers for the collection of food. If there is a sufficient amount of food, the bee makes many journeys between its hive and the flowers. The food is then transported to onlooker bees through "waggle dance". The dance is an efficient way of sharing information and communication among the bees. Onlooker bees observe the dance and hence the know the position of food and it's amount. This mechanism assists the colony to pleat food rapidly and efficiently.

3. Artificial Bee Colony Algorithm

Employee bees go find food and their number is directly proportional to the amount of food. The scout bee is the superfluous food source of employed bee. The search space is searched by the bees. This information is then shared and passed on to the onlooker bees which then chose a food source in the neighborhood. In this iterative algorithm, the total number of bees, onlooker or employed is the same as number of solutions. For each and every iteration, every single bee employed will discover a source of food which is in the neighborhood of its present source of food and then assesses the measure of nectar it has i.e., appropriateness to the purpose.

1. Initializing all the values.
2. The employed bees evaluate the amount of nectar it has i.e., suitability to the purpose, by moving from one source to another
3. The onlooker bees are positioned based on the quantity of nectar attained by bees employed.
4. Scout bees discover new food sources
5. Good sources of food are memorized
6. Go to step 2, if termination criteria is not fulfilled; otherwise pause the procedure and demonstrate the best source of food found so far (Fig. 3).

This algorithm firstly normalizes the database according to homogenous data after the database is prepared. Then analysis is done with the metaheuristics algorithm where GA and ABC are both implemented simultaneously and evaluated to give performance analysis. We then implement the proposed Artificial bee optimization using a comparative study with GA and back propagation to find the optimal solution.

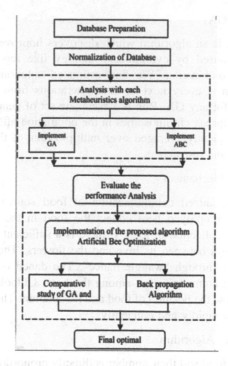

Fig. 3. Artificial bee optimization algorithm [4]

2.2 Port Business Intelligence Framework Design

This framework design consists of four layers namely data extraction layer, data warehouse layer, data mining layer and front layer:

1. Data Extraction layer: The port production system takes its data from huge oracle databases and stores it. Data extraction simply selects the source of the data, cleans the empty and duplicate data, it clears any errors that exists and makes the data reliable and more accurate.
2. Data Warehouse layer: Data mining of the stored data is carried on along with the cleaning of data. A warehouse model is build which consists of fact and dimension model.
3. Data mining layer: This will be integrated to the data warehouse layer using the clustering analysis method. The main idea of this method is to group similar things based on the similarity variable attributes on the foundation of multidimensional space information.
4. The front layer: It divides the business reports and data analysis into customer policies and financial risk control. According to this analysis, it gives direction and a path of collaboration for the business executives (Fig. 4).

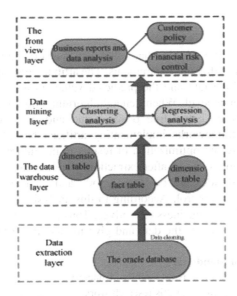

Fig. 4. Port business intelligence system [6]

3 Inference

The above values have been obtained from [4]. On comparing the genetic algorithm with the artificial bee algorithm, we have obtained Table 1. ABC works better with True variables whereas genetic algorithm works better with false variables.

Table 1. Comparison of genetic algorithm and artificial bee colony

Analysis variable	Genetic algorithm	Artificial bee colony
True +ve	233	374
False +ve	237	75
False -ve	274	127
True +ve rate	0.47	0.74
False +ve rate	0.18	0.15

Swarm intelligence has very less constraints on its applications and hence used by the U.S. military for supervising unmanned vehicles. It is also now researched by NASA for its implementation in planetary mapping.

Port business decision making on the other hand is restricted by external factors, except when it is under the stimulus of internal data. Formation of heterogenous data like data, audio, text, and external enterprise web page information may also be different.

4 Conclusion

The paper summarizes and concludes that different frameworks in the business systems are suitable according to different requirements. The artificial bee algorithm gives better results in the true positive rate and the predicted values proving that computers can exceed human intelligence level. When general algorithm and artificial bee colony are compared, ABC gives us better true positive and and positive predicted values. Business intelligence frameworks implementing artificial bee algorithms should be implemented to increase the optimization of results in the future. In port intelligence which consists of four main layers, after extraction, warehousing and mining of data is done, the front layer divides the business reports and data analysis into customer policies and financial risk control. According to this analysis, it gives direction and a path of collaboration for the business executives. Thus each of these frameworks help us in a different way to analyze the data and give the required predictions.

Compliance with Ethical Standards
✓ All authors declare that there is no conflict of interest
✓ No humans/animals involved in this research work.
✓ We have used our own data.

References

1. Quan, X.I., Sanderson, J.: Understanding the artificial intelligence business ecosystem. IEEE Eng. Manage. Rev. **46**(4), 22–25 (2018)
2. Kiruthika, J., Khaddaj, S.: Impact and challenges of using of virtual reality & artificial intelligence in businesses. In: 2017 16th International Symposium on Distributed Computing and Applications to Business, Engineering and Science (DCABES), Anyang, pp. 165–168 (2017)
3. Bai, S.A.: Artificial intelligence technologies in business and engineering (2011)
4. Anusha, R., Nallaperumal, K.: Business intelligence: an artificial bee colony optimization approach. In: 2014 IEEE International Conference on Computational Intelligence and Computing Research, pp. 1–4. IEEE, December 2014
5. Wu, J.Y.: Computational intelligence-based intelligent business intelligence system: concept and framework. In: 2010 Second International Conference on Computer and Network Technology, pp. 334–338. IEEE, April 2010
6. Lei, H., Yifei, H., Yi, G.: The research of business intelligence system based on data mining. In: 2015 International Conference on Logistics, Informatics and Service Sciences (LISS), pp. 1–5. IEEE, July 2015
7. Denić, N., Nešić, Z., Radojičić, M., Vasović, J.V.: Some considerations on business intelligence application in business improvement. In: 2014 22nd Telecommunications Forum Telfor (TELFOR), pp. 1142–1145. IEEE, November 2014
8. Gang, T., Kai, C., Bei, S.: The research & application of business intelligence system in retail industry. In: 2008 IEEE International Conference on Automation and Logistics, pp. 87–91. IEEE. September 2008

9. Muthusamy, V., Slominski, A., Ishakian, V.: Towards enterprise-ready AI deployments minimizing the risk of consuming AI models in business applications. In: 2018 First International Conference on Artificial Intelligence for Industries (AI4I), pp. 108–109. IEEE, September 2018

10. Qihai, Z., Tao, H., Tao, W.: Analysis of business intelligence and its derivative-financial intelligence. In: 2008 International Symposium on Electronic Commerce and Security, pp. 997–1000. IEEE, August 2008

11. Valter, P., Lindgren, P., Prasad, R.: Artificial intelligence and deep learning in a world of humans and persuasive business models. In: 2017 Global Wireless Summit (GWS), pp. 209–214. IEEE, October 2017

12. Rosenberg, L., Pescetelli, N.: Amplifying prediction accuracy using swarm AI. In: 2017 Intelligent Systems Conference (IntelliSys), pp. 61–65. IEEE, September 2017

Image Completion of Highly Noisy Images Using Deep Learning

Piyush Agrawal and Sajal Kaushik[✉]

Bharati Vidyapeeth's College of Engineering, New Delhi, India
sajalkaushik17@gmail.com

Abstract. Generating images from a noisy or blurred image is an active research problem in the field of computer vision that aims to regenerate refined images automatically in a content-aware manner. Various approaches have been developed by academia and industry which includes modern ones applying convolution neural networks and many other approaches to have more realistic images. In this paper, we present a novel approach that leverages the combination of capsule networks and generative adversarial networks that is used for generating images in a more realistic manner. This approach tries to generate images which are locally and globally consistent in nature by leveraging the low level and high-level features of an image.

Keywords: Capsule networks · Generative Adversarial Networks · Convolutional Neural Networks

1 Introduction

Generating images from highly noisy images is an active research problem. Various Deep learning techniques both convolutional and non-convolutional Networks have been previously used by the academia and industry to tackle this Problem. With the advent of Generative Adversarial Networks it is Possible to generate images from highly noisy images.

An architecture of the generator and the discriminator is very important in determining the results. The training stability and performance heavily depends on architecture. Various techniques such as batch normalization, Stacked Architecture and multiple generators and discriminators.

Generative Adversarial network proved to be a game changer in deep learning. It uses two components, Generator and discriminator. The Generator inputs a random data and outputs a sample of data. Then the generated data is fed into the Discriminator Network which compares it with the real data which the discriminator have and then determines whether it is real or fake. Ultimately Generator tries to minimize the Min-Max function and discriminator does opposite of it by maximizing the function.

DCGAN's (Deep Convolutional Generative Adversarial Networks) uses deep convolutional layers in addition to the basic generative and discriminative network models. DCGAN have high similarity to Gans, but specifically focuses on

© Springer Nature Switzerland AG 2020
S. Smys et al. (Eds.): ICCVBIC 2019, AISC 1108, pp. 1031–1043, 2020.
https://doi.org/10.1007/978-3-030-37218-7_108

leveraging Deep Convolutional networks in place of those fully-connected networks. Spatial Correlations are determined by the conv-nets in various areas of the images. DCGAN would likely be more fitting and accurate for image datasets, whereas the acceptability of the gans over the large domains.

This paper presents a brief comparison between capsule gans and dcgans architecture. There are various non learning techniques which try to encapsulate neighboring pixels information and thus to regenerate images from noisy images. Various variants of gans such as Conditional Generative Adversarial Networks, Laplacian Pyramid of Adversarial Networks, Deep Convolutional Generative Adversarial Networks, Adversarial Auto encoders, Information Maximizing Generative Adversarial Networks and Bidirectional Generative Adversarial Networks played a major role in image generation. Motivation behind this paper is to experiment with the results of capsule networks and gans to remove noisy images. This Paper uses capsule Gans to somehow recover from the disadvantages of the Gans. Converging of gans is a big challenge with respect to the state of the art problems of reconstructing images. Combination of capsule networks and generative adversarial networks may give higher accuracy.

The paper is organized as: The Related Work has been described in Sect. 2. Further, Sect. 3 describes the methodology of the approach. Section 4 shows our experiment process, describes about the different datasets used and how they are trained. Section 5 shows the results of the model and graphs generated. And lastly, Sect. 6 concludes the paper and discuss about the results.

2   Related Work

Various Methods have been developed previously in order to complete the noisy blurred Images [1]. GANs proved to be the most influential one to Complete the Blurred Images [2]. Before Gans, Convolutional Networks played a Major role in Classifying and Generating Images [3,4]. Various Methods such as patch Matching and various Non-Learning Approaches such as that rely on propagating appearance information from neighboring pixels to the target region using some mechanisms [5].

Image reconstruction from highly noisy images is an active research problem in the area of computer vision [?]. Basic image processing techniques have been used but the results are not much satisfactory. Deep learning approaches leverages out the obtained correlation between the various patterns obtained from noisy images and real images.

2.1   Generative Adversarial Networks

The Generative Adversarial network model uses latent variables z that used to generate the data that needs to belong to the data x [6–8]. The probability (distribution) of generator pg, is defined over the data with a prior on p(z). The Generator function $G(z, g)$ represents a mapping from the latent variable space to the space of the data. The Discriminator function $D(x, d)$ maps the data space to a single scalar which represents the probability that x belongs to the actual dataset x or to pg. The Mapping of the data to a single scalar that represents whether the data belongs to the actual dataset x or to p(g) The Training of discriminator is to maximize the probability of classifying an image whether coming from an generator or actual datasets.

$$P(G) = E(x)[LogD(x)] + E(z)[log(1 - D(G(z))]$$ (1)

The Eq. 1 E(x) represents the expectation of that point in the real distribution, D(x) represents the output of that point in the discriminator. Z is the latent which is fed to the generative adversarial network. G(z) the output of the variable z from the generative networks is passed to the discriminator as D(g(z)).

2.2 DCGAN

Deep Convolutional Generative Adversarial Network is a class of Convolutional Neural Networks which improves the performance of Gans and make Gans more stable. It is a Technique to generate more realistic images from the noisy images. DCGAN leverages the convolutional networks to get the spatial relationship between various parts of the images and thus use this information to improve the image generation using generative adversarial Networks [9]. DCGANS works better for large-scale scene understanding data. Original Gans are difficult to scale using Convolutional neural networks, thus DCGAN proposed an optimized network for Image Generation.

2.3 Conditional GAN

Conditional Gan is an extension of standard gan which generates samples conditioned on specific variables. An additional variable in the model, such that we generate and discriminate samples conditioned on that variable [10]. Main aim of the generator is to generate samples that are both realistic and match the context of the additional variables.

2.4 Info GAN

For image completion [11], several modifications such as optimal patch search have been proposed [12,13]. In particular [14], proposed a global-optimization based method that can obtain more consistent.

2.5 WGAN

With several reasons to support, GANs are very [difficult] to train. Wasserstein GAN (WGAN), an alternative function [15,16] shows that the convergence properties are better when compared with standard value as a result of value function. This work uses the Wasserstein distance W(q, p), q and p being distributions, which can be understood as the lowest cost to move mass when the q is transformed to p. The value function is:

$$\min_{G} \max_{D \in \mathcal{D}} \mathbb{E}_{\boldsymbol{x} \sim \mathbb{P}_r}[D(\boldsymbol{x})] - \mathbb{E}_{\hat{\boldsymbol{x}} \sim \mathbb{P}_g}[D(\hat{\boldsymbol{x}})] \qquad (2)$$

Above different approaches has been used differently approaches have been studied to improvise the state of the art models which can solve the presented problem.

3   Approach

3.1   Capsule Gans

The paper proposes an architecture for creating these capsules which will replace the neurons in a neural network form image generation. The capsules of nested neurons. V_j is the activation vector outputted by layers. V_{jis} defined such that its length describes the probability of the existence of the entity represented by its capsule. The longer ones are valued a little under 1 and the shorter ones are squashed near zero with the help of the nonlinear activation function:-

$$\mathbf{v}_j = \frac{||\mathbf{s}_j||^2}{1 + ||\mathbf{s}_j||^2} \frac{\mathbf{s}_j}{||\mathbf{s}_j||} \qquad (3)$$

Where V_j is the output vector and S_j is the input of the activation function. Let U_i be the output of the previous capsule layer, which acts as the input for the current layer j. Now, the outputs of the previous layers are multiplied by a weight matrix W_{ij} which results in what is called a prediction vector $U_{j|i}$.

Now, this prediction vector is then multiplied by the coupling coefficients between the capsules of the layers 'i' and 'j': C_{ij}, to give the resultant S_j.

Here, the sum of all coupling coefficients for a capsule is equal to 1. It is determined by the process of 'routing softmax' where the logits, b_{ij} are the log probabilities that a capsule 'i' couples with a capsule 'j'. These log priors are then learned discriminatively at the same time as that of the other parameters. The coupling coefficients are then learned iteratively by monitoring the agreement between the capsules of the different layers. This agreement is measured by the scalar multiplication of the output of the capsule of the given layer (V_j) and that of the input prediction vector coming from the previous layer $(U_{j|i})$. This value of a_{ij} is treated as log likelihood and is added to the initial value of b_{ij}. This is then used to calculate the new values of coupling coefficients using the routing softmax method.

Let u_i be the output of the previous capsule layer, which acts as the input for the current layer j. Now, the outputs of the previous layers are multiplied by a weight matrix W_{ij} which results in what is called a prediction vector $u_{j|i}$. Now, this prediction vector is then multiplied by the coupling coefficients between the capsules of the layers 'i' and 'j': c_{ij}, to give the resultant s_j.

Here, the sum of all coupling coefficients for a capsule is equal to 1. It is determined by the process of 'routing softmax' where the logits, b_{ij} are the log probabilities that a capsule 'i' couples with a capsule 'j'. These log priors are then learned discriminatively at the same time as that of the other parameters. The coupling coefficients are then learned iteratively by monitoring the agreement between the capsules of the different layers. This agreement is measured by the scalar multiplication of the output of the capsule of the given layer (v_j) and that of the input prediction vector coming from the previous layer $(u_{j|i})$.

$$c_{ij} = \frac{\exp(b_{ij})}{\sum_k \exp(b_{ik})} \qquad (4)$$

This value of a_{ij} is treated as log likelihood and is added to the initial value of b_{ij}. This is then used to calculate the new values of coupling coefficients using the routing softmax method.

3.2 Network Architecture and Implementation

This experiment consists of two architectures, first the Generator [refer] which is responsible for generating images from the noise and Discriminator based on Capsule Networks [17] which validates the generated images.

The generator is inspired from DCGAN framework [18] consists of 5 convolutional layers. These layers work for up-sampling a 100-dimensional vector whose each element is sample independently with 1 standard deviation on a normal distribution of mean 0.

To regularize the network, batch normalization [19] has been applied in between layers so that stability could increase. Except for final layer, rectified linear units [20] are used as nonlinearity between all layers and the output is applied with nonlinearity prior to batch normalization. In the end, 'tanh' [21] as activation is used (Fig. 1).

Figure 1: DCGAN generator used for LSUN scene modeling.

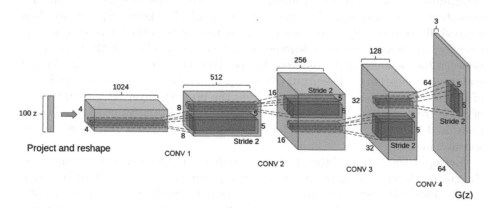

Fig. 1. Generator architecture [9]

The discriminator of the model is based on capsule network which comprises of dynamic routing. The architecture has final output as the 16-dimensional capsule. Input layer consists of 64×64 dimension which is then passed to 18×18 kernels resulting in $256 \times 24 \times 24$ layers, which is then passed to another kernel of size 18×18 and the resulting layer on dynamic routing gives one 16-Dimensional capsule. This process of dynamic routing occurs between the output digit capsule and the primary layer of the capsule. An illustration is provided below (Fig. 2).

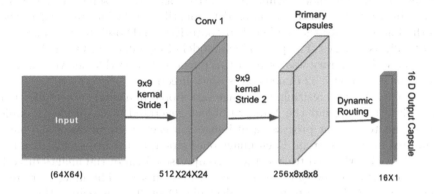

Fig. 2. Discriminator architecture

4   Experiments

4.1   Dataset

In this work, the model prepared is examined on several datasets. The motive of the examination is to evaluate the model on different datasets such that how accurate they can complete images from noisy images. The developed model is evaluated on three different datasets in this work. These three datasets are: CelebA [22], the Street View House Numbers (SVHN) [23], Facescrub [24]. The CelebA dataset consists of more than 200,000 images each having 40 different attributes. There are 10,177 unique identities in the dataset. Having so many unique identities, this dataset played very crucial role in understanding of the facial attributes on different postures for the model. The other dataset, Face-scrub with over 100,000 images of more than 500 unique faces is also focused on improving facial recognition research. The high quality images and manually cleaned images played a very important role for training and providing better diversity. From both datasets, images were cropped and resized to 64 × 64 to keep only the face from each image. SVHN being traned separately, the images were cropped into 32 × 32 for the input to the model.

4.2 Training Process

Data Preprocessing: The two datasets, CelebA and Facescrub are used together to train the model while the SVHN was trained separately. For former combination, Openface [25] is used to identify the facial area and further crop the image. It uses the mode called "outerEyesAndNose" to crop and then save the aligned images. The pre-trained model of Openface on OpenCV (Open Computer Vision) library was used for this purpose. For SVHN, the version with cropped images of 32 × 32 dimension is used directly.

After cropping and resizing, the images are stored in a single object file. This object file is nothing but the Numpy[] object generated using pickle[] module. This is done to prevent processing of images on each re-iteration of the batch in a limited memory system. Each image once stored to the object file then this object file is directly used to load the uncompressed form of the images directly into memory without pre-processing then to un-compress. The images are read in the batch of 32 images from this object file. Overall, This helped the model to alleviate the training process by minimizing the time to process image pixels and convert them to numpy array every-time when we train.

Training: The actual training consists of loading the object file of stored images, and reading the images from it in the batches of 32. The generator is trained by providing the batch of 64 noises [26,27].

4.3 Visual Comparison

We compared the Capsule Network [17] based GAN with the DCGAN [11] based Semantic Inpainting [28,29] and the results are quite interesting to compare how good both models are for image generation task.

Below is the results obtained by DCGAN model on Facescrub dataset. The results are based on Semantic Inpainting. The amount of noise sent to the generator is pretty much high. Due to the above fact it is expected that the results may degrade further. As shown in Fig. 6 capsgan on celebA performed with a very low accuracy of generating the image. Capsule networks leverages the information of capsules and due to lots of feature mixing the information at each [11] (Figs. 3, 4 and 5).

Fig. 3. CapsGAN on Facescrub

Fig. 4. DCGAN on Facescrub

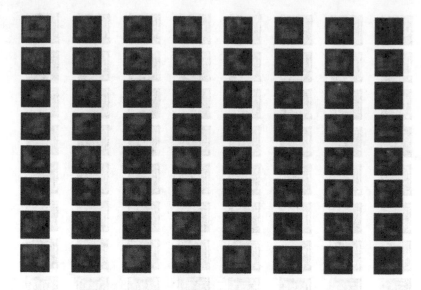

Fig. 5. Noise sent to generator

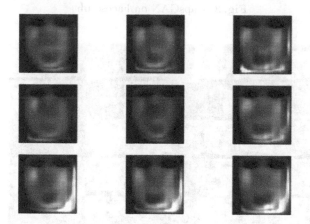

Fig. 6. CapsGAN on CelebA

5   Experimental Results

As described in the training process, the model is trained on 3 different datasets. While the dataset is trained on SVHN dataset as well, our primary focus of evaluation is on the results obtained from the Celeba and Facescrub datasets. The Fig. 7 is a graph on Generator loss for the epochs.

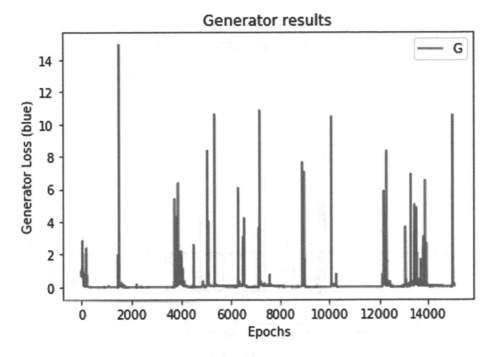

Fig. 7. Generator loss versus epochs

The facescrub dataset is trained on NVIDIA GTX 1060 6 GB GPU until 30 epochs where each epoch is trained for approximately 1000 batches of images from the dataset. While the CelebA [?] Instead of choosing the epochs arbitrarily, they were chosen by visual result convergence.

Figure 7 shows the results from generator which were calculated on loss values over the 15000 epochs. Our Generator is based on the architecture provided here in the Fig. 1.

Figure 8 below shows the results obtained from loss values of Discriminator. This discriminator is based on the Capsule Network [17] whose implementation is presented here. Generator basically trying to generate the images from noisy images. Figure 7 shows a uniform rise at regular intervals of loss which is quite evident in terms of the model's robustness. The peak's of the loss is gradually decreasing which indicates the model is learning gradually. Due to higher number of epochs, the time taken to train the model is around 3 days. The generator and discriminator follows a parallel mechanism due to which training times increases exponentially.

Discriminator Accuracy is constant most of the epochs. Minor reductions in the accuracy is due to the robustness of the model. Independently the Cumulative mean of the loss is increasing which proves as the data increases on the training side the models capability to learn the pattern decreases.

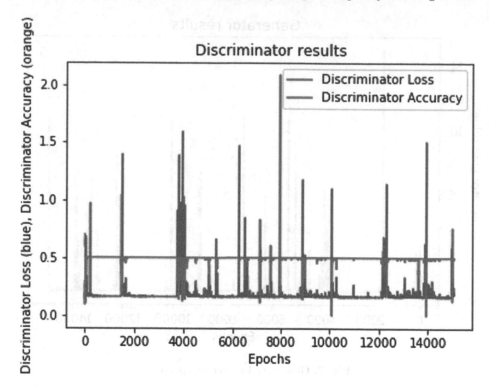

Fig. 8. Discriminator loss and accuracy versus epochs

6   Discussion and Extension

General Adversarial Networks are powerful tools and its variant DCGAN is very popular with different other variants launch recently including infoGAN [30]. Inspired from the success of Capsule Networks [17] over a conventional CNN, this work involved using Capsule Network with GANs as a discriminator on different datasets to examine how the results are different from one of the popular method, DCGAN. The results are generated on different datasets of public faces in this work. We hope that our work helps out future researches to improve Generative models allowing computers to automate the process of Art making.

We (authors) declare that there is no conflict of interest. No humans/animals involved in this research work. We have used our own data.

References

1. Lagendijk, R.L., Biemond, J., Boekee, D.E.: Identification and restoration of noisy blurred images using the expectation-maximization algorithm. IEEE Trans. Acoust. Speech Signal Process. **38**(7), 1180–1191 (1990)

2. Gregor, K., Danihelka, I., Graves, A., Wierstra, D.: DRAW: a recurrent neural network for image generation. CoRR abs/1502.04623, arXiv:1502.04623

3. Krizhevsky, A., Sutskever, I., Hinton, G.E.: ImageNet classification with deep convolutional neural networks. In: Advances in Neural Information Processing Systems, pp. 1097–1105 (2012)

4. Im, D.J., Kim, C.D., Jiang, H., Memisevic, R.: Generating images with recurrent adversarial networks. CoRR abs/1602.05110, arXiv:1602.05110

5. Zeiler, M.D., Fergus, R.: Visualizing and understanding convolutional networks. CoRR abs/1311.2901, arXiv:1311.2901

6. Goodfellow, I.J., Pouget-Abadie, J., Mirza, M., Xu, B., Warde-Farley, D., Ozair, S., Courville, A., Bengio, Y.: Generative adversarial networks. CoRR abs/1602.05110. https://arxiv.org/abs/1406.2661

7. Theis, L., van den Oord, A., Bethge, M.: A note on the evaluation of generative models (2015). arXiv:arXiv:1511.01844

8. Dosovitskiy, A., Springenberg, J.T., Brox, T.: Learning to generate chairs with convolutional neural networks, CoRR abs/1411.5928, arXiv:1411.5928

9. Radford, A., Metz, L., Chintala, S.: Unsupervised representation learning with deep convolutional generative adversarial networks, CoRR abs/1511.06434. arXiv:1511.06434

10. Mirza, M., Osindero, S.: Conditional generative adversarial nets. arXiv preprint arXiv:1411.1784

11. Amos, B.: Image Completion with Deep Learning in TensorFlow. http://bamos.github.io/2016/08/09/deep-completion

12. Gorijala, M., Dukkipati, A.: Image generation and editing with variational info generative adversarial networks. arXiv preprint arXiv:1701.04568

13. Sun, J., Yuan, L., Jia, J., Shum, H.-Y.: Image completion with structure propagation. In: ACM Transactions on Graphics (ToG), vol. 24, pp. 861–868. ACM (2005)

14. Darabi, S., Shechtman, E., Barnes, C., Goldman, D.B., Sen, P.: Image melding: combining inconsistent images using patch-based synthesis. ACM Trans. Graph. 31(4), 82-1 (2012)

15. Arjovsky, M., Chintala, S., Bottou, L.: Wasserstein GAN. arXiv preprint arXiv:1701.07875

16. Nagarajan, V., Kolter, J.Z.: Gradient descent GAN optimization is locally stable. In: Advances in Neural Information Processing Systems, pp. 5585–5595 (2017)

17. Sabour, S., Frosst, N., Hinton, G.E.: Dynamic routing between capsules, CoRR abs/1710.09829. arXiv:1710.09829

18. Suárez, P.L., Sappa, A.D., Vintimilla, B.X.: Infrared image colorization based on a triplet DCGAN architecture. In: 2017 IEEE Conference on Computer Vision and Pattern Recognition Workshops (CVPRW), pp. 212–217. IEEE (2017)

19. Ioffe, S., Szegedy, C.: Batch normalization: accelerating deep network training by reducing internal covariate shift. arXiv preprint arXiv:1502.03167

20. Maas, A.L., Hannun, A.Y., Ng, A.Y.: Rectifier nonlinearities improve neural network acoustic models. In: Proceedings of ICML, vol. 30, p. 3 (2013)

21. Wazwaz, A.-M.: The tan h method: solitons and periodic solutions for the Dodd-Bullough-Mikhailov and the Tzitzeica-Dodd-Bullough equations. Chaos Solitons Fractals 25(1), 55–63 (2005)

22. Liu, Z., Luo, P., Wang, X., Tang, X.: Large-scale CelebFaces Attributes (CelebA) dataset (2018). Accessed 15 Aug 2018

23. Goodfellow, I.J., Bulatov, Y., Ibarz, J., Arnoud, S., Shet, V.: Multi-digit number recognition from street view imagery using deep convolutional neural networks. arXiv preprint arXiv:1312.6082

24. Ng, H.W., Winkler, S.: A data-driven approach to cleaning large face datasets. In. IEEE International Conference on Image Processing (ICIP 2014), pp. 343–347 (2014). https://doi.org/10.1109/ICIP.2014.7025068

25. Amos, B., Ludwiczuk, B., Satyanarayanan, M.: OpenFace: a general-purpose face recognition library with mobile applications. Technical report, CMU-CS-16-118, CMU School of Computer Science (2016)

26. Salimans, T., Goodfellow, I., Zaremba, W., Cheung, V., Radford, A., Chen, X., Chen, X.: Improved techniques for training GANs. In: Lee, D.D., Sugiyama, M., Luxburg, U.V., Guyon, I., Garnett, R. (eds.) Advances in Neural Information Processing Systems 29, pp. 2234–2242. Curran Associates, Inc. (2016)

27. Dziugaite, G.K., Roy, D.M., Ghahramani, Z.: Training generative neural networks via maximum mean discrepancy optimization. arXiv:1505.03906 (2015)

28. Yeh, R.A., Chen, C., Lim, T., Hasegawa-Johnson, M., Do, M.N.: Semantic image inpainting with perceptual and contextual losses, CoRR abs/1607.07539, arXiv:1607.07539

29. Hinton, G.E., Sabour, S., Frosst, N.: Matrix capsules with EM routing. In: International Conference on Learning Representations (2018). https://openreview.net/forum?id=HJWLfGWRb

30. Chen, X., Duan, Y., Houthooft, R., Schulman, J., Sutskever, I., Abbeel, P.: Info-Gan: interpretable representation learning by information maximizing generative adversarial nets. CoRR abs/1606.03657, arXiv:1606.03657

Virtual Vision for Blind People Using Mobile Camera and Sonar Sensors

Shams Shahriar Suny$^{(\boxtimes)}$, Setu Basak,
and S. M. Mazharul Hoque Chowdhury

Department of Computer Science and Engineering, Daffodil International
University, Dhaka, Bangladesh
{shams.cse, setu.cse, mazharul2213}@diu.edu.bd

Abstract. In this modern era of technology, almost all the real life challenges are solved by using the innovative technological solutions. We know there are many blind people, who still remain outside from the touch of modern and digitized technologies. Blind people cannot get proper direction for their destination path and it is still hard to find out what object that he/she will interact with. The proposed model will not only help in their easy movement but also give them a virtual vision. In this implementation we will use NodeMCU and mobile camera which will be guided by OpenCv library that uses Python to detect an object. It will also analyze and report the object type to the user, which will create a virtual vision to the user. On the other hand, there will be some sonar sensors which will include vibration motor to help the blind persons to sense that there is an obstacle ahead in that direction. The future work of this project will provide the blind person a proper direction to his/her way using Google maps.

Keywords: Virtual vision · YOLO · Physically challenged · Mobile camera · Sensors

1 Introduction

There are many visually impaired individuals around us. They have confronted numerous issues to move from one spot to another. The proposed model will assist them by detecting the hindrance present before them. Then again, they can likewise comprehend what is the name of the article that before them. So they can have a virtual vision with deterrent evasion framework. In this paper we have presented a project that will help the blind people by giving them a virtual vision. This project has been created by the picture handling innovation and Arduino. Picture preparing will take the video picture from the camera, at that point it will discover the article before it is analyzed by Yolo model in python OpenCv. It has likewise capacity to recognize different item from the video feed. Yet, we keep it just one for the particular item acknowledgment which is nearer to the camera. In this adaptation we have utilized versatile camera which associate with the picture handling server remotely. Then again, this undertaking will likewise assist the visually impaired individuals with overcoming the obstacle before them by hearing the notice sound. At first there is just a single sensor has been

© Springer Nature Switzerland AG 2020
S. Smys et al. (Eds.): ICCVBIC 2019, AISC 1108, pp. 1044–1050, 2020.
https://doi.org/10.1007/978-3-030-37218-7_109

utilized. In any case, it very well may be executing in various edge so it can likewise identify if there is some enormous gap before the client or not. So at once they can have a virtual vision with along to an impediment evasion framework.

2 Literature Review

Jain along with his team members have worked on a hand gloves that can help a visually hampered person [1]. This glove had a camera included which connected to a raspberry pi device. The camera had object detection feature. Which will help the user to know the object that in front of the gloves. The author used DNN method to detect the object. There were also five vibrators which can help users to get direction to the object. Similar work that can be found in [2] where the Arora et al. have discussed about the measurement of the blind people's problems in everyday life. So they developed some kind of device that can be wearable. This work also implements the real time object detection system. So that the user can know the object in front of it. The author used same DNN method to detect the object. It also used combination of single shot-multiple box detection technology. This work also include speech to tell the user what is closer to him. Another work for the blind people where Frankel have tried to find all the object from an image that is previously captured by the camera [3]. In this case mobile camera is used for taking the pictures. Then the pictures are passed through the algorithm to make a 3D representation. This 3D representation will help to locate all the objects with all the possible angles. After the process of the image is finished then the user can get the image from it. A touch screen will provide to touch in the picture and if there is any object that detected by the algorithm. Bai along with his team members tried to implement both obstacle avoidance and object detection. This device has a camera that attached to a sunglass but the user need to hold on the main device. This system used CNN method to detect the object in front of it. The author trying to make an obstacle avoidance with the camera also. The camera they used RGB-D which used to take image and then it sends it to the device to process and result. Then the user can get acknowledged about the obstacle. User get the obstacle in-front of him/her then the system will beep to alert the user. Islam et al. have compared many systems variation [5]. There are wide details that display over a table. All the sensors and actuators that needed for the projects that are listed nicely. Hogle and his team describes about a patent of a new wearable device that can connect wirelessly to help visually hampered people [6]. In this work camera will take the pictures of the current state and send them to the connected server. The outcome of the process will come to the device and make the user know what type of object that the user facing right now. It has also voice command control so that the user can interact with the device properly. Rodríguez et al. Have presented an assisting system for blind people that will help them avoiding the obstacles [7]. José along with his team proposed a navigation system for blind people [8]. They have proposed a model that will help the user navigating the path with different types of algorithms. Giudice and Legge have discussed about different sensory devices, optical technologies to assist the blind people [9]. They have proposed a method and also compared different models. Cardin et al. Have proposed a technique for the visually challenged people [10]. They proposed a wearable obstacle detection device. Shoval along with his team proposed a mobile robot to assist the blind people to move along like a normal human [11]. They trained the system and then made a nave belt to avoid obstacles. Lim et al. Have proposed a vest for blind people to avoid obstacles in front of user [12]. Pradeep and his team have used robotic vision to help the blind people [13]. Borenstein and his team proposed a navbelt to assist blind persons in moving around easily [14]. Borenstein et al. Proposed a framework guideCane for blind people to move easily [15].

3 Methodology

At the beginning, the camera will begin functioning not surprisingly. Then the continuous video feed will go into the python library called "Yolo". The library will get the picture feed and searches for the information, which then sends us the recognized item name. At that point the name will be passed into the server to refresh the information. The second part is the server part, which will deal with the information among camera and server and furthermore with primary NodeMCU board to the server. It will remains as the fundamental correspondence connect between two channels. The model will dependably stay up with the latest information obtained from the client. If the client needs to think about the item, at that point the server refreshes its information. So all the associated parts will be refreshed. The last part of this flow is the main NodeMCU board. Some sensors are connected with the main board but other sensors are associated with the principle board. They will take constant incentive from the required sensors (Fig. 1).

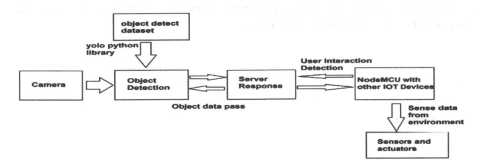

Fig. 1. Work flow of the system

4 Design and Implementation

This is a system with both IOT and the image processing model but we have different elements. At first let's take a look on the proposed system design.

The mobile camera which will be connected wirelessly to the trained database mode. The mobile camera will grab the image from the video feed and sent it the python library. Then the library will identify the object. Initially the python is connected to the computer, so if the camera detects any object, it will be informed to the user by voice. If the user has pressed the active state button. In the main board, it will connect to the Wi-Fi. So that it can connected to the internet to initiation an API communication. It will continuously check for the data arrival. Whenever the user presses the button to change the state, it means that he user wants to know the object type additionally. We can see all these from Fig. 2.

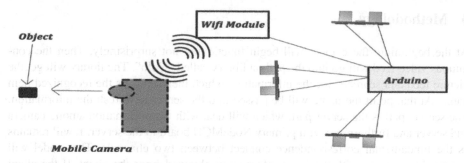

Fig. 2. Full system design

This is the primary stage assignment of the venture camera working rule that appeared in Fig. 3. Then again, it is an IP based camera. So we can get the live stream on what the camera has detected. The camera must be associated with the system from where we are additionally associated with the NodeMCU board. So the camera will take the video feed and after that it will send it to the Yolo python library. Then the object can be detected by the mobile phone.

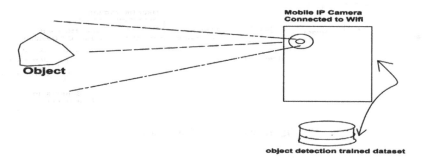

Fig. 3. Object detection.

The server will be controlled by the phpMyadmin database. An API hit will originate from the camera with the information observed by camera. At that point the server information will have refreshed with it. Figure 4 demonstrates that. Be that as it may, then again there will be other solicitation as well. To the primary board end, there will be catch that will press by the client. So the API call will proceed with check whether the condition of the camera change or not. In the event that the client presses the catch and need to comprehend what is on front of them, at that point an API hit will go to the server and change its condition. In this point the camera will proceed with a hunt on whether the client needs to comprehend on the identified object. When the camera detects the front object. It will continuously tell the name of that particular object.

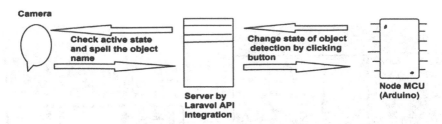

Fig. 4. Server connection

The Fig. 5 shows the detailed design process. In this case we are going to use NodeMCU which is a nice replacement of the Arduino. Because we need another Wi-Fi module that will add to the Arduino. And the configuration of this is pretty much complex. So that we have used the NodeMCU. Because it has an integrated Wi-Fi module in it. On the other hand, it is small in size.

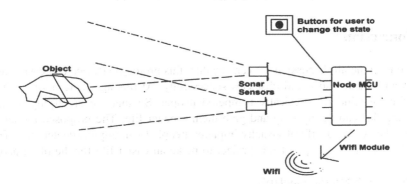

Fig. 5. Main board connection

5   Results and Accuracy

With the information obtained from the image, the Yolo model will execute it with python algorithm. So the portable camera's information is shared in response to popular URL demand. At that point the item identification model will pre display the information. Hence we can get this kind of picture where the items are appeared with the outskirts. The recognized items are combined into a cluster. What's more, the cluster request will be the nearest article recognized. That implies which article is greater, so it will identify first at that particular point. On the off chance that there is no part distinguished from the picture, at that point it will pass the void exhibit as indicated below. Figure 6 represents the detection of different objects. So as far as all the possible scenario that we have tested. We can get all the possible outcomes successfully. We have tested several of them and get at most 90% accuracy in object detection. As objects are larger, it will become closer to the first position of the array.

Fig. 6. Mobile camera object detection [7]

6 Conclusion

At the end of the all discussion we can see that, this prototype device can easily detect objects and it can also avoid obstacles successfully. At this age there is only limited work for the blind or physically hampered people. So that we are trying to make something that can surely help and give them a better life. The proposed model will enhance the quality of life of visually impaired people to an unprecedented rate. There is much more future scope of this project to make an easier life for the blind people.

Compliance with Ethical Standards

✓ All authors declare that there is no conflict of interest.
✓ No humans/animals involved in this research work.
✓ We have used our own data.

References

1. Jain, S., Varsha, S.D., Bhat, V.N., Alamelu, J.V.: Design and implementation of the smart glove to aid the visually impaired. In: 2019 International Conference on Communication and Signal Processing (ICCSP) (2019)
2. Arora, A., Grover, A., Chugh, R., Reka, S.: Real time multi object detection for blind using single shot multibox detector. Wirel. Pers. Commun. **107**, 651 (2019)
3. Frankel, M.: Systems and methods for blind and visually impaired person environment navigation assistance. US20190026939A1. https://patents.google.com/patent/US20190026 939A1
4. Bai, J., Liu, Z., Lin, Y., Li, Y., Lian, S., Liu, D.: Wearable travel aid for environment perception and navigation of visually impaired people. Electronics **8**, 697 (2019)
5. Islam, M.M., Sadi, M.S., Zamli, K.Z., Ahmed, M.M.: Developing walking assistants for visually impaired people: a review. IEEE Sens. J. **19**(8), 2814–2828 (2019)
6. Hogle, R., Beckman, R.: Object detection, analysis, and alert system for use in providing visual information to the blind. US20190070064A1. https://patents.google.com/patent/US20190070064A1
7. Rodríguez, A., Yebes, J.J., Alcantarilla, P.F., Bergasa, L.M., Almazán, J., Cela, A.: Assisting the visually impaired: obstacle detection and warning system by acoustic feedback. Sensors **12**, 17476–17496 (2012). https://doi.org/10.3390/s121217476
8. José, J., Farrajota, M., Rodrigues, J.M.F., du Buf, J.M.H.: The SmartVision local navigation aid for blind and visually impaired persons. JDCTA (2011)
9. Giudice, N.A., Legge, G.E.: Blind navigation and the role of technology. In: The Engineering Handbook of Smart Technology for Aging, Disability, and Independence. Wiley (2008)
10. Cardin, S., Thalmann, D., Vexo, F.: Wearable obstacle detection system for visually impaired people. In: Haptex 2005 (2005)
11. Shoval, S., Borenstein, J., Koren, Y.: Mobile robot obstacle avoidance in a computerized travel aid for the blind, 1050-4729194 $03.00 0 1994 IEEE (1994)
12. Shin, B.-S., Lim, C.-S.: Obstacle detection and avoidance system for visually impaired people. In: Oakley, I., Brewster, S. (eds.) HAID 2007. LNCS, vol. 4813, pp. 78–85. Springer, Heidelberg (2007)
13. Pradeep, V., Medioni, G., Weiland, J.: Robot vision for the visually impaired, 978-1-4244-7030-3/10/$26.00 ©2010 IEEE (2010)
14. Shoval, S., Borenstein, J.: The Navbelt—a computerized travel aid for the blind based on mobile robotics technology, 0018–9294/98$10.00 ã 1998 IEEE (1998)
15. Borenstein, J., Ulrich, I.: The GuideCane - a computerized travel aid for the active guidance of blind pedestrians. In: Proceedings of the 1997 IEEE International Conference on Robotics and Automation Albuquerque, New Mexico, April 1997

Study on Performance of Classification Algorithms Based on the Sample Size for Crop Prediction

I. Rajeshwari[1]([⊠]) and K. Shyamala[2]

[1] Computer Science, Queen Mary's College, Chennai, Tamilnadu, India
rajeshwari_i@yahoo.com
[2] Computer Science, Dr. Ambedkar Government Arts College, Chennai,
Tamilnadu, India
shyamalakannan2000@gmail.com

Abstract. An important issue in agricultural planning is the crop selection for cultivation. This work presents an approach which uses Data Mining classification techniques to predict the crops to be cultivated based on the availability of soil nutrient. The dataset considered in this work has been taken from the soil test centres of Dindigul district, Tamilnadu. The parameters are the various nutrients present in the soil samples collected from the different regions of Dindigul district. The properties of soils are analyzed to find the crop suitable for cultivation in the region. The dataset is taken in 9 different sample sizes. Seven different classification algorithms are deployed on these soil datasets. The performance of the algorithms are analyzed based on certain metrics and by using the Confusion matrix and Classification report are used for analysis. The Decision Tree is found to be the best algorithm for the soil dataset.

Keywords: Crop prediction · Classification · Decision tree · Naïve Bayes · KNN

1 Introduction

Agriculture plays an important role in Indian economy. Although selection of crop for cultivation has been linked to a number of factors such as soil, weather and water, this work investigated the soil nutrient availability. The aim of this work was to determine the classification algorithm suitability for difference sizes of data sets. As nutrition is important for the human health, soil nutrients are important for plant growth. Proper soil nutrition will prevent nutrient-related plant stress and crop losses through pests, diseases, and poor post-harvest quality. From the soil, the roots of a plant receive nutrients which help it grow. Knowledge about soil health and its maintenance is very important for sustaining crop productivity [1]. A soil test gives a brief idea about soil structure and mineral compositions rations. Indian Government has established soil test centers at every district head quarters [2]. Soil testing is a chemical method for estimation of nutrient supplying power of a soil or soil fertility evaluation [3]. The soil test provides information about the various chemical properties such as EC, pH, along with micro nutrients N, P, K and macro nutrients Zn, B, Cu, Fe [4].

© Springer Nature Switzerland AG 2020
S. Smys et al. (Eds.): ICCVBIC 2019, AISC 1108, pp. 1051–1058, 2020.
https://doi.org/10.1007/978-3-030-37218-7_110

The main aim of the paper is to assist farmers in selecting crop for cultivation. Data mining classification techniques are applied over the soil nutrient data sets to predict the crops to be cultivated. For this, the given dataset is divided into two sets., viz., a training set and test set. The training sets are randomly sampled from the dataset. The remaining form the test set. The classifier is built in the learning step by the classification model by analysing the training set. Prediction of the class labels are done in the classification step. The predictive accuracy of a classifier is estimated from the test set. The percentage of test sets that are correctly classified by the classifier is the accuracy of a classifier. The best way to achieve higher accuracy is to test out different algorithms. The best one can be selected by cross-validation. Classification algorithms like decision tree, KNN, Linear SVM, Kernal SVM, Logistic regression, Naive Bayes and Random forest are deployed on the soil datasets and their performance are analysed based on certain metrics like Accuracy score, Cohen's Kappa, Precision, Recall And F-Measures, Hamming Loss, Explained Variance Score, Mean Absolute Error, Mean Squared Error and Mean Squared Logarithmic Error. The Confusion matrix and Classification report are used for analysis. The main problem is comparative study of these classification algorithms, by applying them on soil datasets with 9 different sizes.

The organization of this paper is as follows. In Sect. 2, the literature review conducted is described and the study to be done is introduced. Section 3 describes the dataset collection and its preprocessing activities. Section 4 discusses the results of the classification algorithms for predicting the crop to be cultivated. Finally, Sect. 5 includes conclusion and future work.

2 Literature Review

Deshmukh et al. [5] analysed and compared the ADTree, Bayes Network, Decision Table, J48, Logistic, Naive Bayes, NBTree, PART, RBFNetwork and SMO algorithms on five datasets using WEKA. Correctly Classified Instances (CCI), Incorrectly Classified Instances (ICI), Kappa Statistic (KS), Mean Absolute Error (MAE) and Root Mean Squared Error (RMSE) were the metrics taken for analysis. The paper concluded that no single classification algorithm could provide the best predictive model for all datasets. Ramesh et al. [6] analysed and compared the performance of Navie Bayes, Bayesian Network, Naive Bayes updatable, J48 and Random Forest on soil data set based on Relative Absolute Error, Mean Absolute Error, Correctly classified instances and Cohen Kappa Statistics. They have concluded that Naïve Bayes classifier is the efficient classification technique among the remaining classification techniques. Manjula et al. [1] had used Naïve Bayes, Decision Tree and Hybrid approach of Naïve Bayes and Decision Tree for classifying the soil dataset and their performance was analysed based on the two factors, accuracy and execution time. It was concluded that the hybrid classification algorithm had performed well over the other two methods.

Kumar et al. [7] analysed and compared the behavior of Naïve Bayes, J48(C4.5) and JRip on soil profile data and concluded that JRip model is the best classifier for soil sample data analysis. The performance was analysed based on correctly classified instances, incorrectly classified instances, accuracy and mean absolute error. Baskar et al. [8] had used Naïve Bayes, J48(C4.5) and JRip on samples and concluded J48 model turned out to be the best classifier based on correctly classified instances, incorrectly classified instances, accuracy and mean absolute error. Ten fold validation was used. Hemageetha et al. [9] analysed Naïve Bayes, J48, Bayesian network and JRip classifiers on soil data set based on PH value to predict the suitability of soil for crop cultivation using WEKA. The metrics used were correctly classified instances, incorrectly classified instances, accuracy, Cohen Kappa Statistics and Mean Absolute Error. Ten fold validation was done. It was concluded that J48 outperformed all other classifier models.

Pahwa et al. [10] analyzed and compared the behaviour of different kinds of classification algorithms on medical dataset taken from literature. Naive Bayesian, Decision Tree Induction, Multilayer Perceptron and logistic regression were applied on three different datasets. The classification techniques were compared based on the TPrate, FPrate, precision, recall, classification accuracy and the execution time in seconds. They have concluded that MLP is more accurate and better than others but it takes more time. Shah et al. [11] had compared the performance of Decision tree, Bayesian Network and K-Nearest Neighbor algorithms for predicting breast cancer with the help of WEKA, based on the parameters lowest computing time and accuracy. It was concluded that Naïve Bayes is a superior algorithm. Nabi et al. [12] had evaluated the performance of Naïve Bayes, Logistic Regression, Decision tree and Random forest for predicting the diabetic patients based on correctly classified instances, incorrectly classified instances, Mean Absolute Error (MAE), Root Mean Square Error (RMSE), Relative Absolute Error(RAE), Root Relative squared Error (RRSE) and Confusion Matrix. It was concluded, that Logistic Regression has more correctly classified instances with low RMSE and RRSE whereas Navie Bayes has low RAE and MAE.

From the review, it was found that most of the authors have used few techniques and analysed them based on few metrics. In this study more techniques are applied on datasets of 9 different sizes, and analysed on more metrics. Classification algorithms like decision tree, KNN, Kernal SVM, Linear SVM, Logistic regression, Naive Bayes and Random forest are deployed on the soil datasets and their performance are analysed based on certain metrics like Accuracy score, Cohen's Kappa, Precision, Recall And F-Measures, Hamming Loss, Explained Variance Score, Mean Absolute Error, Mean Squared Error and Mean Squared Logarithmic Error using PYHTON. The Confusion matrix and Classification report are used for analysis.

3   Research Methodology

For this work, data was collected from the soil test centers in Dindigul district. Samples from various villages covering almost the entire district were collected. Nearly 44000 samples with 12 nutrients, viz., EC, pH, N, P, K, Fe, Zn, Cu, B, OC, S and Mn were collected. The collected data was preprocessed by removing unwanted data, noisy data and blank data. Extreme values in each attributes are treated as noisy data. 35700 samples from the processed data were taken. The data was initially coded in EXCEL and finally converted to comma delimited.csv. Using Principal Component Analysis on the soil data in PYTHON, components that has up to 80% of variance were considered and hence the first eight components: pH, N, P, K, S, Zn, Fe and Mn were taken as principal soil components. With the 8 features, a target class was added which gives the 11 set of crops that can be cultivated. The class label is given in the number form. From this dataset, 9 datasets of different sizes, viz., 110, 220, 550, 1100, 2200, 5500, 11000, 22000 and 33000 were formed with balanced class. The various classification algorithms were applied on the datasets with 0.1% of the data taken as test data.

4   Result, Analysis and Review of Outcomes

The algorithms were executed on the processed 9 datasets in.CSV (Comma Separated format) file using PYTHON and the various metrics were analysed. The performance of the algorithms were analysed in two ways, viz., the first one, analyzing how each algorithm works on the different datasets and the second one, how the various algorithms perform on each dataset. Table 1 shows the outcome of the various algorithms based on the metrics that need to be the maximum and Table 2 for the minimum metrics on the dataset of size 110. Similarly for all datasets the results were obtained. For sample size 110, the corresponding comparison charts for Tables 1 and 2 are given in Fig. 1.

Table 1. Performance of various algorithms on dataset of size 110

Algorithm	ATR	AT	AS	AM	AP	AR	AF	EVS	CKS
DT	1	0.929	0.9286	0.9818	1	0.93	0.95	0.9949	0.9195
KerSVM	0.384	0.364	0.3636	0.3008	0.41	0.36	0.36	0.6085	0.3125
KNN	0.586	0.545	0.5455	0.3769	0.65	0.55	0.56	0.4213	0.4811
LinSVM	0.596	0.636	0.6364	0.4587	0.53	0.64	0.57	0.8915	0.5769
LogREG	0.556	0.455	0.4545	0.4133	0.41	0.45	0.42	0.3766	0.3774
NaiveBayes	0.98	0.818	0.8182	0.8148	0.82	0.82	0.79	0.9532	0.7905
RF	0.96	0.545	0.5455	0.5841	0.48	0.55	0.46	0.6096	0.4860

Table 2. Performance on dataset of size 110

Algorithm	ASD	MAE	MSE	MSLE	HL	ET
DT	0.05455	0.0714	0.0714	0.0013	0.0714	14
KerSVM	0.13754	2.0000	6.9091	0.2905	0.6364	18
KNN	0.11823	1.5455	9.0000	0.3374	0.4545	15
LinSVM	0.14990	0.8182	2.0909	0.0655	0.3636	43
LogREG	0.12247	1.8182	9.8182	0.3733	0.5455	18
NaiveBayes	0.13899	0.3636	0.7273	0.0473	0.1818	16
RF	0.20582	1.6364	8.1818	0.3948	0.4545	25

From both the charts of Fig. 1, it can be seen that Decision tree has resulted in the best metrics, followed by NaiveBayes and linear SVM. Similarly for all the datasets, the outperformed algorithms are listed in Table 3.

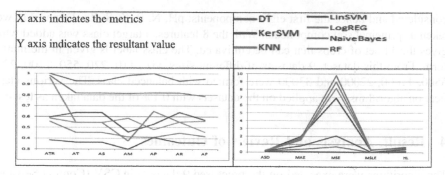

Fig. 1. Performance of various algorithms on datasets of size 110

Table 3. Algorithms that tops in the performance based on the resultant metrics

Dataset size	Maximum metrics	Minimum metrics
110	DT, NB, LinearSVM	DT, NB, LinearSVM
220	DT, NB, RF, KNN	DT, NB, KNN, kerSVM
550	DT, NB, RF, KNN	DT, NB, KNN
1100	DT, RF, NB, KNN	DT, NB, RF, KNN
2200	DT, RF, NB, KNN	DT, RF, NB, KNN
5500	DT, RF, NB, KNN	DT, KNN, RF
11000	DT, RF, KNN, NB	DT, RF, KNN
22000	DT, RF, KNN, NB	DT, RF, KNN
33000	DT, RF, KNN, NB	DT, RF, KNN

From the Table 3, it is clear that the decision tree algorithm results in the highest accuracies and lowest error rates for all datasets. Followed by, when the sample size is smaller i.e. lesser than 1100, Naive Bayes can be used, for larger sample size, random forest can be deployed.

Figure 2 shows the performance of algorithm individually on various datasets. As far as Decision Tree is concerned, accuracies are more than 0.92, accuracies increase with the size of the datasets and reach the maximum when the size is 11000 and deteriorates then. Also it results in null error when the size is between 2200 and 22000. For Kernal SVM, as size increases, accuracy increases and error rate decreases. Maximum resultant accuracy is less than 0.8. Linear SVM results in accuracies less than 0.63, and there is no direct correlation between the dataset size and accuracies. It gives a better result when the size is 11000. It is found that for KNN, the performance increases with the size and the accuracies less than 0.95. Logistic Regression results in accuracies less than 0.62. It shows a good performance when the sample size is 5500. Naïve Bayes performance is not directly correlated with the dataset size. But its accuracies were good when the sample size is 33000. But the error rate are more. As far as Random Forest is concerned, accuracies increase with the size of the datasets and reach the maximum when the size is 22000 and deteriorate then. Also its error rate decreases as the size increases with a minimum when the size is 22000.

Table 4 shows the execution time of the various algorithms on the datasets. Linear SVM, Kernal SVM and logistic regression are having a lengthy time. The remaining algorithms are compared in Fig. 3. Random forest algorithm executes faster when the sample size is less than 1100. As sample size increases, Naïve Bayes algorithm executes faster. Random forest and KNN are the next faster ones.

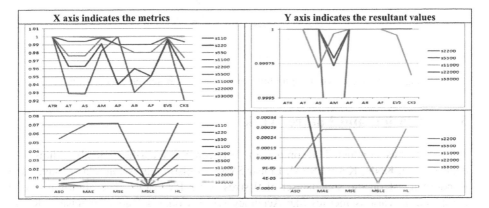

Fig. 2. Performance of decision tree algorithm on the datasets

The outcome of the study is that Decision Tree is the best algorithm for the soil datasets of any size among the considered 7 algorithms with a little compromise in the execution time. Other than decision tree, If execution time is to be taken into consideration, Naive Bayes can be used for sample size less than 1100 and Random Forest for larger datasets.

Table 4. Execution time in ms

Algo	Size								
	110	220	550	1100	2200	5500	11000	22000	33000
DT	14	16	17	21	29	46	62	94	145
KerSVM	18	21	31	63	147	696	23200	7640	17800
KNN	15	17	20	25	34	76	170	421	908
LinSVM	43	131	335	688	14300	42100	93300	24300	51000
LogREG	18	19	26	38	68	148	305	915	2140
NaiveBayes	16	18	17	17	26	36	48	75	116
RF	25	26	27	33	42	71	108	167	256

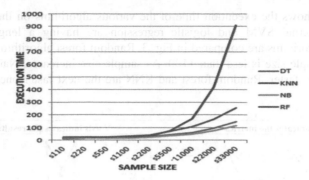

Fig. 3. Performance of algorithms on the datasets based on execution time

5   Conclusion and Future Work

This study has conducted a comparison between the classification algorithms like decision tree, KNN, Kernal SVM, Linear SVM, Logistic regression, Naive Bayes and Random forest on the soil dataset of 9 sizes using PYTHON. The algorithms were compared on various metrics like Accuracy score, Cohen's Kappa, Precision, Recall And F-Measures, Hamming Loss, Explained Variance Score, Mean Absolute Error, Mean Squared Error and Mean Squared Logarithmic Error. The Confusion matrix and Classification report were used for analysis. It is concluded that Decision tree is found suitable for these datasets with little compromise in the execution time for predicting the crops to be cultivated. The other algorithms that top in the rank are NaiveBayes, when the sample size is lesser than 1100, and Random Forest, for larger sample size. When execution time is given importance, NaiveBayes can be used, followed by kNN for sample size less than 5500 and Random Forest for larger datasets. The work can be extended further by comparing these algorithms for different sizes of various datasets other than soil dataset, still larger datasets and varying the parameters for the algorithms. For example, the number of trees in random forest can be changed and analysed. Also the percentage of data taken for testing and training can be changed and analysed.

References

1. Manjula, E., Djodiltachoumy, S.: Data mining technique to analyze soil nutrients based on hybrid classification. Int. J. Adv. Res. Comput. Sci. **8**(8) (2017). http://dx.doi.org/10.26483/ijarcs.v8i8.4794, ISSN 0976-5697
2. RamaKrishna, B.V., Satyanarayana, B.: Agriculture soil test report data mining for cultivation advisory. Int. J. Comput. Appl. (2250-1797), **6**(2), 11–16 (2016)
3. Singh, A.K., Singh, S.R.K., Tomar, K.: Soil testing towards sustainable agriculture and land management: farmer beliefs and attitudes. Asian J. Soil Sci. **8**(2), 290–294 (2013)
4. Dogra, A.K., Wala, T.: A comparative study of selected classification algorithms of data mining. IJCSMC **4**(6), 220–229 (2015). ISSN 2320-088X

5. Deshmukh, B., Patil, A.S., Pawar, B.V.: Comparison of classification algorithms using WEKA on various dataset. Int. J. Comput. Sci. Inf. Technol. (IJCSIT) 4(2), 85–90 (2011)
6. Ramesh, V., Ramar, K.: Classification of agricultural land soils: a datamining approach. Agric. J. 6(3), 82–86 (2011). ISSN 1816-9155
7. Kumar, V.P., Velmide, L.: Datamining plays a key role in soil analysis of Warangal region. Int. J. Sci. Res. Publ. 4(3), 1–3 (2014). ISSN 2250-3153
8. Baskar, S.S., Arockiam, L., Charles, S.: Applying data mining techniques on soil fertility prediction. Int. J. Comput. Appl. Technol. Res. 2(6), 660–662 (2013)
9. Hemageetha, N., Nasira, G.M.: Analysis of soil condition based on pH value using classification techniques. IOSR-J. Comput. Eng. 18(6), 50–54 (2016). e-ISSN 2278-0661, p-ISSN 2278-8727
10. Pahwa, P., Papreja, M., Miglani, R.: Performance analysis of classification algorithms. IJCSMC 3(4), 50–58 (2014). ISSN 2320-088X
11. Shah, C., Jivani, A.G.: Comparison of data mining classification algorithms for breast cancer prediction. In: Conference Paper, July 2013. https://www.researchgate.net/publica-tion/269270867, https://doi.org/10.1109/ICCCNT.2013.6726477
12. Nabi, M., Wahid, A., Kumar, P.: Performance analysis of classification algorithms in predicting diabetes. Int. J. Adv. Res. Comput. Sci. 8(3), 456–461 (2017). ISSN 0976-5697

Global Normalization for Fingerprint Image Enhancement

Meghna B. Patel[1($\boxtimes$)], Satyen M. Parikh[1], and Ashok R. Patel[2]

[1] A.M.Patel Institute of Computer Studies, Ganpat University, Gujarat, India
meghna.patel@ganpatuniversity.ac.in,
parikhsatyen@yahoo.com
[2] Florida Polytechnic University, Lakeland, FL, USA
apatel@floridapoly.edu

Abstract. The enhancement in fingerprint image remains as a vital step to recognize or verify the identity of person. The noise is influenced during the acquisition of fingerprint image. The poor quality images are captured and leads to inaccurate levels of discrepancy in values of gray level beside the ridges and furrows because of non-uniformity of ink and contact of finger on scanner. This poor quality images are affect to the minutiae extraction algorithm which may extract incorrect minutiae and affect to the fingerprint matching during post-processing. Normalization is the pre-processing step for increase the quality of images by removing the noise and alters the range of pixel intensity values. The mean and variance are used in process to reduce variants in gray-level values along ridges and valleys. In this paper show the result of applied famous global normalization and prove by empirical analysis of an image that normalization is remain good for enhancement process and make the noise free image which is useful for next step of fingerprint recognition.

Keywords: Fingerprint · Normalization · Global normalization · Block normalization · Image enhancement

1 Introduction

Due to the wide-spread use of internet, the information is electronically transferred through the connected network. The security of information is become major issue at the time of data transmission. Identification of true person and protect the information from opponent play vital role [1, 2].

The identity of person can be recognized or verified using biometric authentication. The biometric authentication is categorized into physiological and behavioral features of an individual like fingerprints, hand geometry, face, iris, voice, signature, key stroke [3]. Based on the bcc market research report [4] fingerprint is the most popular biometric in the world up to 2015. See the below Fig. 1.

As well as in the latest report of BCC research titled as "Biometrics: Technologies and Global Markets" computed $14.9 billions in 2015, estimated to extended upto $41.5 billions by 2020 [5].

The fingerprint provide highest degree of reliability amongst other biometric modalities like hand, voice, iris, face etc. [6]. The fingerprints are distinctive and

© Springer Nature Switzerland AG 2020
S. Smys et al. (Eds.): ICCVBIC 2019, AISC 1108, pp. 1059–1066, 2020.
https://doi.org/10.1007/978-3-030-37218-7_111

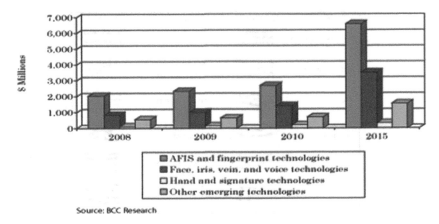

Fig. 1. The global market of biometric technologies from 2008–2015 [5]

remain unchanged during entire life [7]. Also based on study of research paper [8] the possibilities of two fingerprints are similar is very low around 1 in 1.9×10^{15}.

Fingerprints are the marks or pattern that is generated after pressing the tip of finger on any surface. Fingerprint holds ridges and valleys patterns. The papillary lines of finger are called ridges. The gaps exist between ridges are called valley patterns [1]. The ridges are displayed as a dark lines and the white area display between ridges are indicated as valleys. See in below Fig. 2. The Fingerprint Recognition done by first extracting minutiae and matching the minutiae. Minutiae are the features of fingerprint see in below Fig. 3.

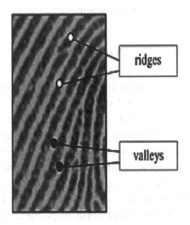

Fig. 2. Ridges and furrows (valleys) of fingerprint [4]

The recognition of fingerprint is done in two parts: (1) Pre-processing (2) Post-processing. The processes called normalization, image enhancement, orientation estimation, binarization and thinning are considered as Pre-processing phases while

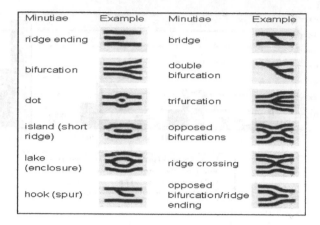

Fig. 3. Minutiae of fingerprint

post-processing includes extraction of minutiae, detect and remove false minutiae, extract core point and minutiae matching [9–12].

The improvement in finger image quality is challengeable task because poor quality finger image are generated due to dry or wet skin or damaged due to age, scratches, worn down, etc. [6]. These factors affect to ridge structure, for that connected ridges become disconnected and disconnected ridges become connected. The result of false ridges structure extracts the false minutiae and gives false match of fingerprint.

The paper formed in different sections. The Sect. 2 include previous work done on normalization Sect. 3 include proposed work on global normalization using mean and variance Sect. 4 include Performance Measurement and Result Discussion Sect. 5 include Conclusion and Future Enhancement.

2 Previous Work Done on Normalization

The pixel-wise enhancement is used in an image processing. In pixel-wise enhancement the novel pixel value is depend on the earlier pixel value but it can't affect the clarity of ridge and furrows structure. Many methods are used at the initial stage of image enhancement are like histogram equalization, contrast stretching, normalization and wiener filtering [6]. Various techniques of finger image enhancement are proposed in paper [13, 14]. The most of techniques are enhance the image from binary image whereas some techniques enhance finger image on gray scale image. Normalization is the first phase to apply enhancement on gray scale image. To increase quality of finger images, the researchers proposed block-wise implementation. They do adaptive normalization based on block processing. Using each block's statistics, input finger image is divided in sub-block and normalized the process [13, 14]. In the research paper [15] states that in pre-processing fingerprint image enhancement can be done in two phases: (1) Use conventional filter for enhancing the image without changing filter parameters. (2) Use conventional filter to enhance the structure of ridges and furrows (valleys) along changes in filter parameter based on ridge orientation and frequency. The author introduces the comparison of combination of local, global and block-local normalization at first phase for analysis of efficiency. In proposed work, followed the widely accepted normalization method established by Hong et al. [16].

3 Global Normalization Using Mean and Variance

The images are captured by scanner during the image acquirement, at that time poor quality images are captured and lead to inaccurate levels of discrepancy in values of gray level beside the ridges and furrows because of non-uniformity of ink and contact of finger on scanner.

Normalization is an important pre-processing step for increase the quality of images by altering the range of pixel intensity values for removing the noise. Sometimes, it is called as contrast stretching. This process reduces variants in gray level values along ridges and furrows using mean and variances.

The following Eq. (1) is used for normalization process [6, 16] to determine the new intensity value of each pixel.

$$
\mathcal{G}(i,j) = \begin{cases} M_0 + \sqrt{\dfrac{VAR_0(\mathcal{I}(i,j)-M)^2}{VAR}} & \text{if } \mathcal{I}(i,j) > M \\ M_0 - \sqrt{\dfrac{VAR_0(\mathcal{I}(i,j)-M)^2}{VAR}} & \text{otherwise} \end{cases} \tag{1}
$$

Here, I(i,j) describe gray-level value of pixel (i, j). M_0 describe desired means value and VAR_0 describe desired variance values.

After applying Eq. (1) and make try and error concept for finding out desired mean and variance value perfect for databases. Then finally decided $M_0 = 100$ and $VAR_0 = 100$ is perfect value for fingerprint database. The steps of algorithm for proposed work in normalization process is describe in Fig. 4.

4 Performance Measurement and Result Discussion

The proposed work is implemented in java language. To check result of normalization process, the experiment is done using two databases FVC2000 [17, 18] with subset DB1,DB2,DB3 and DB4 and FingerDOS [19, 20]. Tables 1 and 2 show detail description of databases.

The proposed work is evaluated using different measurement parameters such as: (i) MSE and PSNR value (ii) Computational Time (iii) Quality of an Image.

(i) MSE and PSNR value

PSNR stands for Peak Signal-to-Noise Ratio. The below Eq. (3) used to measure quality within original and reconstructed finger image. Higher PSNR and lower MSE prove that reconstructed image has better quality.

MSE stands for Mean Squared Error. To calculate PSNR, first calculate the MSE using following Eq. (2) then calculate PSNR using Eq. (3):

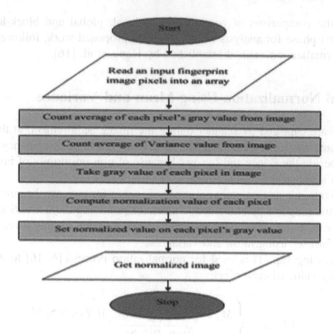

Fig. 4. Flowchart of normalization process

Table 1. Database FVC2000 [17, 18]

FVC2000				
Set B	Sensor type	Image size	No. of impression	Resolution
DB1	Low-cost optical sensor	300 × 300	10 × 8	500 dpi
DB2	Low-cost capacitive sensor	256 × 364	10 × 8	500 dpi
DB3	Optical sensor	448 × 478	10 × 8	500 dpi
DB4	Synthetic generator	240 × 320	10 × 8	About 500 dpi

Table 2. FingerDOS database [19, 20]

Finger DOS			
Sensor type	Image size	No. of impression	Resolution
Optical sensor (secuGen iD-USB SC)	260 × 300	3600 = 60 × 6 × 10 i.e. No. of subjects = 60 No. of fingers = 6 (index, middle and thumb of right and left hand) No. of impression = 10	500 PPI

$$MSE(x,y) = \frac{1}{N}\sum_{I=1}^{N}(x_i - y_i)^2 \qquad (2)$$

Where, N stands for all pixels in input image.

$$PSNR = 10\,\log_{10}\frac{L^2}{MSE} \qquad (3)$$

Where, L stands for number of discrete gray level.

The Table 3 shows comparative study of input image and normalized image with PSNR and MSE values. The comparison shows that normalized image contain higher PSNR and lower MSE value compared to input image. The comparison prove that proposed work present better result in specification of image brightness and contrast.

Table 3. Comparative study of MSE, PSNR and computational time for original image and normalized image

Images	Original image			Normalized image		
	MSE	PSNR	Execution time	MSE	PSNR	Execution time
FVC2000_101	10.634	37.365	176	5.584	39.342	156
FVC2000_102	09.551	38.438	126	6.542	38.469	113
FingerDOS_101	13.096	36.468	138	7.049	40.634	118
FingerDOS_101	12.765	37.287	135	5.045	40.332	116

(ii) **Computational Time**

The calculation time taken in milliseconds for original and normalized image. The result shown in Table 3.

(iii) **Quality of an image**

The visualize appearance of fingerprint images are improved after applying global normalization to converting image into grayscale image as well as noise are also reduced. To show the result two images are selected from FVC2000 database and FingerDOS database respectively. In image red mark show that noise is removed which is occurred due to pressure given on scanner. The implemented result is shown in Fig. 5.

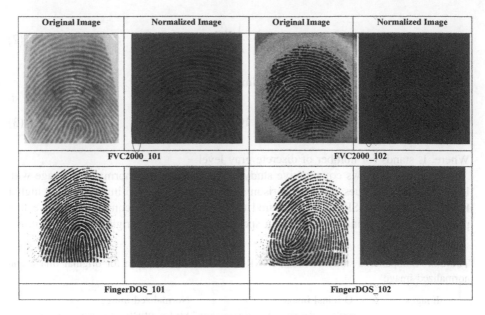

Original Image	Normalized Image	Original Image	Normalized Image
	FVC2000_101		FVC2000_102
	FingerDOS_101		FingerDOS_102

Fig. 5. Visualize appearance of an image [7]

5   Conclusion and Future Work

The proposed work shows implementation of global normalization. The proposed work find out the noise which are occur due to pressure given on scanner and remove noise after altering range of pixel intensity values. The performances of original images and enhanced normalized images figure out using MSE and PSNR value. Also consider the execution time and visualize appearance of an image for comparative study. The experiential results demonstrate that enhanced normalized images provide better result compare to original image.

An another fingerprint recognition phases like image enhancement, orientation estimation, binarization and thinning are done after removing the noise using normalization. To improve overall accuracy of fingerprint recognition, in future work, try to enhance each phase with applying different algorithms.

Compliance with Ethical Standards

✓ All authors declare that there is no conflict of interest.
✓ No humans/animals involved in this research work.
✓ We have used our own data.

References

1. Lee, H.C., Gaensslen, R.E.: Advances in Fingerprint Technology. CRC Press, Boca Raton (2001)
2. Ratha, N.K., Bolle, R.M.: Fingerprint image quality estimation. IBM Computer Science Research Report (1999)

3. Jain, A.K., Ross, A., Prabhakar, S.: An introduction to biometric recognition. IEEE Trans. Circuits Syst. Video Technol. **14**(1), 4–20 (2004)
4. Biometrics: Technologies and Global Markets. Published: November 2010 Report Code: IFT042C. http://www.bccresearch.com/market-research/information-technology/biometrics-technologies-markets-ift042c.html. Accessed Apr 2015
5. Biometrics: Technologies and Global Markets. Published: January 2016 Report Code: IFT042E. https://www.bccresearch.com/market-research/information-technology/biometrics-technologies-markets-report-ift042e.html. Accessed Oct 2017
6. Maltoni, D., Maio, D., Jain, A.K.: Handbook of Fingerprint Recognition, 3rd edn. Springer, New York (2009)
7. Carlson, D.: Biometrics - Your Body as a Key, 2 June 2014, http://www.dynotech.com/articles/biometrics.shtml
8. Mario, D., Maltoni, D.: Direct gray-scale minutiae detection in fingerprints. IEEE Trans. Pattern Anal. Mach. Intell. **19**(1), 27–40 (1997)
9. Patel, M.B., Patel, R.B., Parikh, S.M., Patel, A.R.: An improved O'Gorman filter for fingerprint image enhancement. In: 2017 International Conference on Energy, Communication, Data Analytics and Soft Computing (ICECDS), pp. 200–209. IEEE, August 2017
10. Patel, M.B., Parikh, S.M., Patel, A.R.: Performance improvement in preprocessing phase of fingerprint recognition. In: Satapathy, S., Joshi, A. (eds.) Information and Communication Technology for Intelligent Systems, pp. 321–330. Springer, Singapore (2019)
11. Patel, M.B., Parikh, S.M., Patel, A.R.: An improved approach in fingerprint recognition algorithm. In: Published in International Conference on Computational Strategies for Next Generation Technologies (NEXTCOM 2017), Organized by CT Institute of Engineering Management & Technology Shahpur Jalandhar, 25–26 November 2017. Proceeding in Springer CCIS Series (2017). ISSN No. – 1865-0929
12. Patel, M., Parikh, S.M., Patel, A.R.: An improved approach in core point detection algorithm for fingerprint recognition. In: 3rd International Conference on Internet of Things and Connected Technologies. Elsevier (2018)
13. Kim, B.G., Kim, H.J., Park, D.J.: New enhancement algorithm for fingerprint images. In: Proceedings of 16th International Conference on Pattern Recognition, vol. 3, pp. 879–882. IEEE (2002)
14. Shi, Z., Govindaraju, V.: A chaincode based scheme for fingerprint feature extraction. Pattern Recogn. Lett. **27**(5), 462–468 (2006)
15. Kocevar, M., Kacic, Z.: Efficiency analysis of compared normalization methods for fingerprint image enhancement. ARPN J. Syst. Softw. **3**(3), 40–45 (2013)
16. Hong, L., Member, S., Wan, Y., Jain, A.: Fingerprint image enhancement: algorithm and performance evaluation. IEEE Trans. Pattern Anal. Mach. Intell. **20**(8), 777–789 (1998)
17. Maio, D., Maltoni, D., Cappelli, R., Wayman, J.L., Jain, A.K.: FVC2000: fingerprint verification competition. IEEE Trans. Pattern Anal. Mach. Intell. **24**(3), 402–412 (2002)
18. FVC2000 (n.d.). http://bias.csr.unibo.it/fvc2000/databases.asp. Accessed June–July 2018
19. Bong, F.F.L.D.: FingerDOS: a fingerprint database based on optical sensor. WSEAS Trans. Inf. Sci. Appl. **12**(29), 297–304 (2015)
20. Fingerprint Database Based on Optical Sensor (FingerDOS) (n.d.). https://fingerdos.wordpress.com/. Accessed June–July 2018

Recurrent Neural Network for Content Based Image Retrieval Using Image Captioning Model

S. Sindu[1][✉] and R. Kousalya[2]

[1] Department of Computer Science, Dr. N.G.P. Arts and Science College,
Coimbatore, India
sindupradeep2019@gmail.com
[2] Department of Computer Applications, Dr. N.G.P. Arts and Science College,
Coimbatore, India

Abstract. With the tremendous growth in the collection of digital images by social media, ecommerce applications, medial applications and so on, there is a need for Content based image retrieval. Automatic retrieval process is one of the main focus of CBIR, whereas traditional keyword based search approach is time consuming. Semantic based image retrieval is performed using CBIR where the user query is matched based on the perception of the contents of the image rather than the query which is in text format. One of the main research issue in CBIR is the semantic gap which can be reduced by deep learning. This paper focuses on image captioning model for image retrieval based on the content using recurrent and convolutional neural network in deep learning.

Keywords: Content based image retrieval · Deep learning · Convolutional neural network · Recurrent neural network

1 Introduction

Traditional method for retrieving content based on the images uses text based query to retrieve the images from the database. Since large number of images are generated from different applications, text based image search is more tedious. In content based image retrieval, search is performed taking into consideration on the contents, rather than metadata of the image such as text description, tags, annotation etc. [1]. The term content refers to the low level features such as color, shape, texture or it can be higher level feature which can be derived from the image and thus ensures semantic image retrieval. One of the research challenging issues in CBIR is that the low level features such as color, texture and shape are sufficient to describe the higher level semantic information and to bridge the research gap the Recurrent Neural Network (RNN) and Convolutional Neural Network (CNN) are used.

1.1 Convolutional Neural Network for CBIR

Content based image retrieval uses convolutional neural network to detect and recognize images. Convolutional neural network are special type of neural network which

© Springer Nature Switzerland AG 2020
S. Smys et al. (Eds.): ICCVBIC 2019, AISC 1108, pp. 1067–1077, 2020.
https://doi.org/10.1007/978-3-030-37218-7_112

works very well with the data that is spatially connected and it is used for image processing and image classification. Using Convolutional neural network objects can be detected and recognized. Images are made of pixels, but individual pixels by themselves do not represent any useful information. But a group of pixels preeatures sent adjacent to each other could represent features like shapes (eyes, nose etc.). Normal neural network will not be able to accurately understand the image features correctly.

CNN is composed of many layers and it automatically learns the features like colors, shapes, textures, relationships between shapes etc. The various layers in convolutional neural network are input layer, convolution layer, pooling layer, fully connected layer and output layer. Convolution and pooling are the two special type of layers helpful for image recognition in CNNs. Input layer is the first layer in the Neural network which is fed with normalized Image Pixel data. Raw image pixel data is converted to a grey scale image, and the pixel data are normalized by dividing the pixel intensities by 255 to convert pixel information to be in range 0 to 1. Output Layer is at the end of the Neural network which gives the final classification. Convolution layer convolutes the images by using filters. Pooling layers removes unnecessary parts from the image.

1.2   Recurrent Neural Network

Sequential characteristic of data are recognized by RNN and also predicts the pattern of next likely scenario. It is commonly used in natural language generation, text generation, voice recognition and text generation. RNN can handle dynamic temporal data in a better way [3]. The input to RNN is a time series data and the output can also be a time series data. RNN can handle different types of input and output such as varying input or fixed input, varying output or fixed output. The structure of RNN is shown in Fig. 1 along with the normal feed forward neural network. It has an input layer, hidden layers and an output layer. Recurrent Neural network have loops which allow information to persist and it is recursive in structure [4].

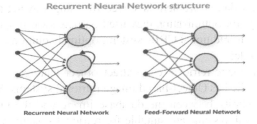

Fig. 1. Structure of recurrent neural network

Some of the applications based on the input type are Text recognition, Sentiment classification, chatbots, image captioning and so on. Image captioning are used in content based image retrieval.

Based on the object and actions of the image, generating textual description for an image is called image captioning. Generating captions for image is a challenging research issue. The role of image captioning for retrieving images based on the content is to achieve semantic similarity by capturing the textual data associated with the images. The caption generated from an image is shown in Fig. 2 and the corresponding caption generated for the below image is "A small aeroplane flying on the sky". The rest of this paper is organized as follows. In Sect. 2 a brief literature survey is made on the existing works. Section 3 describes the proposed method followed by the experimental evaluation with the final conclusion.

Fig. 2. Caption generated from RNN [7]

2 Literature Survey

Zhu et al. [5] used unsupervised algorithm defined by a semantic meaning with hashing functions. The purpose of their research is to enhance the capability of making unique distinctions of hash codes by extracting semantics automatically from the noisy associated texts. In traditional approach, the learning is by maintaining similarity of images visually, which is differentiated by the hash code formulation using a unified unsupervised framework.

In [6] the authors proposed a CBIR schema. To improve the accuracy of the image retrieval system, color, texture and shape features are fine tuned and optimized. The noise and image translation problem is solved by using wavelet and curvelet features for textures. Finally, for optimization they used particle swarm optimization algorithm. When compared with traditional CBIR system, optimal solution and better accuracy is reached in their model.

Content Based image retrieval can extract object features from the image in the database which is called as Object-based image retrieval systems. Segmentation is used for finding the similarity between the database image and the user query image [7]. Object based retrieval systems are suitable for retrieval systems where the objects are easily separated from the environment and that have unique identification for colors or textures [8].

In [9], remote sensing images are used for region based retrieval where a graph based theoretical approach with unsupervised learning is used in their framework. Graph model is used in images and this graph based model can easily find location based feature distribution. Here the resemblance with respect to the image provided by the user and the image in database is based on the graph similarity. In [10], a template based caption generation method are used by the authors which uses fixed template sentence. In this method explicit annotation is required for each class. Image is linked to a sentence and a score is computed. The importance is given for the meaning obtained from the image with the sentence and the score is calculated. Strong enough captions are not generated using this method because it uses a rigid templates.

Image retrieval based caption generation is proposed by the authors in [11]. Similar images based on the content are retrieved and the query images are also compared based on the captions. Transfer based caption generation strategies are used in their work. To compose image description, tree based method is used. The database image content is shown in the form of tree fragments and are used to denote the expressive phrases. The fragments of a tree for a new description composed are pruned.

Neural network approach for caption generation in used by the authors in [12]. Their aim is framing sentence, which is the description for the given image. In their work they presented a learning approach which is sequential and minimal assumption is made on the structure which follows a sequence. Two LSTMs are used in their work. One LSTM is used to map the vector with the input of fixed dimensionality, and another LSTM which is deep is used for decoding from vector to target sequence. To understand the text a log based multimodal approach is proposed by Kiros et al. [13]. In their approach, a word contained on a image is compared with the previous word by probability distribution. Convolutional neural network is used for image-text modeling, and it is used to learn word representations and image features by model training.

In [3, 4] recurrent neural network is used for caption generation. A multimodal neural work approach is used here. Given a previous word, next word is generated by probability. A deep recurrent neural network for text generation and a convolutional neural network for image identification, are the two different neural network types used in their model. The interaction between two model is performed in a multimodal layer to form a new mRNN model. Xu et al. in [14], used an model which automatically learns to describe the content of the image. A deterministic manner is used in their model and the model is trained.

3   Proposed Methodology

Existing CBIR methodologies use pixel level feature similarity which results in large semantic gap. The main disadvantage with existing system is the object type mismatch in the query image and retrieved images. To overcome these difficulties, the proposed model uses CNN and RNN for the content image retrieval. By combining both CNN and RNN, semantic gap can be reduced further by giving importance total number of objects, relative position of objects, actions and so on. The steps to perform image Captioning using CNN and RNN are listed below and the flow chart is shown in Fig. 3.

The Neural Image Captioning model works as follows

- A Standard Pre Trained CNN models like VGG16, VGG19, ResNet or Inception is used as a Image Feature Extractor

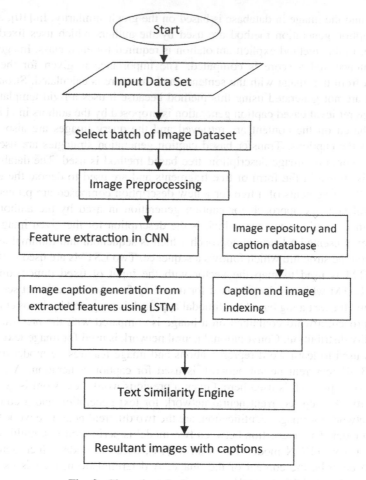

Fig. 3. Flow chart for image captioning model

- The output of CNN Layer which is filtered feature vectors of the Input image is fed into a RNN
- The Image Caption is encoded into a vectorised form using Word to Vector process
- The RNN is trained to Generate Image captions by matching the Image Features from the CNN network corresponding caption of the image from the Image Database
- There is a Attention Mechanism which helps the RNN learn to pay selective attention and focus only on a part of the Image to generate caption.

3.1   The Methodology Is Proposed in Three Steps

- Step 1: Object detection using convolution neural network
- Step 2: Image captioning phase to generate text captions for images

- Step 3: Text similarity based ranking phase on image captions to identify similar images.

Step 1: Object detection using Convolutional neural network

The most important layer in CNN is the Convolution layer, and it is the first hidden layer which performs convolutions and it is a matrix multiplication operation with specific types of filters. Filters are usually a 3 × 3 matrices which convolves the image into processed forms like edge detection, depth detection etc. Each 3 × 3 sub section of original matrix are multiplied with the filter and result is stored as output and fixed stride is taken to start next matrix multiplication operation. This also reduces the input matrix dimension. Pooling layer helps in extracting the important features and removing unessential parts of an image. Pooling helps in dimension reduction and also performs rotational variance to some extent. Some of the pooling operations are max, min, sum and average.

Fully connected layers are the last part of a Convolution Neural Network before the Output Layer and gets the input from the previous pooling layer which also acts as normal input layer. Image data is Stacked or Flattened before feeding the input to the Fully connected layer. By the time image reaches fully connected layers, the image data is dimensionally reduced from the original image size and also important features are highlighted and extracted. The activation function used is the Relu activation function.

The use of convolution neural network bridges the semantic gap for object type matching and ranking on within other low level features. Further semantic gap exists like number of objects, relative position of objects, actions etc., which can be resolved using Recurrent Neural Network.

Step 2: Image captioning phase to generate text captions for images

Recurrent Neural Network is used to generate caption for the images. The output of CNN which is a filtered feature vectors of the input image is fed into RNN. The problem with RNN is the vanishing gradient descent and the memory is short. Hence RNN uses a architecture called Long Short-Term Memory (LSTM) networks as shown in Fig. 4, and the purpose of LSTM is to extend the memory for the required period of time. LSTM has two types of cells, short term memory cells and the long term memory cells, from which a decision is made to retain the information for a long time in memory by looping back into the network or to discard it. The memory in LSTM is in the form of gated cells, where gated means that the cell decides whether to store the information in the memory or to discard it. All these are performed by using different gates such as learn gate, forget gate, remember gate. The caption generated using CNN and LSTM is shown is Fig. 5.

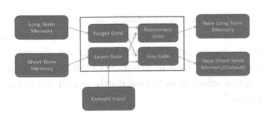

Fig. 4. LSTM architecture

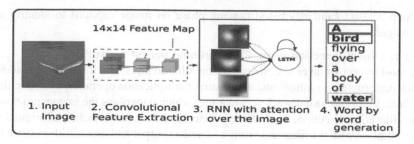

Fig. 5. Image captioning using CNN and RNN(16)

To further improve the performance of the LSTM, an Attention mechanism is used and in this mechanism there is an encoder LSTM which works as a normal LSTM. The intermediate outputs from the encoder LSTM is taken as input. For each input sequence the model is trained to learn to pay selective attention and are related to items in the output sequence. Filtering is performed for each input sequence item based on some conditions and the output sequence is generated. This type of attention based sequence mechanism is useful for applications like image caption generation to make best use of the convolutional neural network and the recurrent neural network. The sequence-based attention mechanism can be applied to computer vision problems to help get an idea of how to best use the convolution neural network to pay attention to images when outputting a sequence, such as a caption.

Step 3: Text similarity based ranking phase on image captions to identify similar images

The content based image retrieval model uses image captioning model to generate captions for the user query image. The database has images along with captions. Text similarity based ranking phase is applied on image captions to identify similar images. The query image caption is compared with the database image caption. If the caption are similar the relevant image is retrieved. Cosine similarity is used for the text similarity processing.

To convert a image into captioning model and to find similarities between text, natural language processing is used. To understand the context of the image for caption generation and to perform text similarity Part of Speech and BLEU (BiLingual Evaluation Understudy) score are used. To measure the BLUE score N-Gram is used.

3.2   Part of Speech (POS) Tagging

POS tagging is used to tag each words based on the grammer and it is the process of marking up a word in a text as similar to a particular part of speech based on both definitions and speech. In the proposed work to reduce the semantic gap in retrieved images higher importance is given to noun and verb of the generated caption. In this proposed work, we have selected to "mandatorily match the noun" and also given next "higher priority to verb".

3.3 N Gram

N-gram refer to the process of combining the nearby words together for representation purposes where N represents the number of words to be combined together. N-Gram combines words together from a sentence for representation purposes and N represents the number of words to be combinded.

Example: A group of girls sitting under a tree

1-Gram - 'A', 'group', 'of', 'girls', 'sitting', 'under', 'a', 'tree'

2-Gram - 'A group', 'group of', 'of girls', 'girls sitting', 'sitting under', 'under a', 'a tree'

3-Gram - 'A group of', 'group of girls', 'of girls sitting', 'girls sitting under', 'sitting under a', 'under a tree'.

Breaking down a natural language into n-grams is essential for maintaining counts of words occurring in sentences which forms the backbone of traditional mathematical processes used in Natural Language Processing.

4 Experimental Evaluation

The dataset used for evaluating the performance model of the proposed system is the MSCOCO dataset. It is implemented using tensorflow and keras. Tensorflow is one of the best library to implement deep learning and keras is built on top of tensorflow. The metrics used for evaluating the image captioning model is the BLUE metrics. All the images from validation dataset is taken for calculating BLEU score and the average score is 66.5 which shows that the sentence generated from query image are very similar compared to the sentence generated from pretrained captioning model.

The custom metrics used to evaluate the performance of image captioning model for Content based image retrieval is Mean Average Precision. No of images used for calculating metrics is limited to 5000 images from validation set of MSCOCO dataset because of performance constraints. To better understand the closing of the semantic gap the results of the new proposed metrics is shown below in Table 1 and the corresponding images are shown from Figs. 6, 7, 8 and 9.

Table 1. Mean Average Precision

Mean Average Precision for Noun matched between generated and top 5 result captions	Mean Average Precision for Verb matched between generated and top 5 result captions
61.71	46.18

The input image given for the proposed image caption generator is shown in Fig. 6 and the corresponding relevant image retrival is shown in Fig. 7. The caption generated for the input image is "a couple of birds sitting on top of a tree" and the results shows that the importance is given to the noun and verb and the related top five images are retrieved. It shows that the image is retrieved with the relevant content and thus reducing the semantic gap.

Fig. 6. Input bird image

Fig. 7. Relavent top 5 images [5]

The input image with action is shown in Fig. 8 and the corresponding output image is shown in Fig. 9. The input image is "A baseball player swinging a bat". The output image generated also shows the relevant image with the same action. It shows that while retrieving the image from the database, it searches for the image with similar action also.

Fig. 8. Input image with action [6]

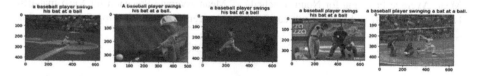

Fig. 9. Output image with action

5 Conclusion and Scope for Further Research

From a large database, Content based image retrieval is used to search a most relevant image based on the content of the image. To reduce the semantic gap in the retrieval, image captioning using recurrent neural network and convolution neural network is proposed in this work. Object classification is performed using CNN and the output of CNN is send as input to RNN. RNN uses LSTM to persist the needed information in the memory. The overall summary of the proposed work is that the object detection is performed using convolution neural network, and the image captioning phase is used to generate text captions for images and RNN is used in the image captioning phase and finally text similarity based ranking is performed on image captions to identify similar images. Non textual features are not captured in the current work and it is a future scope. Features like texture patterns can also be matched to further bridge the semantic gap in the retrieved results.

Compliance with Ethical Standards

✓ All authors declare that there is no conflict of interest.

✓ No humans/animals involved in this research work.

✓ We have used our own data.

References

1. Yasmin, M., Sharif, M., Mohsin, S.: Use of low level features for content based image retrieval: survey. Res. J. Recent Sci. **2**(11), 65–75 (2013)
2. Zhou, W., Li, H., Tian, Q.: Recent advance in context-based image retrieval: a literature survey, electronic edition@arxiv.org (2017)
3. Mao, J., Xu, W., Yang, Y., Wang, J., Yuille, A.L.: Deep captioning with multimodal recurrent neural networks. In: ICLR (2015)
4. Karpathy, A., Fei-Fei, L.: Deep visual-semantic alignments for generating image descriptions. In: CVPR (2015)
5. Zhu, L., Shen, J., Xie, L.: Unsupervised visual hashing with semantic assistant for content-based image retrieval. IEEE Trans. Knowl. Data Eng. **29**, 472–486 (2016)
6. Fadaei, S., Amirfattahi, R., Ahmadzadeh, M.R.: New content-based image retrieval system based on optimised integration of DCD, wavelet and curvelet features. IET Image Process. **11**, 89–98 (2016)
7. Danezis, G., Gürses, S.: A critical review of 10 years of privacy technology. In: Proceedings of the 4th Surveillance and Society Conference (2010)
8. Hoiem, D., Sukthankar, R., Schneiderman, H., Huston, L.: Object based image retrieval using the statistical structure of images? In: Proceedings of the IEEE Conference on Computer Vision and Pattern Recognition (2004)
9. Rashtchian, C., Hockenmaier, J., Forsyth, D.A.: Every picture tells a story: generating sentences from images. In: ECCV (4) (2010)
10. Chaudhuri, B.: Region-based retrieval of remote sensing images using an unsupervised graph-theoretic approach. IEEE Geosci. Remote Sens. Lett. **13**(7), 987–991 (2016)
11. Kuznetsova, P., Ordonez, V., Berg, T., Choi, Y.: Treetalk: composition and compression of trees for image descriptions. TACL **2**, 351–362 (2014)

12. Sutskever, I., Vinyals, O., Le, Q.V.: Sequence to sequence learning with neural networks. In: NIPS (2014)
13. Kiros, R., Salakhutdinov, R., Zemel, R.S.: Multimodal neural language models. In: ICML (2014)
14. Xu, K., Ba, J., Kiros, R., Cho, K., Courville, A.C., Salakhutdinov, R., Zemel, R.S., Bengio, Y.: Show, attend and tell: neural image caption generation with visual attention. In: ICML (2015)

Using Deep Learning on Satellite Images to Identify Deforestation/Afforestation

Apurva Mhatre, Navin Kumar Mudaliar, Mahadevan Narayanan[(✉)],
Aaditya Gurav, Ajun Nair, and Akash Nair

Department of Computer Engineering, SIES Graduate School of Technology,
Navi Mumbai, India
{apurva.mhatre16, navin.mudaliar16,
mahadevan.narayanan16, aaditya.gurav16, ajun.nair16,
akash.nair16}@siesgst.ac.in

Abstract. As of 2018 the area covered by the forests was determined to be 30.6% of the world's land surface which is roughly below 3.8 billion hectares. With the exponential increase in population, the pressure on the natural resources increases which results in cutting the trees for agriculture and for industrial purposes thereby leading to deforestation. In this paper, we investigate the spatial distribution of vegetation cover by using Convolutional Neural Network (CNN) and satellite images. The features of the object identified by the CNN is extremely complex because of the number of filters used to identify various patterns. Supervised learning (learning with annotated data) is used to train the CNN model as it results in better accuracy while determining the forest cover. We leverage this technology of instance segmentation to correctly determine the forest cover in a particular satellite image that radically improves the accuracy.

Keywords: Deforestation · Instance segmentation · Computer vision · Convolutional Neural Network

1 Introduction

India is the home to 1.37 billion people [8], being the 2nd most populous country on the planet with just 3.287 million km^2 of land, there is always a burden on the country's national resources which includes the forests. Indian sub-continent has all types of geographical terrain right from ice clad mountains to hot deserts which makes India home to a wide variety of animal species. The exponential growth of human population has surged the need for space to build civilizations. In India the forest cover was 21% by the end of the last decade [7, 9], but it is constantly increasing in this decade with an average of 0.2% per annum according to Government data.

The main challenges that are faced leading to above mentioned problems are-

- In India, the infrastructure needed to survey forest land constantly and periodically does not exist, and we lack human resources in the Ministry of Environment, Forest and Climate change.

© Springer Nature Switzerland AG 2020
S. Smys et al. (Eds.): ICCVBIC 2019, AISC 1108, pp. 1078–1084, 2020.
https://doi.org/10.1007/978-3-030-37218-7_113

- Much of the Himalayan ranges are geographically challenging to monitor.
- Changes usually come to notice after a substantial portion of forest has been lost.

This paper consists of related works, methodology used, implementation details and the results. The proposed solution consists of 3 phases which are preparing the dataset, feature extraction and object detection with instance segmentation as we want to determine the approximate area of our region of interest. The satellite images will contain forest areas at certain locations in the image that the model has to classify correctly.

Convolutional Neural Network is used here which is a deep learning algorithm that takes image as an input, extract relevant features of objects by using learnable weights and biases and can distinguish between various objects. The pre-processing task which is an important aspect in traditional machine learning techniques is minimal here as the network is able to learn the characteristics with enough training.

Mask Region-CNN (RCNN) [1, 2, 6] is an instance segmentation technique that not only identifies the bounding boxes but also the location of each pixel of the desired object. Its explanation is twofold: region identification and then classifying the said regions and generating masks and bounding boxes. This is done by using another Full Convolution Network (FCN) with input as feature map that gives 0 if the pixel isn't associated with the object and 1 otherwise.

Our model is able to segment the forest and non-forest covers with accuracy over 90% which has various applications. It can be used for comparing the images of previous dates to find out the deforestation rate over time. The model can be improved by training with a dataset of bigger size and variety. The model was tested on images unseen by the model which resulted in results with substantial accuracy.

2 Related Works

Similar work [3] has been done in a competition organized by Planet. The objective was to detect deforestation in the Amazon Basin. Dataset was already provided by Planet with both JPEG and GeoTIFF (it provides info about the weather conditions inside an image) images but the major difference was the approach, the winners of the competition used Custom CNN model at first. Around 30 labels were determined which included mining, lakes, human habitation, etc. They compared the frequency of detection of labels and inferred whether there was significant change in the topology of the land.

Another work [4] which used an Artificial Neural Network (ANN) and GIS to determine the relationship between deforestation and socioeconomic factors. It simulates the deforestation rate over a set of parameters which is preemptively defined and predicts the part of the forest that was destroyed.

Similarly, to establish relation between forest structural features, stem density and basal area, an ANN [10] and remote sensing was used. The ANN generated a map of basal area which provided climatic influences on the forest structure along with structural heterogeneity.

Similar work [11] has been done using Land Change Modeler of TerrSet, a software analytical tool based on GIS that analyses historical data of forests and generates risk maps as well as future forest loss in the Democratic Republic of Congo.

Similarly, Efficient Bayesian Multivariate Classifier (EBMC) [12] and Bayesian networks were used to identify the factors that are associated with direct deforestation drivers and to predict the risk of deforestation.

3 Methodology

3.1 Collection and Annotation of Dataset

Collection of dataset is always a challenge and satellite images are not easy to obtain. We collected images from Kaggle dataset and the rest were acquired from Planet.org, a website which provides satellite images over a timeline. Satellite images have an accuracy up to 3 m. High resolution image was essential to capture the details of our region of interest. The details in a high resolution make feature detection possible using the deep learning algorithm. Annotation of images has been done by pixel annotation tool [5] which is an open-source tool.

3.2 Feature Extraction and Detection

As shown in the Fig. 1, the model extracts thousands of features using the filters which gets better during epoch of training. The layers in the beginning of the network extract features of less relevance whereas the latter layers extract features of very high relevance. It forms a feature map of lower dimension compared to the image. As it is a Feature Pyramid Network (FPN) consisting of 2 pyramids, one receives the features of high relevance and the other receives the features of low relevance.

The localization of the object is done by prediction where the Regional Proposal Network (RPN) scans the FPN. Two outputs are generated by the RPN, for the anchor class and bounding boxes. The anchor class is either background class or a foreground class.

Then the image is further passed on to the hidden layers followed by max pooling and normalization. There are two stages of Mask RCNN, depending on the input image, it performs localization of the object then it predicts the class of the object, confines the bounding box and generates a mask. Both stages are connected to the backbone structure.

The model accepts output of the Regional Proposed Network which are the ROIs and classification and bounding box are its output. With the help of these ROIs 28×28 pixel masks are generated as outputs. During testing, these masks are scaled up.

In Fig. 2, the model depicts the mask formed by it on the input image. Once the model has localized the vegetation via bounding boxes, it has localized it pixel by pixel by forming a mask on it. The mask is then highlighted using OpenCV and the background is transformed to black and the mask is transformed to white.

The architecture of the system is given in Fig. 3. The average binary cross-entropy loss:

$$L_{reg} = \sum_{i \in \{x,y,w,h\}} (t_i - d_i(p))_2 + \lambda \|W\|_2$$

Loss minimization formula

$$-\frac{1}{m^2} \sum_{1 \le i,j \le m} \left[y_{ij} \log \hat{y}_{ij}^k + (1 - y_{ij}) \log\left(1 - \hat{y}_{ij}^k\right) \right]$$

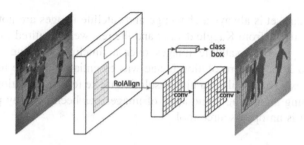

Fig. 1. Masked RCNN framework for instance segmenttion [5]

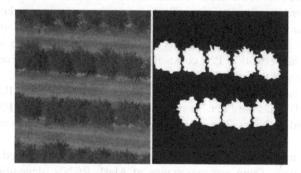

Fig. 2. Masked RCNN pixel-level labelling [9]

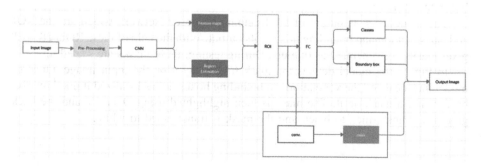

Fig. 3. Architecture of the system

4 Implementation

4.1 Training

The model was trained for over 12 h on a machine with GTX 1070 graphics card. The model was given annotated images from the training dataset. The model generates segmentation masks for each instance of an object in the image. The model progressively was trained for over 15 epochs. The error in classifying forest from human vegetation cover is still above 62%. This can cause inconsistent results and give us misleading information. But this can be solved by training the model with a huge and varied dataset. The output is bounding box around detected features and masked feature with edge detection. Since we had no use of the bounding box it was ignored in the output and only the instant segmentation output is used to generate mask.

4.2 Testing

The model was then given inputs from the test dataset that had an accuracy of over 90%. It wasn't able to classify few images that had less details in it. This could be improved by changing the learning rate and taking more variety in dataset during training process. In Fig. 4, the model is able to classify and perform instance segmentation when the dataset was balanced and could detect the vegetation and non-vegetation patches with an accuracy over 90%. During the earlier stages of training, the model is able to detect only the vegetation present in the image with an accuracy over 90% by forming the mask only on that part of the image with vegetation as shown in Fig. 5.

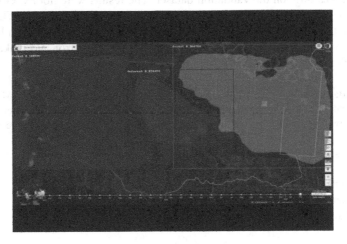

Fig. 4. Result on test image

Fig. 5. Result on test image

5 Conclusion

The goal of the project was to segment forest and non-forest cover using satellite images of any forest/region of interest. The solution was implemented as a proof of concept. The model was tested on a validation dataset of 120 images. The accuracy of the model was 90% on the validation dataset. The results were not extremely accurate on few unseen images but the positive deviation shown was true. This could be improved further by including more images in the training dataset with a lot of variety and also by further training the model we can reduce this error and make the system highly efficient and reliable.

Acknowledgements. We would like to offer our special thanks to Prof. Pranita Mahajan for her valuable and constructive suggestions during the planning and development of this research work. Her willingness to give her time so generously has been very much appreciated.

Compliance with Ethical Standards

✓ All authors declare that there is no conflict of interest
✓ No humans/animals involved in this research work.
✓ We have used our own data.

References

1. Tang, C., Feng, Y., Yang, X., Zheng, C., Zhou, Y.: The object detection based on deep learning. In: 2017 4th International Conference on Information Science and Control Engineering (ICISCE), Changsha, pp. 723–728 (2017)
2. Zhang, X., An, G., Liu, Y.: Mask R-CNN with feature pyramid attention for instance segmentation. In: 2018 14th IEEE International Conference on Signal Processing (ICSP), Beijing, China, pp. 1194–1197 (2018)
3. Kan, W.: Planet: Understanding the Amazon from Space (2017). https://www.kaggle.com/c/planet-understanding-the-amazon-fromspace/. Accessed 10 June 2017
4. Ahmadi, V.: Using GIS and artificial neural network for deforestation prediction. Preprints 2018, 2018030048. https://doi.org/10.20944/preprints201803.0048.v2
5. Abreheret: Pixel Annotation tool (2017). https://github.com/abreheret/PixelAnnotationTool
6. Ren, S., He, K., Girshick, R., Zhang, X., Sun, J.: Object detection networks on convolutional feature maps. IEEE Trans. Pattern Anal. Mach. Intell. **39**(7), 1476–1481 (2017)
7. FAO: The State of the World's Forests 2018 - Forest pathways to sustainable development, Rome. Licence: CC BY-NC-SA 3.0 IGO (2018)
8. India Population, 12 July 2019. http://worldpopulationreview.com/countries/india. Accessed 30 July 2019
9. Forest Cover of India from 1987 to 2015. https://community.data.gov.in/forest-cover-of-india-from-1987-to-2015/
10. Ingram, J., Dawson, T., Whittaker, R.: Mapping tropical forest structure in southeastern Madagascar using remote sensing and artificial neural networks. Remote Sens. Environ. **94**, 491–507 (2005). https://doi.org/10.1016/j.rse.2004.12.001
11. Goldman, E., Harris, N., Maschler, T.: Predicting future forest loss in the democratic republic of the Congo's CARPE landscapes. Technical Note, World Resources Institute, Washington, D.C. (2015)
12. Dlamini, W.M.: Analysis of deforestation patterns and drivers in Swaziland using efficient Bayesian multivariate classifiers. Model. Earth Syst. Environ. **2** (2016). https://doi.org/10.1007/s40808-016-0231-6

An Application of Cellular Automata: Satellite Image Classification

S. Poonkuntran[1(✉)], V. Abinaya[1], S. Manthira Moorthi[2], and M. P. Oza[2]

[1] Velammal College of Engineering and Technology, Madurai, Tamilnadu, India
s_poonkuntran@yahoo.co.in, abino92@gmail.com
[2] Indian Space Research Organization, Space Application Centre, Ahmedabad, India

Abstract. Cellular automata (CA) is an emerging area of research which is being applied in many areas. Satellite image classification is one such area where CA is mainly used for addressing declassification, misclassification and uncertainty. The paper presents an application of CA in satellite image classification. The parallelepiped algorithm in supervised classification have been taken as a base and CA rules were created and added to it in the proposed scheme. In comparative analysis, a test sample of Indian Pines have been experimented with proposed scheme and parallelepiped. The results show 6.96% of improvement in accuracy of proposed classifier with respect to parallelepiped classifier.

Keywords: Image classification · Cellular automata · Satellite image classification · Remote sensing

1 Introduction

The Cellular Automata (CA) is an emerging concept in biological system to study the diseases and its patterns through evolutionary computing model. After the success of CA in biological systems, it has been used in other areas such as gaming, cryptography, classification and prediction. A cellular automata consists of cells and it is a basic structuring element on which cellular automata is performed. It is a two dimensional block forming the portion of an object on which CA is applied such as Image and Texture. Each cell will have its own values from the property of object and it is assigned with particular state initially. This state of cell will be changed through a set of rules that governs the CA [10–12].

One of the recent applications of cellular automata is Satellite Image Classification in which the information about an object is acquired without having physical contact with an object [14]. Such classification becomes important in remote sensing and its results are demanded by Geographic Information Systems (GIS), Global Positioning System (GPS) and Remote Sensing (RS) for better understanding of the data and its relations. This will accelerate the classification to detect the area of interest in the image such as water resources, land, agriculture and roadways [1, 4, 10, 11].

© Springer Nature Switzerland AG 2020
S. Smys et al. (Eds.): ICCVBIC 2019, AISC 1108, pp. 1085–1093, 2020.
https://doi.org/10.1007/978-3-030-37218-7_114

The satellite image classifier performance is based on the number of training sets and its quality. The uncertainty is an issue in any classifier where pixels will be classified into more than one classes. It is an essential to address the uncertainty to improve the accuracy of classifier [1–5]. The other issues include declassification where the pixels will not be classified to any classes due to the limitations of classifier and its properties. The misclassification is another issue where the pixels are classified in to wrong classes. Cellular Automata (CA) has been found as best technique in addressing above mentioned issues [1–14].

This paper discusses an application of cellular automata in satellite image classification against parallelepiped algorithm which is a conventional and widely used technique under supervised classification. The paper is organized as follows. The first section introduces the cellular automata, satellite image classification and discusses the reasons for using cellular automata in classification. Section 2 discusses the related work and its challenges. Section 3 introduces proposed CA based satellite image classification method and its algorithm. Section 4 discusses the experimental results and comparative analysis of proposed scheme. Section 5 discusses conclusions and future enhancement.

2 Related Works

The classification process is categorized in to two classes. They are supervised and unsupervised. The supervised classification requires input data and number of output classes well in advance. The user will supply the samples for all the output classes initially. It is known as training set. The classifier will then classify the input data using training set given by the user. The output classes defined here is non-overlapping classes. However, it is allowed to have minimum overlapping based on application. Many algorithm have been proposed in the literature for performing supervised classification on satellite images. The parallelepiped algorithm is one which is a conventional and widely used in remote sensing application [10–14].

Parallelepiped: Parallelepiped algorithm uses n-dimensional data space where the multispectral satellite data is presented. The mean and standard deviation are taken as base elements for every output classes specified in n-dimensional covering all the bands of the given satellite image. For every spectral value in the given band of satellite image, it is checked against the ranges of threshold formed by mean and standard deviation of the class. If it lies in the range, the spectral value is assigned to the class, other wise it is not classified. This process may classify the pixel into more than one classes that leads to uncertainty in classification. Again, pixels may not be classified in to any classes. This creates declassification issues. Because, it checks with ranges of threshold formed by mean and standard deviation [16]. The threshold ranges contains spectral values and it does not uses contextual values of the pixel in an image. The parallelepiped algorithm is given by the Eq. 1.

$$M(Class, Band)$$
$$- SD(Class, Band) \leq SV(x, Band) \leq M(Class, Band) + SD(Class, Band) \quad (1)$$

Here, M *(Class, Band)* refers the mean of the particular output class at given band. The term SD *(Class, Band)* refers the standard deviation of the particular output class at given band. The spectral value of the pixel x at given band given by SV *(x, Band)*. Here *Band* = *1, 2, 3...b*. Where b is the total number of bands in the satellite image. Similarly, *Class* = *1, 2, 3...n*. Where n is the total number of output classes in the satellite image.

However, all the classification techniques are mainly attempts to understand the pixels in an image and its relationship between them. This understanding plays an important role in classifier accuracy. The pixels will have spectral and contextual information. The spectral information is intensity values described by the method of image acquisition. It is also commonly referred as pixel values. The contextual information is further level of understanding the pixels with its neighbors. It is useful to know how a pixel is located in an image and what objects refer to it.

The common issues in parallelepiped classifier are declassification and uncertainty. In few cases, the pixels may be classified in wrong classes that lead to the misclassification. The major reason for these issues are only spatial information are used in classification. It is observed that when spatial information is combined with contextual information, it improves the accuracy. It also addresses the uncertainty, declassification and misclassification [4, 16].

A cellular automata (CA) is an emerging concept which is being applied in many fields such as physics, chemistry, computer science, biology and mathematics etc [12]. It is an iterative procedure that changes the state value of every cell defined initially over the iteration. A cell is a pattern and it is defined by the user initially. The cell is given with initial state value at the beginning. Later, these state values of cells will change through the iteration based on set of conditions that forms the rule set. The cell patterns, number of states and set of rules are basic components in CA and it is defined by the user based on the application. The cell is a pattern which is having central component and surrounded by neighbors. This structure brings the contextual information of the central component to the application. The cell pattern is not changes in iteration, instead its state value changes. This step addresses the uncertainty related to the central component in a cell. Thereby, CA becomes popular in solving uncertainty in classification.

3 Proposed Work

3.1 Cellular Automata

As per above discussion, the cellular automata can be applied to the data in any dimension. This paper focusses the cellular automata on image classification where the image is a 2-dimensional data and a grid of pixels will form a cell in CA. The image is first divided in to number of cells that forms the basic element in CA. The cell of an image is defined by neighborhood of pixel in an image. The rule set of CA is formed based on neighborhood of pixels. Each cell is assigned a state values that represents initial class assigned to it in classification. This state value will then be modified based on the rule set over iteration. The number of iteration is decided based on the application. According to the literature Von-Neumann is a first person to use this model in to universal constructor application. In 1950, the model were explored for biological system development. In 1970, John Convay invented a "Game of Life" using cellular automata where each cell will have two states namely alive and dead. The mathematical definition of cellular automata is given by a function c: $Z^d \rightarrow S$. Where Z^d is a cellular space on d-dimension, c is a element of Z^d called as cells and S is a set of states that describes the number of output classes in the classification. The function assigns the cell to the state. This paper focusses on 2 dimensional space in CA.

The following CA interpretation are used in the work.

Spatial Dimension: The spatial dimension of CA specifies the dimension in which CA is used. Since, our classification is for images, the spatial dimension is 2. The shape will be a grid.

Size of Neighborhood: It specifies how many neighbor pixels need to be considered for CA operation. The common neighborhood patterns are Von-Neumann that considers 4 neighbors, Moore that consider 8 and Extended Moore that considers 16 neighbors.

States: The state of each cell is defined by two values. 1. Class to which the pixel belongs. 2. Type of the Pixel. Here the pixel will be in any one of four types.

1. Classified – The pixel will be classified to a single class.
2. Border – The pixel will be at border of the classes.
3. Noisy – The pixels are corrupted and not classified to any classes. Simply, the pixels that not belong to any of above mentioned three types.
4. Uncertain – The pixel will be classified to more than one classes.

Transition Function: The set of rules that defines movement of cell from one state to another in CA. It is an important component that brings evolution in CA

Iteration: It specifies the number of iteration used in CA for evolution.

3.2   Proposed Work

To showcase an application of cellular automata, we have developed a classifier that classify the images using parallelepiped algorithm initially. Then, we apply below given four rules created using cellular automata. The proposed classifier classifies the image using parallelepiped classifier with a trained samples given by the user. Then, the results are fed to cellular automata for further improvements.

Rule 1: If the number of spectral output classes of pixel of an image is 1, and the spectral output classes of neighborhood pixels of an image is Not Classified or the same as actual pixel, then Pixel is classified as "Classified"

Rule 2: If the number of spectral output classes of a pixel of an image is 1, and the spectral output classes of neighborhood of pixel of an image are different than actual pixel class, then pixel is classified as "Border"

Rule 3: If the number of spectral output classes of a pixel of an image is 1 and the spectral output class is Noisy, then pixel is classified as "Noisy"

Rule 4: If the number of spectral output classes of a pixel of an image is greater than 1, then the pixel is classified as "Uncertain"

The CA uses set of rules to fine tune the results. The rule set combines the spectral values with contextual details. The contextual details of an image is taken through moore neighborhood pattern on the results of parallelepiped. The contextual information of a pixel is then processed in proposed cellular automata classifier by using the above mentioned four rules.

The proposed CA_Classifier algorithm is as below.

Proposed CA_Classifier (I, T, G)
// I – Input Satellite Image in size of MxNxK: M-No.of Rows, N-No. of Columns &
K- No. of Bands
// T- Training Sample
// G- Ground Truth
1. *Read I , T, G*
2. *Set Threshold as Thr.*
3. *For Iter= 1 to Number of Iteration*
    For k = 1 to K
        *For i= 1 to M*N*
        // Perform Parallelepiped Classifier
            CL= Parallelepiped Classifier (I (i, k), T, Thr)
            // The output of conventional classifier is updated through CA.
            CA=CA_Classifier (CL, Neigbours)
    // Update the Thr
    Thr=Thr+1
4. *Return CA*

To quantitatively measure the performance of the proposed CA classifier, the parallelepiped classifier is taken as base and its results are compared to the proposed scheme.

4   Experimental Results

The proposed CA classifier has been implemented in Matlab 2016 b using sample hyper spectral images taken from Computational Intelligence Group (CIG) of the University of the Basque Country (UPV/EHU), Spain [13]. The sample images are satellite images taken from AVRIS sensors for earth observation. The Indian Pines is a sample image used in the experiment. For the quantitative analysis of results, the following parameters were used as metrics [15].

1. True Positive (TP): A true positive is an outcome where the classifier correctly classifies the positive class.
2. True Negative (TN): A true negative is an outcome where the classifier correctly classifies the negative class.
3. False Positive (FP): A false positive is an outcome where the model incorrectly classifies the positive class.
4. False Negative (FN): A false negative is an outcome where the model incorrectly classifies the negative class.
5. Total Positive (P): It is calculated by TP+FN.
6. Total Negative (N): It is calculated by FP+TN.
7. Sensitivity: True Positive Rate and it is calculated by TP/P.
8. Specificity: True Negative Rate and it is calculated by TN/N.
9. Accuracy: The accuracy specifies how accurate the classifier classifies the pixels. It is given by TP+TN/ (P+N).
10. Error: The error is the difference between the predicted classification and actual classification. It is given by 1-Accuracy.

4.1   Indian Pines

This image is taken on Indian Pines, a sample site in north-western Indiana. The image is taken through AVRIS sensor and specified in 224 spectral bands in 145×145 pixel resolution. The band wavelength is from $0.4–2.5 \ 10^{-6}$ m. The spectral band contains mainly agriculture, forest and vegetation. It also contains rail line, highways, housing and other building structures. The final image contains 200 bands after removing water absorption bands. The ground truth were given for 16 classes which is not mutually exclusive as listed in Table 1. The sample band image and ground truth are shown in Fig. 1.

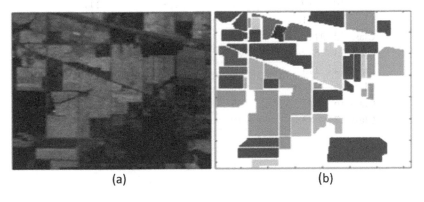

(a)   (b)

Fig. 1. (a) Sample band of Indian pines (b) Ground truth for 16 classes.

4.2 Results and Comparative Analysis

As mentioned in Sect. 4, the experiment has been done on two approaches. First we experimented the proposed scheme that uses both parallelepiped and CA Rules for classification. Second we use only parallelepiped to compare results of the proposed scheme. The positive and negative class portfolio of the proposed scheme is tabulated in Table 2 and parallelepiped is tabulated in Table 3. It is found that the True Positive Rate of the proposed scheme is improved by 12.78% compared to Parallelepiped. At the same time, True negative rate of the proposed scheme is improved only by 0.5% compared to Parallelepiped. The proposed approach (Approach 1) outperforms the parallelepiped (Approach 2) by 6.96% of improvement in overall accuracy of the classification. The sensitivity and specificity of the proposed scheme is improved by 8% and 0.50% respectively with respect to the parallelpiped. The comparative analysis is tabulated in Table 3.

Table 1. Positive/Negative values of Indian pines using approach 1 (parallelepiped with CA rules)

Class	True positive	False positive	False negative	True negative
Class 1	9293	1483	83	10166
Class 2	38	8	178	20801
Class 3	1247	181	178	19419
Class 4	739	91	162	20033
Class 5	212	25	183	20605
Class 6	410	73	163	20379
Class 7	633	97	185	20110
Class 8	22	6	161	20836
Class 9	416	62	175	20372
Class 10	19	1	188	20817
Class 11	841	131	183	19870
Class 12	2112	343	139	18431
Class 13	512	81	158	20274
Class 14	175	30	187	20633
Class 15	1087	178	168	19592
Class 16	337	49	181	20458
Class 17	83	10	177	20755

Table 2. Positive/Negative values of Indian pines using approach 2 (parallelepiped only)

Class	True positive	False positive	False negative	True negative
Class 1	8721	2055	137	10112
Class 2	36	10	242	20377
Class 3	1184	244	229	19368
Class 4	706	124	215	19980
Class 5	202	35	255	20533
Class 6	393	90	246	20296
Class 7	587	143	231	20064
Class 8	22	6	240	20757
Class 9	389	89	218	20329
Class 10	18	2	244	20761
Class 11	791	181	249	19804
Class 12	2001	454	231	18339
Class 13	483	110	209	20223
Class 14	167	38	252	20568
Class 15	1033	232	217	19543
Class 16	314	72	251	20388
Class 17	78	15	234	20698

Table 3. Comparative analysis

Parameter	Approach 1	Approach 2
Accuracy	86.45	79.49
Error	13.55	20.51
Sensitivity	62.63	54.62
Specificity	98.84	98.32

5   Conclusions

This paper presented an application of cellular automata in satellite image classification. The proposed CA method uses parallelepiped, a conventional supervised algorithm as a base and CA rule set is applied on the results of it. The CA improves the accuracy of the parallelepiped classifier by combining the spectral values with contextual values. The experimental result shows that 6.96% of improvement in overall accuracy of proposed classifier compared to parallelepiped. The true positive rate and true negative rate of the proposed classifier is improved by 12.78% and 0.5% respectively. It is concluded that the overall accuracy of the classifier is improved, when parallelepiped classifier is combined with CA. Thereby, CA will become an important component in satellite image classification. The proposed CA scheme is further extended with KNN, CNN classifiers to check its sustainable performance.

Compliance with Ethical Standards

✓ All authors declare that there is no conflict of interest.
✓ No humans/animals involved in this research work.
✓ We have used our own data.

References

1. Aponte, A., Moreno, J.A.: Cellular automata and its application to the modelling of vehicular traffic in the city of Caracas. In: Proceedings of the 7th International Conference on ACRI, vol. 4173, pp. 502–511 (2006)
2. Avolio, M.V., Errera, A., Lupiano, V., Mazzanti, P., Di Gregorio, S.: Development and calibration of a preliminary cellular automata model for snow avalanches. In: Proceedings of the 9th International Conference on ACRI, vol. 6350, pp. 83–94 (2010)
3. Sikdar, B.K., Paul, K., Biswas, G.P., Yang, C., Bopanna, V., Mukherjee, S., Chaudhuri, P. P.: Theory and application of GF(2P) cellular automata as on-chip test pattern generator. In: Proceedings of 13th International Conference on VLSI Design India, pp. 556–561 (2000)
4. Espínola, M., Piedra-Fernández, J.A., Ayala, R., Iribarne, L., Wang, J.Z.: Contextual and hierarchical classification of satellite images based on cellular automata. IEEE Trans. Geosci. Remote Sens. **53**(2), 795–809 (2015)
5. Balzter, H., Braun, P., Kühler, W.: Cellular automata models for vegetation dynamics. Ecol. Modell. **107**(2/3), 113–125 (1998)
6. Bandini, S., Bonomi, A., Vizzari, G.: A cellular automata based modular illumination system. In: Proceedings of the 9th International Conference on ACRI, vol. 6350, pp. 334–344 (2010)
7. Das, S., Chowdhury, D.R.: Generating cryptographically suitable non linear maximum length cellular automata. In: Proceedings of the 9th International Conference on ACRI, vol. 6350, pp. 241–250 (2010)
8. Doi, T.: Quantum cellular automaton for simulating static magnetic fields. IEEE Trans. Magn. **49**(5), 1617–1620 (2013)
9. Dzwinel, W.: A cellular automata model of population infected by periodic plague. In: Proceedings of the 6th International Conference ACRI, vol. 3305, pp. 464–473 (2004)
10. Espínola, M., et al.: ACA multiagent system for satellite image classification. In: Proceedings of the PAAMS, vol. 157, pp. 93–100. AISC (2012)
11. Espínola, M., et al.: A hierarchical and contextual algorithm based on cellular automata for satellite image classification. In: Proceedings of the CiSE, pp. 1–4 (2011)
12. An Application of Cellular Automata in Hermetic Systems: https://www.hermetic.ch/pca/pca.htm
13. López-Fandiño, J., Priego, B., Heras, D.B., Argüello, F.: GPU projection of ECAS-II segmenter for hyperspectral images based on cellular automata. IEEE J. Sel. Top. Appl. Earth Obs. Remote. Sens. **10**(1), 20–28 (2017). https://doi.org/10.1109/jstars.2016.2588530
14. Nichele, S., Ose, M.B., Risi, S., Tufte, G.: CA-NEAT: evolved compositional pattern producing networks for cellular automata morphogenesis and replication. IEEE Trans. Cogn. Dev. Syst. **10**(3), 687–700 (2018). https://doi.org/10.1109/TCDS.2017.2737082
15. The UPV/EHU Hypersepctral Images Data Set. http://www.ehu.eus/ccwintco/index.php/Hyperspectral_Remote_Sensing_Scenes
16. Xiang, M., Hung, C.-C., Pham, M., Kuo, B.-C., Coleman, T.: A parallelepiped multispectral image classifier using genetic algorithms. In: Proceedings of the 2005 IEEE International Geoscience and Remote Sensing Symposium, IGARSS 2005, Seoul (2005), pp. 4-pp. https://doi.org/10.1109/igarss.2005.1526216

Morlet Wavelet Threshold Based Glow Warm Optimized X-Means Clustering for Forest Fire Detection

B. Pushpa$^{(\boxtimes)}$ and M. Kamarasan

Computer and Information Science, Annamalai University, Chidambaram, India
pushpasidhu@gmail.com, smkrasan@yahoo.com

Abstract. Forest fire detection is a significant problem to be resolved for the prevention of life and property safety. Clustering performance of existing fire detection algorithm was poor. In order to resolve the above limitations, Morlet Wavelet Threshold-based Glow Warm Optimized X-Means Clustering (MWT-GWOXC) the technique is proposed. Initially, this technique takes a number of video frames as input and initializes a number of clusters. Then, the proposed approach initializes the glow warm populations with a number of video frames. Then, the technique calculates the fitness function of all clustered video frames and identifies a pre-fire stage or fire stage or critical fire stage. If any video frame not clustered, this technique employs Bayesian probability criterion which determines a higher probability of frame to become a cluster member and improves clustering accuracy. The simulation result demonstrates that the technique is able to increase fire detection accuracy and also minimize fire detection time.

Keywords: Bayesian probability criterion · Color · Fitness function · Intensity · Spatio-temporal energy · Texture · Video frames and X-means clustering

1 Introduction

Forest fire is a serious hazard in different places around the world. Recently, the large fires and the guard of life goods require effective prevention and it is achieved based on the video surveillance - based fire detection. But, a false positive rate of the conventional technique was more. Therefore, the technique is introduced to achieve higher accuracy for finding forest fires in the video. K-medoids Clustering was designed in [1] to get better fire flame discovery performance. But, false-positive detections were very higher. Dirichlet Process Gaussian mixture model-based approach was introduced in [2] for autonomous flame recognition depends on color, dynamics, and flickering traits of flames with higher accuracy. But, the time complexity of fire flame identification was more. Video fire detection was presented in [3] with the application of Gaussian Mixture Model using multi-colour features. However, reducing the rate of the false alarm was not considered. Based on the color, spatial and temporal information, Covariance matrix–based fire and flame discovery method was introduced in [4]. But, flame detection accuracy was poor. Remote identification of forest fires from video

© Springer Nature Switzerland AG 2020
S. Smys et al. (Eds.): ICCVBIC 2019, AISC 1108, pp. 1094–1105, 2020.
https://doi.org/10.1007/978-3-030-37218-7_115

signals was presented in [5] with the help of classifiers using K-SVD learned dictionaries. However, the computational complexity involved during forest fire discovery was not minimized. Spatio–Temporal Flame Modeling and Dynamic Texture examination was introduced in [6] for video-based fire recognition. But, the average frame rate was not efficient. ASTER Digital Elevation Model (DEM) was developed in [7] for determining the fire hazard with improved accuracy. However, processing time taken for analyzing fire risks was not solved. Robust approach for smoke recognition was presented in [8] with application of deep belief networks. But, fire detection was not attained in this approach. A visual analysis method was introduced in [9] to increase the speed of fire flame recognition in video with help of logistic regression and temporal smoothing. But, false positive ratio of fire recognition was higher. An efficient forest fire detection method was designed in [10] with help of background subtraction and color segmentation. However, forest fire identification performance was not efficient. In order to address the above mentioned existing issues, MWT-GWOXC technique is designed with the main contribution is described in below.

To achieve enhanced fire detection accuracy for video sequence through clustering, MWT-GWOXC technique is developed. Morlet Wavelet Transformation is applied for reducing the time complexity of fire detection process. If intrinsic nature of flame pixels, Wavelet signals easily expose the random traits of a given signal and technique to attain more robust detection of flames in video with a lower time. To minimize the false positive rate of fire discovery as compared to existing work, if any video frame does not group into a cluster, X-Means Clustering utilize Bayesian probability criterion that computes a maximum probability to become a cluster member. This helps the technique to improve fire detection accuracy.

This paper is formulated as follows. Section 2 portrays the related works. Section 3 presents the detailed process of MWT-GWOXC technique. In Sect. 4, an experimental setting of proposed technique is demonstrated. Section 5 explains the results and discussion of certain parameters. Finally, in Sect. 6 conclusion of the research work is discussed.

2   Related Works

Video based fire discovery with help of rule method based on wavelet using RGB and HSV color space and spatial analysis based on wavelet was presented in [11]. However, fire detection accuracy was not enhanced. Support Vector Machine classifier was designed in [12] fire and/or smoke during the earliest stages of a fire with a real-time alarm system is identified. But, time and space complexity of this algorithm was very higher. A novel method was developed in [13] to discover fires through analyzing videos obtained by surveillance cameras according to color, shape disparity, and motion analysis. However, true positive rate of fire identification was not enough. A set of motion traits based on motion determination was presented in [14] to discover fire flame in videos. But, the ratio of number of incorrect detection was higher. Real-time multi-characteristic based fire flame identification was introduced in [15] to increase the reliability. However, specificity of fire flame detection was poor. A Saliency-Based

Method was introduced in [16] for enhancing the real-time fire detection performance in video frames. But, processing time needed for fire detection was very higher.

An enhanced probabilistic approach was designed in [17] with application of two trait representations to carry out color-based fire detection. A probability-based framework was introduced in [18] to find out flame based on robust features and randomness testing. But, computational cost was very minimal. A Cumulative Geometrical Independent Component Analysis (C-GICA) model was presented in [19] with aim of enhancing the video fire detection rate. However, the ratio of number of video frames that are clustered was not sufficient. Multi-characteristic fusion based fast video flame recognition was introduced in [20]. But, computational time of flame detection was remained an open issue. The flame detection approach was introduced in [21] for improving the classification accuracy. However, this method was not involving the complex rules. The video-based fire detection was designed in [22] with combined saliency detection and convolutional neural networks to extract the features for the classification. But, false alarm rate does not contain fire detection. A flame and smoke detection system was introduced in [23] to detect the fire incidents using Quick Blaze. However, the frame error rate was not considered. A novel video smoke detection method was designed in [24] with both color and motion features. But the false alarm rate was not considered.

3   Proposed Technique

The Morlet Wavelet Threshold-based Glow Warm Optimized X-Means Clustering technique is introduced with the aim of enhancing the forest fire detection performance with minimal time complexity. Glow Warm Swarm Optimization employed in the technique that provides an optimal solution to effectively cluster the different types of flame (i.e. pre-fire stage, fire stage and critical fire stage) in an input video sequence with higher accuracy. Figure 1 shows the architecture diagram of this technique to attain enhanced forest fire flame identification performance. As presented in Fig. 1 MWT-GWOXC technique initially takes video database as input which contains more number of a video file and it converted into video frames. Then, the technique defines a number of clusters and initializes the glow warm population with a number of video frames. Consequently, the technique applies Morlet Wavelet Transformation for each video frames where it partitions each video frames into a number of sub-blocks. Next, this technique determines fitness function for each frame based on multi-characteristics such as color, spatio-temporal energy, intensity, texture. By using the measured fitness function, then the technique group video frames into diverse clusters (i.e. pre-fire stage, fire stage and critical fire stage). The elaborate process of this technique is described below.

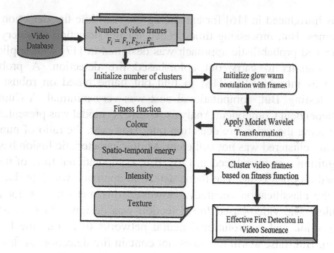

Fig. 1. Architecture diagram of MWT-GWOXC technique

Let us consider an input video dataset includes of many videos represented as '$V_i = V_1, V_2, .., V_n$' where 'n' denotes total number of videos available in given dataset. The videos are initially converted into number of frames denoted as '$F_i = F_1, F_2, ..F_m$'. Here, 'm' represents the no. of frames. The proposed technique initializes the 'x' no. of clusters.

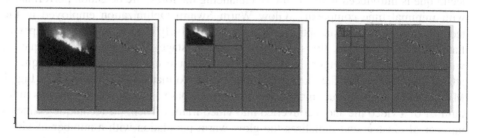

Fig. 2. Morlet's Wavelet Transformation of video frame

After that, this technique initialize the population with help of number of glow warms (i.e. video frames). Then, Morlet Wavelet Transformation is applied in MWT-GWOXC technique to split the images into a number of sub blocks as depicted in above Fig. 2, presents the process involved in Morlet's Wavelet Transformation. During this process, sub block with lower frequency is constantly separated into a number of blocks to increase the fire flame detection performance with lower time consumption. From that, Morlet Wavelet Transformation of a video frame at time '$F_i(t)$' is formulated as follows,

$$F_{wavelet(u,v)} = \frac{1}{\sqrt{|u|}} \int_{-\infty}^{\infty} F_i(t) \, \psi^* \left(\frac{t-v}{u}\right) dt \tag{1}$$

From (1), the video frame is divided into a number of sub-blocks whereas '$F_{wavelet(u,v)}$' wavelet transformation of a frame. After completing the transformation process, the technique determines the fitness value for each frame based on color, spatio-temporal energy, intensity and texture of fire flame.

Color is one of the most considerable characteristics to identify the fire flame. Let us consider $p_1, p_2, ..p_n$ is a fire-colored training RGB samples of the distribution to be approximated. Then, the pixels probability density function 'p_t' is mathematically obtained using kernel 'φ' as,

$$Pr(p_t) = \frac{1}{n} \sum_{i=1}^{n} \varphi(p_t - p_i) \tag{2}$$

From (2), 'φ' refers Gaussian kernel. Here, '$\varphi = n(0, DC)$' whereas 'DC' denotes the diagonal Covariance matrix through diverse standard deviation 'μ_j' with respect to each color channel 'j'. For each pixel the fire color probability is mathematically evaluated as,

$$p_C(i,j) = \frac{1}{n} \sum_{i=1}^{n} \sum_{j=1}^{3} \frac{1}{\sqrt{2\pi\mu_j^2}} e^{-\frac{1}{2}\frac{(p_t-p_i)^2}{\mu_j^2}} \tag{3}$$

For each video frame 'F_i', the average fire color probability is of each pixel $p(i,j)$ in the block is determined and it represented as the total fire color.

$$C_{F_i} = \frac{1}{n} \sum_{i,j} p_C(i,j) \tag{4}$$

From (4), 'n' represents the no. of pixels in the block and '$p_C(i,j)$' indicates the color of each pixel.

The spatio-temporal energy is another significant feature to discriminate the real fire and fire colored objects in video frames. Due to wind or kind of burning material the shape of flame is continuously varied. Therefore, a fire colored object causes lower spatial distinctions within given time interval than a real fire. Thus, the temporal difference of the spatial energy at pixel '$p_{ST}(i,j)$' within a last frame of temporal window 'T'.

$$p_{ST}(i,j) = \frac{1}{T} \sum_{t=0}^{T-1} (\varepsilon_t(i,j) - \overline{\varepsilon_t}(i,j))^2 \tag{5}$$

From (5), 'ε_t' denotes the spatial energy of the pixel at 't' time instance, '$\overline{\varepsilon_t}$' indicates the total value of spatial energy. From that, total spatio-temporal energy is calculated by averaging the individual energy of pixels in blocks using below,

$$ST_{F_i} = \frac{1}{n} \sum_{i,j} p_{ST}(i,j) \tag{6}$$

From (6), for each video frame the total spatio-temporal energy is measured. Followed by, for each block mean value of fire pixels' intensity is mathematically obtained as,

$$I_{F_i} = \frac{1}{n} \sum_{i,j} p_I(i,j) \tag{7}$$

From (7), 'n' refers the total no. of pixels in the block and '$p_I(i,j)$' point outs the fire intensity of pixel. The average of the individual flickering contributions of the each pixels p (i, j) in the block represents the fire texture feature of each frame and it measured using below equation,

$$T_{F_i} = \frac{1}{N} \sum_{i,j} p_f(i,j) \tag{8}$$

From (8), 'n' denotes the no. of pixels in the block and '$p_I(i,j)$' signifies the flickering characteristic of pixel. Based on the measured fire color, intensity and texture, then the technique computes fitness function using below mathematical expression,

$$\delta_{F_i} = \{C_{F_i}, ST_{F_i}, I_{F_i}, T_{F_i}\} \tag{9}$$

From (9), fitness function is measured for each video frame 'F_i'. Followed by, the technique defines threshold fitness function where threshold value is assigned for color, spatio-temporal energy, and intensity, texture. From that, MWT-GWOXC technique clusters the input video frames using below mathematical expression,

$$z = \begin{cases} \textit{If } \delta_{F_i} < \delta_{MinT}, \textit{ then group } 'F_i' \textit{ into prefire stage} \\ \textit{If } \delta_{MinT} < \delta_{F_i} > \delta_{MaxT}, \textit{ then Group } 'F_i' \textit{ into fire stage} \\ \textit{If } \delta_{F_i} < \delta_{MaxT}, \textit{ then Group } 'F_i' \textit{ into critical fire stage} \end{cases} \tag{10}$$

From (10), 'z' represents a clustering output, 'δ_{F_i}' denotes a fitness function value of video frame. Here, 'F_i' refers an 'i^{th}' frame in given video sequence whereas 'δ_{MinT}' refers a minimum threshold fitness function and 'δ_{MaxT}' indicates the minimum threshold fitness function.

After calculating the fitness value, this technique verifies if the fitness function of input video frame 'δ_{F_i}' is lesser than a minimum threshold fitness function 'δ_{MinTh}'. If the above condition is true, the technique clusters the video frame into a pre-fire stage. Otherwise, this technique checks if the fitness function of video frame is greater than a minimum threshold fitness function 'δ_{MinTh}' and lesser than a maximum threshold fitness function 'δ_{MaxTh}'. When the above condition is satisfied, the technique clusters the input video frame into a fire stage. If fitness function of video frame 'δ_{F_i}' is greater than a maximum threshold fitness function 'δ_{MaxTh}', then the technique group the frame into a critical fire stage.

If a video frame does not belong to any of the cluster, this technique improves cluster accuracy with help of Bayesian probability criterion. This helps to cluster the all video frames into the particular cluster with minimal error rate. From that, the Bayesian probability criterion is mathematically estimated as follows,

$$BP_{F_i} = \frac{\delta_{F_2}(t) - \delta_{F_1}(t)}{\sum \delta_{F_n}(t) - \delta_{F_1}(t)} \tag{11}$$

From (11), 'BP_{F_i}' is a Bayesian probability criterion which gives higher probability of video frame becomes a member of the particular cluster. Here '$\delta_{F_2}(t)$' point outs a fitness function of the video frame 'F_2' at time 't'. Here, '$\delta_{F_1}(t)$' represents the fitness function of the video frame 'F_1' at the time 't' and 'δ_{F_n}' denotes a fitness function of the video frame 'n' at the time 't'. This process of this technique is continual until the all video frames in the video are grouped. Through efficient clustering of video frames with a lower time, the technique increase fire detection accuracy as well as reduce fire detection time as compared to existing works.

Input: Number of videos '$V_i = V_1, V_2, .., V_n$'
Output: Improved fire detection accuracy
Step 1: Begin
Step 2:   For each input video 'V_i'
Step 3:   Split 'V_i' into number of frames 'F_i'
Step 4:   Initialize 'x' number of clusters
Step 5:   Initialize the number of frames 'F_i'
Step 6:   For each 'F_i'
Step 7:   Decompose 'F_i' into number of sub blocks
Step 8:   Determine fitness function 'δ_{F_i}'
Step 9:   If '$\delta_{F_i} < \delta_{Min\ Th}$', then
Step 10:   Cluster 'F_i' into pre-fire stage
Step 11:   else if ($\delta_{Min\ Th} < \delta_{F_i} < \delta_{Max\ Th}$)
Step 12:   Cluster 'F_i' into fire stage
Step 13:   else ($\delta_{F_i} > \delta_{Max\ Th}$)
Step 14:   Cluster 'F_i' into critical fire stage
Step 15:   **End If**
Step 16:   If any video frame is not clustered, **then**
Step 17:   Measure Bayesian probability
Step 18:   else
Step 19:   Stop the clustering process
Step 20:   **End if**
Step 21:   **End For**
Step 22: End

Algorithm 1 Morlet Wavelet Thresholding Glow Warm Optimized X-Means Clustering

Algorithm 1 explains the step by step process of MWT-GWOXC technique to attain enhanced clustering performance.

4   Experimental Settings

The experimental evaluation of proposed, MWT-GWOXC technique is implemented in MATLAB with the help of FIRESENSE database [25]. This FIRESENSE database comprises of many forest fire videos. For conducting the simulation process, this technique considers various numbers of video frames in the range of 25–250. The simulation of this technique is conducted for many instances with respect to different number of video frames. The performance of MWT-GWOXC is compared with two existing works namely K-medoids Clustering [1] and Dirichlet Process Gaussian mixture model [2].

5   Performance Analysis

The performance analysis of effectiveness of MWT-GWOXC technique is determined by using following parameters such as fire detection accuracy, fire detection time and false positive rate with aid of tables and graphs.

5.1   Performance Result of Fire Detection Accuracy

Fire Detection Accuracy 'FDA' is determined as the ratio of number of frames that are correctly clustered to the total number of video frames as input. The 'FDA' is estimated in terms of percentage (%) and mathematically calculated as,

$$FDA = \frac{M_{CC}}{m} * 100 \qquad (12)$$

From (12), the accuracy of fire flame detection in video sequence is estimated with respect to a different number of frames. Here, 'M_{CC}' denotes number of frames that are correctly clustered and 'm' represents a total number of video frames considered for simulation work.

Sample Mathematical Calculation

- **Proposed MWT-GWOXC technique:** number of frames correctly clustered is 220 and the total number of frames is 250. Then fire detection accuracy is obtained as,

$$FDA = \frac{220}{250} * 100 = 88\%$$

As presented in below Fig. 3, MWT-GWOXC technique increases the fire detection accuracy by 8% as compared to K-medoids clustering [1] and 16% as compared to Dirichlet Process Gaussian mixture model [2].

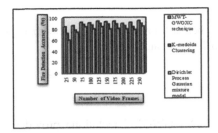

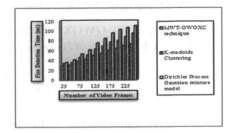

Fig. 3. Simulation results of fire detection accuracy

Fig. 4. Simulation results of fire detection time

5.2 Performance Result of Fire Detection Time

Fire Detection Time '*FDT*' measures the amount of time required for clustering video frames as pre-fire, fire, critical fire stage. The FDT is determined in terms of milliseconds (ms) and mathematically obtained using below,

$$FDT = m * t(CSF) \tag{13}$$

From (13), the time needed for detecting the forest fire flame is evaluated. Here, '*m*' point outs the number of input video frames whereas '*t(CSF)*' indicates the time utilized in single video for finding fire flame through clustering.

Sample Mathematical Calculation

- **Proposed MWT-GWOXC technique:** time taken for clustering single video frame is 0.09 ms and the total number of video frame is 250. Then FDT is estimated as,

$$FDT = 250 * 0.09 = 23 \text{ ms}$$

As illustrated in Fig. 4, as a result, MWT-GWOXC technique decreases the fire detection time by 19% as compared to K-medoids clustering [1] and 31% while compared to Dirichlet Process Gaussian mixture model [2].

5.3 Performance Result of False Positive Rate

False Positive Rate 'FPR' evaluated as the ratio of number of frames that are incorrectly clustered to the total number of video frames as input. The FPR is determined in terms of percentage (%) and mathematically estimated as,

$$FPR = \frac{M_{IC}}{m} * 100 \tag{14}$$

From (14), the false positive rate of fire flame detection is measured. Here, 'M_{IC}' represents number of frames that are inaccurately clustered and '*m*' refers a total number of video frames.

Sample Mathematical Calculation

- **Proposed MWT-GWOXC technique:** number of frames wrongly clustered is 30 and the total number of frame is 250. Then FPR is calculated as,

$$FPR = \frac{30}{250} * 100 = 12\%$$

As demonstrated in Fig. 5, the proposed MWT-GWOXC technique provides a lower false positive rate when compared to existing K-medoids clustering [1] and Dirichlet Process Gaussian mixture model [2].

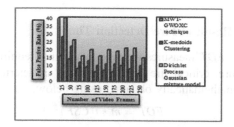

Fig. 5. Simulation result of false positive rate

MWT-GWOXC technique minimizes the false positive rate by 40% as compared to K-medoids clustering [1] and 59% while compared to Dirichlet Process Gaussian mixture model [2].

6 Conclusion

An efficient MWT-GWOXC technique is designed with the goal of attaining higher fire detection accuracy via clustering. The goal of this technique is obtained with the aid of Morlet Wavelet Transformation, Glow Warm Swarm Optimization and X-Means Clustering algorithm. The efficiency of this technique is evaluated by measuring the fire detection accuracy, fire detection time, and false-positive rate and compared with two conventional works. The experimental result shows that this technique gives better performance with an enhancement of fire detection accuracy and reduction of fire detection time as compared with conventional work. In future work, we concentrate on reducing energy consumption for fire detection. In addition, future work is focused to analyze more parameters to get better performance of the proposed technique.

References

1. Khatami, A., Mirghasemi, S., Khosravi, A., Lim, C.P., Nahavandi, S.: A new PSO-based approach to fire flame detection using K-medoids clustering. Expert Syst. Appl. **68**, 69–80 (2017)
2. Li, Z., Mihaylova, L.S., Isupova, O., Rossi, L.: Autonomous flame detection in videos with a Dirichlet process gaussian mixture color model. IEEE Trans. Ind. Inform. **14**(3), 1146–1154 (2018)
3. Han, X.-F., Jin, J.S., Wang, M.-J., Jiang, W., Gao, L., Xiao, L.-P.: Video fire detection based on Gaussian Mixture Model and multi-color features. Signal Image Video Process. **11**(8), 1419–1425 (2017)
4. Habiboğlu, Y.H., Günay, O., Çetin, A.E.: Covariance matrix-based fire and flame detection method in video. Mach. Vis. Appl. **23**(6), 1103–1113 (2012)
5. Rosas-Romero, R.: Remote detection of forest fires from video signals with classifiers based on K-SVD learned dictionaries. Eng. Appl. Artif. Intell. **33**, 1–11 (2014)
6. Dimitropoulos, K., Barmpoutis, P., Grammalidis, N.: Spatio-temporal flame modeling and dynamic texture analysis for automatic video-based fire detection. IEEE Trans. Circuits Syst. Video Technol. (TCSVT) **25**(2), 339–351 (2015)
7. Bahu, K.V.S., Roy, A., Prasad, P.R.: Forest fire risk modeling in Uttarakhand Himalaya using TERRA satellite datasets. Eur. J. Remote. Sens. **49**(1), 1–16 (2016)
8. Pundir, A.S., Raman, B.: Deep belief network for smoke detection. Fire Technol. **53**(6), 1943–1960 (2017)
9. Kong, S.G., Jin, D., Li, S., Kim, H.: Fast fire flame detection in surveillance video using logistic regression and temporal smoothing. Fire Saf. J. **79**, 37–43 (2016)
10. Mahmoud, M.A.I., Ren, H.: Forest fire detection using a rule-based image processing algorithm and temporal variation. Math. Probl. Eng. **2018**, 1–8 (2018). Article ID 7612487
11. Gupta, A., Bokde, N., Marathe, D., Kishore, : A novel approach for video based fire detection system using spatial and texture analysis. Indian J. Sci. Technol. **11**(19), 1–17 (2018)
12. Ho, C.-C.: Nighttime fire/smoke detection system based on a support vector machine. Math. Probl. Eng. **2013**, 1–7 (2013). Article ID 428545
13. Foggia, P., Saggese, A., Vento, M.: Real-time fire detection for video surveillance applications using a combination of experts based on color, shape and motion. IEEE Trans. Circuits Syst. Video Technol. **25**(9), 1545–1556 (2015)
14. Mueller, M., Karasev, P., Kolesov, I., Tannenbaum, A.: Optical flow estimation for flame detection in videos. IEEE Trans. Image Process. **22**(7), 2786–2797 (2013)
15. Chi, R., Zhe-Ming, L., Ji, Q.-G.: Real-time multi-feature based fire flame detection in video. IET Image Process. **11**(1), 31–37 (2017)
16. Jia, Y., Yuan, J., Wang, J., Fang, J., Zhang, Q., Zhang, Y.: A saliency-based method for early smoke detection in video sequences. Fire Technol. **52**(5), 1271–1292 (2016)
17. Zhang, Z., Shen, T., Zou, J.: An improved probabilistic approach for fire detection in videos. Fire Technol. **50**(3), 745–752 (2014)
18. Wang, D.-C., Cui, X., Park, E., JinHakilKim, C.: Adaptive flame detection using randomness testing and robust features. Fire Saf. J. **55**, 116–125 (2013)
19. Rong, J., Zhou, D., Yao, W., Gao, W., Chen, J., Wang, J.: Fire flame detection based on GICA and target tracking. Opt. Laser Technol. **47**, 283–291 (2013)
20. Chen, J., He, Y., Wang, J.: Multi-feature fusion based fast video flame detection. Build. Environ. **45**(5), 1113–1122 (2010)

21. Prema, C.E., Vinsley, S.S., Suresh, S.: Efficient flame detection based on static and dynamic texture analysis in forest fire detection. Fire Technol. **54**(1), 255–288 (2015)
22. Shi, L., Long, F., Lin, C., Zhao, Y.: Video-based fire detection with saliency detection and convolutional neural networks. In: International Symposium on Neural Networks. LNCS, vol. 10262, pp. 299–309 (2017)
23. Qureshi, W.S., Ekpanyapong, M., Dailey, N.M., Rinsurongkawong, S., Malenichev, A., Krasotkina, O.: QuickBlaze: early fire detection using a combined video processing approach. Fire Technol. **52**(5), 1293–1317 (2016)
24. Chunyu, Y., Jun, F., Jinjun, F., Yongming, Z.: Video fire smoke detection using motion and colour features. Fire Technol. **46**(3), 651–663 (2010)
25. FIRESENSE database. https://zenodo.org/record/836749

Advance Assessment of Neural Network for Identification of Diabetic Nephropathy Using Renal Biopsies Images

Yogini B. Patil$^{(\boxtimes)}$ and Seema Kawathekar

Department of Computer Science and Information Technology,
Dr. Babasaheb Ambedkar, Marathwada University,
Aurangabad, Maharashtra, India
yoginibpatil@gmail.com, seema_babrekar@yahoo.co.in

Abstract. The current era of medical science computer plays an important role in disease identification. The accuracy and robustness of the medical test is improved with the help of computer. The significant cause of chronic kidney dieses is the diabetic nephropathy. The diabetic nephropathy has the end stage of renal failure. The research has been done to identification of causes of diabetic nephropathy. The computational technology enhances its usability towards the diabetic nephropathy research. The research has been executed on dataset of 226 images. The classifier has been applied over the geometric features like area, perimeter, eccentricity, perimeter, mean, standard deviation and correlation etc. The neural network used for classification of the dataset. The classification has been done for 10, 15 ad 20 hidden layer variation. The performance is calculated using True positive and true negative statistical measure. The classier is extracted 0.1133% error rate and 88.66% accuracy over the two class classification as normal and diabetic nephropathy dataset. From the above experimental results authors recommended that the neural network is the strong and dynamic classifier for the diabetic nephropathy identification.

Keywords: Diabetic nephropathy · Otsu · Edge · Neural network · Classification · Perimeter

1 Introduction

In the medical image processing the diabetics causes injury to small blood vessels of excretory organ. The diabetes affect on the functionality of the kidney which is responsible for the multigenic disease of human body. The diabetic nephropathy is the type of advancement of the diabetes which is responsible for the illness of the blood vessels. This disease affect the size of blood capillaries and its functionality. This is a type of a microangiopathy and its damage the functionality of the thickening the blood vessel and structure of eye in the body structure. This type is responsible for the micro vascular sickness of the blood structure and differentiation of the patients in the first and second types [1]. There are numerous ways for observant diagnostic assay pictures such as light microscopy research, Immune fluorescence and negatron research studies are essential to differentiate diabetic kidney disease will facilitate to indicate structural

© Springer Nature Switzerland AG 2020
S. Smys et al. (Eds.): ICCVBIC 2019, AISC 1108, pp. 1106–1115, 2020.
https://doi.org/10.1007/978-3-030-37218-7_116

changes like Mesangial enlargement, diffuse glomerular basement membrane thickening and Nodular glomerulosclerosis [1]. The representation of the graphical manner is easily understand and easy to share the convenient information. Image is the most suitable graphical structure which allowed to transmit message with them. The image transmit data from the basic features like shape, size, area and resolution. They depict spatial information such that it can be recognized as an object. The information deriving from the image is very easy and robust task which has been carried by the human being in day to day life. Maximum information has been shared and transmit through the image and it has observable feature. On the basis of observable feature of human eyes the image is classified for the specific reason [2].

For this experiment the data is normalized as per the resolution, size and shape of the image. The edge detection, preprocessing has been done over it using the structural features of the image, like points, lines, curves and surface. The features can be corner point; edge point and salient point of the image object [4].

2   Related Work

This section is provide the detail explanation and highlights the contribution for the work done by the researcher in this domain. Diabetic nephropathy is a small vascular illness of the capillary that affects patient with sorts one and a couple of polygenic disease. The structural and pathological changes in the glomerular affect the thickening in the size of GBM. The structural changes in the GBM has been very poorly understand by the pathology expert. It is correlated to another disease and associated for abnormal treatment for the patients [4].

Glomerular diameter and Bowman's area breadth square measure terribly necessary for detection varied excretory organ connected diseases, for this urinary organ corpuscle and renal objects like capillary wall, glomerular capillary wall etc. It can be helpful for histo-pathological analysis of excretory organ pictures, particularly capillary. The automation in the glomerular detection is not easy task and its a challenging work has been done using the dynamical intensity. The analyzing techniques based on the median filter and it morphological operation has been applied over the capillary vessel diameter extraction. Bowman's area square is measure using this technique. This solution was tested on urinary organ corpuscle pictures of twenty one rats [4].

The pathos potter-k is the special assistance for the pathology expert who is responsible for the automation based results analysis.

The classical image processing and pattern recognition methodology is applied over capillary vessel identification which extract accuracy eighty eight. $3 \pm 3.6\%$. The result indicate that the approach can be applied to the event of systems designed to coach pathology students and specialist additionally in morphological analysis [5].

Here 34 TEM pictures of urinary organ diagnostic test samples square measure taken. Accuracy of segmentation remains challenging task even though' a great deal of researcher has been done on this subject. There is no universal solution for this downside here we've got projected methodology that is that the modification of the initial Chan-Vese algorithm [7].

3 Methodology

For the experiment analysis the database plays a key role. Till data in the era of medical image processing no standard database is available. This research have been collect the dataset which already preprocessed by the medical community for the different application.

3.1 Database Collection

For obtaining the experimental database collection or creation is the challenging task. We have tried several Literatures and on-line sources to search out normal images of excretory organ Glomerular. Various on-line pathological sources are additionally searched, some online pathology, educational sites, medical portals are additionally visited however there was no single resource. Database for this research work has been collect from online medical library and sources like National center for Biotechnology information (NCBI) and Kidney pathology.com [8–13].

3.2 Normalization

Images collected from on top of totally different sources was not in same format that's dimension, height, width, pixels, horizontal and vertical resolution, bit depth etc. For analysis of disease from these images they should be in same format. This was main task in this work. Using mat-lab program first all these images are created in 512 * 512 dimension, width and height 512 pixels, horizontal and vertical dimension 96 dpi, bit-depth 32. These images are hold on in MySQL information. Using ODBC data connection these pictures was load in mat-lab for more experiment.

3.3 Segmentation

The image segmentation is the process which divides the images into sub section no adaptive region. The structure of the segmented image is the use to identify the solution. In the medical image processing the various segmentation methodology is applied. For this research otsu segmentation is employed. This segmentation method work on the clustering based thresholding. For this image has been convert from grayscale to binary. This is threshold based algorithm which extract the minimal distance of pixel and class variance in the image sub section. The algorithm assumes that the image contains 2 categories of pixels following bi-modal bar chart (foreground pixels and background pixels), then calculates the optimum threshold separating the two categories so their combined unfold (intra-class variance) is least, or equivalently (because the sum of try wise square distances is constant). The inter-class variance is supreme. Consequently, Otsu's method is roughly a one-dimensional, discrete analog of Fisher's Discriminate Analysis. The steps of this segmentation is shown in the Fig. 1. Here result step n = 4 are taken for feature extraction. The quality of Otsu segmentation is evaluated using peak signal to noise ratio (PSNR) and signal to noise ration (SNR) applied math approach. SNR is 90%. The PSNR based quality of segmentation is 100% [14].

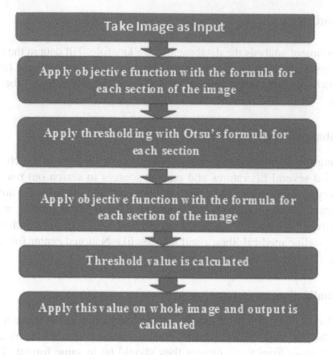

Fig. 1. Detail steps of otsu image segmentation

3.4 Classification

In the era of medical image processing and machine learning the Neural network is the robust and dynamic classifier for the recognition. Neural network has the basic functionality for statistical learning approach. The classification phenomenon has been applied over the character, handwriting digit, text recognition and more recently to satellite image classification [15]. The Artificial Neural Networks and other nonparametric classifiers have a functionality for the input image name for being study SVMs function by nonlinearly jutting the coaching information in the input area to a feature space of higher (infinite) dimension by use of a kernel perform. This results in a linearly separable dataset that will be separated by a linear classifier. This process permits the classification of image datasets which is typically nonlinearly divisible in the input area. In many instances, classification in high dimension feature areas results in over-fitting within the input space, however, in SVMs over-fitting is controlled through the principle of structural risk minimization the empirical risk of misclassification is minimized by maximizing the margin between the data points and also the call boundary. In practice this criterion is softened to the minimization of a price issue involving each the complexness of the classifier and also the degree to which marginal points are misclassified. The tradeoff between these factors is managed through a margin of error parameter (usually designated C) that is tuned through cross-validation procedures. The functions used to project the information from input area to feature space are typically known as kernels (or kernel machines), examples of which embody polynomial, Gaussian (more commonly referred to as radial basis functions) and quadratic functions [16, 17].

4   Experimental Analysis

This experiment is tested over the 226 collected and normalized dataset. The images are normalized in 512 by 512 and 256 by 256 dimensional size. The flowchart of the experimental work is shown in the Fig. 2. For this challenging research the experimental flow diagram is divided into two main modules. The Module one explains the collection of the images from various medical sources and normalizes them to prepare the standard dataset. Our dataset is standardized or not is verified by the medical expert. In the second module the research is done using the computational image processing method. In that the data is passed through the preprocessing, segmentation, feature extraction classification and recognition of the results.

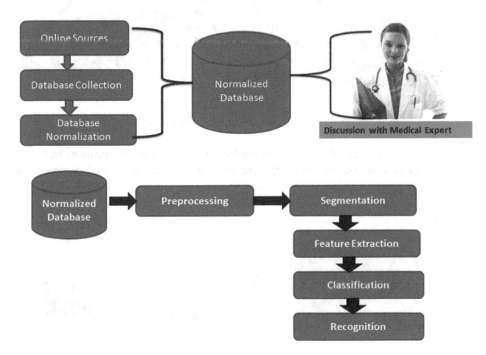

Fig. 2. Flow diagram of the experimental work [7]

The original two samples of the collected images from internet sources are shown in Fig. 3.

The normalize image sample of same size, width and resolution is described in Fig. 4.

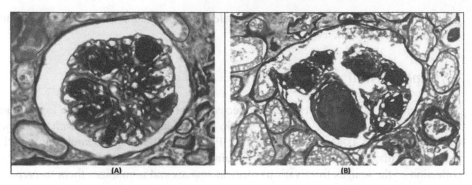

Fig. 3. Sample of two original databases collected from online sources [8]

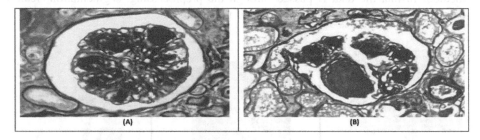

Fig. 4. Normalized image set of sample 2 from collected dataset [9]

The preprocessing of the collected normalized images has been done using histogram and histogram equalization approach. The graphical representation of preprocessing using image processing histogram of above two sample images is shown in Fig. 5.

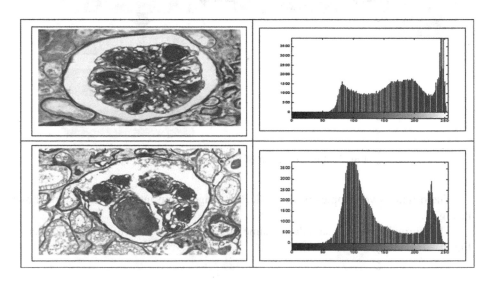

Fig. 5. Preprocessing of the collected normalized two sample image [11]

The otsu segmentation is effectively applied over the 256 image dataset. The Otsu segmentation is applied for the threshold 2, 3 and 4 level. The graphical representation of the threshold 2 and threshold 4 is shown in Figs. 6 and 7 respectively.

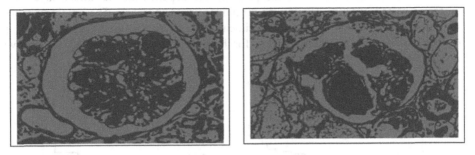

Fig. 6. Otsu segmentation of two sample image at threshold

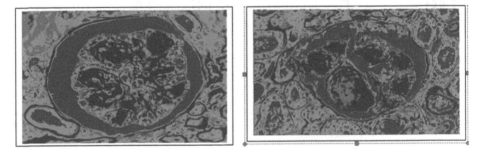

Fig. 7. Otsu segmentation of two sample image at threshold

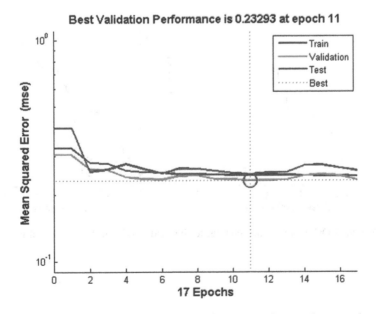

Fig. 8. ROC curve of the performance for the neural network approach

The classification of the neural network is applied over the collected feature of the dataset. The system extracts the 70% training, 15% testing and 15% validation approach. The performance of the neural network is tested over the hidden layer. From the observation the variation of the performance depends on the hidden layer. The system is tested over the 10, 15 and 20 hidden layer approach. The graphical representation of the neural network experiment such as ROC curve for the testing is described in the Fig. 8. The ROC curve for the testing, training and validation is shown in the Fig. 9.

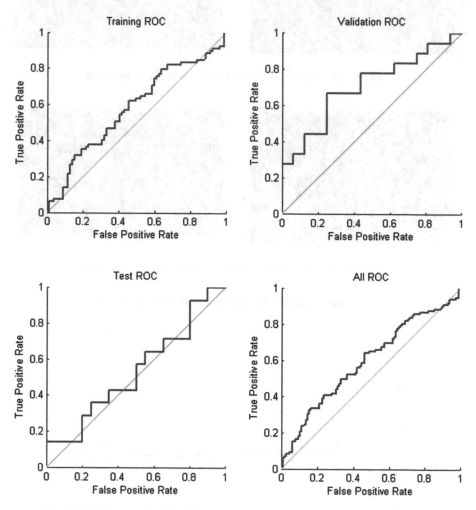

Fig. 9. ROC curve for all training, testing and validation for the neural network

The classification results of the neural network with respective to hidden layer as 10, 15 and 20 is shown in the Table 1. The performance of the neural network is calculated on the basis of true positive, true negative, false positive, and false negative performance ratio. The neural network proposed the **88.66**% accuracy towards the classification experiment.

Table 1. Experimental classification results of neural network

Sr. No	Parameter for classification	Hidden layer 10	Hidden layer 15	Hidden layer 20	Average
1	True positive ratio (TPR)	0.61	0.541	0.5671	**0.573**
2	True negative ratio (TNR)	0.4286	0.285	0.98	**0.565**
3	False negative ratio (FNR)	0.5714	0.7143	0.625	**0.637**
4	False positive ratio (FPR)	0.3323	0.3212	0.3169	**0.323**
5	Accuracy	**88.8286**	**88.7302**	**88.4375**	**88.665**

5 Conclusion

From this research the experiment has been tested on 226 image dataset which are collected from various medical online sources. The database has been normalized by mysql and Matlab tool. The preprocessing of the dataset done using edge detection and histrogram method. The tsu segmentation is applied over the resultant image. The Geometrical feature has been extracted from the dataset. The Neural network of the hidden layer variation of 10, 15, and 20 has been applied over the resultant images. Performance of the classifier is calculated using True positive, True negative statistical measure. The system proved the 88.66% accuracy for the identification of the diabetic nephropathy. The two class classification has been done using neural network. The author recommended that neural network is the robust and dominant methodology for diabetic nephropathy identification.

Compliance with Ethical Standards
✓ All authors declare that there is no conflict of interest
✓ No humans/animals involved in this research work.
✓ We have used our own data.

References

1. Alsaad, K.O., Herzenberg, A.M.: Distinguishing diabetic nephropathy from other causes of glomerulosclerosis: an update. J. Clin. Pathol. **60**(1), 18–26 (2007)
2. Gupta, S., Nijhawan, S.: Comparative analysis of text detection using SVM, KNN and NN. International Journal of Science and Research (IJSR) ISSN (Online), 2319-7064 (2015)
3. Tervaert, T.W.C., Mooyaart, A., Amann, K., Cohen, A., Cook, H.T., Drachenberg, C., Ferrario, F., Fogo, A., Haas, M., De Heer, E., Joh, K., Laure, H.N., Radhakrishnan, J., Seshan, S., Ingeborg, M.B., Bruijn, J.: Pathologic classification of diabetic nephropathy. J. Am. Soc. Nephrol. JASN **21**, 556–563 (2010). https://doi.org/10.1681/asn
4. Kotyk, T., et al.: Measurement of glomerulus diameter and Bowman's space width of renal albino rats. Comput. Methods Prog. Biomed. **126**, 143–153 (2016). https://doi.org/10.1016/j.cmpb
5. Barros, G.O., et al.: PathoSpotter-K: a computational tool for the automatic identification of glomerular lesions in histological images of kidneys. Sci. Rep. **7**, 46769 (2017)
6. Rangayyan, R.M., Kamenetsky, I., Benediktsson, H.: Segmentation and analysis of the glomerular basement membrane in renal biopsy samples using active contours: a pilot study. J. Digit. Imaging **23**(3), 323–331 (2009). https://doi.org/10.1007/s10278-009-9188-6
7. Ravi, M., Hegadi, R.S.: Detection of glomerulosclerosis in diabetic nephropathy using contour-based segmentation. In: International Conference on Advanced Computing Technologies and Applications (ICACTA-2015), pp. 244–249 (2015)
8. Image database. http://medpics.ucsd.edu
9. Image database. http://peir.path.uab.edu
10. Image database. https://library.med.utah.edu
11. Image database. www.ajkd.org
12. Image database. www.ncbi.nlm.nih.gov
13. Image database. www.kidneypathology.com
14. Patil, Y.B., Kawathekar, S.S.: Automated analysis of diabetics nephropathy using image segmentation techniques for renal biopsy images. Int. Multidiscip. J. Pune Res. **3**(4) (2017). Impact Factor: 2.46, ISSN 2455-314X
15. Vapnik, V.N.: The Nature of Statistical Learning Theory. Springer, New York (1995)
16. Huang, C., Davis, L.S., Townshed, J.R.G.: An assessment of support vector machines for land cover classification. Int. J. Remote Sens. **23**, 725–749 (2002)
17. Foody, M.G., Mathur, A.: Toward intelligent training of supervised image classifications: directing training data acquisition for SVM classification. Remote Sens. Environ. **93**, 107–117 (2004)

Advanced Techniques to Enhance Human Visual Ability – A Review

A. Vani[1(✉)] and M. N. Mamatha[2]

[1] Sapthagiri College of Engineering, VTU, Belgaum, India
vaninagesh44@gmail.com
[2] BMS College of Engineering, VTU, Belgaum, India
mamatha222@yahoo.co.in

Abstract. Human eye is one of the most complex organs in human body. Eyes, the integral part of Visual system receives signals and brain computes the obtained visual details. The brain helps in detection and interpretation of information in the visible light. This Complex process includes reception of light, binocular perception, estimation of distance and categorization of objects. Any anomaly in such process causes Blindness or Visually Impairment. According to WHO survey approximately 1.3 billion people live with some form of vision impairment. The need for visual aid is essential to enhance visual ability for visually impaired. In this paper a detailed review of different types of advance techniques to enhance vision is discussed.

Keywords: Neuroprosthesis · BioMEMs · tACS · IoT · Machine learning

1 Introduction

The International organization WHO estimates that around 1.3 billion people around the world are suffering from visual impairment. This can be broadly classified as distance and near vision impairment [1]. It includes mild, moderate, severe impairment and blindness. Globally, there are many causes for visual impairment which includes uncorrected refractive errors, age related macular degeneration, cataract, glaucoma, diabetic retinopathy, trachoma, corneal opacity. More than 80% of the visual impairment are unavoidable [2].

There is a need for technological intervention for avoiding these circumstances. The solution many be simple with refractive error correction using glasses, while some solution includes surgeries like cataract disease. For irreversible vision problem rehabilitation can be achieved with visual neuroprosthesis [3]. The technological intervention can be broadly classified as invasive and non invasive techniques. Invasive techniques requires surgeries to restore vision. The non invasive techniques uses modern tools and techniques to help restore vision. The Fig. 1 shows the classification of advance techniques involved in restoring vision for visually impaired.

© Springer Nature Switzerland AG 2020
S. Smys et al. (Eds.): ICCVBIC 2019, AISC 1108, pp. 1116–1124, 2020.
https://doi.org/10.1007/978-3-030-37218-7_117

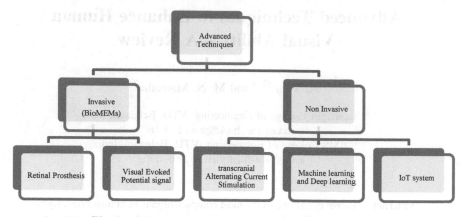

Fig. 1. Advanced techniques for enhancing visual ability

2 Invasive Techniques

2.1 Retinal Prosthesis

The prostheses devices are the electronic devices which substitute the organ function. The Retinal prostheses is the implantable device which substitute phototransduction process of retina which is affected significantly by the retinal disease called retinitis pigmentosa [3]. The normal eye has photoreceptors in outer layer of retina which triggers the phototransduction to generate electrical signal in the presence of light stimulus. These complex signals are processed by retina before ganglion cells. Axonal process from Retinal ganglion cell forms a optical nerve to send signals to visual cortex. In diseases such as retinal pigmentosa the outer layer of retinal are gradually lost causing progressive visual loss, however the inner layer is still working. The restoration of vision is possible with suitable devices which receive the light sinal and process to form electrical impulses to the remaining layers of visual functions [4].

Figure 2 shows the Alpha-IMS subretinal microphotodiode array retinal prosthesis. The size of the device is about 3 mm * 3 mm * 70 mm operating at 5–7 Hz. The energy is provided by external coil placed behind ears which magnetically provides energy and communication wirelessly. Similarly there are other specific designs for epiretinal implants such as Argus II retinal prosthesis system, EPI- RET3, Intelligent Retinal Implant System (IRIS V2), Suprachoroidal Retinal Prosthesis and Subretinal Retinal Prosthesis are available. The only limitation of the system is, invasive techniques. This technology is suitable only for the eye where only retinal layers are affected and rest all the region such as visual nerve path, temporal lobe and occipital lobe of the brain should be normal for better vision.

2.2 VEP Signal Technique

Whenever we see any object our eyes generate an electrical signal due to visual stimulus which is referred as Visual Evoked Potential (VEP) signal. This signal is

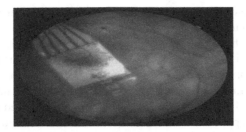

Fig. 2. Fundus photo of Retina Implant Alpha-IMS device [3]

considered to be universal, just like a ECG signal from heart. In most of the cases people lose their vision during injury, the injuries causes damage in the frontal lobe and eye whereas the actual vision process takes place in the occipital lobe. The occipital lobes computes the images based on the VEP signal. It means that the computation block of the brain is still working. So it is possible to generate artificially evoked potential which can be applied to the brain. This evoked potential fires the required neuron in the occipital lobe of the brain [5].

Figure 3 shows the block diagram of Visual Evoked Potential signal with object detection system. The visual stimulus for various images are displayed on the screen. The subject is allowed to see the images displayed on the screen. The EEG electrodes are placed on the brain according to 10–20 international electrode placement system. The EEG signals for different images like standard checker box, face and objects are recorded. Visual Evoked Potential can be extracted from the EEG signals. These VEP signal forms the database.

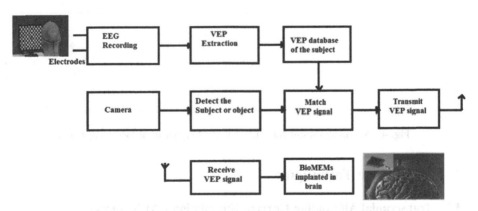

Fig. 3. Block diagram of VEP signal extraction with object detection system [5].

The Camera will be placed on the visually impaired. Simple object and face recognition techniques can be applied to recognize face or objects. Once the object or face is recognized the corresponding VEP signal can be sent to the BioMEMs placed on the occipital lobe of the human brain. The VEP signal helps in partial vision on the visually impaired individual. The placement of BioMEM which is micro electrode, fires the required neuron on the brain in occipital lobe.

Figure 4 shows the simulation results of VEP signal extraction from EEG signal for face recognition and Fig. 5 shows VEP signal extraction for standard checker board image. OpenVibe simulation software is multiplatform used for signal processing and other applications. In the figure image stored in the database and corresponding EEG signal is displayed. The output averaged VEP signal is extracted from EEG signal [6]. This signal can be applied to BioMEMs which can be implanted on brain. Further the BioMEMs fires the neurons in the occipital lobe creating partial image in the brain.

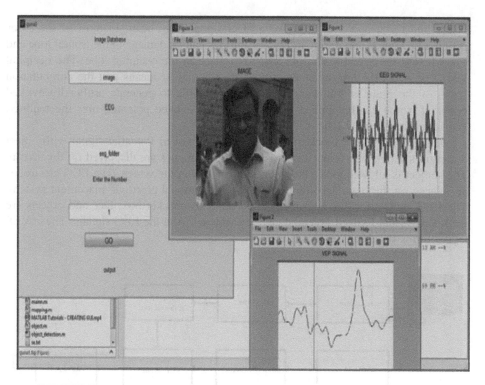

Fig. 4. Simulation result for VEP extracted signal from face recognition.

3 Noninvasive Techniques

3.1 transcranial Alternating Current Stimulation (tACS) Device

Most of the blind aided devices available uses audio cues and other devices involves surgery for implant. There is need for visual feedback with non invasive device. The non invasive visual cues are more intuitive and appealing than auditory feedback or invasive systems. This can be achieved by transcranial Alternating Current Simulation tACS device [7].

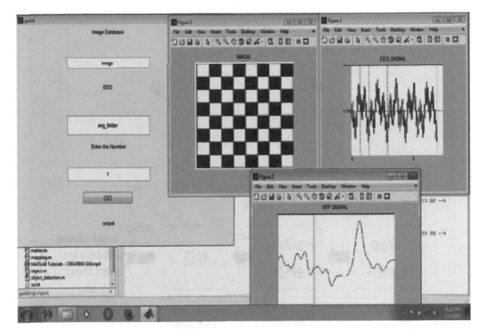

Fig. 5. Simulation result for VEP extracted signal from standard checker board

The tACS device provides very low electrical simulation (μA) to the posterior part of the brain. The applied electrical simulation can be perceived as flashes of light for visually impaired. The change in electrical characteristics causes change in rate of flashes of light. The rate of flash of light can be used as visual feedback for visually impaired. As a proof of concept [7], the test was carried out for 9 blind folded individual with the age group 20-35 years mimicking visual impaired. Three electrical simulation parameters slow, medium and fast are presented for the individuals. These electrical parameters are perceived as a flashes of light at different rates.

The electrodes used can be conventional Ag/AgCl adhesive foam with electrolyte gel. As shown in the Fig. 6 Fp1 and A1 bipolar combination is used to stimulate left side of the brain. Fp2 and A2 is used for right side of the brain. The flickering rate in the brain can be varied by changing frequency and current intensity of tACS device.

The individuals were able to distinguish slow (8 Hz), medium (10 Hz) and fast (20 Hz) flickering rate as flashes of light in brain. This visual cues can be used for visually impaired to detect proximity of the object by making tACS device sensor triggered as shown in the Fig. 7. The sensors like ultrasound sensor which gives the information of the distance can be interfaced to tACS device. Based on the proximity, the tACS generate electrical signal which causes changes in flashes of light. Thus it helps in visual cues for visually impaired with non invasive device.

Figure 8 shows the brain stimulation results for both left and right side of the brain. The frequency is kept constant at 8 Hz. The current intensity is varied in steps of 100 μA from 200 μA to 1200 μA. The simulation result is shown for 9 subjects with

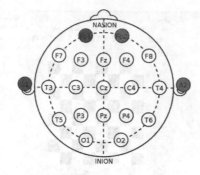

Fig. 6. Electrode placement for tACS device

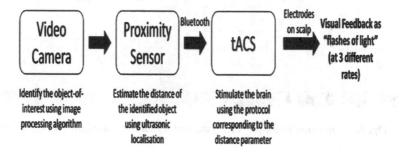

Fig. 7. Block diagram of sensor triggered tACS

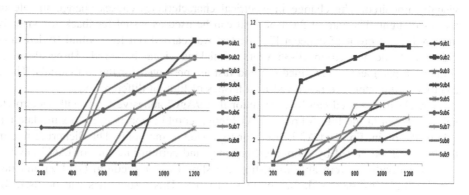

Fig. 8. Left and Right side brain stimulation for frequency 8 Hz for different current intersity

blind folded mimicking visual impairment. According to the graph, subjects can feel the change in flickering rate as the current intensity is varied.

Figure 9 shows the brain stimulation results for both left and right side of the brain. The frequency is kept constant at 10 Hz. The current intensity is varied in steps of 100 µA from 200 µA to 1200 µA. Similar result was obtained for 20 Hz frequency by varying current from 200 µA to 1200 µA.

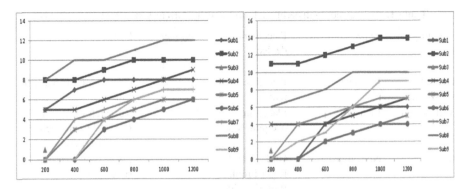

Fig. 9. Left and Right side brain stimulation for frequency 10 Hz for different current intersity

3.2   Image Processing, Machine Learning and Deep Learning Tools

Eye disorders is a major health problem globally [8]. There are many types of age related eye diseases such as diabetic retinopathy, glaucoma and age related macular degeneration and many more [9, 10]. There is need for proper Computer aided analysis and classification tool for eye diseases for restoring vision. The features are extracted from the images of eye diseases. The classification strategies are applied for the eye data set. The new technologies like machine learning and deep learning tools can be used to diagnose eye diseases. Millions of medical images are used to train [9]. With this techniques the tool will be able to categories the healthy and unhealthy signs of eyes. The algorithm will also be able to detect some of the conditions like glaucoma, macular degeneration and fluid in retina as accurately as six expert Ophthalmologist [9, 10].

3.3   IoT Devices

Internet of Things is the device which connects all the devices together. IoT technology plays important role health sector. The sensor will be connected to eye glasses. When the sensor detect threat it immediately sends the message to the nearest hospital [11]. Also through GPS the location of patient can also be detected. IoT with such monitoring can improve health and prevents diseases [12].

4   Conclusion and Future Work

In this paper recent technologies related to enhance visual ability is discussed in detail. Table 1 describes the current status of all the technologies.

Improvement of all the advanced techniques both invasive and non-invasive techniques is very essential. In invasive techniques biocompatability is very crucial for safety as it may cause tissue damage so optimal material for simulation and implantation, long term implantation studies are essential. In tACS device provide visual feedback as flashes as light. This information is very limited for visually impaired individual. Further there is a need for partial vision to get more information. ML and

Table 1. Advanced engineering solution and its current status of major initiatives

Techniques	Type	Placement of device	Number of electrodes	Status
Retinal prosthesis	Invasive	Placed in retina	25–65	Commercially available in Europe
VEP signal	Invasive	Placed in occipital lobe	6	Research of placing BioMEMs on occipital lobe in progress
tACS	Non-invasive	On Left and Right Frontal Posterior side of brain	2	Pilot study is found successful
Machine learning	Non-invasive	Only images of eyes are taken	0	Early detection of eye diseases such as glaucoma, age-related macular degeneration and diabetic retinopathy found successful
IoT system	Non-invasive	Placed on Eye glasses	0	IoT with smart sensor integration helps to contact doctor or guardian

deep learning techniques are provided with millions of images for better diagnosis of few eye diseases. This techniques should be further extended for all types of eye diseases. Smart sensor based vision and threat detection can be integrated to form complete eye solution through smart sensors.

Compliance with Ethical Standards
  ✓ All authors declare that there is no conflict of interest.
  ✓ No humans/animals involved in this research work.
  ✓ We have used our own data and author photo.

References

1. World Health Organization visual impairment report 2018. https://www.who.int/news-room/fact-sheets/detail/blindness-and-visual-impairment
2. Age-Related Eye Disease Study Research Group. Risk factors associated with age-related macular degeneration: a case-control study in the age-related eye disease study: age-related eye disease study report number 3. Ophthalmology **107**(12), 2224–2232 (2000)
3. https://eyewiki.aao.org/Retina_Prosthesis#Background
4. Liu, J., Sun, X.: A survey of vision aids for the blind. In: 2006 6th World Congress on Intelligent Control and Automation, Dalian, pp. 4312–4316 (2006)

5. Suraj, H., Yadav, A.V.B.: The use of VLSI to generate vision in blind people using a suitable neuroprosthesis implant of BIOMEMS in brain. In: 2011 IEEE International Conference on Computer Science and Automation Engineering, Shanghai, pp. 179–183 (2011)
6. Vani, A., Mamatha, M.N.: Artificial generation of visual evoked potential to enhance visual ability. World Acad. Sci. Eng. Technol. Int. J. Med. Health Biomed. Bioeng. Pharm. Eng. 10(9), 443–447 (2016)
7. Vani, A., Mamatha, M.N.: Visual feedback system for visually impaired using tACS – a proof of concept study, vol. 6, no. 1, pp. 738–743, January 2019. ISSN 2349-5162. http://www.jetir.org/
8. https://www.nature.com/articles/d41586-018-02431-1
9. Selvathi, D., Suganya, K.: Support vector machine based method for automatic detection of diabetic eye disease using thermal images. In: 2019 1st International Conference on Innovations in Information and Communication Technology (ICIICT), pp. 1–6. IEEE (2019)
10. Bhatia, K., Arora, S., Tomar, R.: Diagnosis of diabetic retinopathy using machine learning classification algorithm. In: 2016 2nd International Conference on Next Generation Computing Technologies (NGCT), pp. 347–351. IEEE (2016)
11. Prouski, G., Jafari, M., Zarrabi, H.: Internet of things in eye diseases, introducing a new smart eyeglasses designed for probable dangerous pressure changes in human eyes. In: 2017 International Conference on Computer and Applications (ICCA), pp. 364–368. IEEE (2017)
12. Bhatt, Y., Bhatt, C.: Internet of things in HealthCare. In: Bhatt, C., Dey, N., Ashour, Amira S. (eds.) Internet of Things and Big Data Technologies for Next Generation Healthcare. SBD, vol. 23, pp. 13–33. Springer, Cham (2017). https://doi.org/10.1007/978-3-319-49736-5_2

Cross Media Feature Retrieval and Optimization: A Contemporary Review of Research Scope, Challenges and Objectives

Monelli Ayyavaraiah[1(✉)] and Bondu Venkateswarlu[2]

[1] AITS-Rajampet, Dayananda Sagar University, Bangalore, India
ayyavaraiah50@gmail.com
[2] Dayananda Sagar University, Bangalore, India
bonduvenkat-cse@dsu.edu.in

Abstract. Predictive analytics that learns from cross-media is one among the significant research objectives of the contemporary data science strategies. The cross-media information retrieval that often denotes as cross-media feature retrieval and optimization is the crucial and at its infant stage. The traditional approaches of predicative analytics are portrayed in the context of unidimensional media such as text, image, or signal. In addition, the ensemble learning strategies are the alternative, if the given learning corpus is of the multidimensional media (which is the combination of two or more of test, image, video, and signal). However, the contributions those correlates the information of divergent dimensions of the given learning corpus is still remaining in the nascent stage, where it is termed as cross media feature retrieval and optimization. This manuscript is intended to brief the recent escalations and future research scope in regard to cross-media feature retrieval and optimization. In regard to this, a contemporary review of the recent contributions has been portrayed in this manuscript.

Keywords: Cross-media retrieval · Media-gap · UCCG · Multi-modal embedding · DTRNN

1 Introduction

It is vital for several domains such as healthcare, associating diversified media standards such as text, signals, and images, which often together denotes as cross-media. The methods of processing the cross-media to extract information (often denotes as features) has significant role in learning process of predictive analysis.

The methods such as pattern recognition and computer vision are closely bonded to the cross-media that evinced huge significance in performance of corresponding domains [1]. Cross Media Feature Retrieval and Optimization is challenging due to the formation of the cross media learning corpus by interacting with manifold aspects on diverse conditions, comprising of changes in the illumination, stop-words, background interference. The work [2, 3] presents that many of retrieval methods concentrated on the single media like retrieval of image, text, audio, video and signal.

© Springer Nature Switzerland AG 2020
S. Smys et al. (Eds.): ICCVBIC 2019, AISC 1108, pp. 1125–1136, 2020.
https://doi.org/10.1007/978-3-030-37218-7_118

Nevertheless, the challenge is intricate. The unavoidable occurrence of the heterogeneity gap among the information of cross media that is deliberated extensively as fundamental barrier towards feature retrieval of multi-model is the main reason for the challenge to be intricate. The usual model for bridging the gap will be for mapping the data of cross-media into normal feature space where required data is much closer to every other such that queries were carried out and predicted to get desired outcomes. Moreover, the challenge needs machine towards understanding the descriptions in the texts, images & their connections instead of just prominence in the verbs or images & nouns in texts.

Through the improvement of network application & information technology, instant release of the information is developing and spreading quickly. Furthermore, the kind of information improved from the text to the multimedia. Voluminous quantity of the multi-media information & the formats of unstructured data make the retrieval of information, monitoring and analyzing the topic more intricate. It is significant aimed at information retrieval method for precisely examining the entire type of related data and combining the topic relevant to the results of cross-media from diverse sources. Hence, it is required for adopting the topic of cross-media analysis & the model of information retrieval for extracting the features from entire probable media data that are combined to identical space of semantic aimed at processing.

The work [4] presents that the topic of analyzing the cross media and the information retrieval model can enhance the capacity of decision making and its effectiveness. Moreover, the knowledge will be extracted from dataset of cross-media to reflect contemporary burning topics and the event status is evaluated accurately. Through the increase of the multimedia data like image, audio, 3D model, text, video, the CMR became the significant domain, where the users are attaining the outcomes through several media kinds by submitting 1 query of media-type.

The "media-gap" is the challenging issue of CMR, where the depictions of diverse media kinds are unreliable and present in diverse spaces of features. Therefore, it is hugely challenging for assessing the similarities amid them. Here there are numerous models proposed for addressing this problem through examining the correlations comprised in the data of cross-media. Many of the contemporary models are planned for the retrieval of 2 media types mostly text & images, yet CMR emphasizes the variety of media.

The objective of CMR is identifying the most related texts in contemporary data for video or image in contemporary data & conversely. The other significant variance among them is caption of video or image concentrates only on video or image and the text that will not be simple for extending to other types of media.

The CMR is better for 2 reasons. Primary one is that, providing the related outcomes with entire media types that is more convenient and resourceful for the consumers. Next is that diverse kinds of media cases are utilized always for describing the similar concept of semantic from diverse factors. Hence, in CMR, diverse media types can enhance over each other that is comparatively strong towards noise. Besides, users or consumers might submit any media data type as query that is useful and convenient.

2 Review of Literature

The resources of online multi-media were rich in the visual information, text & audio. Mapping the semantics from low range to high range is often indefinite. For resolving such ambiguity in mapping, some of the researchers carried out joint analysis of the Cross-Media data, related to content & context. The work [5] suggested multimodality depictions of the web videos comprising features of latent semantic, audio, surrounding & visual features and implement later-fusion method to train the classifiers separately for diverse modalities and merged these classifiers outputs for estimating the anonymous video category. Further, analyzing the content based on the attention of videos or images is utilized for approximating the regions, events & objects of consumer interest [6]; the structural information of web page like neighboring context, links were also discussed for facilitating the CMR [7–9].

The work [10] predicted that representation of image can be viewed in the form of analogous towards "cross-lingual retrieval issue" and suggested "CMRM (cross-media relevance model)". The work [1] built a "UCCG (uniform cross-media correlation graph)" as per the features of media objects and their information. For performing the CMR, the positive score will be allotted to query instance, and score will be spread along graph & objects of the aimed modality with maximum scores were returned. For improving the performance of retrieval, the relevant short term & long-term feedbacks are utilized for UCCG optimizing.

The work [1, 11] implements the CMR on the level of semantic based on ontology. The work [12] presents the combined speculative method of mining the multi-media semantic, which utilizes the fundamental multimedia speculation characteristics for attaining the CMR. The work [13] proposes a method which integrated semantic description spatial temporal context with CMR, for implementing the prototype method. The work [14] suggests multi-information method related with semantic representation between retrieval map & multimedia information for completing the CMR. The work [15] presents that contemporary CMR algorithm concentrates mostly on extracting the feature & classification of information of the cross-media.

Contemporarily, besides from existing retrieval system based on the content, the novel method came into survival known as CMR. Extracting the correlation amid the documents of heterogeneous multimedia, where modality of document of the input media & retrieved outcomes does not required being similar. The work [16, 17] presents that several systems of cross-media concentrated on combining & handling the data of heterogeneous multimedia. The work [18] proposed "dual cross-media relevance method" for annotating and retrieving the images automatically that deliberates both text-text & text-image associations. The work [15, 19] presents that; gathering of several modalities of media objects with huge amount of similarity amid them is called MMD (Multimedia document). The work [16] presents that generating the MMDs practically might lead to performance problems due to the association among existing images & texts differences.

In the recent years, several image annotation models for the images of internet are proposed as there is a rapid growth in the image data of internet. The work [20] presents that same images were searched from web through extracting the common descriptions & representations from surrounding explanations of same images in the form of annotation aimed at image query. The work [21] presents that new "kNN-sparse graph-based semi-supervised learning" method is proposed for consecutively harnessing the unlabeled & labeled data that exploits the issue of annotating huge scale corpus of image through label propagation above boisterously web images which are tagged. The work [22] introduced novel algorithm called scalable learning for "multi-class image classification" on the basis of regularization penalty of trace-norm & "multi-nominal logistic loss". The work [23] presents that utilizing "stochastic gradient descent optimization" might scale learning towards thousands of the classes and millions of the images.

The tagging of image is utilized extensively in the website of image sharing in the internet, since the users can simply search, manage and generate the images with the assistance of tags. The work [24] established "large-scale automated photo tagging" model through discovering social-images & represented a method on the basis of retrieval for tagging the photo automatically. The work [25] suggested machine learning structure for extracting the social images & examining its application for tagging the image automatically. The work [26] presents that utilizing the information of image tagging for improving the understanding capabilities of image, where they suggest a new "multi-task multi-feature metric learning method" that models information sharing technique among diverse tasks of learning. This method is capable of learning simultaneously through semantic information & multi task learning structure on the basis of social tagging; hence both of them get assistance from the information offered through each other.

Generally, the social-images comprise tags & other literal descriptions, therefore CMR models are resourceful for improving the understandings of the image. The work [27] presents the notion of the semantic propagation on the basis of relevant feedback & query expansion of cross modality that enhances the performances of retrieval when the query will be initialized either through instance image or keyword. The work [28] suggested cross-modality method that will not require entire data trained in the form of pairs of respective text & image. Therefore, their method will exploit accessibility of few pairs of the image-text, along with the text to text & image to image similarities for learning intrinsic associations among text & images.

The "Multi-modal embedding" models map the data of cross-media into feature space in respect to search identical or related objects of cross media. The work [29] presents that "kernelized canonical correlation analysis (kCCA)" is utilized for mapping the visual & textual information and formed breakthrough in the segmentation & annotation. The work [30] developed multi-method deep Boltzmann machines for depicting the visual & textual information jointly. The work [31] learned the depiction of cross-media model in the view of optimizing the problem of list wise grading.

The objective of this contribution is bi-directionally ranking the data of cross-media. Contemporarily, there will be developing domain of contribution, which automatically generates the images caption. Feng et al. integrated the language methods, multi-model identical methods and visual detectors for yielding the images captions. The work [32] presents that the modal performance is predicted for having better or equal quality. The work [33] presents that the generative method is on the basis of deep recurrent structure, which integrated the upgrading of machine translation & computer vision. The work [34] introduced "deep visual-semantic embedding" method for detecting the visual objects utilizing data of labeled image and semantic information collected from the text which is unlabeled.

The work [35] proposed "DTRNN (Dependency Tree Recursive Neural Network)" towards textual information process. The work [36] suggested a method that embeds the fragments of the sentences & images into common space. Amid the models mentioned above, all of them utilized RNN for processing the textual information & used inner products among data features of cross media to explore the similarity. Except the method mentioned in above contribution, remaining methods extracted the global-features. Aimed at images, generally the background takes huge amount even though it is not significant. But when it turns to sentences, the keywords which are mined will match hardly the prominence in image. Therefore, inner product through global features might cause unavoidable mistake. Nevertheless, as stated in contribution [36], the fragments of the sentence were not suitable, specifically when comes to manifold adjectives for 1 noun. Moreover, it will be intricate for corresponding every fragment in 1 image with every phrase or word in related sentence.

The work [15] explored regarding the understanding of "Harmonizing hierarchical manifolds for multimedia document semantics" & CMR. The work [15] presents tri-space & ranking based model, which will be a heterogeneous measure of similarity for CMR, whereas other existing models concentrates on 3rd common space and their suggested tri-space model concentrates on entire features. Learning the joint representation of cross-media with the semi-supervised and sparse regularization and measuring the similarity content among diverse media will be main challenge.

The work [37] proposed novel method called deep learning aimed at retrieving the related cross media. The suggested method focused on 2 stages like distance detection & feature extraction. After attaining the information of feature, they sorts and eradicates the redundant feature vectors.

The work [38] suggested new "cross-media distance metric learning framework on the basis of multi-view matching & sparse feature selection". The model engaged "sparse feature selection" for choosing subset of related features and eradicates the unnecessary features for audio features & "high-dimensional image features". Further, the model built "Multi-modal Semantic Graph" for finding the "embedded manifold cross-media correlation". Further, canonical correlations are merged with multi-information into "multi-view matching" & constructed a "Cross-media Semantic Space" aimed at "cross-media distance measure". The comparisons & simulations verified the superiority, applicability and validity of our method from diverse factors. The important confine is that database will be relatively smaller due to insufficient standard database & audio data.

The work [16] applied "supervised coarse-to-fine semantic hashing model" for effective search of similarity over the datasets for learning sharing space. Generally, by leveraging semantic similarity among samples and to build coarse-fine similarity-matrix, the hamming space is learned. Later, iterative upgrading method is implemented for deriving optimal solutions which are local. Finally, to develop further the hashing codes discrimination, learning the "orthogonal rotation matrix" by reducing the loss of quantization while storing the undisturbed solution optimality. The simulations extensively utilized NUS-WIDE & Wiki datasets showed that suggested model surpassed the contemporary model. The cost of training affects sophisticated learning methods performance like kernel learning.

The work [39] introduced "SCH (Semantic Consistency Hashing)" model for the retrieval of cross-modal. The learned SCH shared simultaneously the semantic space by taking into account both the intra-modal & intermodal semantic correlations. The similar depiction is learned utilizing "non-negative matrix factorization" aimed at samples with diverse modalities in respect to store the semantic consistency of inter-modal. In the mean-time, the algorithm of neighbor preserving is adopted for preserving each modality semantic consistency. The productive optimal algorithm is suggested for lessening the complexity of time and also extended easily towards classification of image, annotation and other domains of learning. The cost of training the "CMFH (Collective Matrix Factorization Hashing)" is more than SCH model in huge training dataset. Hence, SCH model might not be utilized effectively in the small-scale implementations.

The work [40] discussed efficiency of CMR models in "SBIR (Sketch-based image retrieval)" that have been implemented in matching image-text successfully. Contemporary model could not tackle cross modal issue of SBIR. The simulation results displayed that subspace learning will model effectively the domain gap of sketch photo. Computational intricacy of analysis model is very intricate in the procedure of SBIR.

The proposed model "HSNN (Heterogeneous Similarity measure with Nearest Neighbors)" [41] is for extracting the correlation of intra-media. Heterogeneous similarity is achieved through computing the possibility of 2 media objects appropriate to similar semantic class. The "CMCP (Cross-Media Correlation Propagation)" method is to deal simultaneously with negative & positive correlation among media objects of diverse types of media. The negative correlation is very significant due to the offering of effective special information. Moreover, the method is capable of propagating correlation among the heterogeneous-modalities. Lastly, both CMCP & HSNN are adaptable, such that any conventional similarity measure might be included. The effective ranking method is learned through the further fusion of manifold similarity measures by AdaRank aimed at CMR (Table 1).

Table 1. Contemporary contributions and their properties.

Reference no.	Technique employed	Performance measure	Dataset	Benefits	Confines
[42]	Multi-graph cross modal hashing	Inter-media hashing, cross-view hashing, composite hashing through manifold information sources	NUS-WIDE Dataset, Wikipedia	The approximation of Nystrom method is utilized for constructing the productive graph and it displays the improved performance when it is compared with 2 unsupervised cross-modal hashing model	Designing the productive graph with enhancing intricacy in construction of graph, whose time might be linear is more intricate

(continued)

Table 1. (*continued*)

Reference no.	Technique employed	Performance measure	Dataset	Benefits	Confines
[43]	Semantic correlations	Joint representation learning, canonical correlation analysis & cross modal factor analysis	NUS-WIDE, Wikipedia, & X-Media datasets	Among 3 databases, this has greater efficiency than contemporary model	The data weight value is not utilized, which assists to increase the system performance
[44]	Correlation among attributes	Ranking score	Shoes dataset	This model takes the expedient of structural sparsity of demands & simulation outcomes display the updated performance when it is compared with contemporary model of the image retrieval method with identical queries	The image retrieval performance need to be enhanced & only image will be retrieved in this model
[45]	Semi-supervised semantic factorization hashing	MAP score of cross-modal hash value, recall & precision	NUS-WIDE, Wikipedia dataset	This technique alters semantic tag into hash codes, which stores the better semantic values. The simulation outcome displays that its effectiveness is more than contemporary model	The decay of value reasons in eigenvalue, whereas orthogonally there is a limit of hash code
[46]	Link confidence measure	Hit ratio	Local dataset	This model achieves the intermediates match outcomes and estimates potential similarities & this assists the greater connectivity of graph aimed at the confined computational resources	The assessment will not be provided because of its intricacy to build the criteria of objective

(*continued*)

Table 1. (*continued*)

Reference no.	Technique employed	Performance measure	Dataset	Benefits	Confines
[47]	Semantic entity projection	MAP score	Wikipedia dataset	The suggested model utilizes this projection model and simulation outcome display that system attains better outcomes when compared with another contemporary model	The correlation among the levels of entity are not exploited fully and provides low result for lesser quantity of data
[48]	Semantic information	Precision, MAP score, & recall	X-Media & Wikipedia dataset	The algorithm of iterative optimization resolves the issue of resulting optimization and displays the improved outcome in both cross-media & single-media retrieval through state-of-art model	Many types of correlation amid diverse media will be utilized to enhance the performance
[49]	Joint graph regularized heterogeneous metric learning	CCA+SMN CCA, JGRHML & CFA	X-Media, & Wikipedia dataset	This assists for learning greater level of semantic metric with the label propagation	The "jointly modeling manifold modality" is not utilized in this model, which assists for utilizing in several applications
[50]	A cross-modal self-taught learning (CMSTL)	Precision, MAP & recall	NUS-WIDE & Wikipedia articles	Hierarchical generation is utilized for achieving multi-modal theme and it provides the improved performance in the cross-modal retrieval	The method is enlarged towards image hashing & also over other media.
[51]	Bayesian personalized ranking based heterogeneous metric learning	Precision, MAP & Recall	Core 15k image, Wikipedia dataset	They combine the heterogeneous & homogeneous graph regularization into the required value & generated the enhanced performance	This model is carried out on images and is required for testing other media such as video, text & audio

(*continued*)

Table 1. (*continued*)

Reference no.	Technique employed	Performance measure	Dataset	Benefits	Confines
[52]	Deep multimodal learning model	MAP score	NUS Wide 10 k dataset, Wikipedia	The simulation outcome show that suggested method is more effective when it compared with the other models	This process needs to be applied for utilizing the data which is unlabeled in the method for enhancing the performance of system

3   The Issues and Challenges in Cross-media Feature Retrieval Optimization

The scope of data accumulation in recent era of computer aided automation and networks is massive. This tempo of massive data collection often leads to the accumulated corpus with unstructured, dynamic formats, large in volume, and multidimensional factors. The semantic gaps are the significant constraints of the cross-media feature retrieval and optimization, and the semantic gaps noted in sources of cross-media are

(i) between the corpus tuples framed under temporal barriers,
(ii) vivid dimensions of data representation formats,
(iii) distribution diversity of the data projections
      a. within the format
      b. between the formats of the cross media.

Hence, the critical research objectives of the cross-media centric predictive analytics by machine learning strategies are entails to defining significant methods of feature retrieval and optimization from cross-media with semantic gaps stated above.

The methodological approaches those surpasses the semantic gaps of the cross-media in regard to feature retrieval and optimization are

Features of diversified mediums of the cross media are being transformed from multivariate to univariate. This is often done by considering the features of divergent dimensions of the cross-media data having significant correlation.

Adapting the significance of the diversified features of the vivid mediums from the cross media. This is scaled by the significant associability between the features from vivid mediums.

Identifying diversity or similarity of horizontal and vertical distributions of the values projecting corresponding features is another significant strategy. This enables to achieve maximal learning accuracy by partitioning the features in to diversified tuples, such that the entries of each tuple evince the high correlation.

4 Conclusion

The contribution of this manuscript is a contemporary review that intended to identify the significance, scope and challenges of the cross-media information retrieval and optimization. The term "information" often denotes as features in regard to predictive analytics using machine learning. The contemporary review that portrayed in this manuscript is evincing the considerable research scope in cross media feature retrieval and optimization. The possible methodological approaches, which can be adapted by future research also concluded.

Compliance with Ethical Standards

✓ All authors declare that there is no conflict of interest.
✓ No humans/animals involved in this research work.
✓ We have used our own data.

References

1. Zhuang, Y.-T., Yang, Y., Fei, W.: Mining semantic correlation of heterogeneous multimedia data for cross-media retrieval. IEEE Trans. Multimed. **10**(2), 221–229 (2008)
2. Ma, Q., Nadamoto, A., Tanaka, K.: Complementary information retrieval for cross-media news content. Inf. Syst. **31**(7), 659–678 (2006)
3. Mao, X., et al.: Parallel field alignment for cross media retrieval. In: Proceedings of the 21st ACM International Conference on Multimedia. ACM (2013)
4. Wu, Y.: Information management and information service in emergency forecast system. J. Doc. Inf. Knowl. **5**, 73–75 (2006)
5. Yang, L., et al.: Multi-modality web video categorization. In: Proceedings of the International Workshop on Workshop on Multimedia Information Retrieval. ACM (2007)
6. Liu, H., et al.: A generic virtual content insertion system based on visual attention analysis. In: Proceedings of the 16th ACM International Conference on Multimedia. ACM (2008)
7. Gao, B., et al.: Web image clustering by consistent utilization of visual features and surrounding texts. In: Proceedings of the 13th Annual ACM International Conference on Multimedia. ACM (2005)
8. Noah, S.A.M., et al.: Exploiting surrounding text for retrieving web images. J. Comput. Sci. **4**(10), 842–846 (2008)
9. Rege, M., Dong, M., Hua, J.: Graph theoretical framework for simultaneously integrating visual and textual features for efficient web image clustering. In: Proceedings of the 17th International Conference on World Wide Web. ACM (2008)
10. Jeon, J., Lavrenko, V., Manmatha, R.: Automatic image annotation and retrieval using cross-media relevance models. In: Proceedings of the 26th Annual International ACM SIGIR Conference on Research and Development in Information Retrieval. ACM (2003)
11. Hu, T., et al.: Ontology-based cross-media retrieval technique. Comput. Eng. **8** (2009)
12. Xu, D.-Z., Yi, T.: Multi-ontology query based on ontology matching. Comput. Technol. Dev. **18**(11), 13–17 (2008)
13. Yi, Y., Tongqiang, G., Yueting, Z., Wenhua, W.: Cross-media retrieval based on synthesis reasoning model. J. Comput.-Aided Des. Comput. Graph. **9** (2009)

14. Yang, L., Fengbin, Z., Baoqing, J.: Research of cross-media information retrieval model based on multimodal fusion and temporal-spatial context semantic. J. Comput. Appl. **29**(4), 1182–1187 (2009)
15. Guo, Y.-T., Luo, B.: Image semantic annotation method based on multi-modal relational graph. J. Comput. Appl. **30**(12), 3295–3297 (2010)
16. Yang, Y., et al.: Harmonizing hierarchical manifolds for multimedia document semantics understanding and cross-media retrieval. IEEE Trans. Multimed. **10**(3), 437–446 (2008)
17. Yang, J., Li, Q., Zhuang, Y.: Octopus: aggressive search of multi-modality data using multifaceted knowledge base. In: Proceedings of the 11th International Conference on World Wide Web. ACM (2002)
18. Kim, H.H., Park, S.S.: Mediaviews: a layered view mechanism for integrating multimedia data. In: International Conference on Object-Oriented Information Systems. Springer, Heidelberg (2003)
19. Liu, J., et al.: Dual cross-media relevance model for image annotation. In: Proceedings of the 15th ACM International Conference on Multimedia. ACM (2007)
20. Zhuang, Y., et al.: Manifold learning based cross-media retrieval: a solution to media object complementary nature. J. VLSI Signal Process. Syst. Signal Image Video Technol. **46**(2–3), 153–164 (2007)
21. Martinec, R., Salway, A.: A system for image–text relations in new (and old) media. Vis. Commun. **4**(3), 337–371 (2005)
22. Wang, X.-J., et al.: Annotating images by mining image search results. IEEE Trans. Pattern Anal. Mach. Intell. **30**(11), 1919–1932 (2008)
23. Li, X., Snoek, C.G.M., Worring, M.: Learning social tag relevance by neighbor voting. IEEE Trans. Multimed. **11**(7), 1310–1322 (2009)
24. Harchaoui, Z., et al.: Large-scale image classification with trace-norm regularization. In: 2012 IEEE Conference on Computer Vision and Pattern Recognition. IEEE (2012)
25. Perronnin, F., et al.: Towards good practice in large-scale learning for image classification. In: 2012 IEEE Conference on Computer Vision and Pattern Recognition. IEEE (2012)
26. Wu, L., et al.: Distance metric learning from uncertain side information with application to automated photo tagging. In: Proceedings of the 17th ACM International Conference on Multimedia. ACM (2009)
27. Wu, P., et al.: Mining social images with distance metric learning for automated image tagging. In: Proceedings of the fourth ACM International Conference on Web Search and Data Mining. ACM (2011)
28. Wang, S., et al.: Multi-feature metric learning with knowledge transfer among semantics and social tagging. In: 2012 IEEE Conference on Computer Vision and Pattern Recognition. IEEE (2012)
29. Zhang, H.J., Su, Z.: Improving CBIR by semantic propagation and cross modality query expansion. In: Proceedings of the International Workshop on Multimedia Content-Based Indexing and Retrieval, Brescia (2001)
30. Jia, Y., Salzmann, M., Darrell, T.: Learning cross-modality similarity for multinomial data. In: 2011 International Conference on Computer Vision. IEEE (2011)
31. Socher, R., Fei-Fei, L.: Connecting modalities: semi-supervised segmentation and annotation of images using unaligned text corpora. In: 2010 IEEE Computer Society Conference on Computer Vision and Pattern Recognition. IEEE (2010)
32. Srivastava, N., Salakhutdinov, R.R.: Multimodal learning with deep boltzmann machines. In: Advances in Neural Information Processing Systems (2012)
33. Wu, F., et al.: Cross-media emantic representation via bi-directional learning to rank. In: Proceedings of the 21st ACM International Conference on Multimedia. ACM (2013)

34. Fang, H., et al.: From captions to visual concepts and back. In: Proceedings of the IEEE Conference on Computer Vision and Pattern Recognition (2015)
35. Vinyals, O., et al.: Show and tell: a neural image caption generator. In: Proceedings of the IEEE Conference on Computer Vision and Pattern Recognition (2015)
36. Frome, A., et al.: Devise: a deep visual-semantic embedding model. In: Advances in Neural Information Processing Systems (2013)
37. Hochreiter, S., Schmidhuber, J.: Long short-term memory. Neural Comput. 9(8), 1735–1780 (1997)
38. Karpathy, A., Joulin, A., Fei-Fei, L.F.: Deep fragment embeddings for bidirectional image sentence mapping. In: Advances in Neural Information Processing Systems (2014)
39. Jiang, B., et al.: Internet cross-media retrieval based on deep learning. J. Vis. Commun. Image Represent. 48, 356–366 (2017)
40. Zhang, H., et al.: A cross-media distance metric learning framework based on multi-view correlation mining and matching. World Wide Web 19(2), 181–197 (2016)
41. Yao, T., et al.: Supervised coarse-to-fine semantic hashing for cross-media retrieval. Digit. Signal Process. 63, 135–144 (2017)
42. Yao, T., et al.: Semantic consistency hashing for cross-modal retrieval. Neurocomputing 193, 250–259 (2016)
43. Xu, P., et al.: Cross-modal subspace learning for fine-grained sketch-based image retrieval. Neurocomputing 278, 75–86 (2018)
44. Zhai, X., Peng, Y., Xiao, J.: Cross-media retrieval by intra-media and inter-media correlation mining. Multimed. Syst. 19(5), 395–406 (2013)
45. Xie, L., Zhu, L., Chen, G.: Unsupervised multi-graph cross-modal hashing for large-scale multimedia retrieval. Multimed. Tools Appl. 75(15), 9185–9204 (2016)
46. Peng, Y., et al.: Semi-supervised cross-media feature learning with unified patch graph regularization. IEEE Trans. Circuits Syst. Video Technol. 26(3), 583–596 (2016)
47. Cao, X., et al.: Image retrieval and ranking via consistently reconstructing multi-attribute queries. In: European Conference on Computer Vision. Springer, Cham (2014)
48. Deng, J., Du, L., Shen, Y.-D.: Heterogeneous metric learning for cross-modal multimedia retrieval. In: International Conference on Web Information Systems Engineering. Springer, Heidelberg (2013)
49. Kim, K., et al.: Match graph construction for large image databases. In: European Conference on Computer Vision. Springer, Heidelberg (2012)
50. Huang, L., Peng, Y.: Cross-media retrieval via semantic entity projection. In: International Conference on Multimedia Modeling. Springer, Cham (2016)
51. Zhai, X., Peng, Y., Xiao, J.: Learning cross-media joint representation with sparse and semisupervised regularization. IEEE Trans. Circuits Syst. Video Technol. 24(6), 965–978 (2014)
52. Zhai, X., Peng, Y., Xiao, J.: Heterogeneous metric learning with joint graph regularization for cross-media retrieval. In: Twenty-Seventh AAAI Conference on Artificial Intelligence (2013)
53. Xie, L., et al.: Cross-modal self-taught learning for image retrieval. In: International Conference on Multimedia Modeling. Springer, Cham (2015)
54. Wang, J., et al.: Semi-supervised semantic factorization hashing for fast cross-modal retrieval. Multimed. Tools Appl. 76(19), 20197–20215 (2017)
55. Qi, J., Huang, X., Peng, Y.: Cross-media retrieval by multimodal representation fusion with deep networks. In: International Forum of Digital TV and Wireless Multimedia Communication. Springer, Singapore (2016)

Conversion from Lossless to Gray Scale Image Using Color Space Conversion Module

C. S. Sridhar[1]($\boxtimes$), G. Mahadevan[2], S. K. Khadar Basha[3], and P. Sudir[3]

[1] Bharathiar University, Coimbatore, India
sridhar_cs@yahoo.com
[2] ANNAI College, Kumbakonam, India
g_mahadevan@yahoo.com
[3] SJCIT, Chickballapur, India
basha_skb@rediffmail.com, sudirhappy@gmail.com

Abstract. Compression plays a vital role in better utilization of available memory space and bandwidth. Hence it is widely used for image archiving, image transmissions, multimedia applications, cloud-computing, E-commerce, etc. For effective presentation and utilization of digital image specific techniques are required to reduce the number of bits. There are two types of image namely Lossy and Lossless. Conversion from lossless to lossy images requires specialized and licensed software.

A software prototype module has been developed using color space conversion module using OpenCV which converts lossless image to lossy image after resizing, further convert's lossy image to gray scale image and vice-versa. The lossless images such as TIFF, BITMAP, and PNG are initially resized, converted to lossy image and further converted to Gray scale. On conversion to gray-scale it has been found that the number of bits has been considerably reduced. Performance analysis of software proto type is obtained by comparing the Decompressed image quality with respect to compression ratio, compression factor, PSNR (Power to Signal Noise Ratio) and MSE (Maximum Signal Error). The results of the developed color space module are compared and presented in the form of tables and snapshots.

Keywords: Color space module · OpenCV · Bitmap · PNG · Tiff · JPEG

1 Introduction

The compression of digital images has been largely focused in research area in recent years. New algorithms are developed for compression and compact representation of images. Digital image processing exploits the problem of redundancy by taking the exact amount of information required. It is a process that intends to yield a compact representation of an image and redundant data. The compression occurs by taking the benefits of redundancy information of the image. This property is exploited for achievement of better storage space and bandwidth. Compression of image is achieved by exploiting the properties of inter pixel, coding redundancy, psycho visual redundancy [4]. There are two types of image compression, namely lossless and lossy image

© Springer Nature Switzerland AG 2020
S. Smys et al. (Eds.): ICCVBIC 2019, AISC 1108, pp. 1137–1145, 2020.
https://doi.org/10.1007/978-3-030-37218-7_119

compression [6]. Bandwidth plays a very important role for transmission of images over wireless network. Hence it becomes very crucial for reducing the size of the image before transmission.

2 Methodology

2.1 Image Compression

Basically there are two types of image compression (Fig. 1).

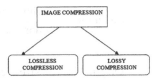

Fig. 1. Compression types

Lossless image compression reduces the size of an image without any quality loss. Hence it is used to retain the highest quality of the image by using more storage space. In case of Lossy image compression the size of the image reduces significantly by losing some data from the original. It plays a very important role for bloggers, photography, E-commerce and cloud computing.

2.2 Parameters for Measurement

It mainly depends on compressed file, algorithms and exploitation of redundancy in the image. The parameters considered for measurement are compression ratio, compression factor, percentage of saving, mean square error and peak signal to noise ratio.

Figure 2 shown below, describes the Block Diagram of the software prototype module developed using open source tool. Initially the incoming image is resized using

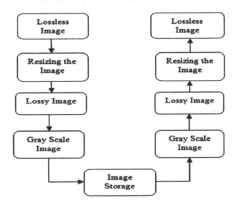

Fig. 2. Block diagram for lossless to Gray scale and vice-versa

interpolation technique. Conversion of Lossless Image to Lossy Image after Resizing [2, 3] is done. Lossy image is further converted to gray scale Image using color space conversion module. The converted Gray scale image is stored. Gray scale image will be converted into the original in the reverse process using decompression algorithm [9].

2.3 Equations

1. $$\text{compression ratio} = \frac{\text{compressed image size}}{\text{original image size}}$$

2. $$\text{compression factor} = \frac{\text{original image size}}{\text{compressed image size}}$$

3. $$\text{Percentage of saving in bits} = \frac{\text{Original image size - compressed image size}}{\text{original image size}} \times 100$$

4. $$MSE = \frac{1}{mn} \sum_{i=0}^{m-1} \sum_{j=0}^{n-1} [I(i,j) - K(i,j)]^2$$

Where I denotes Original image and K denotes decompressed image

5. $$\begin{aligned} PSNR &= 10 \cdot \log_{10}\left(\frac{MAX_I^2}{MSE}\right) \\ &= 20 \cdot \log_{10}\left(\frac{MAX_I}{\sqrt{MSE}}\right) \\ &= 20 \cdot \log_{10}(MAX_I) - 10 \cdot \log_{10}(MSE) \end{aligned}$$

Qualitative measurements are taken to compare different image enhancement algorithms. The parameters to be considered for measurements are chosen according to their requirements (Fig. 3).

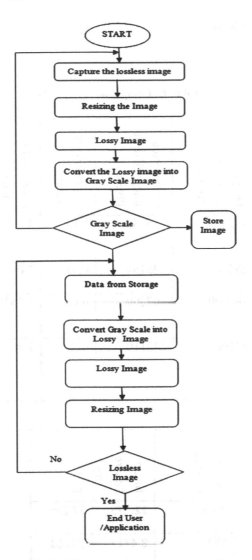

Fig. 3. Flow chart for proto type model

Compression ratio is a dimensionless quantity. It is measured as a ratio of compressed image to original image. Compression factor is a reciprocal of compression ratio. Percentage of saving is the number of bits saved after compression [6, 8].

The methodology of the flow chart for conversion from lossless to grayscale is as same as that explained in the block diagram.

2.4   Comparative Results

See Table 1.

Table 1. Comparisions of results

Image Format	Original Image Size in Pixels	Compressed Image Size in Pixels	Compression Ratio	Compression Factor	Percentage of saving
PNG	432000	276480	0.64	1.5625	36%
TIFF	262144	128164	0.48	2.045	51.11%
BITMAP	480000	307200	0.64	1.5625	36%
JPEG (Lossy)	341000	242110	0.71	1.408	29%

Image Format	Mean Square Error	PSNR
PNG	12.175	37.276
TIFF	8.4717	38.851
BITMAP	6.842	39.778
JPEG (Lossy)	19.8113	35.161

TIFF format

Original Image

Resized Imag

Lossy Image (Jpeg Image)

Gray Scale Image   De-Compressed Image

PNG Image

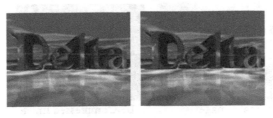

Original Image   Resized Image

Lossy Image (Jpeg Image)

Gray Scale Image   De-Compressed Image

BMP Image

Original Image   Resized Image

Lossy Image (Jpeg image)

Gray Scale Image   De-Compressed Image

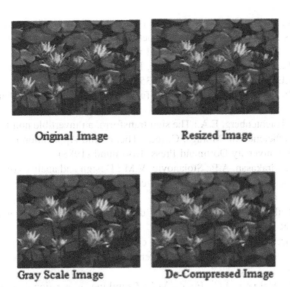

Original Image   Resized Image

Gray Scale Image   De-Compressed Image

3 Conclusion and Future Enhancement

This software prototype module can convert lossless image to gray scale image using Open source. From the results it is evident that there is a great savings in number of bits after converting to gray scale and highest quality of the De-compressed image has been maintained.

In future Color space conversion module can be implemented using hardware [7].

Acknowledgement. We kindly acknowledge NCH software for using trail version of pixillion image converter for comparison of results.

Compliance with Ethical Standards
   ✓ All authors declare that there is no conflict of interest.
   ✓ No humans/animals involved in this research work.
   ✓ We have used our own data.

References

1. Sridhar, C.S., Mahadevan, G., Khadar basha, S.K.: Design and implementation of a color space conversion module using open- computer vision. J. Adv. Res. Dyn. Control. Syst. Scopus SJR, vol. 12-special issue, 2321–2329 (2017). ISSN 1943-023X
2. Sridhar, C.S., Mahadevan, G., Khadar basha, S.K.: Compression of image using hybrid combination of haar wavelet transform and color space model in noisy environment. International journal of Engineering and technology **7**(1.5), 283–285 (2018). 12-special issue

3. Sridhar, C.S., Mahadevan, G., Khadar Basha, S.K.: Transmission of images using image compressions for low powered wireless networks. J. Adv. Res. Dyn. Control. Syst. 2158–2165 (2018). ISSN 1943-023x. 12-special issue
4. Donoho, D.L., Vetterli, M., Devore, R.A., Daubechies, I.: Data Compression and harmonic analysis. IEEE Trans. Inf. Theory **44**, 2435–2476 (1998)
5. Haar, A.: Zur theorieder orthogonalen funktionensysteme. Math. Ann. **69**(3), 331–371 (1910)
6. Besslich, P., Trachtenberg, E.A.: The sign transform: an invertible non-linear transform with quantized coefficients. In: Moraga, C. (ed.) Theory and Application of Spectral Technique, pp. 150–160. University Dortmund Press, Dortmund (1988)
7. Chen, F., Chandrakasan, A.P., Stojanovic, V.M.: Design and analysis of a hardware-efficient compressed sensing architecture for data compression in wireless sensors. IEEE J. Solid-State Circuits **47**(3), 744–756 (2012)
8. Díaz, M., et al.: Compressed sensing of data with a known distribution. Appl. Comput. Harmon. Anal. **45**(3), 486–504 (2018)
9. Duda, K., Turcza, P., Marszałek, Z.: Low complexity wavelet compression of multichannel neural data. In: 2018 International Conference on Signals and Electronic Systems (ICSES). IEEE (2018)
10. Nassar, M.R., Helmers, J.C., Frank, M.J.: Chunking as a rational strategy for lossy data compression in visual working memory. Psychol. Rev. **125**(4), 486 (2018)
11. Sahin, A., Martin, O., O'Hare, J.J.: Intelligent data compression. U.S. Patent Application No. 10/013,170
12. Parvanova, R., Todorova, M.: Compression of images using wavelet functions. Appl. Comput. Technol. ACT, 90 (2018)
13. Zemliachenko, A.N., Abramov, S.K., Lukin, V.V., Vozel, B., Chehdi, K.: Preliminary filtering and lossy compression of noisy remote sensing images. In: Image and Signal Processing for Remote Sensing XXIV, 10789, p. 107890 V. International Society for Optics and Photonics, October 2018
14. Gregor, K., Besse, F., Rezende, D. J., Danihelka, I., Wierstra, D.: Towards conceptual compression. In: Advances In Neural Information Processing Systems, pp. 3549–3557 (2016)

Iris Recognition Using Bicubic Interpolation and Multi Level DWT Decomposition

C. R. Prashanth[1(✉)], Sunil S. Harakannanavar[2], and K. B. Raja[3]

[1] Dr. Ambedkar Institute of Technology, Bangalore, Karnataka, India
Prashanthcr.ujjani@gamil.com
[2] S. G. Balekundri Institute of Technology, Belagavi, Karnataka, India
Sunilsh143@gmail.com
[3] University Visvesvaraya College of Engineering, Bangalore, Karnataka, India

Abstract. Iris recognition is endorsed as the most important one as it takes a live tissue verification. Iris recognition model is presented having multiple descriptors viz., DWT, ICA and BSIF for producing hybrid features. The 2nd level DWT is performed to produce unique iris features. The high frequency bands of DWT are employed by performing bicubic interpolation to generate new sub-bands. Inverse DWT is applied on low frequency and higher frequency subbands of DWT and interpolation techniques for reconstruction of images. The features are selected based on ICA in collabration with BSIF for extraction of features. Test image and database features are matched in the presence of ED to accept or reject the sample on CASIA database. The iris model performed better accuracy compared with other iris biometric system.

Keywords: Biometrics · Physiological · Bicubic interpolation · Behavioral · Euclidean distance

1 Introduction

Biometric trait "iris" falls under physiological characteristics of humans to authenticate an individual. Unique features of iris texture provide a better for authentication of person. Compared to other biometric traits, iris recognition is considered as most important one as it takes a live tissue verification. Reliability of system of the authentication depends on the acquired signal and the recorded signal of a live person, which cannot be manufactured as a template. The human eye consists of pupil, iris and sclera [3]. The texture of iris is categorized as quite complex and unique, so it is considered as one of the secured biometric trait to identify an individual. It is more reliable and accurate for identification. So it is widely accepted in national identification programs for processing of millions of peoples data. It is a muscular portion within pupil. It is observed that no two irises are same even for the twins of parents. It represents point between sclera and pupil and having unusual structure that generates numerous features (crypts, collarets, freckles, corona etc). Here, iris recognition model is presented. The multistage decomposition of DWT is performed for extraction of features. Bicubic interpolation is applied on higher components of 1st stage DWT to produce new set of sub bands. Inverse DWT is used to reconstruct the image. The BSIF

© Springer Nature Switzerland AG 2020
S. Smys et al. (Eds.): ICCVBIC 2019, AISC 1108, pp. 1146–1153, 2020.
https://doi.org/10.1007/978-3-030-37218-7_120

and ICA are adopted for obtaining hybrid features. The features are matched using ED classifier.

The structure of paper continues in such a way that the existing methodologies of iris recognition are explained in Sect. 2. Iris model is described in Sect. 3. Performance analysis is explained in Sect. 4. Proposed iris work is concluded in Sect. 5.

2   Related Work

Neda et al. [1] adopted multilayer perceptron NN to recognize the human iris. The two dimensional Gabor kernel technique was employed to extract the detailed coefficients of iris and experiments are conducted on CASIA v-3. The model needs to work with the combination of fuzzy systems and MLPNN-PSO methods with different classification methods. Syed et al. [2] described HT and optimized light version technique to recognize the iris samples in smart phones. The generated iris features are compared with test features using HD on CASIA v4 dataset to measure accuracy. Bansal et al. [4] designed adaptive fuzzy filter to remove the noise. The extracted features using Haar transformation are compared with test features using HD on IITD database. However, the model was developed only for the images corrupted with impulse noise and the method needs to be extended for other types of noises to improve the recognition rate. Li [5] applied HT and calculus techniques to obtain the location of iris from the eye samples. Using Gabor filter bank iris features are extracted. The coefficients of test and database are matched in presence of ED. The experiments are tested on own iris database consisting of 8 individuals and 16 different samples of iris. However the model is tested only on limited samples of iris, the recognition rate decreases for large scale sample of database. Muthana et al. [6] adopted PCA and Fourier descriptors to obtain genuine features of image. The significant details of the iris patterns were represented by high spectrum coefficients whereas low spectrum coefficients provide the general description of iris pattern. The performance of iris model was evaluated on three classifiers viz., cosine, Euclidean and Manhattan classifiers to accept or reject the image samples. Some of the existing algorithms for iris recognition is tabulated in Table 1.

Table 1. Summary of different Existing methodologies for Iris Recognition

Authors	Techniques	Limitations
Kiran et al. [7]	BSIF and SIFT	Iris model needs to be improved for larger database and new deep method should be developed to get accuracy
Kavita et al. [8]	LGW and HAAR Wavelet.	Work to be carried for other artificial intelligence techniques to improve FRR
Aparna et al. [9]	Combinations of HAAR transform and sum block algorithms are used.	Algorithms should develop to robust in its results
Ximing et al. [10]	2D Gabor Wavelet transform	Work to be carried for other artificial intelligence techniques to improve accuracy

3   Proposed Methodology

In proposed model, multistage decomposition of DWT is performed for extraction of unique iris features. Bicubic interpolation is applied on higher components of 1st stage DWT for the production of new subbands. Further, inverse DWT is used to reconstruct the image. The final features of iris are generated using BSIF and ICA filter. Finally features are matched using ED classifier on CASIA database. The iris model is given in Fig. 1.

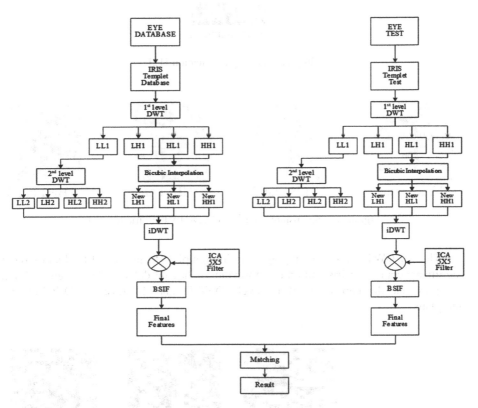

Fig. 1. Block diagram of proposed model

The sample of CASIA iris images are considered to extract iris region. The morphological operations are carried out on eye image to obtain segmented iris of regions (left and right), each side is having a size of 40 pixels. The obtained regions are concatenated to form a template and is shown in Fig. 2.

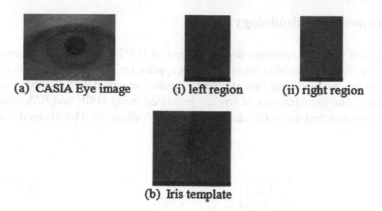

(a) CASIA Eye image   (i) left region   (ii) right region

(b) Iris template

Fig. 2. Iris template creation [12]

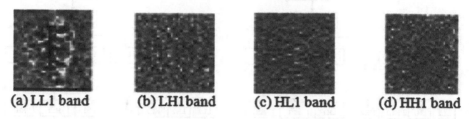

(a) LL1 band   **(b) LH1 band**   **(c) HL1 band**   **(d) HH1 band**

Fig. 3. Decomposition of 1st level 2D-DWT on iris template

The 2D DWT [9, 10] is employed on iris template to produce LL1 band having significant content of iris and other information is having detailed data. Figures 3 and 4, shows the decomposition of 1st level 2D-DWT and 2nd level 2D-DWT on iris template.

(a) LL2 Band   **(b) LH2 Band**   **(c) HL2 Band**   **(d) HH2 Band**

Fig. 4. Decomposition of 2nd level 2D-DWT on iris template

Bicubic interpolation is employed on detailed region of DWT for generation of new sub-band images. The IDWT is considered for reconstruction of images from sub-bands having smaller bandwidths. ICA having 5X5 with size 10 are adopted to filter iris template inturn produces binary images [11]. Binariezed features are extracted using BSIF and the sample image of BSIF is given in Fig. 5.

Fig. 5. BSIF output

Finally the hybrid features are matched in presence of ED classifier for identification of an individual using Eq. 1 [12].

$$\text{Euclidean Distance} = \sum_{i=1}^{M}(P_i - q_i)^2 \qquad (1)$$

Where, 'M' is no. of coefficients, P_i denotes coefficient values of database and q_i is related to test images.

4   Experimental Results

The accuracy of iris model is evaluated by considering CASIA iris database [13–15]. First 6 images of the database are used to calculate FAR, FRR and TSR [14]. The remaining image of the database is used as out of database to evaluate FAR [16–19]. Table 2 gives the OTSR and MTSR for different combinations of PID and POD's. It is seen that, the values of OTSR and EER varies and doesn't maintain constant for increasing in PID's when POD kept constant. The MTSR for different combinations of [20, 30, 40, 50, 60]:[30] PIDs and POD results 100%.

Figure 6 shows the plot of performance parameters for 40:30 combination of PID and POD. For the iris model, as the threshold value varies the FRR decreases and FAR, TSR increases respectively. The TSR value of 95% and EER of 5% is recorded for the proposed iris model for 40:30 of PID and POD values. Table 3 gives the OTSR and MTSR for different values of PID and POD's. It is seen that, the values of OTSR and EER remains 93%, and 6.67% respectively for increasing in POD's when PID kept constant. The MTSR for different combinations of [30]:[20, 30, 40, 50, 60] PIDs and POD results 100%.

Table 2. EER, OTSR and MTSR for different values of PID and POD's

PID	POD	EER	OTSR	MTSR
20	30	10	90	100
30	30	6.67	93.33	100
40	30	5	95	100
50	30	7	93	100
60	30	10	90	100

Figure 7 shows the plot of performance parameters for 30:40 combination of PID and POD. For the iris model, as the threshold value varies the FRR decreases and FAR, TSR increases respectively. The TSR value of 93.33% and EER of 6.67% is recorded for the proposed iris model for 30:40 combinations of PID and POD.

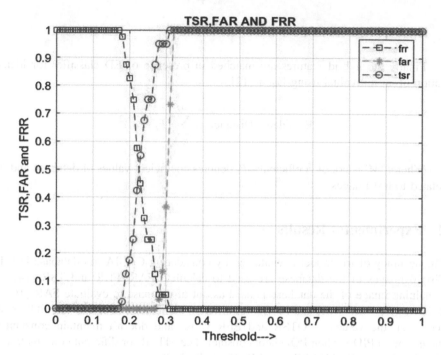

Fig. 6. Simulation of Performance parameters for 40:30 (PID:POD)

Table 3. EER, OTSR and MTSR for different values of PID and POD's

PID	POD	EER	OTSR	MTSR
30	20	6.67	93.33	100
30	30	6.67	93.33	100
30	40	6.67	93.33	100
30	50	6.67	93.33	100
30	60	6.67	93.33	100

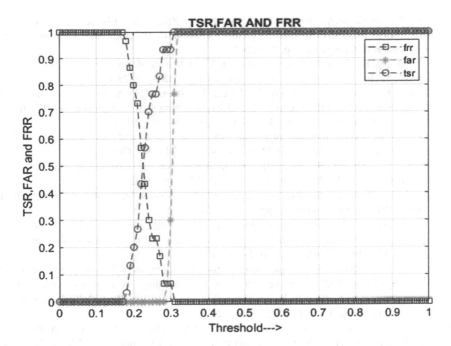

Fig. 7. Simulation of Performance parameters for 30:40 (PID:POD)

5 Conclusion

In this paper, 2nd level DWT is performed to produce unique iris features. The high frequency bands of DWT are employed by performing bicubic interpolation to generate new sub-bands. Inverse DWT is applied on lower and higher frequency subbands of DWT and interpolation techniques for reconstruction of images. The features are selected based on ICA in collabration with BSIF for extraction of features. Test image and database features are matched in the presence of ED to accept or reject the sample on CASIA database. In future, model needs to be adopted with fusion of different domains viz spatial and transform domains to improve its recognition.

Compliance with Ethical Standards

✓ All authors declare that there is no conflict of interest.
✓ No humans/animals involved in this research work.
✓ We have used our own data.

References

1. Ahmadi, N., Akbarizadeh, G.: Hybrid robust iris recognition approach using Iris image pre-processing, two-dimensional Gabor features and multi-layer perceptron neural network/PSO. IET Biometrics **7**(2), 153–162 (2018)

2. Ali, S., Shah, M.A., Javed, T., Abdullah, S., Zafar, M.: Iris recognitionsystem in smartphones using light version recognition algorithm. In: IEEE International Conference on Automation & Computing, pp. 1–6 (2017)
3. Choudhury, B., Then, P., Issac, B., Raman, V., Haldar, M.K.: A survey on biometrics and cancellable biometrics systems. Int. J. Image Graph. 18(1), 1–28 (2018)
4. Bansal, R., Juneja, A., Agnihotri, A., Taneja, D., Sharma, N.: A fuzzified approach to Iris recognition for mobile security. In: IEEE International Conference on Electronics, Communication and Aerospace Technology, pp. 246–251 (2017)
5. Li, Z.: An Iris recognition algorithm based on coarse and fine location. In: IEEE International Conference on Big Data Analysis, pp. 744–747 (2017)
6. Hamd, M., Ahmed, S.: Biometric system design for Iris recognition using intelligent algorithms. Int. J. Modern Educ. Comput. Sci. 10(3), 9–16 (2018)
7. Joshi, K., Agrawal, S.: An Iris recognition based on robust intrusion detection. In: Annual IEEE India Conference, pp. 1–6 (2016)
8. Arunalatha, J.S., Rangaswamy, Y., Shaila, K., Raja, K.B., Anvekar, D., Venugopal, K.R., Iyengar, S.S., Patnaik, L.M.: IRHDF: Iris recognition using hybrid domain features. In: Annual IEEE India Conference, pp. 1–5 (2015)
9. Gale, A.G., Salankar, S.S.: Evolution of performance analysis of Iris recognition system by using hybrid method of feature extraction and matching by hybrid classifier for Iris recognition system. In: IEEE International Conference on Electrical, Electronics and Optimization Techniques, pp. 3259–3263 (2016)
10. Alkassar, S., Woo, W.L., Dlay, S.S., Chambers, J.A.: Robust sclera recognition system with novel sclera segmentation and validation techniques. IEEE Trans. Syst. Man Cybern. Syst. 47(3), 474–486 (2017)
11. Zhao, D., Fang, S., Xiang, J., Tian, J., Xiong, S.: Iris template protection based on local ranking. Int. J. Secur. Commun. Netw. 2018, 1–9 (2018)
12. Kaur, B., Singh, S., Kumar, J.: Robust Iris recognition using moment invariants. Wireless Pers. Commun. 99(2), 799–828 (2018)
13. Bhagat, K., Deshmukh, R., Patil, P., Kirange, D.K., Waghmare, S.: Iris recognition using radon transform and GLCM. In: IEEE International Conference on Advances in Computing, Communications and Informatics, pp. 2257–2263 (2017)
14. Wei, X., Wang, H., Wan, H., Scotney, B.: Multiscale feature fusion for face identification. In: IEEE International Conference on Cybernetics, pp. 1–6 (2017)
15. Wang, Q., Liu, Z., Tong, S., Yang, Y., Zhang, X.: Efficient Iris localization via optimization model. Hindawi Math. Probl. Eng. 2017, 1–9 (2017)
16. Joshi, K., Agrawal, S.: An Iris recognition based on robust intrusion detection. In: IEEE Annual India Conference, pp. 1–6 (2016)
17. Vishwakarma, D.K., Jain, D., Rajora, S.: Iris detection and recognition using two-fold techniques. In: International Conference on Computing, Communication and Automation, pp. 1046–1051 (2017)
18. Zhang, W., Xue, P., Wang, C.: Application of convolution neural network in Iris recognition technology. In: IEEE International Conference on Systems and Informatics, pp. 1169–1174 (2017)
19. Samant, P., Agarwal, R., Bansal, A.: Enhanced discrete cosine transformation feature based iris recognition using various scanning techniques. In: IEEE International Conference on Electrical, Computer and Electronics, pp. 660–665 (2017)

Retinal Image Classification System Using Multi Phase Level Set Formulation and ANFIS

A. Jayachandran[1]([⊠]), T. Sreekesh Namboodiri[2], and L. Arokia Jesu Prabhu[3]

[1] Department of CSE, Presidency University, Bangalore, India
ajayachandran@presidencyuniversity.in
[2] Department of CSE, PSN College of Engineering and Technology, Tirunelveli, India
tsreekeshn@yahoo.com
[3] Department of CSE, Sri Sakthi Institute of Engineering and Technology, Coimbatore, India
arokiajeruprabhu@gmail.com

Abstract. Computerized systems for eye maladies distinguishing proof are significant in the ophthalmology field. Conservative systems for the detection of eye disease depend on labour-intensive awareness of the retinal segments. This work exhibits another directed technique for hemorrhages discovery in advanced retinal images. This strategy utilizes an ANFIS plot for pixel association and registers a 5-D vector made out of dim dimension and Cross Section Profie (CSP) Study-constructed highlights for pixel portrayal. Classification of diseases is a crucial aspect in eye disease categorization through image processing techniques. The categorization of diseases according to pathogen groups is a significant research domain and potentially a challenging area of work. Various classification techniques for single as well as multiple diseases is identified. Classification and detection are very similar, but in classification primary focus is on the categorization of various diseases and then the classification according to various pathogen groups.

Keywords: DR · CSP · ANFIS · Classification · Retinal imaging · Feature extraction

1 Introduction

Ongoing advances in PC innovation have empowered the advancement of various sorts of fully automatic computer based diagnosis Presently, restorative picture investigation is an exploration zone that pulls in a ton of worry from the two researchers and doctors. Electronic restorative imaging and examination techniques utilizing different modalities have encouraged early determination [1, 2]. The presence of hemorrhages is commonly used to analyze DR or hypertensive retinopathy by utilizing the arrangement plan of Scheie. Notwithstanding distinguishing microaneurysms, it is troublesome for ophthalmologists to discover them in noncontrast fundus pictures. The unpredictability found in a microaneurysm picture is extraordinarily low; thus, ophthalmologists for the

© Springer Nature Switzerland AG 2020
S. Smys et al. (Eds.): ICCVBIC 2019, AISC 1108, pp. 1154–1161, 2020.
https://doi.org/10.1007/978-3-030-37218-7_121

most part recognize microaneurysms by utilizing fluorescein angiograms. In any case, it is hard to utilize fluorescein as a difference mode for diagnosing all the medicinal examinees exposed to mass screening.

A few robotized methods have been accounted for to evaluate the adjustments in morphology of retinal pots demonstrative of retinal ailments. A portion of the procedures quantity the pot morphology as a normal esteem speaking to the whole vessel organize, e.g., normal tortuosity [3]. Anyway as of late, vessel morphology estimation explicit to supply routes or veins was observed to be related with malady. Contours usually contain key visual information of an image. In computer vision, contours have been widely used in many practical tasks. Although quite a few contour detection methods have been developed over the past several decades, contour detection is still a challenging problem in the image field. Among the non-learning approaches, many early methods, such as the famous Canny detector, find contours by extracting edges where the brightness or color changes sharply. However, such methods usually employ regular kernels, e.g., Gaussian filter and Gabor filter, to measure the extents of local changes, and thus can hardly deal with textures. To address this problem, many texture suppression methods have been proposed. Examples are the method based on non-classical receptive field inhibition, the method based on sparseness measures, the method based on surround-modulation, etc. It has been validated that texture suppression can help improve contour detection performance. Nonetheless, these methods still mainly use low-level local features. Moreover, some of them are computationally heavy, which leads to difficulties in practical applications [4–6].

In classification, SVM is a popular supervised machine learning technique which is designed primarily to solve binary classification tasks. In an SVM, the classification task is performed by finding the optimum separating hyperplane with maximum margin using structural risk minimization (SRM) principle [7]. Since the present study deals with four binary classification tasks, therefore SVM is chosen in this work. In the case of non-linear SVMs, a method to map the training sample to a high dimension feature space exists. For mapping purpose, different kernel functions are used and the mapping operation is done satisfying Mercer's theorem. The different kernel functions present in an SVM are linear, polynomial and radial basis function. k-nearest neighbour (kNN) is another well-known machine learning technique widely used for solving different classification problems. The advantages of using kNN algorithm are its robustness and relatively ease of implementation. A kNN assigns class to the test data by majority voting technique, depending on the most frequent class of its nearest neighbours [8–10]. In a kNN algorithm, there are two important parameters. The first one is the distance parameter and the second one is the kvalue. In this study, the performance of the classifier is evaluated considering Euclidean distance parameter and by varying the k value from 2 to 10. One downside to these methodologies is their reliance upon strategies for finding the beginning stages, which should dependably be either at the optic nerve or at in this way recognized branch focuses. Veins were distinguished by methods for scientific morphology [11, 12]. Coordinated channels were connected related to different methods. The planned discharge identification strategy is displayed in area 2, the test results are exhibited in segment 3 and the end in segment 4.

2 Hemorrhages Detection System

In this work, we propose a system for optic disc localization and its segmentation. The method can be used as a initial step in the design of a CAD system for many retinal diseases. The proposed method is based on ANFIS system. The optic disc is located using the brightest pixel in the green frequency of the retinal image, and the region of detection algorithm is employed to precisely extract the optic disc. The segmented optic disc is approximated to the nearest possible circle to calculate its diameter in terms of numbers of pixels. The method is designed and implemented using a hospital fundus database and it is quantitatively evaluated by the opthalmology domain experts on seven publicaly available fundus databases namely, STARE and DRIVE.

In this study, for designing an optimized CNN to segment the ROI areas from the background, GOA is utilized. The duty of the GOA here is to justify the number of hyper-parameters of CNN to achieve better performance than the manual justification. The candidate solutions in this configuration are a sequence of integers. For preventing the system error, the minimum (min) and the maximum (max) limitations are considered 2 and the size of the sliding window, respectively. The minimum value 2 here represents that the allowed min value for the max pooling so that there does not exist any lower size. It should be noted that this optimization problem has an inequality constraint such that the value of the sliding window should be less than the input data.

The half-value precision of CNN is selected as the cost function on a breast cancer validation process. Because of the presence of CNN and GOA, the general configuration has a high computational cost; because each agent of the swarm of the CNN needs to be trained on the breast cancer dataset by applying backpropagation technique for 1000 iterations. After initializing and evaluating the cost of the agents, the position of the search agents has been updated based on the GOA parameters and the process repeats until the stop criteria have been achieved. The preprocessing state is given in Fig. 1.

3 Extraction of Features

The point of the element extraction arrange is pixel portrayal by methods for an element vector, a pixel portrayal as far as some quantifiable estimations which might be effectively utilized in the order stage to choose whether pixels have a place with a genuine vein or not. A clear broadness first filter count is associated for the calculation of grayscale morphological multiplication. Pixels of the image are taken care of progressively, and diverged from their 8-neighbors. In case all neighbors have a lower control, the pixel itself is a LMR. In case there is a neighboring pixel with higher power, the present pixel may not be a generally extraordinary. Pixels of a LMR are seen as solely as possible hopefuls, and the pixel with the most outrageous last score will address the territory; this strategy is suggested as non most prominent camouflage. To inspect the encompassing of a solitary most extreme pixel in a MA applicant locale, the force esteems along discrete line fragments of various introductions, whose focal pixel is the competitor pixel, are recorded. Along these lines, a lot of cross-sectional power profiles is acquired. On the acquired cross-area profiles a pinnacle discovery step

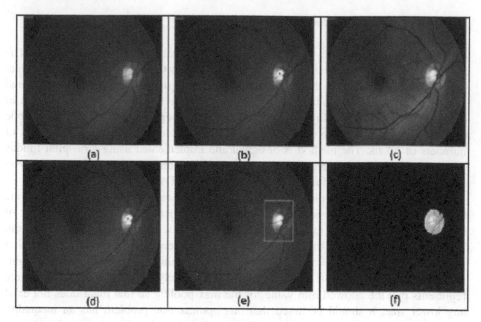

Fig. 1. Processing steps. (a) Input image [4] (b) De-blurred green channel image (c) Contrast enhanced image (d) Brightest pixel detected (marked as a dot in the optic disc) (e) Marking of Region-of-interest (f) Extraction of foreground object

is performed. Our point is to choose whether a pinnacle is available at the focal point of the profile, i.e., at the area of the competitor point for a particular heading. A few properties of the pinnacle are determined, and the last list of capabilities comprises of a lot of factual estimates that show how these qualities shift as the introduction of the cross-segment is evolving. Along these lines, the variety of critical qualities, for example, symmetry and state of the structure, and its distinction from the foundation might be numerically communicated.

According to the two clues, lumen and illumination high-light, we processed RETINAL images with histogram equilibrium to enhance image contrast as the first step. Obviously there are others image manipulation enhanced algorithms, such as linear contrast broadening, retinex model and so on. However all the transformation must be homeomoriphic, since related research [11] showed that in a image, the shape information of objectives was represented through levelset of the image, in another words the shape information of objectives was included in the isolux lines of image. In this paper, we adopted histogram equilibrium to improve image contrast. It makes the illumination of tumors prominent.

Allow for the whole cell's boundary, which imbedding in the black background. We adopt multiphase level set formulation, in theory, two boundary can indicate 4 regions, according to analyzing, that the region of interest is whole included in the bigger boundary, so we use two boundary represent 3 regions in this work. Moreover, this method shows a desirable performance in the presence of intensity inhomogeneities for both synthetic and real images. Therefore, the method is good at intensity inhomogeneities, suitable for the illumination high-light. Simultaneously, the method is appropriate for another clue, the lumen highlight. We consider the three-phase RETINAL image domain is segmented into three disjoint regions 1, 2 and 3. Meanwhile, two level set function 1 and 2 is used to define the three regions.

4 ANFIS System

ANFIS is one of mixture insightful neuro-fluffy induction frameworks and it working under Takagi-Sugeno-type fluffy derivation framework. With respect to the forecast of directing plot for portable robot we accept the fluffy derivation framework under thought of four information sources i.e., Front hindrance remove, Right deterrent separation, Left impediment separate, target point which are gathered from sensors and each information variable has three participation functions(MF) individually, at that point a Takagi-Sugeno-type fuzzy inference framework., if-then rules is set up as per Eq. (1)

$$f_n = p_n X_1 + q_n X_2 + r_n X_3 + s_n X_4 + u_n \tag{1}$$

In ANFIS Classifier, Gradient drop and Backpropagation computations are used to alter the parameters of enlistment limits (fleecy sets) and the heaps of defuzzification (neural frameworks) for cushioned neural frameworks. ANFIS applies two techniques in reviving parameters. The ANFIS is a FIS completed in the arrangement of a flexible cushy neural framework. It combines the express data depiction of a FIS with the learning force of ANNs. The objective of ANFIS is to fuse the best features of fleecy structures and neural framework. The classifier is utilized to recognize the irregularities in the retinal cerebrum images [12–15]. By and large the info layer comprise of seven neurons relating to the seven highlights. The yield layer comprise of one neuron demonstrating whether the retinal is of an ordinary cerebrum or strange and the concealed layer changes as indicated by the quantity of principles that give best acknowledgment rate for each gathering of highlights. The framework is a blend of neural system and fluffy frameworks in which that neural system is utilized to decide the limits of fluffy framework. It to a great extent evacuates the necessity for physical enhancement of limits of fluffy scheme. The framework by the learning abilities of neural system and with the benefits of the standard base fluffy framework can enhance the execution altogether and neuro-fluffy framework can likewise give an instrument to consolidate past perceptions into the grouping process. In neural system the preparation basically fabricates the framework.

5 Experimental Results

5.1 Performance Measures

So as to evaluate the execution of the planned strategy on a experimental image, the subsequent division is contrasted with its relating best quality level picture.

Five performance metrics have been used for the analysis the proposed method as per the standard evaluation metrics such as sensitivity, PPV, NPV, specificity and accuracy and the segmentation results of proposed method in hemorrhages is given in Fig. 2.

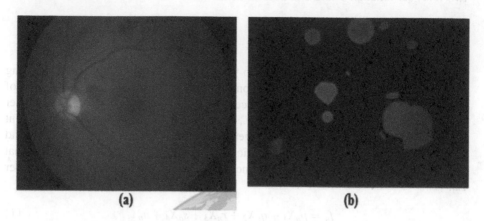

Fig. 2. (a) Fundus image [7] (b) Hemorrhage obtained

5.2   Evaluation of Proposed System

This technique was assessed on DRIVE and STARE database pictures with accessible best quality level pictures. Since the pictures' dim foundation outside the FOV is effectively distinguished. Affectability, explicitness, positive prescient esteem, negative prescient esteem and exactness esteems were figured for each picture considering FOV pixels as it were. Since FOV covers are not accommodated STARE pictures, they were created with a surmised breadth of 650 * 550. The trial consequences of affectability, particularity, PPV, NPV and exactness of DRIVE information base is appeared in Fig. 3 and STARE information base is appeared in Fig. 4.

6   Conclusion

The hemorrhages are difficult to recognize from foundation varieties since it normally low differentiation. Programmed identification of discharge can be befuddled by other dim zones in the picture, for example, the veins, fovea, and microaneurysms. It have a variable size and regularly they are small to the point that can be effectively mistaken for the pictures clamor or microaneurysms and no standard database that characterize discharge by shape. The most false discovery is the situation when the veins are contiguous or covering with hemorrhages. So the powerful recognition of hemorrhages strategy is required. The proposed methodology contains highlight extraction and characterization. The advantage of the framework is to help the doctor to settle on a ultimate choice without vulnerability. Our proposed discharge division strategy does not require any client intercession, and has steady execution in both typical and strange pictures. The proposed drain identification calculation delivers over 96% of division precision in both Databases.

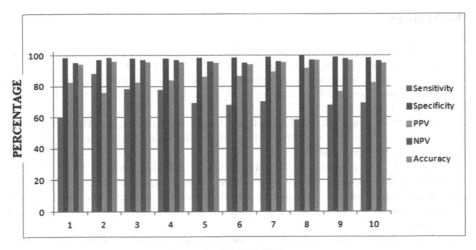

Fig. 3. Classification results of proposed method using DRIVE database

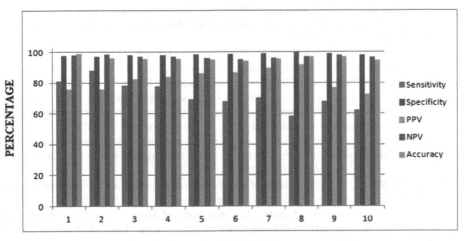

Fig. 4. Classification results of proposed method STARE database

Compliance with Ethical Standards
✓ All authors declare that there is no conflict of interest.
✓ No humans/animals involved in this research work.
✓ We have used our own data.

References

1. Doi, K.: Computer-aided diagnosis in medical imaging: historical review, current status and future potential. Comput. Med. Imaging Graph. **31**, 198–211 (2007)
2. Sundaraj, G.K., Jayachandran, A.: Abnormality segmentation and classification of multi model brain tumor in MR images using Fuzzy based hybrid kernel SVM. Int. J. Fuzzy Syst. **17**(3), 434–443 (2016)
3. Ronald, P.C., Peng, T.K.: A Textbook of Clinical Ophthalmology: A Practical Guide to Disorders of the Eyes and Their Management, 3rd edn. World Scientific Publishing Company, Singapore (2003)
4. Mahiba, C., Jayachandran, A.: Severity analysis of diabetic retinopathy in retinal images using hybrid structure descriptor and modified CNNs. Measurement **135**, 762–767 (2019)
5. Sukkaew, L., Makhanov, B., Barman, S., Panguthipong, S.: Automatic tortuosity-based retinopathy of prematurity screening system. IEICE Trans. Inf. Syst. **91**(12), 2868–2874 (2008)
6. Jayachandran, A., Dhanasekaran, R.: Automatic detection of brain tumor in magnetic resonance images using multi-texton histogram and support vector machine. Int. J. Imaging Syst. Technol. **23**(2), 97–103 (2013)
7. Niemeijer, M., Xu, X., Dumitrescu, A., Gupta, P., Ginneken, B., et al.: Automated measurement of the arteriolar-to-venular width ratio in digital color fundus photographs. IEEE Trans. Med. Imaging **30**(11), 1941–1950 (2011)
8. Vickerman, M., Keith, P., Mckay, T.: VESGEN 2D: automated, user-interactive software for quantification and mapping of angiogenic and lymphangiogenic trees and networks. Anat. Record **292**, 320–332 (2009)
9. Jayachandran, A., Dhanashakeran, R., Sugel Anand, O., Ajitha, J.H.M.: Fuzzy information system based digital image segmentation by edge detection. In: 2010 IEEE International Conference on Computational Intelligence and Computing Research, 28–29 December 2010
10. Leandro, J.J., Cesar, J.R., Jelinek, H.F.: Blood vessels segmentation in retina: preliminary assessment of the mathematical morphology & the wavelet transform techniques. In: Proceding on Computer Graphics and Image Processing, pp. 84–90 (2001)
11. Jayachandran, A., Dhanasekaran, R.: Multi class brain tumor classification of RETINAL images using hybrid structure descriptor and fuzzy logic based RBF kernel SVM. Iranian J. Fuzzy Syst. **14**(3), 41–54 (2017)
12. Zana, F., Kelin, J.C.: Segmentation of vessel-like patterns using mathematical morphology and curvature evaluation. IEEE Trans. Image Process. **10**, 1010–1019 (2001)
13. Al-Rawi, M., Karajeh, H.: Genetic algorithm matched filter optimization for automated detection of blood vessels from digital retinal images. Comput. Methods Programs Biomed. **87**, 248–253 (2007)
14. Jayachandran, A., Dhanasekaran, R.: Severity analysis of brain tumor in RETINAL images using modified multi-text on structure descriptor and kernel-SVM. Arabian J. Sci. Eng. **39**(10), 7073–7086 (2014)
15. Jayachandran, A., Dhanasekaran, R.: Brain tumor detection using fuzzy support vector machine classification based on a texton co-occurrence matrix. J. Imaging Sci. Technol. **57**(1), 10507-1–10507-7 (2013)

FSO: Issues, Challenges and Heuristic Solutions

Saloni Rai and Amit Kumar Garg[✉]

Electronics and Communications Department,
Deenbandhu Chhotu Ram University of Science and Technology,
Murthal, Haryana, India
salonirai4@gmail.com, garg_amit03@yahoo.co.in

Abstract. Most dominant and viable advanced technology for broadband wireless applications is free space communication (FSO) that provides unlicensed spectrum, very high data rate and has no electromagnetic interference. Atmospheric turbulences has great effect on FSO link and quality of light beam gets Detroit while propagating to long distance through the atmosphere. The signal exhibits a random fluctuation in the existence of atmospheric turbulence which degrades functioning of system. Bit Error Rate (BER), Outage Probability (OP) and Quality Factor are most important factors that decide quality of transmission of any data. Free space optical communication concept is introduced in this paper along with its issues, challenges and heuristic solutions. The findings in the paper demonstrate deployment of different methods for enhancing the system performance. It is seen that by using heuristic solutions such as FSO-WDM, hybrid RF/FSO better results with respect to BER can be obtained in comparison to simple FSO for high speed wireless communication.

Keywords: Free space optical communication · Atmospheric turbulences · LOS · BER · FSO/RF · OOK · DPSK · QPSK · NRZ · RZ

1 Introduction

To transmit data wirelessly, free-space optical communication (FSO) is best suited approach which uses light that propagates in free space. It is a recent growing technology that uses point to point laser links and transmits the data from transmitter to receiver and uses direct LOS [3–14]. This technology provides unlicensed spectrum, flexibility, cost effectiveness, bandwidth scalability, speed of deployment and security and is an optimal technology for future generations. FSO link working is similar to fiber optics communication but channel is air [2]. It ranges from about 100 m to few kilometers. With the increase in demand of wireless broadband communications, it has received appreciable attention because it has low cost as compared to fiber optic communication and offers low dispersion with high data rate and very less bit error rate. It is attaining market acceptance as a functional, high band-width access tool. It is preferred over Radio fiber a traditional RF system has limited frequency spectrum. This system is appropriate for regions where optical fiber cables cannot be deployed like in hilly areas [4, 5]. This problem of limited frequency spectrum and cost of laying fibers can be relieved by the

© Springer Nature Switzerland AG 2020
S. Smys et al. (Eds.): ICCVBIC 2019, AISC 1108, pp. 1162–1170, 2020.
https://doi.org/10.1007/978-3-030-37218-7_122

implementation of this communication technology. The various challenges for this system is atmospheric turbulence like rain, snow, dust, haze, fog, Aerosol, gases and various other suspended particles geometric loss, atmospheric loss, blocking, shadowing which provides the significant attenuation to the system. Reliability and functionality of this system is adversely affected by the atmospheric conditions.

To enhance the system reliability and effectiveness and to mitigate all atmospheric impairments several techniques have been deployed [8–11]. Techniques like hybrid RF/FSO, WDM and modulation schemes can be deployed at physical layer and retransmission, reconfiguration and rerouting can be implemented at network layer level.

This paper is arranged as follows. In Sect. 2, the basic architecture of FSO link system is given along with its schematic diagram. Issues and challenges are presented in Sect. 3. Heuristic solutions to enhance the system performance is presented in Sect. 4. Results and discussions are presented in Sect. 5 and conclusion and future scope is presented in Sect. 6.

2 Functional Schematic of FSO

The functional schematic of FSO communication is illustrated in Fig. 1. It constitutes of transmitter, receiver and channel i.e. air. The information is communicated through the source over this system. To modulate the data for transmission, different modulation techniques such as OOK, BPSK, QPSK and PPM can be implemented on this system. This source output is modulated to an optical carrier called laser and then transmitted through the atmospheric channel. Transmitter converts incoming electrical signal into optical form and transmits to atmospheric channel. Atmospheric channel through which the optical beam passes possess a lot of challenges for the transmitted signal like rain, snow, dust, haze, fog, Aerosol, gases and various other suspended particles geometric loss, atmospheric loss, blocking, shadowing which provides the significant attenuation to the system. Thus the signal has to be properly modulated before transmission. The signal is optically collected and detected at the receiving side which converts received optical signal to electrical form [12].

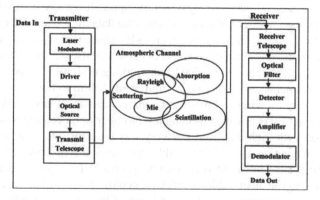

Fig. 1. Functional schematic diagram of FSO [12]

Quality of transmission of data and system performance can be measured by bit error rate (BER). It is the prob. that an error may occur in a bit i.e. '1' bit turns to '0' bit or '0' bit turns to '1' bit. The Typical Value of BER required for commercial optical wireless systems is 10-9 [13]. The system should have less error probability less than acceptable rate with particular data rate.

$$BER = \frac{Number\ of\ errors}{Total\ number\ of\ bitssent} \tag{1}$$

$$BER = \frac{P0}{2} + \frac{P1}{2} \tag{2}$$

P0 is the prob. of receiving '0' instead of '1' and P1 is the prob. of receiving '1' instead of '0'. Multiplication factor here is because '0' and '1' are equally likely events [13].

Figure 2 represents the BER vs. range under different weather conditions. Due to excessive heating, turbulences, particles fluctuations and scintillation atmospheric conditions leads to temporal and random spatial changes.

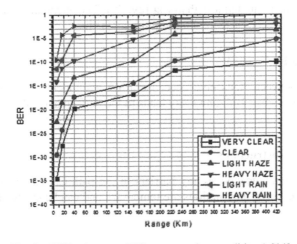

Fig. 2. BER v/s range (different weather conditions) [14]

3 Issues and Challenges

Various challenges occur with the deployment of FSO technology and there is a need to reduce these challenges to harness complete capability of technology.

3.1 Atmospheric Disturbances

Atmospheric turbulence is the major challenge that affects the most in FSO. Fluctuations in refractive index degrade the performance of link due to in homogeneities in temperature [15, 16]. The parameters like length, operating wavelength, fog, rain, gases amongst transmitter and receiver leads in attenuation due to photon absorption and scattering [17]. Gamma–Gamma model is used for atmospheric turbulence i.e. "weak turbulence (WT), moderate turbulence (MT) and strong turbulence (ST) conditions [31]".

When the signal is transmitted over system, it suffers from misalignment and atmospheric turbulence between transmitter and receiver, using Gamma–Gamma model, the PDF of Ia is expressed as

$$P(Ia) = \frac{2(\alpha\beta)^{\frac{\alpha+\beta}{2}}}{\Gamma(\alpha)\Gamma(\beta)} Ia^{\frac{\alpha+\beta}{2}-1} K\alpha - \beta(2\sqrt{\alpha\beta Ia}) \tag{3}$$

Γ (.) denotes Gamma Function. α is no. of larger scale turbulence cell and β denotes no. of small scale turbulence cell.

Most dominant forms of interference for free space optics is fog. It consists of small water droplets, so due to moisture, it may reflect and scatter. Its visibility range lies between 0–2,000 m, so the phrases "advection fog" and "convection fog" are used [18]. Rain also has influence on this link, but its impact is comparatively less than other weather conditions [19]. Attenuation because of snow lies in range 30–350 dB/km which is more than rain and less than fog [29, 30]. The attenuation factor value ranges from 0 to 3 dB/km for clear weather conditions [19].

3.2 Pointing Error (PE)

Error occurs between transceiver because FSO equipment is placed on high rise structure in urban areas. Building sway leads to the performance of FSO links degrades. [20].

3.3 Physical Obstructions

FSO is LOS communication so any barriers in the path can cause interruption in the communication. These interruptions are temporary and can be easily and automatically resumed [21].

3.4 Link Margin (LM)

It can be calculated by the power received at the receiver [27]. It is expressed as:

$$LM = 10\log\frac{Pr}{Ps} \tag{4}$$

Where, Pr = power received, Ps = receiver sensitivity. The value of $Pr > Ps$. The value of Ps is in between −20 dbm to −40 dbm.

3.5   Geometric Attenuation (GE)

It is square of ratio aperture diameter of receiver (dr) by summation aperture diameter of transmitter (dt) and θL [27, 28].

$$GE = (dr/(\varnothing l + dt))^2 \qquad (5)$$

Table 1 indicates various weather conditions with respect to attenuation values in dB/Km. Weather conditions has the major influence on performance of FSO links as explained above.

Table 1. Weather conditions vs. attenuation value [22]

Weather conditions	Attenuation value (dB/Km)
Advection fog	4.28
Convection fog	15.6
Light rain	6.27
Heavy rain	19.8
Haze	20.68

4   Heuristic Solutions to Enhance System Performance

Different mitigation techniques are deployed to increase FSO reliability and its performance. These techniques can be deployed within physical layer or network layer.

At Physical layer, Wavelength division multiplexing (WDM) is the technique that can be deployed to reduce atmospheric attenuation and has high-capacity so high data rate for long distance can be transmitted. It is a multiplexing technique in which the many optical signals are multiplexed on a medium at altered wavelengths to maximize bandwidth usage [1]. The weather dependence is a significant factor in performance of FSO link so WDM system is designed to overcome this.

FSO and RF have some of the common features between them but impact of atmospheric conditions is not same [24]. Fog is the major issue that affects the performance of the FSO link, but RF link has least influence of fog. Similarly, rain does not have any effect on FSO link, but RF link has great influence on performance [25, 26]. RF/FSO hybrid systems are deployed to provide good signal quality. Thus Hybrid system does not experience abrupt changes in turbulence and the weather conditions are steady.

Different modulation schemes can also be deployed to increase the efficiency of system. OOK is simplest scheme and is band-width efficient. The light source transmits logic '1' when it is on and transmits logic '0' at off state. It is also known as non-return to zero (NRZ) modulation. A return to zero (RZ) coding can also be used in which the logic "1" returns to zero in the middle of the sample. RZ is more sensitive than NRZ [35]. Despite from the above schemes, coherent modulation schemes like differential phase shift keying (DPSK) and binary phase shift keying (BPSK) can also be used. Under the weather conditions, the BER performance of BPSK is better than OOK.

Pulse position modulation (PPM) scheme is an orthogonal modulation technique and it needs both slot and symbol synchronization [33, 34].

In Network layer, retransmission of data is one of the solutions. It is useful when there is strong atmospheric turbulence. Repeated transmission of data may increase the probability of correct detection. Rerouting is also one of the solutions and can be handled by Rerouting transmit paths. When direct connection is disabled, networking between nodes can reduce communication problems by enabling alternative routes from sender to receiver. Best route can be selected while considering traffic demands. It provides point to multipoint transceiver.

Table 2 describes various types of mitigation techniques deployment at physical level and network layer in order to increase FSO system reliability and performance.

Table 2. Mitigation techniques

Physical layer	Network layer
Hybrid RF/FSO	Retransmission
WDM	Rerouting
Aperture averaging	Reconfiguration
Modulation and coding	Replay

5 Results and Discussions

Simulation results of different techniques are discussed in this section. The values of BER at different weather conditions are recorded in Table 3. It is detected that value of BER decreases at clear weather conditions and is improved using WDM technique.

Table 3. Weather conditions vs. BER values (WDM)

Weather conditions	BER values (distance = 2.5 km)
Thin fog	10−12
Light fog	10−10
Medium fog	10−4
Heavy fog	10−2
Clear weather	10−34
Haze	10−15

Figure 3 represents the graph between BER vs. range under different weather conditions for WDM system. Graph shows different weather conditions along with its BER values for transmission range. It can be seen that by deploying WDM technique the functionality of system can be improved. Thus, as shown in graph at clear weather it ranges more than 13 km and under heavy haze it last to 1 km [23].

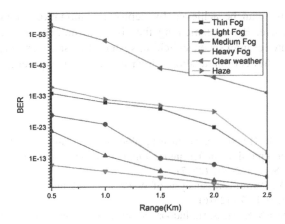

Fig. 3. BER vs. range (WDM FSO) [23]

Figure 4 shows comparison amongst hybrid RF/FSO links and FSO link for moderate and strong turbulences. From the graph it can be illustrated that fso link has more error rate as compared to hybrid RF/FSO network. By using hybrid RF/FSO links, the reliability and system performance is improved which can be seen in the graph [32].

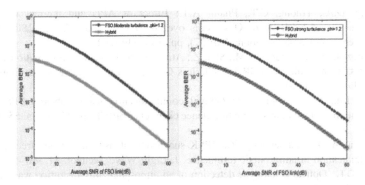

Fig. 4. FSO link vs hybrid RF for strong turbulence and moderate turbulence [32]

6 Conclusion and Future Scope

In this paper a brief survey of FSO its issues, challenges and heuristic solutions is presented. The major advantages of FSO prove that it is good and reliable technology for future generations. The adverse effect on the performance of FSO link results from atmospheric turbulence. To improve the reliability of system certain mitigation techniques are deployed. Different mitigation techniques like Hybrid FSO/RF system, WDM-FSO system and modulation schemes are latest approach to improve the system performance and reliability for faster communication. The obtained results prove that better performance can be achieved by deploying mitigation techniques to enhance system performance. By implementing the above techniques this technology can be deployed in future for high speed switching system, free space exploration and it can be analogous to satellite communication which is the major requirement nowadays.

References

1. Kaur, G., Singh, H.: Simulative Investigation of 32x5 Gb/s DWDM-FSO system using semiconductor optical amplifier under different weather conditions. Int. J. Adv. Res. Comput. Sci. **8**(4), 392–397 (2017)
2. Mazin, A.A.A.: Performance analysis of WDM-FSO link under turbulence channel. World Sci. News **50**, 160–173 (2016)
3. Elgala, H., Mesleh, R., Haas, H.: Indoor optical wireless communication: potential and state-of-the-art. IEEE Commun. Mag. **49**(9), 56–62 (2011). https://doi.org/10.1109/mcom.2011.6011734
4. Esmail, M.A., Ragheb, A., Fathallah, H., Alouini, M.S.: Investigation and demonstration of high speed full-optical hybrid FSO/fiber communication system under light sand storm condition. IEEE Photon. J. **9**(1) (2017). Article no. 7900612
5. Khalighi, M.A., Uysal, M.: Survey on free space optical communication: a communication theory perspective. IEEE Commun. Surv. Tutors. **16**(4), 2231–2258 (2014)
6. Ghassemlooy, Z., Popoola, W., Rajbhandari, S.: Optical Wireless Communications: System and Channel Modeling with Matlab. CRC Press Taylor and Francis Group, Boca Raton (2013)
7. Luong, D.A., Truong, C.T., Pham, A.T.: Effect of APD and thermal noises on the performance of SC-BPSK/FSO systems over turbulence channels. In: 18th Asia-Pacific Conference on Communications (APCC), pp. 344–349 (2012)
8. Ramirez-Iniguez, R., Idrus, S.M., Sun, Z.: Atmospheric transmission limitations. In: Optical Wireless Communications: IR for Wireless Connectivity, p. 40. Taylor & Francis Group, LLC, London (2008)
9. Li, J., Liu, J.Q., Taylor, D.P.: Optical communication using subcarrier PSK intensity modulation through atmospheric turbulence channels. IEEE Trans. Commun. **55**, 1598–1606 (2007)
10. Popoola, W.O., Ghassemlooy, Z.: BPSK subcarrier intensity modulated free space optical communication in atmospheric turbulence. J. Light Wave Technol. **27**, 967–973 (2009)
11. Fried, D.L.: Optical heterodyne detection of an atmospherically distortion wave front. Proc. IEEE **55**, 57–77 (1967)
12. Alkholidi, A., Altowij, K.: Effect of clear atmospheric turbulence on quality of free space optical communications in Western Asia. In: Das, N. (ed.) Optical Communications Systems (2012)
13. Majumdar, A.K., Ricklin, J.C.: Free Space Laser Communications: Principles and Advances. Springer, Heidelberg (2008). ISBN-13 978-0-387-28652-5
14. http://www.laseroptronics.com/index.cfm/id/57-66.htm
15. Andrews, L.: Atmospheric Optics. SPIE Optical Engineering Press, Bellingham (2004)
16. Andrews, L., Phillips, R., Hopen, C.: Laser Beam Scintillation with Applications. SPIE Optical Engineering Press, Bellingham (2001)
17. Gagliardi, R., Karp, S.: Optical Communications. Wiley, New York (1995)

18. Kim, I., Mcarthur, B., Korevaar, E.: Comparison of laser beam propagation at 785 and 1550 nm in fog and haze for optical wireless communications. In: Proceedings of SPIE, vol. 4214, pp. 26–37 (2001)
19. Akiba, M., Ogawa, K., Walkamori, K., Kodate, K., Ito, S.: Measurement and simulation of the effect of snow fall on free space optical propagation. Appl. Opt. **47**(31), 5736–5743 (2008)
20. Achour, M.: Simulating free space optical communication; part I, rain fall attenuation. In: Proceedings of SPIE, vol. 3635 (2002)
21. Bouchet, O., Marquis, T., Chabane, M., Alnaboulsi, M., Sizun, H.: FSO and quality of service software prediction. In: Proceedings of SPIE, vol. 5892, pp. 1–12 (2005)
22. Niaz, A., Qamar, F., Ali, M., Farhan, R., Islam, M.K.: Performance analysis of chaotic FSO communication system under different weather conditions. Trans. Emerg. Telecommun. Technol. (2019). https://onlinelibrary.wiley.com/doi/abs/10.1002/ett.3486
23. Sarangal, H., Singh, A., Malhotra, J., Thapar, S.S.: Performance evaluation of hybrid FSO-SACOCDMA system under different weather conditions. J. Opt. Commun. (2018). https://doi.org/10.1515/joc-2018-0172
24. Chatzidiamantis, N.D., Karagiannidis, G.K., Kriezis, E.E., Matthaiou, M.: Diversity combining in hybrid RF/FSO systems with PSK modulation. In: IEEE International Conference on Communications (2011)
25. Nadeem, F., Kvicera, V., Awan, M.S., Leitgeb, E., Muhammad, S.S., Kandus, G.: Weather effects on hybrid FSO/RF communication link. IEEE J. Sel. Areas Commun. **27**(9), 1687–1697 (2009)
26. Ghoname, S., Fayed, H.A., El Aziz, A.A., Aly, M.H.: Performance evaluation of an adaptive hybrid FSO/RF communication system: impact of weather attenuation. Iran. J. Sci. Technol. Trans. Electr. Eng. (2019). https://doi.org/10.1007/s40998-019-00244-0
27. Alma, H., Al-Khateeb, W.: Effect of weather conditions on quality of free space optics links (with focus on Malaysia). In: 2008 International Conference on Computer and Communication Engineering, Kuala Lumpur, pp. 1206–1210 (2008). https://doi.org/10.1109/ICCCE.2008.4580797
28. Chaudhary, S., Amphawan, A.: The role and challenges of free-space systems. J. Opt. Commun. **35**, 327–334 (2014). https://doi.org/10.1515/joc-2014-0004
29. Kim, I.I., Korevaar, E.: Availability of free space optic (FSO) and hybrid FSO/RF systems. Light pointe Tech report. http://www.opticalaccess.com
30. Kaushal, H., Kaddoum, G.: Optical communication in space: challenges and mitigation techniques. IEEE Commun. Surv. Tutor. https://doi.org/10.1109/comst.2016.2603518
31. Sarkar, D., Mctya, S.K.: Effects of atmospheric weather and turbulence in MSK based FSO communication system for last mile users. Telecommun. Syst. (2019). https://doi.org/10.1007/s11235-019-00602-7
32. Singh, N.: To analyze the aftermath of using hybrid RF/FSO link over FSO link under various turbulence of atmosphere. Int. J. Eng. Res. Technol. (IJERT) **8**(05) (2019). http://www.ijert.org, ISSN 2278-0181 IJERTV8IS050421 (This work is licensed under a Creative Commons Attribution 4.0 International License)
33. Majumdar, A.K., Ricklin, J.C.: Free-Space Laser Communications Principles and Advances. Springer, New York (2008)
34. David, F.: Scintillation loss in free-space optic systems. In: LASER 2004, San Jose, USA, vol. 5338 (2004)
35. Henniger, H., Wilfert, O.: An introduction to free space optical communications. Radio Eng. **19**(2), 203–212 (2010)

Design of Canny Edge Detection Hardware Accelerator Using xfOpenCV Library

Lokender Vashist[(✉)] and Mukesh Kumar

National Institute of Technology Kurukshetra,
Kurukshetra 136119, Haryana, India
lokendervashist2@gmail.com,
mail2mukeshsharma@gmail.com

Abstract. Real time image processing using CPU takes more computational power and thereby processing speed also decreases. FPGA with its parallel processing capability does complex processes with much more speed and less power consumption sharing load of CPU hence it can be used for edge detection in real time where large number of pixels are to be processed. In this paper we developed programmable logic for canny edge detection algorithm intended for Zynq 7000 based Zybo board. xfOpenCV library is used to generate the hardware logic for algorithm after application of a wrapper function so as to make it portable with AXI stream interface of Zynq device. Better edges are detected even for noisy image using xfOpenCV library. Resources utilization of FPGA are very less thus the edge detection done using Zynq 7000 device will be 100x faster and more reliable then CPU's

Keywords: xfOpenCV library · Vivado HLS · Canny edge detection algorithm · Zybo board

1 Introduction

Real time image and video processing has been key element for development of algorithms for object detection, video surveillance, medical imaging, multimedia, space technology and security [4]. Image segmentation, image categorization, object detection, and feature tracking are key functions helpful in above mentioned applications. The video pipeline contains edge detection as its initial phase. Image in general contains large number of edges and detecting them using CPU takes lot of computational time. Moreover, in real time framework processing becomes more complex and it requires high computational power to handle enormous number of pixels associated with the incoming video [9]. At high frame rate, CPU's with little computation capability becomes vulnerable in correct detection of edges in real time framework. FPGA with its parallel processing capability makes it a suitable candidate for the above mentioned applications. FPGA uses hardware accelerators to reduce the amount of processes to be done by CPU thereby helps in reducing speed and energy [1]. Some pre-processing functions like filtering, thresholding which are high speed operations can be implemented using hardware accelerator before sending pixels to CPU which reduces the data to be processed in CPU leaving only regions of interest to be taken

© Springer Nature Switzerland AG 2020
S. Smys et al. (Eds.): ICCVBIC 2019, AISC 1108, pp. 1171–1178, 2020.
https://doi.org/10.1007/978-3-030-37218-7_123

care of. Also control logic can be implemented directly using FPGA without the interaction of CPU which reduces high computational power and increases speed considerably.

Gudis, Lu et al. [12] used hardware accelerators for computer vision application using embedded solution to describe an architecture framework. They provided a structure using a comprehensive library of real-time vision hardware accelerators and software architecture. In [13] the use of OpenCV library and its implementation on Zynq platform is shown. This application was for road sign detection. This algorithm was illustrated by using Zynq Z7102 based zedboard in which the input image was of 1920 × 1080 resolution which was coming from an ON-Semiconductor VITA-2000 sensor attached via the FPGA Mezzanine Card (FMC) slot. The programmable logic in the design was used for preprocessing of image analysis and image segmentation.

In this paper we are going to implement canny edge detection technique on Zynq Z7010 processor based Digilent Zybo board. The canny edge detection hardware accelerator is generated using Vivado HLS by setting constraint and using optimization through directives of HLS tool. xfOpenCV library provided by Xilinx for SDSoC environment is used containing the top-level canny edge detector function for synthesis using HLS. As xfOpenCV function generates the input and output port of IP as pointers hence it cannot be used directly for zynq z7010 which requires their input and output port in accordance with AXI stream format. Hence a wrapper function is designed so as to transform the original function in AXI stream format and then synthesis is done using this wrapper function.

2 Canny Edge Detection Algorithm

It is one of the most widely used algorithm in modern day for detecting edges as its performance is reliable for noisy images [11]. It is based on kernel convolution [10]. The steps involved in this algorithm are described below:

1. Smoothening of image
   The very first step is to remove noise by using gaussian filter which removes high frequency components from image. If input image I (i, j) convolved with gaussian filter G it results in smoothened image [2].

$$F(i, j) = I(i, j) * G \qquad (1)$$

2. Gradients in horizontal and vertical direction
   The sobel operator is used to find gradient in horizontal and vertical direction. Where the change in greyscale intensity is maximum this operator is applied to obtain edges for every pixel. Sobel operator in horizontal and vertical direction:

$$H = \begin{matrix} -1 & 0 & 1 \\ -2 & 0 & 2 \\ -1 & 0 & 1 \end{matrix} \quad \text{and} \quad V = \begin{matrix} 1 & 2 & 1 \\ 0 & 0 & 0 \\ -1 & -2 & -1 \end{matrix}$$

These operators are applied convolved with smoothened image giving gradients in the respective directions.

$$Gh = F\,(i, j) * H \quad \text{and} \quad Gv = F\,(i, j) * V \tag{2}$$

Magnitude and direction of gradient of pixel is given by

$$G = \sqrt{(Gh^{\wedge}2 + Gv^{\wedge}2)} \quad \text{and} \quad \theta = \arctan\frac{Gv}{Gh} \tag{3}$$

3. Non Maximum Suppression
   To obtain thin edges local maxima in the gradient image is kept and rest is deleted for every pixel. Start with gradient direction of 45° for the current pixel K (i, j), then in the positive and negative gradient direction find the gradient magnitude and mark them as P (i, j) and N (i, j) respectively. If the strength of edge of the pixel K (i. j) is more than that of P (i, j) and N (i, j) then mark this as an edge pixel if not discard it or partially suppress it.
4. Hysteris Thresholding
   As noise is still present, there are local maxima found. To supress them choose two threshold THIGH and TLOW. If the gradient magnitude is less then TLOW get rid of edge. If magnitude is high then THIGH preserve it. If it falls in between them and if any of its neighbouring pixel has magnitude greater than THIGH keep that edge otherwise remove it.

3 Implementation of Canny Edge Detection on Vivado HLS Using xfOpenCV Library Targeted for Zybo Board

Most of the emerging embedded processor and SoC devices require hardware logic to be written in Verilog/VHDL which is difficult for complex applications. The automated flow of HLS tool enables the hardware logic to be written in high level language C/C++ [3]. In this work we are using Xilinx Vivado HLS tool to convert the canny edge detection function from xfOpenCV library into HDL implementation (Verilog/VHDL) which is then synthesized using Vivado tool targeted for FPGA (Zybo board) [5].

The IP to be generated for canny algorithm is targeted for Zybo board having Zynq 7000 Z7010 device. The Zynq 7000 contains ARM cortex A9 processor with a single FPGA which provides extreme high speed [6]. The generated IP has to be used in accordance with AXI4 Stream protocol for the input and output ports of IP. Figure 1 shows the architecture of Zynq 7000 SoC.

The algorithm to generate a wrapper function around the original top level canny edge detection function from xfOpenCV library is discussed below so as to generate IP in accord with AXI protocol.

xf::Mat is the class used to define the image object in xfOpenCV library. H - Height of image, W - width of image, input and output are objects of class hls::stream which defines the image objects in AXI format. xf::AXIvideo2xfMat – function which converts Input video in AXI format into xf::mat image format, xf::xfMat2AXIvideo –

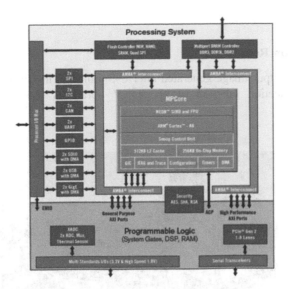

Fig. 1. System architecture of Zynq 7000 SoC [5]

function which converts xf::mat image object back to AXI format video object. Steps for algorithm are described below:

1. Denote Original top-level function from xfOpenCV library for synthesis written in C++ as: Canny_accel (source image, destination image, kernel matrix)
2. Denote wrapper function for synthesis as: xf_ip_accel_app (input, output, H, W, kernel matrix).
3. Use pragma directive of HLS for interface
   #pragma HLS INTERFACE axisport = in
   #pragma HLS INTERFACE axis port = out
4. Define two temporary image objects of xf::mat class as temp_in1, temp_out1.
5. Use pragma directive given below to define the HLS stream variables which are temporary.
   #pragma HLS stream variable = temp_in1.data
   #pragma HLS stream variable = temp_out1.data
6. Inside the data flow pragma perform the below steps:
   a. Use function as xf::AXIvideo2xfMat (input, temp_in1) which convert temp_in1 of mat type in to input object of AXI type.
   b. Now call function Canny_accel to obtain edges:
      Canny_accel (temp_in1, temp_out1, kernel)
   c. After this convert temp_out1 back to AXI video as:
      xf::xfMat2AXIvideo (temp_out1, output)

4 Results

4.1 Simulation Results

The input and output image generated after C-Simulation in Vivado HLS after using canny edge detection function of xfOpenCV library is shown in Figs. 2 and 3 respectively.

Fig. 2. Input image for C-simulation in Vivado HLS [9]

Figure 2 is the still image given as an input to the standalone application developed in Vivado HLS for C-simulation. The input image file is of resolution 1280 × 720. Figure 3 shows the edges detected using the canny edge detection after C-simulatiton. Edges are present at every pixel having a gradient change as visible from Fig. 3.

Fig. 3. Output Image after C-simulation

4.2   IP Generated Before and After Application of C++ Wrapper Function

The IP generated before applying wrapper function having input and output port generated as pointers is shown in Fig. 4 and after applying the wrapper function xf_ip_accel_app is shown in Fig. 5.

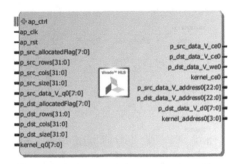

Fig. 4. IP generated before applying the wrapper function on top level C function for canny edge detection from xfOpenCV library.

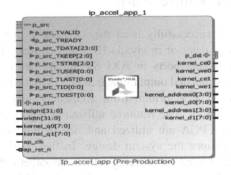

Fig. 5. IP generated after applying wrapper function

The IP generated after C-synthesis and export function in Vivado HLS is depicted in Fig. 4. It contains input ports on left and output ports on right of the block which are of pointer type standard. Figure 5 depicts the IP generated using the C-synthesis and export function on the wrapper function. The generated IP block contains input and output ports having AXI standard protocol which is the port standard used on Zynq processor.

4.3   Resource Utilization

Table 1 gives the estimate of the resources of FPGA that are going to be used by the generated RTL in the system design. Utilization of BRAM_18K is only 5% which is used to implement line buffer for canny edge detection algorithm. Utilization of Flip Flops is 4% which is used to implement window buffer for canny edge detection C++ code.

Table 1. Resource Utilization of the hardware accelerator for canny edge detection.

Utilization Estimates				
Summary				
Name	BRAM_18K	DSP48E	FF	LUT
DSP	-	-	-	-
Expression	-	-	0	46
FIFO	0	-	40	304
Instance	6	-	1489	2419
Memory	-	-	-	-
Multiplexer	-	-	-	54
Register	-	-	9	-
Total	6	0	1538	2823
Available	120	80	35200	17600
Utilization (%)	5	0	4	16

5 Conclusion and Future Work

In this paper we have successfully used the xfOpenCV library having a top level function for canny edge detector in vivado HLS and converted it into a synthesizable hardware accelerator having ports in AXI Stream format with the help of a C++ wrapper function. The simulated output image indicates that edges are better in terms of brightness and clarity. The algorithm from xfOpenCV gives better results than its counterpart OpenCV library. The resource utilization report indicates that very little available resources of FPGA are utilized and hence more complex blocks can be included with it to enhance the system design. In the classic canny edge detection system gaussian filter is used for filtering which smoothens the image without preserving some edges, hence we can use median filter [7] or bilateral filter which smoothens image along with preserving edges in place of gaussian filter and change the C++ code in xfOpenCV library accordingly and generate IP for that. Secondly, in image thresholding the upper and lower threshold are fixed by user which results in the loss of edges in the real time [8]. We can develop automatic setting of threshold values and this part of Canny algorithm can be changed in xfOpenCV C++ function to obtain better edges.

Compliance with Ethical Standards

   ✓ All authors declare that there is no conflict of interest.
   ✓ No humans/animals involved in this research work.
   ✓ We have used our own data.

References

1. Asano, S., Maruyama, T., Yamaguchi, Y.: Performance comparison of FPGA, GPU and CPU in image processing. In: 2009 International Conference on Field Programmable Logic and Applications, pp. 126–131. IEEE (2009)
2. Kumar, M., Saxena, R.: Algorithm and technique on various edge detection: a survey. Sig. Image Process. **4**(3), 65 (2013)
3. Chinnadurai, M., Joseph, M.: High level synthesis tools-an overview from model to implementation. Middle-East J. Sci. Res. **22**, 241–254 (2014)
4. Nayagam, M.G., Ramar, K.: A survey on real time object detection and tracking algorithms. Int. J. Appl. Eng. Res **10**(9), 8290–8297 (2015)
5. Amara, A.B., Pissaloux, E., Atri, M.: Sobel edge detection system design and integration on an FPGA based HD video streaming architecture. In: 2016 11th International Design & Test Symposium (IDT), pp. 160–164. IEEE (2016)
6. Sadri, M., Weis, C., Wehn, N., Benini, L.: Energy and performance exploration of accelerator coherency port using Xilinx ZYNQ. In: Proceedings of the 10th FPGAworld Conference, p. 5. ACM (2013)
7. Dong, Y., Li, M., Li, J.: Image retrieval based on improved Canny edge detection algorithm. In: 2013 International Conference on Mechatronic Sciences, Electric Engineering and Computer (MEC), 20–22 December 2013, pp. 1453–1457 (2013)
8. Bao, Y., Wu, D.: An improved Canny algorithm based on median filtering and adaptive threshold in robot training system. In: 2015 6th International Conference on Manufacturing Science and Engineering. Atlantis Press (2015)
9. Stefaniga, S.-A., Gaianu, M.: Performance analysis of morphological operation in CPU and GPU for medical images. In: 2017 19th International Symposium on Symbolic and Numeric Algorithms for Scientific Computing (SYNASC), pp. 377–384. IEEE (2017)
10. Vincent, O.R., Folorunso, O.: A descriptive algorithm for sobel image edge detection. In: Proceedings of Informing Science & IT Education Conference (InSITE), vol. 40, pp. 97–107. Informing Science Institute, California (2009)
11. Zeng, J., Li, D.: An improved canny edge detector against impulsive noise based on CIELAB space. In: 2010 International Symposium on Intelligence Information Processing and Trusted Computing, pp. 520–523. IEEE (2010)
12. Gudis, E., Lu, P., Berends, D., Kaighn, K., Wal, G., Buchanan, G., Chai, S., Piacentino, M.: An embedded vision services framework for heterogeneous accelerators. In: Proceedings of the IEEE Conference on Computer Vision and Pattern Recognition Workshops, pp. 598–603 (2013)
13. Russell, M., Fischaber, S.: OpenCV based road sign recognition on Zynq. In: 2013 11th IEEE International Conference on Industrial Informatics (INDIN), pp. 596–601. IEEE (2013)

Application of Cloud Computing for Economic Load Dispatch and Unit Commitment Computations of the Power System Network

Nithiyananthan Kannan[1(✉)], Youssef Mobarak[1,2], and Fahd Alharbi[1]

[1] Faculty of Engineering Rabigh, King Abdulaziz University,
Jeddah, Saudi Arabia
{nmajaknap,ysoliman,fahdalharbi}@kau.edu.sa
[2] Faculty of Energy Engineering, Aswan University, Aswan, Egypt

Abstract. The objectives of this research paper is to discuss the impacts of the economic load dispatch solutions and unit commitment solutions for online power system applications in cloud computing environment. The economic load dispatch solutions and unit commitment solutions calculations for multi- area power system implemented on cloud computing environment is to give solutions to the real time monitoring arrangement, it is affected by the inter-operability problems among conventional client – server architecture. The existing load dispatch and commitment system computing cannot able to meet the expectations of the efficient monitoring and accurate delivery of the output with the massive enlargement of power system network in day to day basis. Besides, the cloud computing implementation on economic load dispatch and unit commitment and through threads of single server – multi client architecture is suggested to implement by considering each client as thread instance as power system client. The current methods on solving these problems through high performance computing infrastructures creased more issues on scalability, more maintenance and investments. The proposed method is ideal replacement due to high capacity, very high scalability and less cost in all the services.

Keywords: Cloud computing · High-performance computing · Unit commitment · Economic load dispatch · Power system applications

1 Introduction

Over past twenty years the power systems expanded constantly and reached a state of isolated system to highly interconnected network. This interconnection makes power systems to develop regional and International grids worldwide to achieve maximum possible power system reliability. Recent growth in extraction of power from distributed sources through various renewable energy sources is increased the complexity in integrating the power systems and its management. This network growth brought lot of problems such as large scale storage of data, big level of scale computing, more hardware and software requirements etc. High number of research work has been reported the contribution of computers in to Electrical related applications [1–32]. Among various types of online power system analysis the economic load dispatch and

© Springer Nature Switzerland AG 2020
S. Smys et al. (Eds.): ICCVBIC 2019, AISC 1108, pp. 1179–1189, 2020.
https://doi.org/10.1007/978-3-030-37218-7_124

unit commitment monitoring place wider role is to utilize the resources in an optimized way to produce the power at much cheaper cost per unit as indicates in Fig. 1. Economic Load dispatch (ELD) is the algorithm which allocates the generation levels among power plants in tandem to meet the load demand as cheap as possible. The objectives of the ELD algorithm are to bring out the best optimal power values of a power plant which satisfies all the constraints and should reduce the fuel cost to an optimal value as shown in Fig. 1.

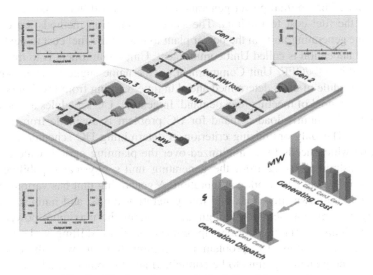

Fig. 1. Display of economic load dispatch concept

Normally the ELD based issues are formed by a quadratic equation and it is find solution by using numerical methods based computer program using the standard optimization algorithms based on standard algorithms lambda iteration, dynamic programming methods and gradient methods. The Eq. 1 indicates the total fuel cost.

$$F_T = \sum_{i=1}^{n} F(P_i) = \sum_{i=1}^{n} \left(a_i P_i^2 + b_i P_i + c_i \right) \tag{1}$$

Where F_T indicates the overall fuel cost. a_i, b_i, and c_i are the coefficient amount, P_i is the power generated. Equation 2 indicates the minimum and maximum power limits.

$$P_{i\,min} \leq P_i \leq P_{i\,max} \quad \text{for } i = 1, 2, \ldots, n \tag{2}$$

The Eq. 3 indicates the calculations for power demand which related with transmission line losses.

$$P_D = \sum_{i=1}^{n} P_i - P_{loss} \tag{3}$$

The Eq. 4 indicates the power loss equation based on B matrix.

$$P_{loss} = \sum_{j=1}^{n} \sum_{j=1}^{n} P_i B_{ij} P_j \tag{4}$$

B_{ij}'s are the elements of coefficient of loss matrix B.

ELD reduces the cost of the energy sources and provides better control over emissions. ELD is also identifies expensive non performing power plants. ELD is also improves power system reliability, efficiency and capable of change in output in a short span of time. The optimal power generation allotment for different power plants helps to reduce the fuel cost to minimal. The same optimization methods are extended to allocate the generating units in the power plant to reduce the fuel cost to next leve. This method of allocation is called Unit commitment. Unit commitment is to find out which unit should be ON/OFF. Unit Commitment, as the name suggests, is an hourly commitment schedule of units determined ahead in time, varying from few hours to a week, with the main goal of meeting load demand. In general, UC schedules are determined a day ahead. The hourly load demand for UC problem is available from precise load forecasting. The only optimizing criterion in determining UC schedule is the cost of generation which needs to be minimized over the planning period while fulfilling all system constraints arising from the generating unit's physical capabilities and the transmission system's network configuration. A generating unit has various limitations such as min up-down time, max and min generation limits, and ramp rate limits etc. Similarly transmission network configuration governs the maximum power flow possible through lines while transformer tap settings and phase shifting angle limits impose the physical constraints. UC problem is a combination of two sub-problems. One determines the generating units to be committed and the other actually focuses on the amount of generation from each of these committed units. Generating units exhibit different operating efficiencies and performance characteristics which reflect on the required inputs. Thus the cost of generation also depends on the amount of output from each committed unit apart from the choice of generating units. Thus, a UC problem is solved in two stages. The combination of generating units yielding the least final production cost is thus chosen as the UC schedule for the hour as shown in Fig. 2.

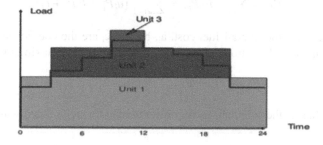

Fig. 2. Combination of generating units

2   Cloud Computing and Its Benefits

Cloud computing is the latest technology uses multiple computer in a network connected with internet to meet out fast growing web based service request. It is capable providing variety of services to the business community and the society. Cloud computing technology able to serve at greater flexibility and able to provide computing resources as minimal cost. This latest technology creates new revolution in the electronic services of various disciplines. In this research work, it has been exploring the features of cloud computing services and its suitability of ELD and UC applications. Large number of cloud services providers available in the market who pays vital role in utilization of resource effectively and make the user friendly approach enables all the small to big enterprices to get benifited [33–42] (Fig. 3).

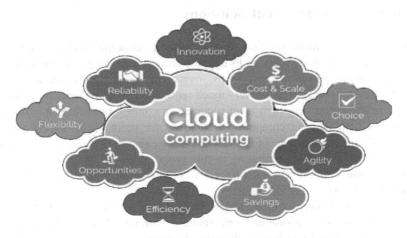

Fig. 3. Cloud computing features

The powerful computational capacity, rapid scalability, efficient, high reliability, flexibility and high cost effectiveness are the major benefits to cloud computing that can be exploited for important power system application such as ELD and UC. The cloud computing offers three important services as shown in Fig. 4. The services are software as service, Platform as service and Hardware as service. In software as service the service is offered via internet the several standard software applications can be assessed by the user as shared resource. The user can access the software without installing any software in his own computer as end user.

In Platform as a service is offered by several service providers. Any one want to upload their own applications, they can upload in the cloud server. Based on the increase on the usage cloud computing architecture supports high level of scalability. It helps developer to code the applications without worrying about the platform infrastructure requirements.

In hardware as a service some service providers in the world offers their infrastructure to the users. Instead of offering physical infrastructure they offer their hardware in a virtualize environment. This service is used for computing or storing large data [42–52].

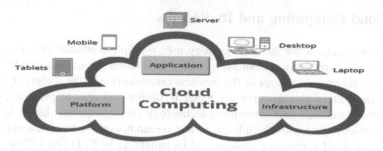

Fig. 4. Services of cloud computing

3 Cloud Computing Based Economic Load Dispatch and Unit Commitment Solutions

To give fast and reliable solutions to the existing economic load dispatch and unit commitment calculations it needs to upgrade the systems, improve the data communications among power system clients, overcome the complex operations and to increase the computing levels are very much required. This complexity in handling data indicates the need of very high level accuracy in control, operation and coordination capabilities with efficient computing techniques for online load dispatch and commitment of units monitoring. In the view of these existing problems and to overcome the difficulties associated computing in this proposal the on-line economic load dispatch and unit commitment calculations through cloud computing environment is proposed as shown in Fig. 5.

This proposed work outlines a new approach to develop a cloud based power system solutions in a distributed environment without additional investment. This application developed will have a character of super scalability achieved through 100

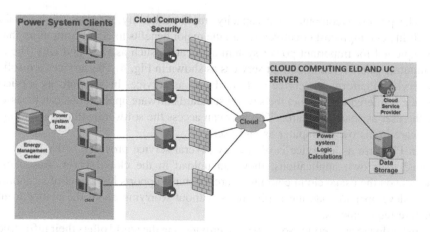

Fig. 5. Proposed architecture of cloud computing based ELD and UC

million servers offered by google cloud computing, and other famous vendors who hold large number of servers. ELD and UC based cloud computing model will become virtualization. The cloud computing model supports the power system clients in various places to have an interface for ELD and UC service. The required ELD and UC calculations results will arrive from the clouds server, not from the typical physical server. The Economic load dispatch calculations are done in different locations somewhere in the cloud which is not known by the individual power system client. This indicates the decentralized computing and strong virtualization. This high reliability model has a strongly able to recover under disaster or abnormal situation. The application and power system information will be back-up in three different positions of the architecture. Since the ELD and UC applications are developed in cloud computing it becomes scalable which cope up with the vast expansion of power system network. Usage of cloud computing technology makes the entire application cheaper because it is not required to buy any additional hardware facilities, natural fault tolerant facilities, effective utilization of computing resources and no maintenance charges. The short comings of the current economic load dispatch and unit commitment calculations will be eliminated in the proposed design of the research work. The system is classified in to four architecture layers as shown in Fig. 6. The physical layer, application layer, service oriented architecture layer and access layer. The Economic Load dispatch service will be deployed in cloud computing platform based SOA layer and various power system clients will access the server with their data through advance access layer. This model indicates five types of interface services such as data, graphics, topology, ELD calculation and web services. The cloud computing platform enables the communications through server, network, storage, and application virtualization.

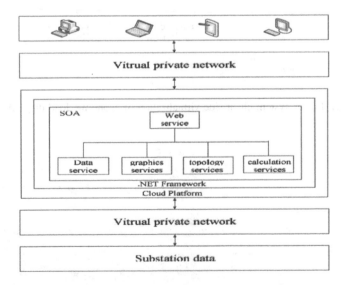

Fig. 6. Virtualization of proposed economic load dispatch system.

Since the legacy power system hardware structure itself utilized for implementation of cloud will not need to have an additional investment. Hence it will have a very attractive market share with other conventional energy management organizations. Even though the initial cost of the proposed cloud based system is costlier. Since the smart grid and micro grid is the future of power systems, the proposed model will be widely utilized by energy management organizations. So the financial risk is very minimal. This proposed cloud based architecture is robust, fault tolerant and highly secure in nature. This makes the implementation of cloud for the legacy power system gives good climate for business model. Cloud based Economic load dispatch calculations implementation creates job for the young electrical and computer engineers for the development, deployment and maintenance area.

4   Results and Conclusion

After successful implementation of the proposed model brings out the cloud based distributed environment model to monitor at regular intervals the online load dispatch and commitment of generating units for a multi- area power system network in a distributed environment as shown in Figs. 7 and 8.

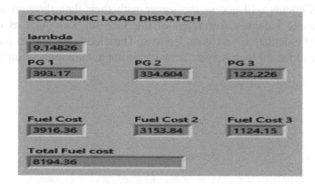

Fig. 7. Power system client side virtualization of economic load dispatch output

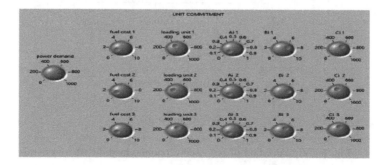

Fig. 8. Power system client side virtualization unit commitment output

The proposed systems developed based on cloud architecture are highly scalable and sustainable. It can have the ability to interoperate with any kind of legacy power systems in a platform independent, language independent and architectural independent way through XMLisation of Power system data. Although the conventional calculations for load dispatch and commitment of units is available, the importance of this research emphasizes its uniqueness of interaction of large number of power systems clients access the service in a distributed environment which enables the strong interoperability nature of the various types of power system clients across various platforms. Each power systems client approaches the ELD cloud service with the dynamic power system data and able to get the results in a distribute environment at regular intervals. The proposed model will be tested and validated through standard IEEE Power system data. Accordingly, the outcome of the proposed work can be implemented for big real time power systems network. After implementing this system, load dispatch and commitment solutions for multi- area power systems network performed effectively to get the advantages in language independent, architectural independent and platform independent way. This proposed model will reduce the complexity in collecting, processing and analyzing the power system related and it also provides solutions to legacy issues. This deployment of new technology implementation will bring a good revolution in the energy management sector especially in smart grid and micro grid applications. It makes easier connection of renewable energy sources in to the grid. Many energy management centers will adopt this technology for the benefit of their service.

Acknowledgements. This project was funded by the Deanship of Research Grant, King Abdulaziz University, Jeddah, Grant No (D-201-829-1440). The authors, therefore, gratefully acknowledge the DSR technical and financial support.

Compliance with Ethical Standards

✓ All authors declare that there is no conflict of interest.
✓ No humans/animals involved in this research work.
✓ We have used our own data.

References

1. Thomas, S., Nithiyananthan, K.: A novel method to implement MPPT algorithms for PV panels on a MATLAB/SIMULINK environment. J. Adv. Res. Dyn. Control Syst. **10**(4), 31–40 (2018)
2. Gowrishankar, K., Thomas, S., Nithiyananthan, K.: Wireless integrated-sensor network based subsea tunnel monitoring system. J. Adv. Res. Dyn. Control Syst. **10**(12), 647–658 (2018)
3. Gowrishankar, K., Nithiyananthan, K., Mani, P., Thomas, S.: GSM based dual power enhanced LED display notice board with motion detector. Int. J. Eng. Technol. **7**(2.8 Special Issue 8), 559–566 (2018)
4. Sing, T.Y., Nithiyananthan, K., Siraj, S.E.B., Raguraman, R., Marimuthu, P.N.: Application of cluster analysis and association analysis model based power system fault identification. Eur. J. Sci. Res. **138**, 16–28 (2016)

5. Nair, P., Nithiyananthan, K., Sing, T.Y., Raguraman, R., Siraj, S.E.B.: Enhanced R package-based cluster analysis fault identification models for three phase power system network. Int. J. Bus. Intell. Data Min. **14**, 106–120 (2019)

6. Nithiyananthan, K., Varma, S.: MATLAB simulations based identification model for various points in global positioning system. Int. J. Comput. Appl. **138**(13), 15–18 (2016)

7. Amudha, A., Nithiyananthan, K., Raja, S.S., Sundar, R.: Unit commitment calculator model for three phase power system network. Int. J. Power Renew. Energy Syst. **3**(1), 37–42 (2016)

8. Gowrishankar, K., Nithiyananthan, K., Mani, P., Venkatesan, G.: Neural network based mathematical model for feed management technique in aquaculture. J. Adv. Res. Dyn. Control Syst. **9**(Special Issue 18), 1142–1161 (2017)

9. Sing, T.Y., Siraj, S.E.B., Raguraman, R., Marimuthu, P.N., Nithiyananthan, K.: Cosine similarity cluster analysis model based effective power systems fault identification. Int. J. Adv. Appl. Sci. **4**(1), 123–129 (2017)

10. Nair, P., Nithiyananthan, K., Dhinakar, P.: Design and development of variable frequency ultrasonic pest repeller. J. Adv. Res. Dyn. Control Syst. **9**(Special Issue 12), 22–34 (2017)

11. Nithiyananthan, K., Sundar, R., Ranganathan, T., Raja, S.S.: Virtual instrumentation based simple economic load dispatch estimator model for 3-phase power system network. J. Adv. Res. Dyn. Control Syst. **9**(11), 9–15 (2017)

12. Amudha, A., Raja, T.S., Nithiyananthan, K., Sundar, R.: Virtual stability estimator model for three phase power system network. Indones. J. Electr. Eng. Comput. Sci. **4**(3), 520–525 (2016)

13. Nithiyananthan, K., Nair, M.P.: An effective cable sizing procedure model for industries and commercial buildings. Int. J. Electr. Comput. Eng. **6**(1), 34–39 (2016)

14. Nair, M.P., Nithiyananthan, K.: Feasibility analysis model for mini hydropower plant in Tioman Island. Malaysia. Distrib. Gener. Altern. Energy J. **31**(2), 36–54 (2016)

15. Umasankar, U., Nithiyananthan, K.: Environment friendly voltage up-gradation model for distribution power systems. Int. J. Electr. Comput. Eng. **6**(6), 2516–2525 (2016)

16. Kumaar, G.S., Nithiyananthan, K., Thomas, S., Karthikeyan, S.P.: MATLAB/SIMULINK simulations based modified SEPIC DC to DC converter for renewable energy applications. J. Adv. Res. Dyn. Control Syst. **11**(4), 285–295 (2019)

17. Venkatesan, G., Nithiyananthan, K.: Static and transient thermal analysis of power factor converters in switched reluctance motor drive. J. Adv. Res. Dyn. Control. Syst. **10**(13), 1656–1662 (2018)

18. Ramachandran, V., Nithiyananthan, K.: Effective data compression model for on-line power system. Int. J. Eng. Model. **27**(3–4), 101–109 (2014)

19. Nithiyananthan, K., Ramachandran, V.: Distributed mobile agent model for multi area power systems automated online state estimation. Int. J. Comput. Aided Eng. Technol. **5**(4), 300–310 (2013)

20. Ramachandran, V., Nithiyananthan, K.: Versioning-based service-oriented model for multi-area power system online economic load dispatch. Comput. Electr. Eng. **39**(2), 433–440 (2013)

21. Ramachandran, V., Nithiyananthan, K.: Location independent distributed model for on-line load flow monitoring for multi-area power systems. Int. J. Eng. Model. **24**(1–4), 21–27 (2011)

22. Nithiyananthan, K., Loomba, A.K.: MATLAB/Simulink based speed control model for converter controlled DC drives. Int. J. Eng. Model. **24**(1–4), 49–56 (2011)

23. Thomas, S., Nithiyananthan, K., Eski, A.: Investigations on transient behavior of an energy conservation chopper fed DC series motor subjected to a change in duty cycle. J. Green Eng. **9**(1), 92–111 (2019)

24. Nithiyananthan, K., Ramachandran, V.: A plug and play model for JINI based on-line relay control for power system protection. Int. J. Eng. Model. **21**(1–4), 65–68 (2008)
25. Ramachandran, V., Nithiyananthan, K.: A distributed model for capacitance requirements for self-excited induction generators. Int. J. Autom. Control **2**(4), 519–525 (2008)
26. Ramachandran, V., Nithiyananthan, K.: Component model simulations for multi-area power system model for on-line economic load dispatch. Int. J. Emerg. Electr. Power Syst. **1**(2). Art no: 1011
27. Elavenil, V., Nithiyananthan, K.: CYMGRD based effective earthling design model for substation. Int. J. Comput. Appl. Eng. Sci. **1**(3), 341–346 (2011)
28. Nithiyananthan, K., Ramachandran, V.: Distributed mobile agent model for multi-area power system on-line economic load dispatch. Int. J. Eng. Model. **17**(3–4), 87–90 (2004)
29. Mobarak, Y.A., Hemeida, A.M., Nithiyananthan, K.: Voltage and frequency based load dependent analysis model for Egyptian power system network. J. Adv. Res. Dyn. Control Syst. **11**(6), 971–978 (2019)
30. Hau, L.K., Kasilingam, G., Nithiyananthan, K.: Development of prototype model for wireless based control pick and place robotic vehicle. TELKOMNIKA Indones. J. Electr. Eng. **14**(1), 110–115 (2015)
31. Nithiyananthan, K., Ramachandran, V.: A distributed model for multi-area power systems on-line dynamic security analysis. In: IEEE Region 10 Annual International Conference, Proceedings/TENCON, vol. 3, pp. 1765–1767 (2002)
32. Nithiyananthan, K., Ramachandran, V.: EJB based component model for distributed load flow monitoring of multi-area power systems. Int. J. Eng. Model. **15**(1–4), 63–67 (2002)
33. Huang, Q., Zhou, M., Zhang, Y., Wu, Z.: Exploiting cloud computing for power system analysis. In: 2010 International Conference on Power System Technology, Hangzhou, pp. 1–6 (2010)
34. Dongxu, Y., Hua, W., Hongbo, W.: Architecture design of power system fault calculation based on cloud computing technology. In: 2017 IEEE Conference on Energy Internet and Energy System Integration (EI2), Beijing, pp. 1–5 (2017)
35. Pepermans, G., et al.: Cloud generation: definition, benefits and issues. Energy Policy **33**(6), 787–798 (2005)
36. Di Santo, M., et al.: A cloud architecture for online power systems security analysis. IEEE Trans. Ind. Electron. **51**(6), 1238–1248 (2004)
37. Li, Z., Liu, Y.: Reactive power optimization using agent based grid computing. In: The 7th International Power Engineering Conference (2005)
38. Huang, Q., Qin, K., Wang, W.: Development of a grid computing platform for electric power system applications. In: Power Engineering Society General Meeting. IEEE (2006)
39. Al-Khannak, R., Bitzer, B.: Load balancing for cloud and integrated power systems using grid computing. In: Clean Electrical Power, ICCEP 2007 (2007)
40. Schainker, R., et al.: On-line dynamic stability analysis using cloud computing. In: Power and Energy Society General Meeting – Conversion and Delivery of Electrical Energy in the 21st Century. IEEE (2008)
41. Huang, Q., et al.: Cloud state estimation with PMU using grid computing. In: Power & Energy Society General Meeting. IEEE (2009)
42. Ali, M., Dong, Z.Y., Zhang, P.: Adoptability of grid computing technology in power systems analysis, operations and control. IET Gener. Transm. Distrib. **3**(10), 949–959 (2009)
43. Di Silvestre, M.L., Gallo, P., Ippolito, M.G., Sanseverino, E.R., Sciumè, G., Zizzo, G.: An energy blockchain, a use case on tendermint. In: 2018 IEEE International Conference on Environment and Electrical Engineering and 2018 IEEE Industrial and Commercial Power Systems Europe (EEEIC/I&CPS Europe), Palermo, pp. 1–5 (2018)

44. Vecchiola, C., Pandey, S., Buyya, R.: High-performance cloud computing: a view of scientific applications. In: 10th International Symposium on Pervasive Systems, Algorithms and Networks (ISPAN), pp. 4–16 (2009)
45. Evangelinos, C., Hill, C.: Cloud computing for parallel scientific HPC applications: feasibility of running coupled atmosphere-ocean climate models on Amazon's EC2. In: Cloud Computing and Its Applications (2008)
46. Feng, Y., Li, P., Liang, S.: The design concepts of power analysis software for smart dispatching in china southern power grid. South. Power Syst. Technol. 4(1), 29–34 (2010). (in Chinese)
47. Geberslassie, M., Bitzer, B.: Cloud computing for renewable power systems. In: International Conference on Renewable Energies and Power Quality (2012)
48. Agamah, S., Ekonomou, L.: A PHP application library for web-based power systems analysis. In: European Modelling Symposium. IEEE (2015)
49. Milano, F.: An open source power system analysis toolbox. IEEE Trans. Power Syst. 20(3), 1199–1206 (2005)
50. Saha, A.K.: Challenges in power systems operations, analyses, computations and application of cloud computing in electrical power systems. In: 2018 IEEE PES/IAS PowerAfrica, Cape Town, pp. 208–213 (2018)
51. Wang, X.Z., Ge, Z.Q., Ge, M.H., Wang, L., Li, L.: The research on electric power control center credit monitoring and management using cloud computing and smart workflow. In: 2018 China International Conference on Electricity Distribution (CICED), Tianjin, pp. 2732–2735 (2018)
52. Venugopal, G., Jees, S.A., Kumar, T.A.: Cloud model for power system contingency analysis. In: 2013 International Conference on Renewable Energy and Sustainable Energy (ICRESE), Coimbatore, pp. 26–31 (2013). https://doi.org/10.1109/ICRESE.2013.6927827

Hamiltonian Fuzzy Cycles in Generalized Quartic Fuzzy Graphs with Girth k

N. Jayalakshmi[1,2], S. Vimal Kumar[3(✉)], and P. Thangaraj[4]

[1] Bharathiar University, Coimbatore 641046, India
[2] Department of Mathematics, RVS Technical Campus-Coimbatore,
Coimbatore 641402, Tamilnadu, India
jayaanand6@gmail.com
[3] Department of Mathematics, Dr.R.K. Shanmugam College of Arts & Science,
Indili, Kallakurichi 606213, Tamilnadu, India
svimalkumar16@gmail.com
[4] Department of CSE, Bannari Amman Institute of Technology,
Sathiamangalam 638401, Tamilnadu, India
proftp@bitsathy.ac.in

Abstract. In this paper, we show that the Hamiltonian fuzzy cycles in generalized quartic fuzzy graph on $n(n \geq 19)$ vertices with girth k, where $k = 10, 11, 12, 13, 14, \ldots$. This, together with the result of Hamiltonian fuzzy cycles in Quartic fuzzy graphs with girth k, $k = 3, 4, 5, 6, 7, 8, 9$. Moreover, we prove that the maximum number of Hamiltonian fuzzy cycles in generalized quartic fuzzy graph on $n(n \geq 6)$ vertices with girth $k(k \geq 3)$.

Keywords: Fuzzy Graph · Quartic fuzzy graph · Hamiltonian cycles · Hamiltonian fuzzy cycles · Girth

1 Introduction

Graphs theory serve as mathematical models to analyse with success several concrete real-world issues. In modern world, fuzzy graph theory is extremely young it's been growing in no time and has various applications modern science and technology particularly within the fields of information technology, neural network and cluster analysis, medical diagnosis. The authors in [8] discussed the total degree for totally regular fuzzy graphs. [5] addressed the concept Fuzzy If-Then Rule in k-Regular Fuzzy Graph. Vimal Kumar et al. [6,12] introduced the concept of domination and dynamic chromatic number for 4-regular graphs with girth 3. Rosenfeld et al. [10] discussed the paths, cycles and connectedness results for fuzzy graphs. The author [9], proposed the concept of Hyper-Hamiltonian generalized Petersen graphs. Recently, Vimal Kumar et al. [3] introduced the concept of Hamiltonian fuzzy cycles in Quartic fuzzy graphs with girth k, $k = 3, 4, 5, 6, 7, 8, 9$. Throughout this paper, we discuss the various aspects of Hamiltonian fuzzy cycles (HFCs) in Quartic fuzzy graphs (QFGs)

© Springer Nature Switzerland AG 2020
S. Smys et al. (Eds.): ICCVBIC 2019, AISC 1108, pp. 1190–1202, 2020.
https://doi.org/10.1007/978-3-030-37218-7_125

with girth k, where $k = 10, 11, 12, 13, 14, \dots$. For more information about regular graphs and fuzzy graphs we refer the reader to [1,2,4,7,11]. In this paper, our aim is to extend the result from [3] and present to find the maximum number (max. no.) of Hamiltonian fuzzy cycles (HFCs) in generalized quartic fuzzy graphs (GQFGs) with girth k for $n \geq 6$ and $k = 3, 4, 5, \dots$

1.1 Preliminaries

We, first give a few definitions are useful for development in the succeeding articles.

Definition 1.1.1. [3]. A graph $G^* : (V, E)$ is said to be a Quartic (4-regular), if each vertex has degree 4 or $d(v) = 4$ for all $v \in V$.

Definition 1.1.2. [3]. A cycle C in a graph $G^* : (V, E)$ is said to be a Hamiltonian cycle, if it covers all the vertices of G^* exactly once except the end vertices.

Definition 1.1.3. [3]. The girth of G^* is the length of a shortest cycle in G^* and it is denoted by k.

Definition 1.1.4. [3]. Let V be a non-empty set. A fuzzy graph $G : (\sigma, \mu)$ is a pair of function $\sigma : V \rightarrow [0,1]$ and $\mu : V \times V \rightarrow [0,1]$ such that $\mu(v_1, v_2) \leq \mu(v_1) \wedge \mu(v_2)$ for all v_1, v_2 in V. Here $\mu(v_1) \wedge \mu(v_2)$ denotes the minimum of $\mu(v_1)$ & $\mu(v_2)$ and $v_1 v_2$ denotes the edge between v_1 and v_2.

Definition 1.1.5. [3]. Let $G : (\sigma, \mu)$ be a fuzzy graph on $G^* : (V, E)$. A graph G is Quartic fuzzy graph, if each vertex $\sigma(v)$ in G has same degree κ, i.e., $d[\sigma(v)] = \kappa$ for all $v \in V$.

Definition 1.1.6. [3]. In a fuzzy graph G a fuzzy cycle C covers all the vertices of G exactly once except the end vertices then the cycle is called Hamiltonian fuzzy cycle.

The GQFGs with girth 3, 4 and 5 are presented in Figs. 1, 2 and 3, respectively.

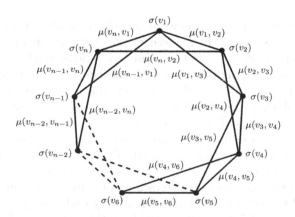

Fig. 1. Fuzzy graphs G with girth 3

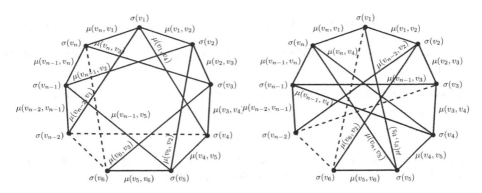

Fig. 2. Fuzzy graphs G with girth 4   **Fig. 3.** Fuzzy graphs G with girth 5

2   Main Results

In this section, we provide the HFCs in GQFGs with girth k, for $k = 10, 11, 12, ..., 20$. Finally, the main theorem, we provide the HFCs in GQFGs with girth k, for $k = 3, 4, 5, ...$

Theorem 2.1. For $n \geq 19$, GQFG with girth 10 is decomposable into
$$\begin{cases} 1, \text{ if } n \text{ is a multiple of 3} \\ 2, \text{otherwise} \end{cases}, \text{HFCs.}$$

Proof. Let G be a GQFG on $n(n \geq 19)$ vertices with girth 10. Let v_1 be an initial and a terminal vertex. By the definition of HFCs as following cases:

Case 1: Assume that n is multiples 3,
   Cycle C_1: $\sigma(v_1)\sigma(v_2)\sigma(v_3)...\sigma(v_{n-1})\sigma(v_n)\sigma(v_1)$.
Therefore, the max. no. of HFCs in GQFGs with girth 10 is one.

Case 2: Assume that n is not a multiples 3,
   Cycle C_1: $\sigma(v_1)\sigma(v_2)\sigma(v_3)...\sigma(v_{n-1})\sigma(v_n)\sigma(v_1)$.
   Cycle C_2: $P_1P_2P_3P_4P_5P_6P_7P_8P_9P_{10}$ or $P_1P_9P_8P_7P_6P_5P_4P_3P_2P_{10}$ or $P_1P_5P_9P_4P_8P_3P_7P_6P_2P_{10}$ or $P_1P_8P_6P_4P_2P_9P_7P_5P_3P_{10}$ or $P_1P_6P_2P_7P_3P_8P_4P_9P_5P_{10}$ or $P_1P_3P_5P_7P_9P_2P_4P_6P_8P_{10}$, where
$P_1 = \sigma(v_1)\sigma(v_{10})\sigma(v_{19})\sigma(v_{28})\sigma(v_{37})...\sigma(v_{9m-8})$,
$P_2 = \sigma(v_2)\sigma(v_{11})\sigma(v_{20})\sigma(v_{29})\sigma(v_{38})...\sigma(v_{9m-7})$,
$P_3 = \sigma(v_3)\sigma(v_{12})\sigma(v_{21})\sigma(v_{30})\sigma(v_{39})...\sigma(v_{9m-6})$,
$P_4 = \sigma(v_4)\sigma(v_{13})\sigma(v_{22})\sigma(v_{31})\sigma(v_{40})...\sigma(v_{9m-5})$,
$P_5 = \sigma(v_5)\sigma(v_{14})\sigma(v_{23})\sigma(v_{32})\sigma(v_{41})...\sigma(v_{9m-4})$,
$P_6 = \sigma(v_6)\sigma(v_{15})\sigma(v_{24})\sigma(v_{33})\sigma(v_{42})...\sigma(v_{9m-3})$,
$P_7 = \sigma(v_7)\sigma(v_{16})\sigma(v_{25})\sigma(v_{34})\sigma(v_{43})...\sigma(v_{9m-2})$,
$P_8 = \sigma(v_8)\sigma(v_{17})\sigma(v_{26})\sigma(v_{35})\sigma(v_{44})...\sigma(v_{9m-1})$,
$P_9 = \sigma(v_9)\sigma(v_{18})\sigma(v_{27})\sigma(v_{36})\sigma(v_{45})...\sigma(v_{9m})$,

$P_{10} = \sigma(v_1)$ and $v_{9m-8}, v_{9m-7}, v_{9m-6}, v_{9m-5}, v_{9m-4}, v_{9m-3}, v_{9m-2}, v_{9m-1}, v_{9m} \leq v_n$, $m = 1, 2, 3, \ldots$

Therefore, the max. no. of HFCs in GQFGs with girth 10 is two. This completes the proof.

Theorem 2.2. For $n \geq 21$, GQFG with girth 11 is decomposable into
$$\begin{cases} 1, & \text{if } n \text{ is even} \\ 2, & \text{otherwise} \end{cases}, \text{ HFCs.}$$

Proof. Let G be a GQFG on $n(n \geq 21)$ vertices with girth 11. Let v_1 be an initial and a terminal vertex. By the definition of HFCs as following cases:

Case 1: Suppose n is even,
   Cycle C_1: $\sigma(v_1)\sigma(v_2)\sigma(v_3)\ldots\sigma(v_{n-1})\sigma(v_n)\sigma(v_1)$.
Therefore, the max. no. of HFCs in GQFGs with girth 11 is one.

Case 2: Suppose n is not even,
   Cycle C_1: $\sigma(v_1)\sigma(v_2)\sigma(v_3)\ldots\sigma(v_{n-1})\sigma(v_n)\sigma(v_1)$.
   Cycle C_3: $P_{11}P_{12}P_{13}P_{14}P_{15}P_{16}P_{17}P_{18}P_{19}P_{20}P_{10}$ or $P_{11}P_{20}P_{19}P_{18}P_{17}P_{16}P_{15}$
$P_{14}P_{13}P_{12}P_{10}$ or $P_{11}P_{14}P_{17}P_{20}P_{13}P_{16}P_{19}$ $P_{12}P_{15}P_{18}P_{10}$ or $P_{11}P_{18}P_{15}P_{12}P_{19}$
$P_{16}P_{13}P_{20}P_{17}P_{14}P_{10}$ or $P_{11}P_{19}P_{17}P_{15}P_{13}P_{20}P_{18}P_{16}P_{14}P_{12}P_{10}$ or $P_{11}P_{13}P_{15}P_{17}$
$P_{19}P_{12}P_{14}P_{16}P_{18}P_{20}P_{10}$, where
$$P_{11} = \sigma(v_1)\sigma(v_{11})\sigma(v_{21})\sigma(v_{31})\sigma(v_{41})\ldots\sigma(v_{10m-9}),$$
$$P_{12} = \sigma(v_2)\sigma(v_{12})\sigma(v_{22})\sigma(v_{32})\sigma(v_{42})\ldots\sigma(v_{10m-8}),$$
$$P_{13} = \sigma(v_3)\sigma(v_{13})\sigma(v_{23})\sigma(v_{33})\sigma(v_{43})\ldots\sigma(v_{10m-7}),$$
$$P_{14} = \sigma(v_4)\sigma(v_{14})\sigma(v_{24})\sigma(v_{34})\sigma(v_{44})\ldots\sigma(v_{10m-6}),$$
$$P_{15} = \sigma(v_5)\sigma(v_{15})\sigma(v_{25})\sigma(v_{35})\sigma(v_{45})\ldots\sigma(v_{10m-5}),$$
$$P_{16} = \sigma(v_6)\sigma(v_{16})\sigma(v_{26})\sigma(v_{36})\sigma(v_{46})\ldots\sigma(v_{10m-4}),$$
$$P_{17} = \sigma(v_7)\sigma(v_{17})\sigma(v_{27})\sigma(v_{37})\sigma(v_{47})\ldots\sigma(v_{10m-3}),$$
$$P_{18} = \sigma(v_8)\sigma(v_{18})\sigma(v_{28})\sigma(v_{38})\sigma(v_{48})\ldots\sigma(v_{10m-2}),$$
$$P_{19} = \sigma(v_9)\sigma(v_{19})\sigma(v_{29})\sigma(v_{39})\sigma(v_{49})\ldots\sigma(v_{10m-1}),$$
$$P_{20} = \sigma(v_{10})\sigma(v_{20})\sigma(v_{30})\sigma(v_{40})\sigma(v_{50})\ldots\sigma(v_{10m}),$$
$P_{10} = \sigma(v_1)$ and $v_{10m-9}, v_{10m-8}, v_{10m-7}, v_{10m-6}, v_{10m-5}, v_{10m-4}, v_{10m-3}, v_{10m-2}, v_{10m-1}, v_{10m} \leq v_n$, $m = 1, 2, 3, \ldots$
Therefore, the max. no. of HFCs in GQFGs with girth 11 is two. This completes the proof.

Theorem 2.3. For $n \geq 23$, GQFG with girth 12 is decomposable into
$$\begin{cases} 1, & \text{if } n \text{ is multiple of } 11 \\ 2, & \text{otherwise} \end{cases}, \text{ HFCs.}$$

Proof. Let G be a GQFG on $n(n \geq 23)$ vertices with girth 12. Let v_1 be an initial and a terminal vertex. By the definition of HFCs as following cases:

Case 1: Assume that n is multiple of 11,
   Cycle C_1: $\sigma(v_1)\sigma(v_2)\sigma(v_3)\ldots\sigma(v_{n-1})\sigma(v_n)\sigma(v_1)$.
Therefore, the max. no. of HFCs in GQFGs with girth 12 is one.

Case 2: Assume that n is not a multiple of 11,

   Cycle C_1: $\sigma(v_1)\sigma(v_2)\sigma(v_3)...\sigma(v_{n-1})\sigma(v_n)\sigma(v_1)$.

   Cycle C_4: $P_{21}P_{22}P_{23}P_{24}P_{25}P_{26}P_{27}P_{28}P_{29}P_{30}P_{31}P_{10}$ or $P_{21}P_{31}P_{30}P_{29}P_{28}P_{27}$ $P_{26}P_{25}P_{24}P_{23}P_{22}P_{10}$ or $P_{21}P_{30}P_{28}P_{26}P_{24}$ $P_{22}P_{31}$ $P_{29}P_{27}P_{25}P_{23}P_{10}$ or $P_{21}P_{23}$ $P_{25}P_{27}P_{29}P_{31}P_{22}P_{24}P_{26}$ $P_{28}P_{30}P_{10}$ or $P_{21}P_{25}P_{29}P_{22}P_{26}P_{30}P_{23}P_{27}P_{31}P_{24}P_{28}P_{10}$ or $P_{21}P_{26}P_{31}P_{25}P_{30}P_{24}P_{29}P_{23}P_{28}P_{22}P_{27}P_{10}$, where

$P_{21} = \sigma(v_1)\sigma(v_{12})\sigma(v_{23})\sigma(v_{34})\sigma(v_{45})...\sigma(v_{11m-10})$,
$P_{22} = \sigma(v_2)\sigma(v_{13})\sigma(v_{24})\sigma(v_{35})\sigma(v_{46})...\sigma(v_{11m-9})$,
$P_{23} = \sigma(v_3)\sigma(v_{14})\sigma(v_{25})\sigma(v_{36})\sigma(v_{47})...\sigma(v_{11m-8})$,
$P_{24} = \sigma(v_4)\sigma(v_{15})\sigma(v_{26})\sigma(v_{37})\sigma(v_{48})...\sigma(v_{11m-7})$,
$P_{25} = \sigma(v_5)\sigma(v_{16})\sigma(v_{27})\sigma(v_{38})\sigma(v_{49})...\sigma(v_{11m-6})$,
$P_{26} = \sigma(v_6)\sigma(v_{17})\sigma(v_{28})\sigma(v_{39})\sigma(v_{50})...\sigma(v_{11m-5})$,
$P_{27} = \sigma(v_7)\sigma(v_{18})\sigma(v_{29})\sigma(v_{40})\sigma(v_{51})...\sigma(v_{11m-4})$,
$P_{28} = \sigma(v_8)\sigma(v_{19})\sigma(v_{30})\sigma(v_{41})\sigma(v_{52})...\sigma(v_{11m-3})$,
$P_{29} = \sigma(v_9)\sigma(v_{20})\sigma(v_{31})\sigma(v_{42})\sigma(v_{53})...\sigma(v_{11m-2})$,
$P_{30} = \sigma(v_{10})\sigma(v_{21})\sigma(v_{32})\sigma(v_{43})\sigma(v_{54})...\sigma(v_{11m-1})$,
$P_{31} = \sigma(v_{11})\sigma(v_{22})\sigma(v_{33})\sigma(v_{44})\sigma(v_{55})...\sigma(v_{11m})$,
$P_{10} = \sigma(v_1)$ and v_{11m-10}, v_{11m-9}, v_{11m-8}, v_{11m-7}, v_{11m-6}, v_{11m-5}, v_{11m-4}, v_{11m-3}, v_{11m-2}, v_{11m-1}, $v_{11m} \leq v_n$, $m = 1, 2, 3,...$ Therefore, the max. no. of HFCs in GQFGs with girth 12 is two. This completes the proof.

Theorem 2.4. For $n \geq 25$, GQFG with girth 13 is decomposable into
$$\begin{cases} 1, & \text{if } n \text{ is even and multiple of 27} \\ 2, & \text{otherwise} \end{cases}, \text{HFCs}.$$

Proof. Let G be a GQFG on $n(n \geq 25)$ vertices with girth 13. Let v_1 be an initial and a terminal vertex. By the definition of HFCs as following cases:

Case 1: Assume that n is even and multiple of 27,

   Cycle C_1: $\sigma(v_1)\sigma(v_2)\sigma(v_3)...\sigma(v_{n-1})\sigma(v_n)\sigma(v_1)$.

Therefore, the max. no. of HFCs in GQFGs with girth 13 is one.

Case 2: Assume that n is odd and not a multiple of 27,

   Cycle C_1: $\sigma(v_1)\sigma(v_2)\sigma(v_3)...\sigma(v_{n-1})\sigma(v_n)\sigma(v_1)$.

   Cycle C_5: $P_{32}P_{33}P_{34}P_{35}P_{36}P_{37}P_{38}P_{39}P_{40}P_{41}P_{42}P_{43}P_{10}$ or $P_{32}P_{43}P_{42}P_{41}$ $P_{40}P_{39}P_{38}P_{37}P_{36}P_{35}P_{34}P_{33}P_{10}$ or $P_{32}P_{39}P_{34}$ $P_{41}P_{36}P_{43}P_{38}P_{33}P_{40}P_{35}P_{42}P_{37}P_{10}$ or $P_{32}P_{37}P_{42}P_{35}P_{40}P_{33}P_{38}$ $P_{43}P_{36}P_{41}P_{34}P_{39}P_{10}$, where

$P_{32} = \sigma(v_1)\sigma(v_{13})\sigma(v_{25})\sigma(v_{37})\sigma(v_{49})...\sigma(v_{12m-11})$,
$P_{33} = \sigma(v_2)\sigma(v_{14})\sigma(v_{26})\sigma(v_{38})\sigma(v_{50})...\sigma(v_{12m-10})$,
$P_{34} = \sigma(v_3)\sigma(v_{15})\sigma(v_{27})\sigma(v_{39})\sigma(v_{51})...\sigma(v_{12m-9})$,
$P_{35} = \sigma(v_4)\sigma(v_{16})\sigma(v_{28})\sigma(v_{40})\sigma(v_{52})...\sigma(v_{12m-8})$,
$P_{36} = \sigma(v_5)\sigma(v_{17})\sigma(v_{29})\sigma(v_{41})\sigma(v_{53})...\sigma(v_{12m-7})$,
$P_{37} = \sigma(v_6)\sigma(v_{18})\sigma(v_{30})\sigma(v_{42})\sigma(v_{54})...\sigma(v_{12m-6})$,
$P_{38} = \sigma(v_7)\sigma(v_{19})\sigma(v_{31})\sigma(v_{43})\sigma(v_{55})...\sigma(v_{12m-5})$,
$P_{39} = \sigma(v_8)\sigma(v_{20})\sigma(v_{32})\sigma(v_{44})\sigma(v_{56})...\sigma(v_{12m-4})$,
$P_{40} = \sigma(v_9)\sigma(v_{21})\sigma(v_{33})\sigma(v_{45})\sigma(v_{57})...\sigma(v_{12m-3})$,

$P_{41} = \sigma(v_{10})\sigma(v_{22})\sigma(v_{34})\sigma(v_{46})\sigma(v_{58})...\sigma(v_{12m-2})$,
$P_{42} = \sigma(v_{11})\sigma(v_{23})\sigma(v_{35})\sigma(v_{47})\sigma(v_{59})...\sigma(v_{12m-1})$,
$P_{43} = \sigma(v_{12})\sigma(v_{24})\sigma(v_{36})\sigma(v_{48})\sigma(v_{60})...\sigma(v_{12m})$,
$P_{10} = \sigma(v_1)$ and $v_{12m-11}, v_{12m-10}, v_{12m-9}, v_{12m-8}, v_{12m-7}, v_{12m-6}, v_{12m-5},$
$v_{12m-4}, v_{12m-3}, v_{12m-2}, v_{12m-1}, v_{12m} \leq v_n$, $m = 1, 2, 3,...$ Therefore, the max.
no. of HFCs in GQFGs with girth 13 is two. This completes the proof.

Theorem 2.5. For $n \geq 27$, GQFG with girth 14 is decomposable into
$\begin{cases} 1, \text{ if } n \text{ is multiple of 13} \\ 2, \text{ otherwise} \end{cases}$, HFCs.

Proof. Let G be a GQFG on $n(n \geq 27)$ vertices with girth 14. Let v_1 be an
initial and a terminal vertex. By the definition of HFCs as following cases:

Case 1: Assume that n is multiple of 13,
   Cycle C_1: $\sigma(v_1)\sigma(v_2)\sigma(v_3)...\sigma(v_{n-1})\sigma(v_n)\sigma(v_1)$.
Therefore, the max. no. of HFCs in quartic fuzzy graphs with girth 14 is one.
Case 2: Assume that n is not a multiple of 13,

   Cycle C_1: $\sigma(v_1)\sigma(v_2)\sigma(v_3)...\sigma(v_{n-1})\sigma(v_n)\sigma(v_1)$.
   Cycle C_6: $P_{44}P_{45}P_{46}P_{47}P_{48}P_{49}P_{50}P_{51}P_{52}P_{53}P_{54}P_{55}P_{56}P_{10}$ or $P_{44}P_{54}P_{51}$
$P_{48}P_{45}P_{55}P_{52}P_{49}P_{46}P_{56}P_{53}P_{50}P_{47}P_{10}$ or $P_{44}P_{52}P_{47}P_{55}P_{50}P_{45}P_{53}P_{48}P_{56}P_{51}P_{46}$
$P_{54}P_{49}P_{10}$ or $P_{44}P_{50}P_{56}P_{49}P_{55}P_{48}P_{54}P_{47}P_{53}P_{56}P_{52}P_{45}P_{51}P_{10}$ or $P_{44}P_{46}P_{48}$
$P_{50}P_{52}P_{54}P_{56}P_{45}P_{47}P_{49}P_{51}P_{53}P_{55}P_{10}$ or $P_{44}P_{53}P_{49}P_{45}P_{54}P_{50}P_{46}P_{55}P_{51}P_{47}P_{56}$
$P_{52}P_{48}P_{10}$ or $P_{44}P_{55}P_{53}P_{51}P_{49}P_{47}P_{45}P_{56}P_{54}P_{52}P_{50}P_{48}P_{46}P_{10}$, where
$P_{44} = \sigma(v_1)\sigma(v_{14})\sigma(v_{27})\sigma(v_{40})\sigma(v_{53})...\sigma(v_{13m-12})$,
$P_{45} = \sigma(v_2)\sigma(v_{15})\sigma(v_{28})\sigma(v_{41})\sigma(v_{54})...\sigma(v_{13m-11})$,
$P_{46} = \sigma(v_3)\sigma(v_{16})\sigma(v_{29})\sigma(v_{42})\sigma(v_{55})...\sigma(v_{13m-10})$,
$P_{47} = \sigma(v_4)\sigma(v_{17})\sigma(v_{30})\sigma(v_{43})\sigma(v_{56})...\sigma(v_{13m-9})$,
$P_{48} = \sigma(v_5)\sigma(v_{18})\sigma(v_{31})\sigma(v_{44})\sigma(v_{57})...\sigma(v_{13m-8})$,
$P_{49} = \sigma(v_6)\sigma(v_{19})\sigma(v_{32})\sigma(v_{45})\sigma(v_{58})...\sigma(v_{13m-7})$,
$P_{50} = \sigma(v_7)\sigma(v_{20})\sigma(v_{33})\sigma(v_{46})\sigma(v_{59})...\sigma(v_{13m-6})$,
$P_{51} = \sigma(v_8)\sigma(v_{21})\sigma(v_{34})\sigma(v_{47})\sigma(v_{60})...\sigma(v_{13m-5})$,
$P_{52} = \sigma(v_9)\sigma(v_{22})\sigma(v_{35})\sigma(v_{48})\sigma(v_{61})...\sigma(v_{13m-4})$,
$P_{53} = \sigma(v_{10})\sigma(v_{23})\sigma(v_{36})\sigma(v_{49})\sigma(v_{62})...\sigma(v_{13m-3})$,
$P_{54} = \sigma(v_{11})\sigma(v_{24})\sigma(v_{37})\sigma(v_{50})\sigma(v_{63})...\sigma(v_{13m-2})$,
$P_{55} = \sigma(v_{12})\sigma(v_{25})\sigma(v_{38})\sigma(v_{51})\sigma(v_{64})...\sigma(v_{13m-1})$,
$P_{56} = \sigma(v_{13})\sigma(v_{26})\sigma(v_{39})\sigma(v_{52})\sigma(v_{65})...\sigma(v_{13m})$,
$P_{10} = \sigma(v_1)$ and $v_{13m-12}, v_{13m-11}, v_{13m-10}, v_{13m-9}, v_{13m-8}, v_{13m-7},$
$v_{13m-6}, v_{13m-5}, v_{13m-4}, v_{13m-3}, v_{13m-2}, v_{13m-1}, v_{13m} \leq v_n$, $m = 1, 2, 3,...$
Therefore, the max. no. of HFCs in GQFGs with girth 14 is two. This com-
pletes the proof.

Theorem 2.6. For $n \geq 29$, GQFG with girth 15 is decomposable into
$\begin{cases} 1, \text{ if } n \text{ is even} \\ 2, \text{ otherwise} \end{cases}$, HFCs.

Proof. Let G be a GQFG on $n(n \geq 29)$ vertices with girth 15. Let v_1 be an initial and a terminal vertex. By the definition of HFCs as following cases:

Case 1: Assume that n is even,

Cycle C_1: $\sigma(v_1)\sigma(v_2)\sigma(v_3)...\sigma(v_{n-1})\sigma(v_n)\sigma(v_1)$.

Therefore, the max. no. of HFCs in quartic fuzzy graphs with girth 15 is one.

Case 2: Assume that n is odd,

Cycle C_1: $\sigma(v_1)\sigma(v_2)\sigma(v_3)...\sigma(v_{n-1})\sigma(v_n)\sigma(v_1)$.

Cycle C_7: $P_{57}P_{58}P_{59}P_{60}P_{61}P_{62}P_{63}P_{64}P_{65}P_{66}P_{67}P_{68}P_{69}P_{70}P_{10}$ or $P_{57}P_{70}P_{69}$ $P_{68}P_{67}P_{66}P_{65}P_{64}P_{63}P_{62}P_{61}P_{60}P_{59}P_{58}P_{10}$ or $P_{57}P_{60}P_{63}P_{66}P_{69}P_{58}P_{61}P_{64}P_{67}P_{70}$ $P_{59}P_{62}P_{65}P_{68}P_{10}$ or $P_{57}P_{62}P_{67}P_{58}P_{63}P_{68}P_{59}P_{64}P_{69}P_{60}P_{65}P_{70}P_{61}P_{68}P_{10}$ or P_{57} $P_{59}P_{61}P_{63}P_{65}P_{67}P_{69}P_{58}P_{60}P_{62}P_{64}P_{66}P_{68}P_{70}P_{10}$ or $P_{57}P_{68}P_{65}P_{62}P_{59}P_{70}P_{67}P_{64}$ $P_{61}P_{58}P_{69}P_{66}P_{63}P_{60}P_{10}$, where

$P_{57} = \sigma(v_1)\sigma(v_{15})\sigma(v_{29})\sigma(v_{43})\sigma(v_{57})...\sigma(v_{14m-13})$,
$P_{58} = \sigma(v_2)\sigma(v_{16})\sigma(v_{30})\sigma(v_{44})\sigma(v_{58})...\sigma(v_{14m-12})$,
$P_{59} = \sigma(v_3)\sigma(v_{17})\sigma(v_{31})\sigma(v_{45})\sigma(v_{59})...\sigma(v_{14m-11})$,
$P_{60} = \sigma(v_4)\sigma(v_{18})\sigma(v_{32})\sigma(v_{46})\sigma(v_{60})...\sigma(v_{14m-10})$,
$P_{61} = \sigma(v_5)\sigma(v_{19})\sigma(v_{33})\sigma(v_{47})\sigma(v_{61})...\sigma(v_{14m-9})$,
$P_{62} = \sigma(v_6)\sigma(v_{20})\sigma(v_{34})\sigma(v_{48})\sigma(v_{62})...\sigma(v_{14m-8})$,
$P_{63} = \sigma(v_7)\sigma(v_{21})\sigma(v_{35})\sigma(v_{49})\sigma(v_{63})...\sigma(v_{14m-7})$,
$P_{64} = \sigma(v_8)\sigma(v_{22})\sigma(v_{36})\sigma(v_{50})\sigma(v_{64})...\sigma(v_{14m-6})$,
$P_{65} = \sigma(v_9)\sigma(v_{23})\sigma(v_{37})\sigma(v_{51})\sigma(v_{65})...\sigma(v_{14m-5})$,
$P_{66} = \sigma(v_{10})\sigma(v_{24})\sigma(v_{38})\sigma(v_{52})\sigma(v_{66})...\sigma(v_{14m-4})$,
$P_{67} = \sigma(v_{11})\sigma(v_{25})\sigma(v_{39})\sigma(v_{53})\sigma(v_{67})...\sigma(v_{14m-3})$,
$P_{68} = \sigma(v_{12})\sigma(v_{26})\sigma(v_{40})\sigma(v_{54})\sigma(v_{68})...\sigma(v_{14m-2})$,
$P_{69} = \sigma(v_{13})\sigma(v_{27})\sigma(v_{41})\sigma(v_{55})\sigma(v_{69})...\sigma(v_{14m-1})$,
$P_{70} = \sigma(v_{14})\sigma(v_{28})\sigma(v_{42})\sigma(v_{56})\sigma(v_{70})...\sigma(v_{14m})$,

$P_{10} = \sigma(v_1)$ and $v_{12m-11}, v_{12m-10}, v_{12m-9}, v_{12m-8}, v_{12m-7}, v_{12m-6}, v_{12m-5}, v_{12m-4}, v_{12m-3}, v_{12m-2}, v_{12m-1}, v_{12m} \leq v_n$, $m = 1, 2, 3,...$ Therefore, the max. no. of HFCs in GQFGs with girth 15 is two. This completes the proof.

Theorem 2.7. For $n \geq 31$, GQFG with girth 16 is decomposable into
$$\begin{cases} 1, \text{ if } n \text{ is multiple of 15} \\ 2, \text{ otherwise} \end{cases}, \text{ HFCs.}$$

Proof. Let G be a GQFG on $n(n \geq 31)$ vertices with girth 16. Let v_1 be an initial and a terminal vertex. By the definition of HFCs as following cases:

Case 1: Assume that n is multiple of 15,

Cycle C_1: $\sigma(v_1)\sigma(v_2)\sigma(v_3)...\sigma(v_{n-1})\sigma(v_n)\sigma(v_1)$.

Therefore, the max. no. of HFCs in GQFGs with girth 16 is one.

Case 2: Assume that n is not a multiple of 15,

Cycle C_1: $\sigma(v_1)\sigma(v_2)\sigma(v_3)...\sigma(v_{n-1})\sigma(v_n)\sigma(v_1)$.

Cycle C_8: $P_{71}P_{72}P_{73}P_{74}P_{75}P_{76}P_{77}P_{78}P_{79}P_{80}P_{81}P_{82}P_{83}P_{84}P_{85}P_{10}$ or $P_{71}P_{85}$ $P_{84}P_{83}P_{82}P_{81}P_{80}P_{79}P_{78}P_{77}P_{76}P_{75}P_{74}P_{73}P_{72}P_{10}$ or $P_{71}P_{84}P_{82}P_{80}P_{78}P_{76}P_{74}P_{72}$

$P_{85}P_{83}P_{81}P_{79}P_{77}P_{75}P_{73}\ P_{10}$ or $P_{71}P_{52}P_{78}P_{74}P_{85}P_{81}P_{77}P_{73}P_{84}P_{80}P_{86}P_{72}P_{83}P_{79}$
$P_{75}P_{10}$ or $P_{71}P_{75}P_{79}P_{83}P_{72}P_{76}P_{80}\ P_{84}P_{73}P_{77}P_{81}P_{85}P_{74}P_{78}P_{82}P_{10}$ or $P_{71}P_{78}P_{85}$
$P_{77}P_{84}P_{76}P_{83}P_{75}P_{82}P_{74}P_{81}P_{83}P_{80}P_{72}P_{79}P_{10}$, where
$P_{71} = \sigma(v_1)\sigma(v_{16})\sigma(v_{31})\sigma(v_{46})\sigma(v_{61})...\sigma(v_{15m-14}),$
$P_{72} = \sigma(v_2)\sigma(v_{17})\sigma(v_{32})\sigma(v_{47})\sigma(v_{62})...\sigma(v_{15m-13}),$
$P_{73} = \sigma(v_3)\sigma(v_{18})\sigma(v_{33})\sigma(v_{48})\sigma(v_{63})...\sigma(v_{15m-12}),$
$P_{74} = \sigma(v_4)\sigma(v_{19})\sigma(v_{34})\sigma(v_{49})\sigma(v_{64})...\sigma(v_{15m-11}),$
$P_{75} = \sigma(v_5)\sigma(v_{20})\sigma(v_{35})\sigma(v_{50})\sigma(v_{65})...\sigma(v_{15m-10}),$
$P_{76} = \sigma(v_6)\sigma(v_{21})\sigma(v_{36})\sigma(v_{51})\sigma(v_{66})...\sigma(v_{15m-9}),$
$P_{77} = \sigma(v_7)\sigma(v_{22})\sigma(v_{37})\sigma(v_{52})\sigma(v_{67})...\sigma(v_{15m-8}),$
$P_{78} = \sigma(v_8)\sigma(v_{23})\sigma(v_{38})\sigma(v_{53})\sigma(v_{68})...\sigma(v_{15m-7}),$
$P_{79} = \sigma(v_9)\sigma(v_{24})\sigma(v_{39})\sigma(v_{54})\sigma(v_{69})...\sigma(v_{15m-6}),$
$P_{80} = \sigma(v_{10})\sigma(v_{25})\sigma(v_{40})\sigma(v_{55})\sigma(v_{70})...\sigma(v_{15m-5}),$
$P_{81} = \sigma(v_{11})\sigma(v_{26})\sigma(v_{41})\sigma(v_{56})\sigma(v_{71})...\sigma(v_{15m-4}),$
$P_{82} = \sigma(v_{12})\sigma(v_{27})\sigma(v_{42})\sigma(v_{57})\sigma(v_{72})...\sigma(v_{15m-3}),$
$P_{83} = \sigma(v_{13})\sigma(v_{28})\sigma(v_{43})\sigma(v_{58})\sigma(v_{73})...\sigma(v_{15m-2}),$
$P_{84} = \sigma(v_{14})\sigma(v_{29})\sigma(v_{44})\sigma(v_{59})\sigma(v_{74})...\sigma(v_{15m-1}),$
$P_{85} = \sigma(v_{15})\sigma(v_{30})\sigma(v_{45})\sigma(v_{60})\sigma(v_{75})...\sigma(v_{15m}),$
$P_{10} = \sigma(v_1)$ and $v_{15m-14},\ v_{15m-13},\ v_{15m-12},\ v_{15m-11},\ v_{15m-10},\ v_{15m-9},\ v_{15m-8},$
$v_{15m-7},\ v_{15m-6},\ v_{15m-5},\ v_{15m-4},\ v_{15m-3},\ v_{15m-2},\ v_{15m-1},\ v_{15m} \leq v_n,\ m = 1,2,3,...$ Therefore, the max. no. of HFCs in GQFGs with girth 16 is two. This completes the proof.

Theorem 2.8. For $n \geq 33$, GQFG with girth 17 is decomposable into
$$\begin{cases} 1, & \text{if } n \text{ is even} \\ 2, & \text{otherwise} \end{cases}, \text{HFCs.}$$

Proof. Let G be a GQFG on $n(n \geq 33)$ vertices with girth 17. Let v_1 be an initial and a terminal vertex. By the definition of HFCs as following cases:

Case 1: Assume that n is even,
   Cycle C_1: $\sigma(v_1)\sigma(v_2)\sigma(v_3)...\sigma(v_{n-1})\sigma(v_n)\sigma(v_1)$.
Therefore, the max. no. of HFCs in GQFGs with girth 17 is one.

Case 2: Assume that n is odd,
   Cycle C_1: $\sigma(v_1)\sigma(v_2)\sigma(v_3)...\sigma(v_{n-1})\sigma(v_n)\sigma(v_1)$.
   Cycle C_9: $P_{86}P_{87}P_{88}P_{89}P_{90}P_{91}P_{92}P_{93}P_{94}P_{95}P_{96}P_{97}P_{98}P_{99}P_{100}P_{101}P_{10}$ or
$P_{86}P_{101}P_{100}P_{99}P_{98}P_{97}P_{96}P_{95}P_{94}P_{93}P_{92}P_{91}P_{90}P_{89}P_{88}P_{87}P_{10}$ or $P_{86}P_{97}P_{92}P_{87}$
$P_{98}P_{93}P_{88}P_{99}P_{94}P_{89}\ P_{100}P_{95}P_{90}P_{101}P_{96}P_{91}P_{10}$ or $P_{86}P_{99}P_{96}P_{93}P_{90}P_{87}P_{100}P_{97}$
$P_{94}P_{91}P_{88}P_{101}P_{98}P_{95}P_{92}P_{89}P_{10}$, or $P_{86}P_{95}P_{88}P_{97}P_{90}P_{99}P_{92}\ P_{101}P_{94}P_{87}P_{96}P_{89}$
$P_{98}P_{91}P_{100}P_{93}P_{10}$, where
$P_{86} = \sigma(v_1)\sigma(v_{17})\sigma(v_{33})\sigma(v_{49})\sigma(v_{65})...\sigma(v_{16m-15}),$
$P_{87} = \sigma(v_2)\sigma(v_{18})\sigma(v_{34})\sigma(v_{50})\sigma(v_{66})...\sigma(v_{16m-14}),$
$P_{88} = \sigma(v_3)\sigma(v_{19})\sigma(v_{35})\sigma(v_{51})\sigma(v_{67})...\sigma(v_{16m-13}),$
$P_{89} = \sigma(v_4)\sigma(v_{20})\sigma(v_{36})\sigma(v_{52})\sigma(v_{68})...\sigma(v_{16m-12}),$
$P_{90} = \sigma(v_5)\sigma(v_{21})\sigma(v_{37})\sigma(v_{53})\sigma(v_{69})...\sigma(v_{16m-11}),$
$P_{91} = \sigma(v_6)\sigma(v_{22})\sigma(v_{38})\sigma(v_{54})\sigma(v_{70})...\sigma(v_{16m-10}),$

$P_{92} = \sigma(v_7)\sigma(v_{23})\sigma(v_{39})\sigma(v_{55})\sigma(v_{71})...\sigma(v_{16m-9}),$

$P_{93} = \sigma(v_8)\sigma(v_{24})\sigma(v_{40})\sigma(v_{56})\sigma(v_{72})...\sigma(v_{16m-8}),$

$P_{94} = \sigma(v_9)\sigma(v_{25})\sigma(v_{41})\sigma(v_{57})\sigma(v_{73})...\sigma(v_{16m-7}),$

$P_{95} = \sigma(v_{10})\sigma(v_{26})\sigma(v_{42})\sigma(v_{58})\sigma(v_{74})...\sigma(v_{16m-6}),$

$P_{96} = \sigma(v_{11})\sigma(v_{27})\sigma(v_{43})\sigma(v_{59})\sigma(v_{75})...\sigma(v_{16m-5}),$

$P_{97} = \sigma(v_{12})\sigma(v_{28})\sigma(v_{44})\sigma(v_{60})\sigma(v_{76})...\sigma(v_{16m-4}),$

$P_{98} = \sigma(v_{13})\sigma(v_{29})\sigma(v_{45})\sigma(v_{61})\sigma(v_{77})...\sigma(v_{16m-3}),$

$P_{99} = \sigma(v_{14})\sigma(v_{30})\sigma(v_{46})\sigma(v_{62})\sigma(v_{78})...\sigma(v_{16m-2}),$

$P_{100} = \sigma(v_{15})\sigma(v_{31})\sigma(v_{47})\sigma(v_{63})\sigma(v_{79})...\sigma(v_{16m-1}),$

$P_{101} = \sigma(v_{16})\sigma(v_{32})\sigma(v_{48})\sigma(v_{64})\sigma(v_{80})...\sigma(v_{16m}),$

$P_{10} = \sigma(v_1)$ and $v_{16m-15},\ v_{16m-14},\ v_{16m-13},\ v_{16m-12},\ v_{16m-11},\ v_{16m-10},\ v_{16m-9},$ $v_{16m-8},\ v_{16m-7},\ v_{16m-6},\ v_{16m-5},\ v_{16m-4},\ v_{16m-3},\ v_{16m-2},\ v_{16m-1},\ v_{16m} \leq v_n,$ $m = 1,2,3,...$ Therefore, the max. no. of HFCs in GQFGs with girth 17 is two. This completes the proof.

Theorem 2.9. For $n \geq 35$, GQFG with girth 18 is decomposable into $\begin{cases} 1, & \text{if } n \text{ is multiple of } 17 \\ 2, & \text{otherwise} \end{cases}$, HFCs.

Proof. Let G be a GQFG on $n(n \geq 35)$ vertices with girth 18. Let v_1 be an initial and a terminal vertex. By the definition of HFCs as following cases:

Case 1: Assume that n is multiple of 17,

   Cycle C_1: $\sigma(v_1)\sigma(v_2)\sigma(v_3)...\sigma(v_{n-1})\sigma(v_n)\sigma(v_1)$.

Therefore, the max. no. of HFCs in GQFGs with girth 18 is one.

Case 2: Assume that n is not a multiple of 17,

   Cycle C_1: $\sigma(v_1)\sigma(v_2)\sigma(v_3)...\sigma(v_{n-1})\sigma(v_n)\sigma(v_1)$.

   Cycle C_{10}: $P_{102}P_{103}P_{104}P_{105}P_{106}P_{107}P_{108}P_{109}P_{110}P_{111}P_{112}P_{113}P_{114}P_{115}$ $P_{116}P_{117}P_{118}P_{10}$ or $P_{102}P_{118}P_{117}P_{116}P_{115}P_{114}P_{113}P_{112}P_{111}P_{110}P_{109}P_{108}P_{107}$ $P_{106}P_{105}P_{104}P_{103}P_{10}$ or $P_{102}P_{104}P_{106}P_{108}P_{110}P_{112}P_{114}P_{116}P_{118}P_{103}P_{105}P_{107}$ $P_{109}P_{111}$ $P_{113}P_{115}P_{117}P_{10}$ or $P_{102}P_{106}P_{110}P_{114}P_{118}P_{105}P_{109}P_{111}P_{113}P_{117}P_{104}$ $P_{108}P_{112}P_{116}P_{103}P_{107}P_{105}P_{10}$ or $P_{102}P_{107}P_{112}P_{117}P_{105}P_{110}P_{115}P_{103}P_{108}P_{113}$ $P_{118}P_{106}P_{111}P_{116}P_{104}P_{109}P_{104}P_{10}$ or $P_{102}P_{105}P_{108}P_{111}P_{114}P_{117}P_{103}P_{106}P_{109}$ $P_{112}P_{115}P_{118}P_{104}P_{107}P_{110}P_{113}P_{116}P_{10}$ or $P_{102}P_{110}P_{118}P_{109}P_{117}P_{108}P_{116}P_{107}$ $P_{115}P_{114}P_{105}P_{113}P_{104}P_{112}P_{103}P_{111}P_{106}P_{10},$ or $P_{102}P_{111}P_{103}P_{112}P_{104}P_{103}P_{105}$ $P_{114}P_{106}P_{115}P_{107}P_{116}P_{108}P_{117}P_{109}P_{118}P_{110}$ P_{10}, where

$P_{102} = \sigma(v_1)\sigma(v_{18})\sigma(v_{35})\sigma(v_{52})\sigma(v_{69})...\sigma(v_{17m-16}),$

$P_{103} = \sigma(v_2)\sigma(v_{19})\sigma(v_{36})\sigma(v_{53})\sigma(v_{70})...\sigma(v_{17m-15}),$

$P_{104} = \sigma(v_3)\sigma(v_{20})\sigma(v_{37})\sigma(v_{54})\sigma(v_{71})...\sigma(v_{17m-14}),$

$P_{105} = \sigma(v_4)\sigma(v_{21})\sigma(v_{38})\sigma(v_{55})\sigma(v_{72})...\sigma(v_{17m-13}),$

$P_{106} = \sigma(v_5)\sigma(v_{22})\sigma(v_{39})\sigma(v_{56})\sigma(v_{73})...\sigma(v_{17m-12}),$

$P_{107} = \sigma(v_6)\sigma(v_{23})\sigma(v_{40})\sigma(v_{57})\sigma(v_{74})...\sigma(v_{17m-11}),$

$P_{108} = \sigma(v_7)\sigma(v_{24})\sigma(v_{41})\sigma(v_{58})\sigma(v_{75})...\sigma(v_{17m-10}),$

$P_{109} = \sigma(v_8)\sigma(v_{25})\sigma(v_{42})\sigma(v_{59})\sigma(v_{76})...\sigma(v_{17m-9}),$

$P_{110} = \sigma(v_9)\sigma(v_{26})\sigma(v_{43})\sigma(v_{60})\sigma(v_{77})...\sigma(v_{17m-8}),$

$P_{111} = \sigma(v_{10})\sigma(v_{27})\sigma(v_{44})\sigma(v_{61})\sigma(v_{78})...\sigma(v_{17m-7})$,
$P_{112} = \sigma(v_{11})\sigma(v_{28})\sigma(v_{45})\sigma(v_{62})\sigma(v_{79})...\sigma(v_{17m-6})$,
$P_{113} = \sigma(v_{12})\sigma(v_{29})\sigma(v_{46})\sigma(v_{63})\sigma(v_{80})...\sigma(v_{17m-5})$,
$P_{114} = \sigma(v_{13})\sigma(v_{30})\sigma(v_{47})\sigma(v_{64})\sigma(v_{81})...\sigma(v_{17m-4})$,
$P_{115} = \sigma(v_{14})\sigma(v_{31})\sigma(v_{48})\sigma(v_{65})\sigma(v_{82})...\sigma(v_{17m-3})$,
$P_{116} = \sigma(v_{15})\sigma(v_{32})\sigma(v_{49})\sigma(v_{66})\sigma(v_{83})...\sigma(v_{17m-2})$,
$P_{117} = \sigma(v_{16})\sigma(v_{33})\sigma(v_{50})\sigma(v_{67})\sigma(v_{84})...\sigma(v_{17m-1})$,
$P_{118} = \sigma(v_{17})\sigma(v_{34})\sigma(v_{51})\sigma(v_{68})\sigma(v_{85})...\sigma(v_{17m})$,
$P_{10} = \sigma(v_1)$ and v_{17m-16}, v_{17m-15}, v_{17m-14}, v_{17m-13}, v_{17m-12}, v_{17m-11}, v_{17m-10}, v_{17m-9}, v_{17m-8}, v_{17m-7}, v_{17m-6}, v_{17m-5}, v_{17m-4}, v_{17m-3}, v_{17m-2}, v_{17m-1}, $v_{17m} \le v_n$, $m = 1,2,3,...$ Therefore, the max. no. of HFCs in GQFGs with girth 18 is two. This completes the proof.

Theorem 2.10. For $n \ge 37$, GQFG with girth 19 is decomposable into
$$\begin{cases} 1, \text{ if } n \text{ is even and multiple of } 39 \\ 2, \text{ otherwise} \end{cases}, \text{ HFCs.}$$

Proof. Let G be a GQFG on $n(n \ge 37)$ vertices with girth 19. Let v_1 be an initial and a terminal vertex. By the definition of HFCs as following cases:

Case 1: Assume that n is even and a multiple of 39,
   Cycle C_1: $\sigma(v_1)\sigma(v_2)\sigma(v_3)...\sigma(v_{n-1})\sigma(v_n)\sigma(v_1)$.
Therefore, the max. no. of HFCs in GQFGs with girth 19 is one.

Case 2: Assume that n is odd and not a multiple of 39,
   Cycle C_1: $\sigma(v_1)\sigma(v_2)\sigma(v_3)...\sigma(v_{n-1})\sigma(v_n)\sigma(v_1)$.
   Cycle C_{11}: $P_{119}P_{120}P_{121}P_{122}P_{123}P_{124}P_{125}P_{126}P_{127}P_{128}P_{129}P_{130}P_{131}P_{132}$ $P_{133}P_{134}P_{135}P_{136}P_{10}$ or $P_{119}P_{136}P_{135}P_{134}P_{133}P_{132}P_{131}P_{130}P_{129}P_{128}P_{127}P_{126}$ $P_{125}P_{124}P_{123}P_{122}P_{121}P_{120}$ P_{10} or $P_{119}P_{132}P_{127}P_{122}P_{135}P_{130}P_{125}P_{120}P_{133}P_{128}$ $P_{123}P_{136}P_{131}P_{126}P_{121}P_{134}P_{129}P_{124}P_{10}$ or $P_{119}P_{130}P_{123}P_{134}P_{120}P_{131}$ P_{124} P_{135} $P_{128}P_{121}P_{132}P_{125}P_{136}P_{129}P_{122}P_{133}P_{126}P_{121}P_{10}$ or $P_{119}P_{124}P_{129}P_{134}P_{121}P_{126}$ $P_{131}P_{136}P_{123}P_{128}P_{133}P_{120}P_{125}P_{130}P_{135}P_{122}P_{127}P_{132}P_{10}$ or $P_{119}P_{126}P_{133}P_{122}$ $P_{129}P_{136}P_{125}P_{132}P_{121}P_{128}P_{135}P_{126}P_{131}P_{120}P_{127}P_{134}P_{123}P_{130}P_{10}$, where
$P_{119} = \sigma(v_1)\sigma(v_{19})\sigma(v_{37})\sigma(v_{55})\sigma(v_{73})...\sigma(v_{18m-17})$,
$P_{120} = \sigma(v_2)\sigma(v_{20})\sigma(v_{38})\sigma(v_{56})\sigma(v_{74})...\sigma(v_{18m-16})$,
$P_{121} = \sigma(v_3)\sigma(v_{21})\sigma(v_{39})\sigma(v_{57})\sigma(v_{75})...\sigma(v_{18m-15})$,
$P_{122} = \sigma(v_4)\sigma(v_{22})\sigma(v_{40})\sigma(v_{58})\sigma(v_{76})...\sigma(v_{18m-14})$,
$P_{123} = \sigma(v_5)\sigma(v_{23})\sigma(v_{41})\sigma(v_{59})\sigma(v_{77})...\sigma(v_{18m-13})$,
$P_{124} = \sigma(v_6)\sigma(v_{24})\sigma(v_{42})\sigma(v_{60})\sigma(v_{78})...\sigma(v_{18m-12})$,
$P_{125} = \sigma(v_7)\sigma(v_{25})\sigma(v_{43})\sigma(v_{61})\sigma(v_{79})...\sigma(v_{18m-11})$,
$P_{126} = \sigma(v_8)\sigma(v_{26})\sigma(v_{44})\sigma(v_{62})\sigma(v_{80})...\sigma(v_{18m-10})$,
$P_{127} = \sigma(v_9)\sigma(v_{27})\sigma(v_{45})\sigma(v_{63})\sigma(v_{81})...\sigma(v_{18m-9})$,
$P_{128} = \sigma(v_{10})\sigma(v_{28})\sigma(v_{46})\sigma(v_{64})\sigma(v_{82})...\sigma(v_{18m-8})$,
$P_{129} = \sigma(v_{11})\sigma(v_{29})\sigma(v_{47})\sigma(v_{65})\sigma(v_{83})...\sigma(v_{18m-7})$,
$P_{130} = \sigma(v_{12})\sigma(v_{30})\sigma(v_{48})\sigma(v_{66})\sigma(v_{84})...\sigma(v_{18m-6})$,
$P_{131} = \sigma(v_{13})\sigma(v_{31})\sigma(v_{49})\sigma(v_{67})\sigma(v_{85})...\sigma(v_{18m-5})$,
$P_{132} = \sigma(v_{14})\sigma(v_{32})\sigma(v_{50})\sigma(v_{68})\sigma(v_{86})...\sigma(v_{18m-4})$,

$P_{133} = \sigma(v_{15})\sigma(v_{33})\sigma(v_{51})\sigma(v_{69})\sigma(v_{87})...\sigma(v_{18m-3})$,
$P_{134} = \sigma(v_{16})\sigma(v_{34})\sigma(v_{52})\sigma(v_{70})\sigma(v_{88})...\sigma(v_{18m-2})$,
$P_{135} = \sigma(v_{17})\sigma(v_{35})\sigma(v_{53})\sigma(v_{71})\sigma(v_{89})...\sigma(v_{18m-1})$,
$P_{136} = \sigma(v_{18})\sigma(v_{36})\sigma(v_{54})\sigma(v_{72})\sigma(v_{90})...\sigma(v_{18m})$,
$P_{10} = \sigma(v_1)$ and v_{18m-17}, v_{18m-16}, v_{18m-15}, v_{18m-14}, v_{18m-13}, v_{18m-12}, v_{18m-11}, v_{18m-10}, v_{18m-9}, v_{18m-8}, v_{18m-7}, v_{18m-6}, v_{18m-5}, v_{18m-4}, v_{18m-3}, v_{18m-2}, v_{18m-1}, $v_{18m} \leq v_n$, $m = 1,2,3,...$ Therefore, the max. no. of HFCs in GQFGs with girth 19 is two. This completes the proof.

Theorem 2.11. For $n \geq 39$, GQFG with girth 20 is decomposable into
$$\begin{cases} 1, & \text{if } n \text{ is multiple of } 19 \\ 2, & \text{otherwise} \end{cases}, \text{ HFCs.}$$

Proof. Let G be a GQFG on $n(n \geq 39)$ vertices with girth 20. Let v_1 be an initial and a terminal vertex. By the definition of HFCs as following cases:

Case 1: Assume that n is multiple of 19,
   Cycle C_1: $\sigma(v_1)\sigma(v_2)\sigma(v_3)...\sigma(v_{n-1})\sigma(v_n)\sigma(v_1)$.
Therefore, the max. no. of HFCs in GQFGs with girth 20 is one.

Case 2: Assume that n is not a multiple of 19,
   Cycle C_1: $\sigma(v_1)\sigma(v_2)\sigma(v_3)...\sigma(v_{n-1})\sigma(v_n)\sigma(v_1)$.
   Cycle C_{12}: $P_{137}P_{138}P_{139}P_{140}P_{141}P_{142}P_{143}P_{144}P_{145}P_{146}P_{147}P_{148}P_{149}P_{150}$ $P_{151}P_{152}P_{153}P_{154}P_{155}P_{10}$ or $P_{137}P_{155}P_{154}P_{153}P_{152}P_{151}P_{150}P_{149}P_{148}P_{147}P_{146}$ $P_{145}P_{144}P_{143}P_{142}P_{141}P_{140}P_{139}P_{138}P_{10}$ or $P_{137}P_{151}P_{146}P_{141}P_{155}P_{150}P_{145}P_{140}$ $P_{154}P_{149}P_{144}P_{139}P_{153}P_{148}P_{143}P_{138}P_{152}P_{147}P_{142}P_{10}$ or $P_{137}P_{150}P_{144}P_{138}P_{151}$ $P_{145}P_{139}P_{152}P_{146}P_{140}P_{153}P_{147}P_{141}P_{154}P_{148}P_{142}P_{155}P_{149}P_{143}P_{10}$ or $P_{137}P_{153}$ $P_{150}P_{147}P_{144}P_{141}P_{138}P_{154}P_{151}P_{148}P_{145}P_{142}P_{139}P_{155}P_{152}P_{149}P_{146}P_{143}P_{140}P_{10}$ or $P_{137}P_{139}P_{141}$ $P_{143}P_{145}P_{147}P_{149}P_{151}P_{153}P_{155}P_{138}P_{140}P_{142}P_{144}P_{146}P_{148}P_{150}P_{152}$ $P_{154}P_{10}$ or $P_{137}P_{149}P_{142}P_{144}P_{147}P_{140}P_{152}P_{145}P_{138}P_{150}P_{143}P_{155}P_{148}P_{141}P_{153}$ $P_{146}P_{139}P_{151}P_{144}P_{10}$, where
$P_{137} = \sigma(v_1)\sigma(v_{20})\sigma(v_{39})\sigma(v_{58})\sigma(v_{77})...\sigma(v_{19m-18})$,
$P_{138} = \sigma(v_2)\sigma(v_{21})\sigma(v_{40})\sigma(v_{59})\sigma(v_{78})...\sigma(v_{19m-17})$,
$P_{139} = \sigma(v_3)\sigma(v_{22})\sigma(v_{41})\sigma(v_{60})\sigma(v_{79})...\sigma(v_{19m-16})$,
$P_{140} = \sigma(v_4)\sigma(v_{23})\sigma(v_{42})\sigma(v_{61})\sigma(v_{80})...\sigma(v_{19m-15})$,
$P_{141} = \sigma(v_5)\sigma(v_{24})\sigma(v_{43})\sigma(v_{62})\sigma(v_{81})...\sigma(v_{19m-14})$,
$P_{142} = \sigma(v_6)\sigma(v_{25})\sigma(v_{44})\sigma(v_{63})\sigma(v_{82})...\sigma(v_{19m-13})$,
$P_{143} = \sigma(v_7)\sigma(v_{26})\sigma(v_{45})\sigma(v_{64})\sigma(v_{83})...\sigma(v_{19m-12})$,
$P_{144} = \sigma(v_8)\sigma(v_{27})\sigma(v_{46})\sigma(v_{65})\sigma(v_{84})...\sigma(v_{19m-11})$,
$P_{145} = \sigma(v_9)\sigma(v_{28})\sigma(v_{47})\sigma(v_{66})\sigma(v_{85})...\sigma(v_{19m-10})$,
$P_{146} = \sigma(v_{10})\sigma(v_{29})\sigma(v_{48})\sigma(v_{67})\sigma(v_{86})...\sigma(v_{19m-9})$,
$P_{147} = \sigma(v_{11})\sigma(v_{30})\sigma(v_{49})\sigma(v_{68})\sigma(v_{87})...\sigma(v_{19m-8})$,
$P_{148} = \sigma(v_{12})\sigma(v_{31})\sigma(v_{50})\sigma(v_{69})\sigma(v_{88})...\sigma(v_{19m-7})$,
$P_{149} = \sigma(v_{13})\sigma(v_{32})\sigma(v_{51})\sigma(v_{70})\sigma(v_{89})...\sigma(v_{19m-6})$,
$P_{150} = \sigma(v_{14})\sigma(v_{33})\sigma(v_{52})\sigma(v_{71})\sigma(v_{90})...\sigma(v_{19m-5})$,
$P_{151} = \sigma(v_{15})\sigma(v_{34})\sigma(v_{53})\sigma(v_{72})\sigma(v_{91})...\sigma(v_{19m-4})$,
$P_{152} = \sigma(v_{16})\sigma(v_{35})\sigma(v_{54})\sigma(v_{73})\sigma(v_{92})...\sigma(v_{19m-3})$,

$P_{153} = \sigma(v_{17})\sigma(v_{36})\sigma(v_{55})\sigma(v_{74})\sigma(v_{93})...\sigma(v_{19m-2})$,
$P_{154} = \sigma(v_{18})\sigma(v_{37})\sigma(v_{56})\sigma(v_{75})\sigma(v_{94})...\sigma(v_{19m-1})$,
$P_{155} = \sigma(v_{19})\sigma(v_{38})\sigma(v_{57})\sigma(v_{76})\sigma(v_{95})...\sigma(v_{19m})$,
$P_{10} = \sigma(v_1)$ and $v_{19m-18}, v_{19m-17}, v_{19m-16}, v_{19m-15}, v_{19m-14}, v_{19m-13}, v_{19m-12}, v_{19m-11}, v_{19m-10}, v_{19m-9}, v_{19m-8}, v_{19m-7}, v_{19m-6}, v_{19m-5}, v_{19m-4}, v_{19m-3}, v_{19m-2}, v_{19m-1}, v_{19m} \leq v_n$, $m = 1, 2, 3, ...$ Therefore, the max. no. of HFCs in GQFGs with girth 20 is two. This completes the proof.

Now, we focus the main theorem, that is to find the HFCs in GQFG on $n(n \geq 6)$ vertices with girth $k(k \geq 3)$.

Theorem 2.12. For $n \geq 6$, GQFG with girth k, $k = 3, 4, 5, ...$ and $k \neq 10$ is decomposable into

$$\begin{cases} 1, & \text{if } n \text{ is even, multiple of } k-1 \text{ and } 2k+1 \\ 2, & \text{otherwise} \end{cases}, \text{ HFCs.}$$

Proof. Let G be a GQFG on $n(n \geq 6)$ vertices with girth $k = 3, 4, 5, ...$ and $k \neq 10$. Let v_1 be an initial and a terminal vertex. By the definition of HFCs as following cases:

Case 1: Assume that n is even, multiple of $k-1$ and $2k+1$,
    Cycle C_1: $\sigma(v_1)\sigma(v_2)\sigma(v_3)...\sigma(v_{n-1})\sigma(v_n)\sigma(v_1)$.
Therefore, the max. no. of HFCs in GQFGs with girth $k = 3, 4, 5, ...$ and $k \neq 10$ is one.

Case 2: Assume that n is odd and not a multiple of $k-12$ and $2k+1$,
    Cycle C_1: $\sigma(v_1)\sigma(v_2)\sigma(v_3)...\sigma(v_{n-1})\sigma(v_n)\sigma(v_1)$.
    Cycle C_2: $\{P_i : 1 \geq i \leq k-1\}\sigma(v_1)$,
where $\{P_i = \sigma(v_i)\sigma(v_{k-1+i})\sigma(v_{2k-2+i})\sigma(v_{3k-3+i})... \sigma(v_{(k-1)m-(k-(i+1))})$ for all $1 \leq i \leq k-1$ and $\sigma(v_i)\sigma(v_{k-1+i})\sigma(v_{2k-2+i})\sigma(v_{3k-3+i})...\sigma(v_{(k-1)m-(k-(i+1))} \leq v_n$, $m \geq 1$, $k \geq 3$. Therefore, the max. no. of HFCs in GQFGs with girth $k = 3, 4, 5, ...$ and $k \neq 10$ is two. This completes the proof.

Remark 2.13. For girth $k = 10$, the maximum number of Hamiltonian fuzzy cycle is 1, if n is a multiple of 3; otherwise the Hamiltonian fuzzy cycle is 2.

3   Conclusion

The paper [2, Theorem 3.1–3.7] is a survey about the result for HFCs in Quartic fuzzy graphs with girth k, $k = 3, 4, 5, 6, 7, 8, 9$. and present paper combining results of these Theorems 2.1, 2.2, 2.3, 2.4, 2.5, 2.6, 2.7, 2.8, 2.9, 2.10 and 2.11, we have completely proved the main Theorem 2.12. From the discussion above we can conclude that the max. no. of HFCs in GQFG on $n(n \geq 6)$ vertices with girth $k(k \geq 3)$. This generalized result is applicable for network applications and graph theory.

Compliance with Ethical Standards. All authors declare that there is no conflict of interest. No humans/animals involved in this research work. We have used our own data.

References

1. Bondy, J.A., Murty, U.S.R.: Graph Theory with Applications. North-Holland, New York (1980)
2. Shan, E., Kang, L.: The ferry cover problem on regular graphs and small-degree graphs. Chin. Ann. Math. Ser. B **39**(6), 933–946 (2018)
3. Jayalakshmi, N., Vimal Kumar, S., Thangaraj, P.: Hamiltonian fuzzy cycles in quartic fuzzy graphs with girth K (K = 3, 4, 5, 6, 7, 8, 9). Int. J. Pharm. Technol. **8**, 17619–17626 (2016)
4. Lyle, J.: A structural approach for independent domination of regular graphs. Graphs Comb. **1**, 1–22 (2014)
5. Kohila, S., Vimal Kumar, S.: Fuzzy if-then rule in k-regular fuzzy graph, karpagam. Int. J. Appl. Math. **7**, 19–23 (2016)
6. Mohanapriya, N., Vimal Kumar, S., Vernold Vivin, J., Venkatachalam, M.: Domination in 4-regular graphs with girth 3. Proc. Nat. Acad. Sci. India Sect. A **85**, 259–264 (2015)
7. Mordeson, J.N., Nair, P.S.: Fuzzy Graphs and Fuzzy Hyper Graphs. Springer, Heidelberg (1998)
8. Nagoor Gani, A., Radha, K.: On regular fuzzy graphs. J. Phys. Sci. **12**, 33–40 (2008)
9. Parimelazhagan, R., Sulochana, V., Vimal Kumar, S.: Secure domination in 4-regular planar and non-planar graphs with girth 3. Utilitas Math. **108**, 273–282 (2018)
10. Rosenfeld, A., Zadeh, L.A., Fu, K.S., Shimura, M.: Fuzzy Graphs, Fuzzy Sets and their Applications, pp. 77–95. Academic Press, New York (1975)
11. Sunitha, M.S., Mathew, S.: Fuzzy graph theory: a survey. Ann. Pure Appl. Math. **4**, 92–110 (2013)
12. Vimal Kumar, S., Mohanapriya, N., Vernold Vivin, J., Venkatachalam, M.: On dynamic chromatic number of 4-regular graphs with girth 3 and 4. Asian J. Math. Comput. Res. **7**, 345–353 (2016)

Effective Model Integration Algorithm for Improving Prediction Accuracy of Healthcare Ontology

P. Monika[1(⊠)] and G. T. Raju[2]

[1] VTU, Dayananda Sagar College of Engineering, Bengaluru, Karnataka, India
monikamanjunath@gmail.com
[2] VTU, RNS Institute of Technology, Bengaluru, Karnataka, India
gtraju1990@yahoo.com

Abstract. In this digital era, the web is filled with enormous data in various formats & sizes resulting in decreased utilization of the data. Healthcare is one of the major domains that contributes data to web on daily basis in large numbers and numerous formats. Semantic web deals intelligently with its ontological data. The concept of Machine Learning coupled with statistical techniques for good prediction accuracy includes tree based algorithms like ID3, C4.5 etc. During the phase of resolving tree under-fitting & over-fitting issues which arises based on the dataset quality, couple of strong rules may be lost due to tree pruning which results in decreasing the prediction accuracy. An Effective Model Integration Algorithm (EMIA) has been proposed as a solution based on the concepts of Hidden Markov Model (HMM) and Stochastic Automata Model (SAM). The presented approach attempts to generate improved decision rules by integrating the rules of chosen models based on threshold in a Healthcare ontology resulting in increased prediction accuracy of 86%.

Keywords: Semantic web · Ontologies · Ontology agents · Ontologies integration · Health care · Diabetology · Domains · Decision support · Decision tree · SPARQL · Rule based reasoning

1 Introduction

Web in general intellect, is a huge collection of data in raw, semi-processed and processed format irrespective of several domains. Semantic web deals with facts and meaning associated within the web. Healthcare is one such domain which contributes towards huge volume of data generation on a daily basis. However the generated data is not promised to be successfully utilized when needed even within the same domain due to semantic limitations with respect to data representation and processing formats. Discovering knowledge from the vast semantic web tosses lot of challenges to be handled like various data patterns, missing data, uncertainty, different class labels etc. Tabular representation of data no more supports for automated knowledge discovery. The data represented in the graphical pattern unlocks lot of opportunities towards achieving machine automated decisions and better prediction accuracy. A way of representing web data: Ontology - formally defined by Gruber [24] as "Explicit

© Springer Nature Switzerland AG 2020
S. Smys et al. (Eds.): ICCVBIC 2019, AISC 1108, pp. 1203–1212, 2020.
https://doi.org/10.1007/978-3-030-37218-7_126

specification of a shared conceptualization" is the one to which today's web is being transformed to. Operating on ontological graph data helps achieve automated decision making by constructing decision rules and applying those rules during querying the graphical representations like.rdf, .ttl, .owl, .trig etc. file formats using SPARQL Protocol And RDF Query Language (SPARQL) queries.

Most common predictive modeling approach - Decision tree algorithms learn from the decision tree built during the process of training the model. Trained model works on the testing datasets predicting the results based on the rules, outputting the confusion matrix for accuracy scores. Researchers often prefer decision tree algorithms for data mining in a semantic web due to their various advantages like simplicity, ability to handle various types of data, statistical validation support, ability to handle large datasets and robustness against co-linearity over the limitations like NP – Complete, over fitting etc. An attempt is made here to produce improved decision rules by following the steps of proposed Effective Model Integration Algorithm [EMIA] and use the generated rules for constructing ontology & derive superior automated instances. Shooting SPARQL queries on so created ontology, demonstrates the improved relevance and accuracy of the search results benefitting medical researchers & practitioners. The proposed work has been experimented on dataset of Diabetology, a subdomain of Healthcare. The work include proposal of an EMIA for generating better decision rules and methodologies for creating ontology based on the generated enhanced rules for improved prediction accuracy. The rest of the paper is organized as follows: Sect. 2 describes the algorithms and techniques employed by other researchers towards efficient rules generation. Section 3 presents the proposed EMI Algorithm. Section 4 elaborates on Experimental Testing and results of the suggested methodology. Applying the enhanced rules onto Ontology and extracting the semantic information is explained in Sect. 5 followed by conclusion in Sect. 6.

2   Related Work

During the phase of learning for knowledge extraction, extracted rules can be applied onto the ontological data. Allegro graph and Owlim [1, 2] are few of such rule based reasoners which are capable of inferring large scale instances of an ontology. Ontology based classification (OBC) is applied more on document classification [3, 4] as it is best suitable for representing contextual knowledge with an advantage of integrating heterogeneous data [5]. The Ontostar [6] ontology works by applying C4.5 learning algorithm on to the training dataset, by extracting the rules in the Semantic Web Rule Language (SWRL). A predictive analysis method for diabetes prediction using C4.5 algorithm [7] method builds a decision tree based on the training dataset and with the learnt experience predicts the future cases with the prediction accuracy of 76% approx. for the given test dataset [8]. Healthcare system working models respond by predicting risk of patients based on the features monitored regularly [9] supported by wrapper feature selection concept with bagging [10]. Random forest method shows promising result of 80.84% improvement during diabetes prediction when all attributes were used during prediction [11]. Similarity measures like Euclidian distance used between the classification response vectors enhances the classification accuracy [12]. In medical practice, the concept of ontology is extensively used for semantic description and data interaction & integration to achieve automated knowledge discovery [13]. Overall, Ontologies can be summarized as knowledge representation tools in decision support systems [14]. At the other hand, for remote assessment of the patients' health, the concept of Hidden Semi Markov Model has been employed [15] where the observations are compared relatively for the judgment. Prediction of diabetes and hypotension remotely using HMM coupled with neural networks and fuzzy sets have resulted in good accuracies [16, 17] concluding that the HMM works better with high prediction accuracies.

3 Effective Model Integration Algorithmic Approach Based on C4.5 Decision Tree Algorithm

The decision tree models with target variable owning discrete set of values are called as classification trees. The most famous decision tree based classification technique extensively used today, C4.5 [18] is a supervised learning algorithm proposed by Quinlan which uses pessimistic method during decision rule generation. The algorithm begins with computation of measuring criteria like *SplitInfo, Entropy, Gain*, and *Gain Ratio* based on the given *Training Dataset (T)* & *Attributes (A)* followed with selection of the best attribute a_{best} as candidate for root node of the decision tree. The process continues on the subsets of the training dataset induced based a_{best} until all the attributes are processed or the stopping criteria is reached. Further at end, the algorithm outputs the constructed decision tree and learnt decision rules which will be used for further predictions with better accuracy. Lot many techniques like *Bagging, Boosting, AdaBoost* are noted in literature to enhance the accuracy of prediction of classification algorithms. Couple of machine learning algorithms itself can be hybrid to improve the prediction accuracy. With the related literature survey it's understood that not much contributions are found towards directly improving the quality of model rules generated. The well-known decision tree algorithm: C4.5 shows less prediction accuracy in some cases due to loss of quality decision rules while fixing the issues of over-fitting and under-fitting of the decision trees.

For modeling the stochastic sequences with an underling Finite Automata structure, Hidden Markov Model (HMM) shown in Fig. 1 is most extensively used. Generally, HMM can be viewed as Non-deterministic Finite Automata (NFA) SAM generalization [19] in which state reductions can be made as part of generalization or as Markovian finite state structure probabilistic generalization forming Markov chains depending on their immediate predecessors with hidden states during emissions.

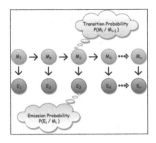

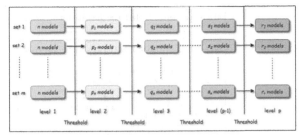

Fig. 1. Hidden Markov Model

Fig. 3. Proposed EMIA models' states

The state variables of HMM depends only on their immediate predecessors thereby forming a Markov chain following greedy strategy. The famous state merging hypothesis [19] of Automata Theory is as follows: *Hypothesis 1: "Two states can be integrated if they are in-distinguishable".* Considering states as models, to integrate any two models, initial model to be generated, an algorithm to be identified to integrate the models, resultant models to be evaluated followed with halting the process with generalization. Based on the theory of SAM & Markov chain perspectives of IIMM and the fact that decision accuracy of C4.5 algorithm depends on the quality of decision rules generated with respect to the training dataset though the model prediction accuracy computed depends on the test dataset predictions, the effective model integration algorithmic approach represented in Fig. 2 has been proposed. Figure 3 depicts the generalized representation of the suggested algorithm with model status at various levels.

Given,

$$\mu = \frac{\sum_{i=1}^{n} a_i}{n} \tag{1}$$

$$\sigma = \sqrt{\frac{\sum_{i=1}^{n} (\mu - a_i)^2}{n}} \tag{2}$$

$$Threshold = \mu + k\sigma \tag{3}$$

$$Sim(x, y) = \frac{q}{q + r + s} \tag{4}$$

Where, m is the number of sets, n is the number of models in each set
$\forall p_i \in P$, $i < n$ *(level-1)*, $\forall q_i \in Q$, $i < n$ *(level-2)*, $\forall s_i \in S$, $i < n$ *(level p-2)*
$\forall r_i \in R$, $i < n$ *(level p-1)*, a_i is the model accuracy of the ith model in that particular set, $k \in N$. μ (1) is the mean and σ (2) is standard deviation of the resultant values.

Algorithm: Effective Model Integration Algorithm (EMIA) to improve prediction Accuracy

---

Input: Training dataset (T), Number of sets (M), Number of Models in each set (N)
Output: Enhanced Model based on C4.5 classification rules

1. Let DS be the Dataset created from T of varied size, where $|DS| = M$
2. Let A be a vector representing prediction
3. Let BSM be a list with 3 data fields representing Binary Similarity Measure
4. Let T_N represent threshold of all models in particular set of M
5. $\forall m \in M, |N| = C$ /* C is a decimal constant > 5 for optimal solution */
6. for $i = 1$ to $|M|$ do
7.   for $j = 1$ to $|N|$ do
8.   Using DS_i, execute C4.5 algorithm
9.   Document rules of model (j)
10.   Document Prediction Accuracy of model(j) in A
11.   end for
12. end for
13. $\forall N \in M$, Compute Threshold (T_N) based on A grouped on M,
14. for $i = 1$ to $|M|$ do
15.   for $j = 1$ to $|N|$ do
16.   if $A(N) \geq T_N$ then
17.   Mark model m_{ij} eligible for integration   /* where $m \in N$ */
18.   end if
19.   end for
20. end for
21. for $i = 1$ to $|M|$ do
22. $\forall (a,b) \in N_i$ /* N_i gets updated with integrated models every iteration */
23.   if models (a and b) are marked as eligible for integration then
24.   Compute Binary Similarity Measure of a and b and store in BSM
25.   Compute Threshold (T_N) based on BSM scores from recent construct
26. $\forall (a,b) \in N_i$ present in BSM /* BSM from recent construct */
27.   if $BSM(a,b) > T_N$ then   /* T_N from recent construct */
28.   Integrate the models a & b and rename as Integrated model
29. end for
30. Repeat the preceding for loop until no more intra models are eligible for integration
31. Consider M itself as N as result of previous step
32. Repeat the preceding for loop until no more inter models are eligible for integration
33. Finally from the resultant integrated models, choose the model with highest prediction accuracy by verifying on C4.5 results as final Enhanced Model
34. Return the Enhanced Model

Fig. 2. Proposed Effective Model Integration Algorithm

The proposed idea starts with m sets with each set having n models equivalent to states of HMM in a SAM form. Instead of probability functions in HMM, the suggested approach considers the threshold values for integration of models if found similar proved upon computation of binary similarity scores between the models for models generalization. To begin with accuracies of all models in all the sets will be documented at level-1. The intra models get selected to further levels for integrating based on the following steps:

$\forall a \in n$ in each model m,

- At Level-1: calculate accuracy, apply threshold (3) to select the model a further.
- From level-2 to level-p, using the Asymmetric Binary Similarity measure method, the concept of clustering is applied for determining the pair of rules to be integrated and the process continues till the last level to get final set of rules so as to attain better decision accuracy. Asymmetric Binary Similarity Measure (BSM) whose coefficients x and y are called as Jaccard coefficient is expressed as in (4) using the documented results of Table 1. The contingency table for binary attributes in Table 1 holds the quantitative values measuring the interrelation between them. Let x and y be 2 models for comparison. The Eq. (4) computes asymmetric binary

similarity value between the rule sets of 2 different models. sim(x, y) = 0 implies no match between the models x and y. whereas Sim(x, y) nearest to value 1 indicates better match between the models. From the resultant matrix of models & respective similarity measures and observations of prediction similarity, the models crossing the threshold based on the similarity scores are integrated at each level within the respective models.

Table 1. Binary attributes contingency matrix

Object x	Object y		
	1	**0**	**Sum**
1	q	r	q + r
0	s	t	s + t
Sum	q + s	r + t	p

The process will be repeated recursively until no further combinations are possible. At the last level, inter model combinations will be attempted for integration resulting in the r models of level p.

4 Experimental Testing and Results

The proposed approach has been experimented on Pima Indian Diabetic dataset of 768 records collected from UCI machine learning repository [20]. The dataset is about diabetic and non-diabetic readings about the women of age above 21 years. After pre-processing the data with Cluster based Missing Value Imputation Algorithm (CMVI) [21], the attributes Plasma glucose concentration, Diastolic blood pressure, Diabetic pedigree function and Age were finalized as most prominent attributes for accuracy enhancement of the prediction results using the proposed EMIA. As proof of concept, a sample computation steps are as follows: The input dataset has been split into 5 sets with varying sizes of 50, 100, 150, 200 and 250 randomly chosen entries. Each set consists of 10 different models. Initially at level 1, prediction accuracies of all the models based on C4.5 algorithmic procedure has been considered for selecting the models eligible for integration based on threshold. At level 2 & 3, the BSM has been computed among the selected models from previous level and all the models satisfying the threshold condition are integrated and fed as input to the next level. The stated procedures to be followed until no further integrations are possible at intra-set level. At level 4, which is last level in the present example, BSM is computed at inter-set level as previous level winded up intra-set combinations and the models of set 4 and set 5 from level 3 were satisfying the threshold condition. Hence integrated and selected as final enhanced model. Further inter set combination of these models couldn't be done as the threshold rule was not satisfied.

$$Precision = \frac{TP}{TP + FP} \tag{5}$$

$$Recall = \frac{TP}{TP + FN} \tag{6}$$

$$Accuracy = \frac{TP + TN}{TP + FP + FN + TN} \tag{7}$$

The so obtained final model was evaluated using *Precision* (5), *Recall* (6) and *Accuracy* (7) metrics. The observed precision, recall and accuracy scores of all the models for a sample dataset of size 150 are as recorded in Table 2. Similarly the observations were tabulated for the rest of the dataset sizes.

Table 2. Sample Precision, Recall and Accuracy scores of existing and proposed techniques

Model no.	Dataset size = 150					
	Existing method			Enhanced method		
	Precision	Recall	Accuracy (%)	Precision	Recall	Accuracy (%)
1	0.46	0.86	73.33	0.78	0.34	80.68
2	0.46	0.86	73.33	0.69	0.33	82.88
3	0.29	0.57	56.66	0.60	0.23	78.04
4	0.50	0.71	76.66	0.83	0.50	83.85
5	0.50	0.71	76.66	0.80	0.50	83.85
6	0.45	0.71	73.33	0.79	0.33	85.68
7	0.50	0.57	76.66	0.70	0.49	84.33
8	0.50	0.57	76.66	0.70	0.48	84.69
9	0.50	0.57	76.66	0.78	0.39	85.09
10	0.50	0.57	76.66	0.48	0.40	76.92
Average →	**0.47**	**0.67**	**73.66**	**0.72**	**0.40**	**82.60**

By observing the mean scores of precision, recall and accuracy against various data sizes (Fig. 4), it's evident that the proposed model precisions are higher, recalls are lower and accuracies are higher for all the cases compared to the existing approach.

The prediction accuracy of the proposed EMIA enhanced model for the complete dataset was compared with the other standard algorithms as witnessed in Fig. 5 justifying that the accuracy of the proposed EMIA approach is better with the value of 86% compared to K-means (71%), CART (73%) and C4.5 (82%) algorithms.

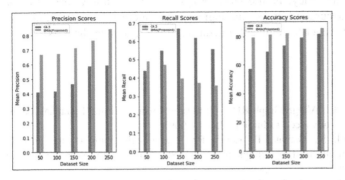
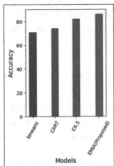

Fig. 4. Metrics comparison between existing and proposed methods

Fig. 5. Accuracy scores

5 Enhanced Rules and Ontology

The enhanced rules obtained using EMIA can be utilized efficiently by following the sequence of methodology starting with enhanced rule generation, followed by building Ontology - by applying the rules defined in .pie file using tools like GraphDB ([22]) or using Cellfie tool available in protege [23] and querying the constructed knowledge graph using SPARQL queries for efficient knowledge retrieval. Observations conclude that the ontology built using the stated methodology is witnessed with the increased derived instances of up to 10% compared to the derived instances of well-known decision algorithms resulting in good prediction rate comparatively.

6 Conclusion

The web today is termed as Semantic web as it operates on the semantics of graph data rather than the syntax being stored to achieve high prediction accuracy. İnspite of lot of technological advancement in the field of Machine Learning, Still the existing algorithms fail to achieve good prediction accuracy due to rule reductions during tree pruning. To overcome this drawback, an Effective Model Integration Algorithm (EMIA) has been proposed based on the concepts of Hidden Markov Model and Stochastic Automata Model (SAM). Depending on the accuracy scores and Binary Similarity Measures, threshold is set. The models from various sets with different dataset sizes undergo integration process by integrating their respective decision rules generated at both intra and inter set levels until the threshold condition is satisfied. The experimental results reveal that the proposed algorithmic technique achieves enhanced precision, recall and accuracy scores compared to existing techniques. The suggested model outperforms with the overall accuracy score of 86% for the present dataset. The observations conclude that the proposed approach performs better irrespective of domains and hence can be applied extensively to various datasets & domains to enhance the prediction accuracy. Future work progresses on extending the work for enhancing prediction accuracy across ontologies upon interaction connecting the semantic data at schema level for efficient utilization of the available datasets.

References

1. Chang, W.W., Miller, B.: AllegroGraph RDF – triplestore evaluation. In: Technical report, Joint study with Adobe advanced technology Labs (2009)
2. Thakker, D.A., Gohil, S., Osman, T., Lakin, P.: A pragmatic approach to semantic repositories benchmarking. In: Proceedings of the 7th Extended Semantic Web Conference 2010, pp. 379–393 (2010)
3. Song, M., Lim, S., Kang, D., Le, S.: Ontology-based automatic classification of web documents. In: International Conference on Intelligent Computing Computational Intelligence, pp. 690–700. Springer (2006)
4. Zhang, X., Hu, B., Chen, J., Moore, P.: Ontology-based context modeling for emotion recognition in an intelligent web. World Wide Web 16(4), 497–513 (2013)
5. Gómez-Romero, J., Serrano, M.A., García, J., Molina, J.M., Rogova, G.: Context-based multi-level information fusion for harbor surveillance. J. Inf. Fusion 21, 173–186 (2014)
6. Liu, B., Yao, L., Han, D.: Harnessing ontology and machine learning for RSO classification. SpringerPlus 5, 1655 (2016)
7. Jain, R.: Rule generation using decision trees. In: Indian Agricultural Statistics Research Institute (IASRI) (2012)
8. Kalyankar, G.D., Poojara, S.R., Dharwadkar, N.V.: Predictive analysis of diabetic patient data using machine learning and hadoop. In: IEEE International Conference on IoT in Social, Mobile, Analytics and Cloud (I-SMAC) (2017)
9. Lakshmi, B.N., Indumathi, T.S., Nandini, R.: A study on C4.5 decision tree classification algorithm for risk predictions during pregnancy. Procedia Technol. 24, 1542–1549 (2016)
10. Lee, S.-J., Xu, Z., Li, T., Yang, Y.: A novel bagging C4.5 algorithm based on wrapper feature selection for supporting wise clinical decision making. J. Biomed. Inf. 88, 144–155 (2018)
11. Zou, Q., Qu, K., Luo, Y., Yin, D., Ju, Y., Tang, H.: Predicting diabetes mellitus with machine learning techniques. Frontiers Genet. 9 (2018). Article 515
12. Wang, H., Wang, J.: An effective image representation method using kernel classification. In: IEEE 26th International Conference on Tools with Artificial Intelligence (2014)
13. Gangemi, A., Pisanelli, D.M., Steve, G.: An overview of the ONIONS project: applying ontologies to the integration of medical terminologies. Data Knowl. Eng. 31, 183–220 (1999)
14. Zhang, X., Hu, B., Ma, X., Moore, P., Chen, J.: Ontology driven decision support for the diagnosis of mild cognitive impairment. J. Comput. Methods Programs Biomed. 113, 781–791 (2014)
15. Capecci, M., Ceravolo, M.G., Ferracuti, F., Iarlori, S., Kyrki, V., Monteriù, A., Romeo, L., Verdini, F.: A hidden semi-markov model based approach for rehabilitation exercise assessment. J. Biomed. Inf. 78, 1–11 (2018)
16. Singh, A., Tamminedi, T., Yosiphon, G., Ganguli, A., Yadegar, J.: Hidden markov models for modeling blood pressure data to predict acute hypotension. In: IEEE International Conference on Acoustics, Speech and Signal Processing (2010)

17. Gill, N.S., Mittal, P.: A Novel hybrid model for diabetic prediction using hidden markov model, fuzzy based rule approach and neural network. Indian J. Sci. Technol. **9**(35), 192–199 (2016)
18. Quinlan, J.R.: Improved use of continuous attributes in C4.5. ACM J. Artif. Intell. Res. **4**(1), 77–90 (1996)
19. Hopcraft, J.E., Ullman, J.D.: Introduction to Automata Theory, Languages and Computation, Text book edition edn. Addison-Wesley, Boston (1979)
20. Dua, D., Graff, C.: UCI machine learning repository. University of California, School of Information and Computer Science, Irvine (2019). [http://archive.ics.uci.edu/ml]
21. Monika, P., Raju, G.T.: Data pre-processing and customized onto-graph construction for knowledge extraction in healthcare domain of semantic web. Int. J. Innovative Technol. Exploring Eng. (IJITEE) **8**(11) (2019)
22. Guting, R.H.: GraphDB: modeling and querying graphs in databases. In: Proceedings of 20th International Conference on Very Large Databases, pp. 297–308 (1994)
23. Musen, M.A.: The protégé project: a look back and a look forward. AI Matters **1**(4), 4–12 (2015). Association of Computing Machinery Specific Interest Group in Artificial Intelligence
24. Gruber, T.R.: A translation approach to portable ontology specifications. J. Knowl. Acquis. Curr. Issues Knowl. Model. **5**(2), 199–220 (1993). https://doi.org/10.1006/knac.1993.1008

Detection of Leaf Disease Using Hybrid Feature Extraction Techniques and CNN Classifier

Vidyashree Kanabur[1]([⊠]), Sunil S. Harakannanavar[1],
Veena I. Purnikmath[1], Pramod Hullole[1], and Dattaprasad Torse[2]

[1] S. G. Balekundri Institute of Technology, Belagavi, Karanataka, India
Vidyashreerk1992@gmail.com, Sunilsh143@gmail.com
[2] KLS Gogte Institute of Technology, Belagavi, Karanataka, India

Abstract. Identification of leaf disease using multiple descriptors is presented. Initially the images are resized to 256×256 to maintain the uniformity throughout the experiment. The Histogram Equalization (HE) technique is employed on resized leaf images to improve their quality. Segmentation is performed using k means clustering. The contour tracing technique is applied on leaf images to trace the boundary of affected areas. Prominent features of leaf image are extracted using DWT, PCA and GLCM techniques. The performance of the system is evaluated using three different classifiers viz., SVM, KNN and CNN using Matlab on bean leaf database. The performance of proposed model on CNN classifier is better when compared with other existing methodologies.

Keywords: Acquisition · Discrete wavelet transform · Grey level co-occurrence matrix · Support vector machine · Convolutional neural network

1 Introduction

The main objective of research in agriculture is to increase the productivity of food with reduced expenditure and enhanced quality of goods. The production of agriculture and its quality depends on soil, seed, and agro chemicals. The agricultural products (vegetables and fruits) are the essential for the survival of human beings. The agricultural productivity decreases when the crops are affected by various diseases. The diseases of plants are nothing but impairments that modifies the normal plants in their vital functions viz., photosynthesis, pollination, fertilization and germination. Due to the effect of adverse environmental conditions, diseases in plants are caused by various fungi, bacteria and viruses. If the disease is not detected at early stage then yield decreases. Figure 1 shows the sample of bacterial disease on plant. It reduces twenty to twenty five percent of its attribute production due to bacterial diseases on plant. In this paper, identification of leaf disease using multiple descriptors is presented. Initially the images are captured using physical devices and database is created containing healthy and unhealthy leaf. The HE is employed on resized images for improving the leaf quality. The significant features of leaf images are extracted using DWT, PCA and

© Springer Nature Switzerland AG 2020
S. Smys et al. (Eds.): ICCVBIC 2019, AISC 1108, pp. 1213–1220, 2020.
https://doi.org/10.1007/978-3-030-37218-7_127

GLCM techniques. The classifiers such as SVM, CNN and KNN are employed to classify and recognize the leaf images.

Fig. 1. Bacterial disease on plant [6]

In paper, rest of the content is arranged as follows. Related work on leaf disease detection is discussed in Sect. 2. Section 3 discusses the proposed methodology for leaf disease detection. The experimentation results on proposed model are explained in Sect. 4. The conclusion of the proposed work is given in Sect. 5.

2 Related Work

Here, the work related to existing methodologies on detection of leaf disease is discussed. Kaur et al. [1] detected the leaf disease in earlier stage of plant. The RGB image of leaf was converted into HSI image in the preprocessing approach to enhance the quality of image. The significant details of the image are extracted using k mean clustering approach. The features of the leaf image are classified using Neural Network, SVM, Fuzzy and K-nearest neighborclassifier. Sujatha et al. [2] adopted k means clustering algorithm for segmentation of leaf image which divides the leaf into different clusters. The diseases of different plant species have recorded using SVM classifier. Kumar et al. [3] applied color transformation approach to convert RGB to HSI image. Then the segmentation process is performed on converted image by masking of pixel and removes the unwanted color using specific threshold. The statistics from SGDM matrices evaluates various plant leaf diseases using neural network, K-nearest neighbors and fuzzy logic classifiers. Khairnar et al. [4] considered input RGB color image of leaf and converted into other color spaces such as HSI and CIELAB. The segmentation of leaf image was performed by k-mean clustering algorithm. After segmentation of leaf image, infected regions are described by extracting features. The features produced based on color, texture and shape are normally used for description of region. The classification is performed using minimum distance criterion and support vector machine. Batule et al. [5] applied the conversion of RGB into HSI color space image. In addition to this, it also converts the leaf image from RGB to HSV format. The model includes image color transform, image smoothing and disease spot segmentation. The features of the leaf image are classified using Neural Network, SVM. Kamlapurkar et al. [6] adopted Gabor filter to segment the leaf image and improve its quality. The features obtained are matched using ANN. Chaware et al. [7] adopted the various classifiers to match the features using SVM, KNN, ANN, and Fuzzy Logic. Rathod et al. [8] converted the leaf image from RGB to CIELAB color components. The features are extracted using K-Medoids algorithm which provides the color and shape descriptions. The produced features are classified using neural Network classifier. Varshney et al. [9] converts the color leaf image from RGB model into HSI model. The color co-occurrence method (CCM) was employed on segmented image where significant details of leaf image are extracted. Finally the obtained features are classified using neural network. Baskaran et al. [10] developed a model to calculate the level of RGB content in an image. The features are extracted using K-mean clustering approach and the features extracted are matched using SVM. Mainkar et al. [11] adopted a combination of GLCM and K-means clustering approach to fetch the useful information of leaf image. Features extracted from GLCM are matched using neural network and radial basis functions. Al-Hiary et al. [12] converted the color leaf image from RGB into HSI format. The features are extracted using K-means clustering and co-occurrence approach for texture analysis of leaf image. The NN is employed to classify the leaf image. Megha et al. [13] applied a method to convert the RGB into Grayscale model for resizing and image filtration. The primary details of the leaf are extracted using K-means and FCM clustering approaches. Further other features of leaf image (color, shape and texture) are extracted and matched using SVM. Sanjana et al. [14] adopted HE, AHE and CLHE to improve its quality. The median filter Gaussian filter and speckle reducing anisotropic diffusion approaches are applied on leaf image to fetch useful features of leaf image. Features extracted are matched using SVM and ED classifiers. Jagtap et al. [15] identified a model for conversion of defected image to HSI color space. Histogram of the intensity image classification was carried out using artificial neural network.

3   Proposed Model

In this section, identification of leaf disease using multiple descriptors is presented. Initially the images are resized to 256×256 to maintain the uniformity throughout the experiment. The HE is employed on resized leaf images to improve their quality. Segmentation is performed using k means clustering which results in a partitioning of the data space into cells. The contour tracing technique is employed on images to extract the boundary of leaf images. The multiple descriptors such as DWT, PCA and GLCM are applied on enhanced leaf images to extract useful details of leaf images. Finally the features generated are classified using SVM, CNN and KNN. Figure 2 shows the proposed model. The main objective of the model is to

- Detect the leaf disease
- Increase accuracy of model
- Decrease the error

Preprocessing includes cropping, resizing, image enhancement and segmentation. In this paper, HE is employed on database leaf image to improve the quality as shown

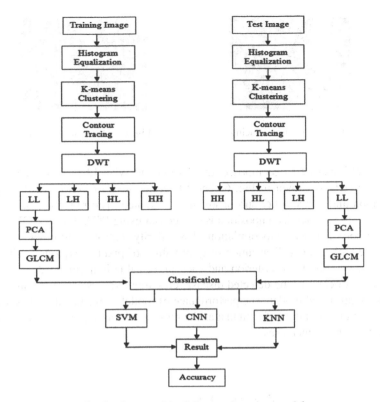

Fig. 2. Proposed leaf disease detection model

in Fig. 3. Segmentation of images is carried out using K means clustering algorithm as shown in Fig. 4. The boundaries of the leaf images are obtained using contour tracing shown in Fig. 5.

Fig. 3. Histogram equalization

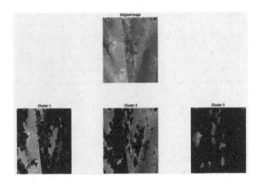

Fig. 4. K-mean clustering

Fig. 5. Contour tracing   **Fig. 6.** DWT decomposition

The leaf images are decomposed into 4 sub bands viz., LL, LH, HL and HH using DWT approach as shown in Fig. 6. The LL component of DWT has the significant information and hence other sub bands are eliminated. The reduction of the features obtained from wavelet decomposition is carried out using PCA. Further, GLCM is used to derive homogeneity, autocorrelation, dissimilarity, entropy and other several properties. The higher-order distribution of gray values of pixel is exploited using specific distance or neighborhood criterion and each pixel value is normalized.

In order to classify the detected disease, classifiers viz., SVM, KNN and CNN are used in the proposed model. The performance of model is evaluated for CNN classifier as shown in Fig. 7 and it is found that as the number of training iterations increases, the detection accuracy increases.

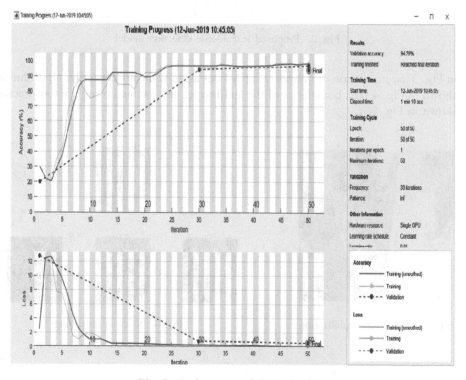

Fig. 7. Performance of CNN classifier

4 Performance Evaluation

The performance of leaf detection model based on the detection of leaf disease is described here. The database was created using bean leaf. Using this database, experimentations are performed on MATLAB to measure the accuracy of model. Figure 8 shows the approach employed for accurate detection of the diseased leaf. The performance of proposed model can be understood using similarity score. In the similarity score, if distance is more then, it will give high percentage of non-similar value and if distance is less then it will shows the low percentage of non-similar values. The similarity score value is depends on the distance or length of the feature database. In Fig. 9, the shows the variations of FAR values with increase in similarity score. FAR gives the percentage of the false acceptance rate of the database. FAR values are depends on the database and neural networks. In neural network if more number of iterations is considered then, FAR decreases for such database. In case of FAR and similarity score graphs, the score of 4000, the value of FAR is 1. It is seen that as the similarity score increases, the value of FAR increases and approaches unity value.. If the distance value of the similarity score decreases then FAR value also reduces and meets to zero.

```
training started...Wait for ~200 seconds...
training started...
Elapsed time is 2.033151 seconds.
Elapsed time is 2.239313 seconds.
...training finished.
testing started....
test error is
Elapsed time is 1.085832 seconds.
CNN Accuracy   =99.0909
CNN Precision  =0.9913
CNN Sensitivity  =0.99091
CNN Specificity  =0.99773
CNN Confutionmatrix  =

confmatrix =

    22    1    0    0    0
     0   21    0    0    0
     0    0   22    0    0
     0    0    0   22    0
     0    0    0    0   22
```

Fig. 8. Overall output of CNN

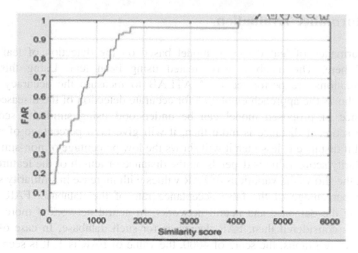

Fig. 9. FAR vs similarity score

The results obtained using different classifiers such as SVM, KNN and CNN for detection of leaf disease is shown in Fig. 10.

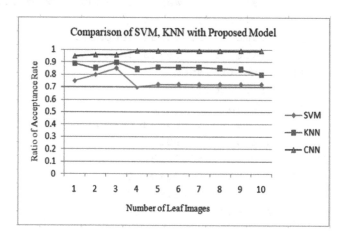

Fig. 10. Comparison of SVM, KNN and CNN

5   Conclusion

In this paper, leaf images are cropped and resized to maintain the uniform dimensions. HE technique is adopted to increase the image quality. Segmentation is performed using k means clustering. The contour tracing technique is used to extract the boundary of leaf images. The multiple descriptors DWT, PCA and GLCM are used to extract useful coefficients of leaf. Lastly, the extracted coefficients are classified using SVM,

KNN and CNN. The performance of the proposed method shows satisfactory accuracy of 99.09% on CNN classifier. In future, a mobile application can be developed and the specific solution should send to the farmer for the disease through short message service or email.

Compliance with Ethical Standards
  ✓ All authors declare that there is no conflict of interest.
  ✓ No humans/animals involved in this research work.
  ✓ We have used our own data.

References

1. Kaur, R., Kaur, M.: A brief review on plant disease detection using image processing. Int. J. Comput. Sci. Mob. Comput. **6**(2), 101–106 (2017)
2. Sujatha, R., Kumar, S., Akhil, G.: Leaf disease detection using image processing. J. Chem. Pharm. Sci. **10**(1), 670–672 (2017)
3. Kumar, S., Kaur, R.: Plant disease detection using image processing-a review. Int. J. Comput. Appl. **124**(16), 6–9 (2015)
4. Khairnar, K., Dagade, R.: Disease detection and diagnosis on plant using image processing-a review. Int. J. Comput. Appl. **108**(13), 36–38 (2014)
5. Batule, V., Chavan, G., Sanap, V., Wadkar, K.: Leaf disease detection using image processing and support vector machine. J. Res. **2**(2), 74–77 (2016)
6. Kamlapurkar, S.: Detection of plant leaf disease using image processing approach. Int. J. Sci. Res. Publ. **6**(2), 73–76 (2016)
7. Chaware, R., Karpe, R., Pakhale, P., Desai, S.: Detection and recognition of leaf disease using image processing. Int. J. Eng. Sci. Comput. **7**(5), 11964–11967 (2017)
8. Rathod, A., Tanawala, B., Shah, V.: Leaf disease detection using image processing and neural network. Int. J. Adv. Eng. Res. Dev. **1**(6), 1–10 (2014)
9. Varshney, S., Dalal, T.: Plant disease prediction using image processing techniques-a review. Int. J. Comput. Sci. Mob. Comput. **5**(5), 394–398 (2016)
10. Baskaran, S., Sampath, P., Sarathkumar, P., Sivashankar, S., Vasanth Kumar, K.: Advances in image processing for detection of plant disease. SIJ Trans. Comput. Sci. Eng. Appl. **5**(3), 8–10 (2017)
11. Mainkar, P., Ghorpade, S., Adawadkar, M.: Plant leaf disease detection and classification using image processing techniques. Int. J. Innov. Emerg. Res. Eng. **2**(4), 139–144 (2015)
12. Al-Hiary, H., Bani-Ahmad, S., Reyalat, M., Braik, M., AlRahamneh, Z.: Fast and accurate detection and classification of plant diseases. Int. J. Comput. Appl. **17**(1), 31–38 (2011)
13. Megha, S., Niveditha, C.R., Soumyashree, N., Vidhya, K.: Image processing system for plant disease identification by using FCM-clustering technique. Int. J. Adv. Res. Ideas Innov. Technol. **3**(2), 445–449 (2017)
14. Sanjana, Y., Sivasamy, A., Jayanth, S.: Plant disease detection using image processing techniques. Int. J. Innov. Res. Sci. Eng. Technol. **4**(6), 295–301 (2015)
15. Jagtap, S., Hambarde, S.: Agricultural plant leaf disease detection and diagnosis using image processing based on morphological feature extraction. IOSR J. VLSI Sig. Process. **4**(5), 24–30 (2016)

Fish Detection and Classification Using Convolutional Neural Networks

B. S. Rekha[1]([✉]), G. N. Srinivasan[2], Sravan Kumar Reddy[3],
Divyanshu Kakwani[3], and Niraj Bhattad[3]

[1] Bharathiar University, Coimbatore, Tamil Nadu, India
rekhabs@rvce.edu.in
[2] Department of Information Science and Engineering,
R.V. College of Engineering, Bengaluru, India
srinivasangn@rvce.edu.in
[3] R.V. College of Engineering, Bengaluru, India
sravankr96@rvce.edu.in, divkakwani@gmail.com

Abstract. About fifty percent of the world relies on seafood as main protein source. Due to this, the illegal and uncultured fishery activities are proving to be a threat to marine life. The paper discusses a novel technique that automatically detects and classifies various species of fishes such as dolphins, sharks etc. to help and protect endangered species. Images captured through boat-cameras have various hindrances such as fluctuating degrees of luminous intensity and opacity. The system implemented, aims at helping investigators and nature conservationists, to analyze images of fishes captured by boat-cameras, detect and classify them into species of fishes based on their features. The system adapts to the variations of illumination, brightness etc. for the detection process. The system incorporates a three phase methodology. The first phase is augmentation. This phase involves using data augmentation techniques on real time images dataset captured by boat-cameras and is passed to the detection module. The second phase is the detection. This phase involves detecting fishes in the image by searching for regions in the image having high probability of fish containment. The third phase is the classification of the detected fish into its species. This step involves the segmented image of fishes to be passed to the classifier model which specifies to which species the detected fish belongs to. CNN (Convolutional neural network) is used at the detection and classification phase, with different architectures, to extract and analyze features. The system provides confidence quotients on each image, expressed on a 0–1 scale, indicating the likelihood of the image belonging to each of the following eight categories ALB, BET, YFT, LAG, DOL, Shark, Other and None. The system provides detection and classification with an accuracy of 90% and 92% respectively.

Keywords: Object detection · Classification · Computer vision · Deep learning · Convolutional neural network

© Springer Nature Switzerland AG 2020
S. Smys et al. (Eds.): ICCVBIC 2019, AISC 1108, pp. 1221–1231, 2020.
https://doi.org/10.1007/978-3-030-37218-7_128

1 Introduction

In order to preserve marine ecosystem, real-time inspection of fisheries is necessary. In coastal areas, where most of the fishes are caught, uncultured practices are proving to be detrimental to marine ecosystem. For the conservation of fish species, the study of size and diversity of each of such species is vital. There has been a lack of automatic methods to perform real-time detection and classification of captured fishes. Traditionally, this task has been achieved by doing manual inspection. This requires having an expert on-board who determines fish species at the time of capture. Other methods involve broadcasting the boat snapshots to remote experts who determine the fish species that the captured fish belongs to.

Each of these methods has their drawbacks. First, having a fish inspector at every boat is not feasible. Second, it is difficult to find experts in the field of ichthyology. Third, transmitting boat snapshots to remote experts is vulnerable to failures. With the nature of human beings turning out to be destructive towards marine eco-system, the conservation of fishes becomes an important aspect. Thus a system which analyses the images of fishes captured by boat cameras and automatically detects and classifies them into their type of species acts as a key to solve such a problem. The current system receives real time input images from a camera installed aboard boats. Its task is to detect and classify the fishes in these images into one of the following 8 categories - ALB, BET, YFT, LAG, DOL, Shark, Others and no fish. The images in the available dataset are shot with varying imaging conditions. These images are highly distorted, cluttered, and in several of them, the fishes are partly occluded. This makes the task of detection and classification challenging. The current state-of-the-art algorithms using the traditional image processing techniques do not perform very well on such images and also has proved to be very specific with the type of environment.

2 Related Work

The system has to be independent of the environment and reliable in the case of noise. This can be achieved by using Deep Learning techniques which involve Convolutional Neural Network (CNN) in the detection and classification tasks. This technique was introduced first by Alex as ALEXNET in the ImageNet competition which gave remarkable results compared to the previous approaches [4]. Similar to all the machine learning algorithms, the method had a training phase and a testing phase. In training phase, the pipeline consists of three parts, data augmentation, convolution network and a fully connect layer at the end combined with a softmax activation. Each part of this design has a specific function. The main function of convolutional layers is to extract features from the image. Each convolutional layer has certain number of filters. These filters are weighted kernels that are passed through out the image to generate corresponding feature maps. These feature maps are passed on to the fully connected network. The fully connected layer weights the features in the final feature maps generated and gets a final probability distribution over all the possible classes. Several hyper parameters are used to tune the network model which include error correction method, learning rate, number of filters in each convolutional layer, number of convolutional layers, input image size etc. Using deeper networks results in better accuracy can be achieved when combined with the techniques like dropout and skip networks which are introduced in [5, 6]. A dropout is the technique where a fraction of connections in the fully connected layers are desperately removed in a random order. A clear explanation is given by the Alex team in [6]. Overfitting is the condition where the grading moves from least error point towards high error rate. This generally happens due to over training a network with same dataset. A simple indication for this is the training accuracy keeps increasing whereas the validation accuracy shows a gradual decrease. The Skips nets also uses a similar approach where few convolutional layers will be desperately skipped and the feature maps are passed on to the further layers. The main idea behind this is to compute a feature map based on two feature maps, one from the direct parent and another from the ancestor. This experimentally showed that the approach reduces overfitting of the network. Using an Ensemble of different models is another technique which gives an edge over using a single model prediction is another popular technique which is used in most of the modern deep learning problems [4, 6–8]. Different approaches like using parallel convolutional neural networks are used by the Google research team [7] which showed improved results but uses a very complex architecture.

Since the success of AlexNet [4], several detection approaches based on CNNs have been proposed. One such successful technique combined region proposals with CNNs, giving rise to R-CNN. The initial version of R-CNN is described in [9]. Several variants of the R-CNN in [9] have been proposed since then. The most successful among them are: Fast R-CNN [10], Faster R-CNN [11], DeepBox [12], R-FCN [13].

The R-CNN described in [9] first generates candidate regions using Selective Search [14], and for each region, generates a feature vector using a CNN, and finally uses an SVM per each category to assign category scores to each region proposal.

R-CNNs were later improved by changing the region proposal generation algorithm, for instance Edgeboxes [15] is used in [16]. For a comparison of region proposal algorithm, see [17]. DeepBox [12] refined the region proposal algorithm by adding a shallow CNN network to filter the candidates. A better version of R-CNN, called Fast R-CNN, was introduced in [10]. It reduced the time by training in single-stage and by sharing the computation of feature vector corresponding to each region proposal. The techniques based on region proposal generation, however, fail in case of cluttered images. In such cases, it is difficult to extract meaningful regions out of an image, mainly because of high-object density and occlusion rates.

3 System Architecture

The proposed system consists of three phase – Data augmentation, detection and classification. As the initial dataset is small and highly imbalanced, the images were augmented to produce a sizeable balanced dataset. Various augmentation techniques were used to increase the size of the data set. In the detection stage, the input image is segmented to obtain image patches that contain fish. The obtained patches are passed to the classifier. The classifier produces a probability distribution of the fish classes of the detected fish (Fig. 1).

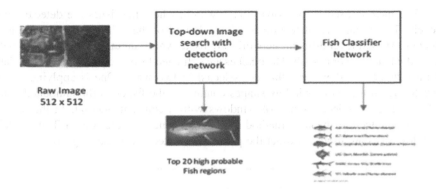

Fig. 1. System architecture.

3.1   Augmentation

Augmentation phase is used to enhance the training dataset in terms of both quality and quantity. Simple affine transforms and preprocessing techniques are used to achieve that

3.2   Detection

The fishes present in the input image are detected in this phase. Different techniques are used to find an optimal approach to extract the fishes from the image. Initially, localization technique was applied in which, the input image is fed to a localization network which regresses coordinates for the area within which fish is present. The limitation of regression bounding is that it cannot detect multiple fishes in a single image. As the dataset consists of multiple fishes in a single image, a better method to find more than one fish was required, hence a custom detection algorithm that best matches the dataset is designed and implemented (Fig. 2).

Fig. 2. Output of the localization algorithm [6]

The method used is top down image search, where the fishes are detected in two levels. The initial level detection is a loose bound of the fish and the latter is a tight bound. At each level different networks are used, whose input is a patch which will be classified into fish or no-fish. The patches are extracted from the raw image in the dataset with different window sizes that are selected in two ways. One is applying k-means clustering on the set of window shapes containing the fishes in the annotated dataset. The other is to select a range of windows with a constant interval. In the proposed system architecture, the latter method is chosen to generalize the system. These obtained window sizes are used to extract the patches for detection phase (Fig. 3).

Fig. 3. Detection algorithm flow diagram.

The detection pipeline consists of a large patch extractor, binary classifier, small patch extractor followed by another binary classifier. The initial large patch extractor extracts large square patches ranging from a scale of 300×300 to 600×600 with an interval of 50. These patches are classified with a binary classifier that is trained with similar image patches manually extracted from the dataset. The binary classifier also outputs the probability of each image patch containing a fish. Based on the classifier output, top 10 high probable image patches are selected and forwarded to next level segmentation. Smaller image patch extractor extracts small patches ranging from a scale of 100×100 to 300×300 with an interval of 50. These patches are classified with a binary classifier that is trained with similar fish patches manually extracted from the dataset. Moreover, computation time depends on model and image crop size, but precision is also affected; usually, time and precision have trade-off relation (Fig. 4).

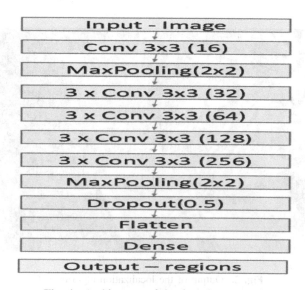

Fig. 4. Architecture of the detection network

The detection phase uses the top down image search technique with sliding window approach in order to initially detect the presence of the fish or no-fish. Figure 5 shows the presence of a fish.

Fig. 5. Output of the detection algorithm.

3.3   Classification

All the input patches obtained in the detection phase are resized appropriately passed to the classifier network. For each input patch, the classifier returns a probability distribution of fish classes. The final probability distribution is computed as follows:

$$P(x) = max\ (pi\ (x)),\ if\ x\ != 'NoF'$$

$$avg\ (pi\ (x)),\ otherwise \quad (1)$$

The Classification network consisted of several convolutional layers followed by a fully connected (FC) layer. VGG-16 network is used for classifier with the input size of (224, 224, 3). This network was chosen because pre-trained VGG networks are available and since our dataset is small, a pre-trained model is indispensable (Fig. 6).

| Input - Image |
| 2 x Conv 3x3 (32) |
| MaxPooling(2x2) |
| 2 x Conv 3x3 (64) |
| MaxPooling(2x2) |
| 2 x Conv 3x3 (128) |
| 2 x Conv 3x3 (256) |
| MaxPooling(2x2) |
| Flatten -> Dense |
| Dropout(0.5) -> Dense |
| Dropout(0.5) -> Dense |
| Softmax->output class |

Fig. 6. Architecture of the classification network.

Once the images pass thru the detection phase are passed thru the classification phase. In the classification phase, if the fish is detected the classifier classifies the detected fish into one of the eight categories. The class with the highest probability is conjectured to be the class of the fish contained in the image. Figure 7 shows the classification of a fish detected in the input image to be of the category BET with a high probability of 1.0. Figure 8 shows that there is No-Fish on the board with a high probability of 1.0.

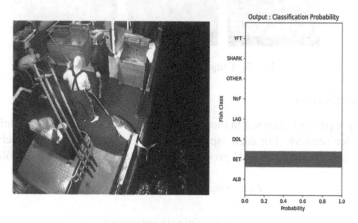

Fig. 7. Output of the classification algorithm identifying BET fish

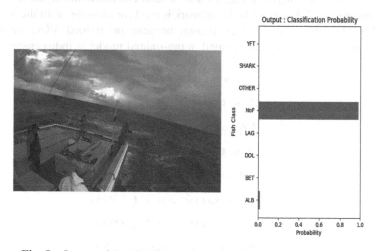

Fig. 8. Output of the classification algorithm identifying No fish [7]

4 Dataset

The original dataset is provided by National Conservancy Organization, who in turn extracted it from camera feeds obtained from boat cameras. The dataset contains a total of 3777 images divided into 8 categories, namely ALB, BET, YFT, SHARK, LAG, DOL, NoF, Other. Since the dataset is relatively small, and contains plenty of variances in imaging conditions, the task of classification is challenging. Following are some of the challenges faced during processing such images:

- Different boat environments: Different environments contain different camera position and orientation.
- Different capture times: Some images are taken in day-light, while some are taken in night-time.
- Distortion: Many images are distorted due to the general hustle typically present in fisheries.
- Occlusion: Several images contain fishes occluded by other objects or fishermen.

Prior to training the networks, the available dataset is heavily augmented with rotation and Gaussian blur. The initial dataset contained 3777 images, and augmentation process resulted in 16000 images. The augmented dataset is fully balanced across all the categories. This dataset is split into training and validation in the ratio of 8:2. Data augmentation increases the data set size which provides better training and also helps in overcoming the neural net getting over fit.

5 Training

In training phase, the pipeline consists of three parts, data augmentation, convolution network and a fully connect layer at the end combined with a softmax activation. Each part of this design has a specific function. The main function of convolutional layers is to extract features from the image. Each convolutional layer has certain number of filters. These filters are weighted kernels that are passed through out the image to generate corresponding feature maps. These feature maps are passed on to the fully connected network. The fully connected layer weights the features in the final feature maps generated and gets a final probability distribution over all the possible classes. Several hyper parameters are used to tune the network model which include error correction method, learning rate, number of filters in each convolutional layer, number of convolutional layers, input image size etc.

The detection and classification networks are trained with the following parameters:

- Optimizer: Adams
- Learning Rate: 1e−3
- Objective: Categorical Cross-entropy
- Epochs: 40

The classifier uses a VGG-16 network pre-trained on the ImageNet dataset, and fine-tuned on tight patches of fishes extracted from our training data. 2000 patches are extracted for each category from the training dataset. For categories that do not have enough samples, data augmentation techniques are used.

6   Results

Initially a single network was used that performed the classification job. The Training Accuracy was 95%, and the Validation Accuracy was poor. When this initial model was tested it was found that there was a dependency on the environment which made us to introduce a detection phase before classification.

The initial approach for detection was localization of the fish in the image. This approach failed in the case where there are multiple fishes in a single image as the localization task was meant to find only one object of interest in an image. This made us to evolve a new method of detecting all the fishes in an image using an image search algorithm. With this final design of network architectures and training parameters, Following are the results obtained:

Detection
Training Accuracy: 94%
Validation Accuracy: 90%
Training Time: About 6 h
Test Time: 5 s
Classification
Training Accuracy: 96%
Validation Accuracy: 92%
Training Time: About 3 h
Test time: 50 ms

7   Conclusion

The usage of CNNs in the detection and classification produced remarkable results. It gave significantly better results than the methods based on manual feature extraction and the traditional image processing techniques. The final networks used are derived from the VGGNet. The CNNs used in the detection module and the classification module achieved a validation accuracy of about 90% and 92% respectively. While the accuracy of the networks has been remarkable, the whole process is a little slow. The detection module takes about 5 s to process the image, while the classification module is almost instantaneous, taking less than a second. The training of the networks has been done on limited datasets. The usage of better aggregation tools could produce much better results in production. The future scope incudes working on R-CNN for better results in the detection and classification phases. The project could be extended to detecting and classification of other species of fishes also.

Compliance with Ethical Standards
✓ All authors declare that there is no conflict of interest.
✓ No humans/animals involved in this research work.
✓ We have used our own data.

References

1. Lowe, D.G.: Distinctive image features from scale-invariant key-points. IJCV **60**, 91–110 (2004)
2. Clara Shanthi, G., Saravanan, E.: Background subtraction techniques: systematic evaluation and comparative analysis. Int. J. Mod. Eng. Res. (IJMER) **3**, 514–517 (2013)
3. Jones, V.: Rapid object detection using a boosted cascade of simple features. In: Computer Vision and Pattern Recognition (2001)
4. Krizhevsky, A., et al.: ImageNet classification with deep convolutional neural networks. In: NIPS (2012)
5. He, K., et al.: Deep residual learning for image recognition. ISLR (2015)
6. Alex, et al.: Dropout: a simple way to avoid overfitting in the network. JMLR (2014)
7. Szegedy, C., Liu, W., et al.: Going deeper with convolutions. In: CVPR. Google Research (2015)
8. Simonyan, K., Zisserman, A.: Very deep convolutional networks for large scale image recognition. In: ICLR. Visual Geometry Group, Department of Engineering Science, University of Oxford (2015)
9. Girshick, R., Donahue, J., Darrell, T., Malik, J.: Rich feature hierarchies for accurate object detection and semantic segmentation. In: The IEEE Conference on Computer Vision and Pattern Recognition (CVPR), pp. 580–587 (2014)
10. Girshick, R.: Fast R-CNN. In: The IEEE International Conference on Computer Vision (ICCV), pp. 1440–1448 (2015)
11. Ren, S., He, K., Girshick, R., Sun, J.: Faster R-CNN: towards real-time object detection with region proposal networks. In: Advances in Neural Information Processing Systems 28, NIPS (2015)
12. Kuo, W., Hariharan, B., Malik, J.: DeepBox: learning objectness with convolutional networks. In: The IEEE International Conference on Computer Vision (ICCV), pp. 2479–2487 (2015)
13. Dai, J., Li, Y., He, K., Sun, J.: R-FCN: object detection via region-based fully convolutional networks. In: Advances in Neural Information Processing Systems 29, NIPS 2016 (2016)
14. Uijlings, J.R.R., van de Sande, K.E.A., Gevers, T., Smeulders, A.W.M.: Selective search for object recognition. Int. J. Comput. Vis. **104**(2), 154–171 (2013)
15. Lawrence Zitnick, C., Dollr, P.: Edge boxes: locating object proposals from edges. In: European Conference on Computer Vision, ECCV 2014, pp. 391–405 (2014)
16. Tang, S., Yuan, Y.: Object detection based on convolutional neural network, Stanford Project report (2016)
17. Hosang, J., Benenson, R.: How good are detection proposals, really? Computer Vision and Pattern Recognition, arXiv (2014)
18. Gidaris, S., Komodakis, N.: LocNet: improving localization accuracy for object detection, arXiv (2016)
19. Sermanet, P., Eigen, D., Zhang, X., Mathieu, M., Fergus, R., LeCun, Y.: OverFeat: integrated recognition, localization and detection using convolutional networks, arXiv (2014)

20. Jaderberg, M., Simonyan, K., Zisserman, A., Kavukcuoglu, K.: Spatial transformer networks, arXiv (2016)
21. Redmon, J., Divvala, S., Girshick, R., Farhadi, A.: You only look once: unified, real-time object detection. In: CVPR (2015)
22. Kingma, D.P., Ba, J.L.: Adam: a method for stochastic optimization. In: ICLR (2015)
23. Srivastava, N., Hinton, G., Krizhevsky, A., Sutskever, I., Salakhutdinov, R.: Dropout: a simple way to prevent neural networks from overfitting. J. Mach. Learn. Res. **15**, 1929–1958 (2014)
24. Ioffe, S., Szegedy, C.: Batch normalization: accelerating deep network training by reducing internal covariate shift, arXiv (2015)
25. Liu, W., Anguelov, D., Erhan, D., Szegedy, C.: SSD: Single Shot MultiBox Detector, arXiv (2016)
26. Mnih, V., Heess, N., Graves, A., Kavukcuoglu, K.: Recurrent models of visual attention. In: ICLR. Google DeepMind (2016)
27. Liang, M., Hu, X.: Recurrent convolutional neural network for object recognition. In: CVPR (2015)

Feature Selection and Ensemble Entropy Attribute Weighted Deep Neural Network (EEAw-DNN) for Chronic Kidney Disease (CKD) Prediction

S. Belina V. J. Sara[✉] and K. Kalaiselvi

Department of Computing Science, School of Computing Science,
Vels Institute of Science, Technology and Advanced Studies (VISTAS),
(Formerly Vels University), Chennai, India
belina_jyotsna@yahoo.co.in,
kalairaghu.scs@velsuniv.ac.in

Abstract. Initial prediction and appropriate medication are the ways to cure Chronic Kidney Disease (CKD) in the early stage of progression. The rate of accuracy in the classification algorithms focuses on the usage of exact algorithms used to select he features in order to minimize the dataset dimensions. The accuracy not only relies on the feature selection algorithms but also on the methods of classification, where it predicts the severities that are useful for the medical experts in the field of clinical diagnosis. To minimize the time for computation and to maximize the classifiers accuracy level, the proposed study, Ensemble Entropy Attribute Weighted Deep Neural Network (EEAw-DNN) classification was aided to predict Chronic Kidney Disease. The rate of accuracy of the EEAw-DNN is surveyed with the help of feature selection using data reduction. Hence Hybrid Filter Wrapper Embedded (HFWE) based Feature Selection (FS) is formulated to choose the optimal subset of features from CKD set of data. This HFWE-FS technique fuses algorithm with filter, wrapper and embedded algorithm. At last, EEAw-DNNbased algorithm used for prediction is used to diagnose CKD. The database used for the study is "CKD" which is implemented using MATLAB platform. The outputs prove that the EEAw-DNNclassifier combined with HFWE algorithm renders greater level of prediction when correlated to other few classification algorithms like Naïve Bayes (NB), Artificial Neural Network (ANN) and Support Vector Machine (SVM) in the prediction of severity of CKD. Datasets were taken from University of California Irvine (UCI) machine learning repository.

Keywords: Chronic Kidney Disease (CKD) · Classification · Ensemble Entropy Attribute Weighted Deep Neural Network (EEAw-DNN) · Feature selection (FS) · Hybrid Filter Wrapper Embedded (HFWE) · University of California Irvine (UCI)

S. Smys et al. (Eds.): ICCVBIC 2019, AISC 1108, pp. 1232–1247, 2020.
https://doi.org/10.1007/978-3-030-37218-7_129

1 Introduction

Chronic kidney disease is a popular disease causing threatening to the public with the maximum occurrence, existence and very expensive treatment. Around 2.5–11.2% of the population in adults across the countries like Europe, Asia, North America, and Australia are likely to be affected with chronic kidney disease [1], In USA alone 27 million individuals are affected with CKD [2].

To attain an innovative knowledge among the huge amount of data, examining the CKD victims, there is a advantage of using data mining techniques [4]. Data mining techniques focuses on machine learning, database management and statistics have been used in the proposed work [5, 6]. The advantage of the usage of the electronic devices is also emphasized and also the software solutions for victims data banks in medical industry, large size of data are extracted every day [7, 8]. At present, there are no broadly accepted diagnostic tools or instruments for the progression of CKD. Hence the medical experts should take ideal decisions about the patients' treatment, delaying risks in the treatments and the patients' progress towards the renal failure and the patients who do not require higher clinical concentration. The level of CKD has been recommended in order to help in treatment focussed actions [9]. Decision making in the diagnosis of the medical industry is a challenging task because of the heterogeneity of kidney diseases, variability in stages of disease progression, and the cardiovascular RISK COVERAGE [10]. The risk is accurately predicted that could help individualized decision making process, enhancing initial and adequate care in patients [11]. Hence for the effective clinical diagnosis, the data mining techniques provides appropriate information from large medical databases which are collected often [12]. Major studies are carried out on the medical databases that are related to the prediction of the cancer disease too. Many methods of classification are carried out on disease prediction and the diagnosis of the medical experts. These classification techniques can reduce the prediction error that exists by the under experienced experts and the outputs can be extracted in no time [12]. But this is not an individual algorithm aided in machine learning criteria and the models for decision making to predict all the varieties of diseases. The high dimensional sets of data acquires a accuracy rate of classification which can be minimal and can also be over-fitting the risks, the computational efforts may be greater and expensive. Usually, the datasets of low dimensions can cause higher level of classification accuracy with the minimal cost for computation and over-fitting risk [13].

This formulated proposes a better prediction system based on the computer vision and the machine learning techniques for aiding in the prediction of CKD under various phases of CKD. Innovative features and Ensemble Entropy Attribute Weighted Deep Neural Network (EEAw-DNN) were used for the process of rapid detection. In this study, many evaluations are carried out on variety of classes were done and correlated as per the estimated rates of glomerular filtration (GFR). The outputs prove that the system might yield consistent prediction process and the medical treatment of CKD victims. Datasets were extracted from the University of California Irvine (UCI) machine learning repository.

2 Literature Review

The parameters of variety of clinical examinations which were examined by Anandanadarajah and Tharmarajah [14] in order identify the data which is helpful for predicting CKD. A repository with many parameters of healthy patients and patients suffering from CKD are investigated using different methods. The initial recognition of the dominant parameters carried out by Common spatial pattern (CSP) filter and linear discriminant analysis could yield positive results in the prediction of CKD. Then the classification techniques are utilized to recognize the feature which seems to be domiant. Also when there is inadequate data based on hypertension and diabetes mellitus, random blood glucose and blood pressure, these parameters could be utilized.

Salekin and Stankovic [16] analysis take into account 24 predictive attributes and construct a machine learning classifier in the diagnosis of CKD. Utilizing machine learning methods attain a prediction accuracy of 0.993 as per the F1-measure with 0.1084 root mean square error. This is a 56% reduction of mean square error correlated to the arts state. The process of feature identification recognizes the most reliable parameters for the prediction of CKD and assesses them for their detection process. At last an analysis is carried out on cost-accuracy tradcoff in order to recognize an innovative CKD prediction process with greater accuracy and at a lower cost.

Luck et al. [17] constructed a summarized and an approach which is easy to understand, which identifies the discriminate local phenomena using the actual exhaustive algorithm that is rule mining, to diagnose two different group of subjects: (1) subjects with least to the medium phases with no renal failure and (2) subjects with medium to severely affected CKD phases suffering from patients having moderate to established CKD stages with renal failure. The prediction algorithm brings out the table containing the m-dimensional variable space to acquire the local over densities of the 2 categories of subjects within the form of rules that are easy for interpretation. Moreover, the basic knowledge of concentration of urinary metabolites classifies the various phase of CKD of the subjects exactly.

Huang et al. [18] formulated a prediction model of chronic diseases prognosis and detection system of combining data mining techniques along with case-based reasoning (CBR). The major procedure of the system adds: (1) acquiring the sensible techniques of data mining to identify the rulc which are implicitly meaningful after examining the health data, (2) with the help of generated rules for the particular chronic disease diagnosis (3) implementing CBR to fix the chronic diseases prediction and medications and (4) extending these progression of work within the system for the easy gaining of knowledge for the prediction of chronic disease, construction, expanding these processes to work within a system for the convenience of chronic diseases knowledge generation, arrangement, cleaning and distribution.

Di Noia et al. [19, 20] formulated software which acts as a tool for exploiting the potential of Artificial Neural Networks (ANNs) in order to classify subjects' health resulting to (ESKD). ESKD infects kidneys in human beings demanding renal replacement medication along with dialysis or even kidney transplantation. The tool which was constructed was made available to both online Web application and as an

Android mobile application. Its medical utility is the development focused on greatest worldwide availability.

Polat et al. [21] formulated an algorithm namely Support Vector Machine (SVM) classification which was used in the prediction of CKD. In the wrapper method, the evaluator which was used as a subset classifier aided with greedy stepwise search engine and wrapper subset evaluator with the Best First search (BFS) engine were also utilized.

Norouzi et al. [22] formulated an adaptive neurofuzzy inference system (ANFIS) for the detection of the renal failure timeframe of CKD focused on actual medical information. A Takagi-Sugeno type ANFIS technique was used in the identification of GFR values. Several variables are identified for the prediction model. This model can accurately detect the variations in GFR, in spite of several uncertainties within the human body and the nature which seems to be dynamic for the progression of CKD for the future period of time.

Tangri et al. [23] constructed a prediction model which is validated for the purpose of CKD progression. The Development phase and the stage of validation of the prediction models, utilizes the information gathered from demographic, medical, and laboratory information from 2 independent Canadian cohorts of subjects suffering from various CKD phases of 3 to 5, were suggested for medication from nephrologists between April 1, 2001, and December 31, 2008. A prediction model utilized by regularly conducted lab investigations can accurately diagnose the progression to the failure of the kidney in the subjects who suffer with CKD phases 3 to 5.

3 Proposed Methodology

This work formulated a detection system focussed on the vision of computer and the techniques involved in Machine Learning process which helps in the prediction of CKD and its various phases. Innovative features and the algorithm namely Novel features and Ensemble Entropy Attribute Weighted Deep Neural Network (EEAw-DNN) were subjected for the purpose of quick prediction process. To diagnose the CKD, 3 significant varieties of algorithms of FS namely wrapper, filter and embedded methods are used in the selection of reduction of features from the CKD dataset. In the proposed work, many validations are done on various classes which were carried out as per the eGPR. The output proved that the system could yield continuous prediction and the medication involved for the CKD cases. From University of California Irvine (UCI) machine learning repository the datasets were obtained. Figure 1 proves that the system of EEAw-DNN classifier used for classification process and CKD identification. The total layout of the formulated system is depicted in Fig. 1.

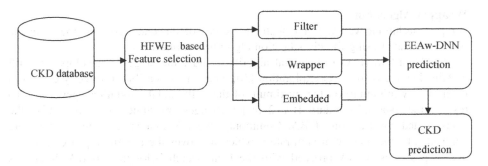

Fig. 1. Proposed EEAw-DNN classifier architecture for CKD prediction

CKD is categorized into phases from I–V as per the evaluated Glomerular Filtration Rate (GFR) [24, 25]. UCI containing CKD data set consists of 24 attributes and one extra attribute class (binary) [26]. It consists of few samples which are 400 in number onto two distinct classes. ("CKD"-.250 cases; "NOTCKD"-.150 cases). Within the 24 attributes, 11 attributes are numeric and 13 attributes are nominal. The set of data consists of some which are missing. Removing the missing valued tuples, around 160 samples were utilized in this proposed work.

3.1 Feature Selection

The filter algorithm is used to select the features whose rank is considered to be the highest within them, and the selected features of subset can be used for any of the prediction based algorithm.

Relief-F
Relief-F is an algorithm of instance-dependent attribute selection algorithm which identifies the finest features having the value which distinct the instances with the reason from many groups that can correlate with each others.

One-R
One-R is a powerful algorithm [27], it constructs a single rule applicable for every feature within the training information and selects the rule with a minimal error. Each and every numerically valued parameters as constant and utilizes the basic algorithm to discrete the values with the range of dislocated intervals.

Gain Ratio
The Gain Ratio is The non-symmetrical evaluation is the Gain ratio constructed with the purpose of balancing the direction formulated got the IG bias [28].

Gini Index (GI)
An algorithm called Gini index [29] is supervised multivariate FS technique which is the filter mechanism for evaluating the capability of a feature to distinct various classes.

Wrapper Algorithm

Wrapper technique validates the features' scores of the data sets that focus on the power estimated using the classification algorithm as black box.

Bat Algorithm (BA) is potential algorithm under the stage of development, still certain shortage is found at exploration [30], hence it is possible to directly chosen in local, a very minimum on the maximum of the multimodal validating operations. To resolve the issue of standard BA, 2 major changes are subjected to maximize the analysis and construction of BA. Commonly, the BA analysis and construction are controlled by the pulse emission rate r, which improves the repetitive process. Step 8 and step 9 lines in BA proceed with the local search belonging to BA. Step 7 is investigated and the algorithm slowly loses its construction potentiality whereas iteration continuing. In order to resolve this issue, the ability of BA is improved by proposing the linear decreasing inertia weight factor [31, 32]. The factor namely inertia weight factor restricts the analysis and growth of BA. Frequency value which is randomly created is allotted for the purpose of feature identification solution within BA and this value of such frequency will have the similar impact to all dimensions of the feature selection solution. All the features with certain variation make no sense in this aspect. This creation minimizes BA's local search results.

Embedded Algorithm

Depending on the SVM the Embedded algorithm is carried out (SVM-t) to select the attributes which seems to be analytical, obtained from the dataset of CKD. The Support vectors' utilize the data of SVM in direction towards the creation of greatest segmentation hyper plane and identification of each CKD dataset classes. The position of nearest CKD dataset points within two classes in SVM, take part an important role for the purpose of selecting the features.

Ensemble Entropy Attribute Weighted Deep Neural Network (EEAw-DNN) Prediction

Customarily, around 10,000 reiterations and there are a huge number of emphasess and furthermore neural system parts in the learning arrangement of DNNs. Considering the adequate calculation multifaceted nature in the testing system, one thing is to rapidly choose a little subset of neural system parts in various preparing cycles. In the meantime, thinking about the high exactness prerequisite, something else is to adaptively consolidate this subset of neural system parts and develop a last arrangement framework. In the following, the unified framework of EEAw-DNN is first formulated. Next, the detail procedure of EEAw-DNN is then described.

To detail the gathering choice [33], the individual classifier choices can be join by dominant part casting a vote, which aggregates the votes in favor of each class and chooses the class that gets the majority of the votes. While the majority voting is the most popular combination rule, a major limitation of majority voting is that only the decision of each model is taken into account without considering the distribution of decisions Specifically, all the conceivable models in the speculation space could be abused by thinking about their individual choices and the connections with different theories.

Therefore, BM assigns the optimal label y to y^* according to the following decision rule, i.e.,

$$P(y^*) = argmax_y P(y|H, x) \qquad (1)$$

where $W(y, h_i(x))$ is a function of y and $h_i(x)$. By multiplying the scaling factor $\lambda > 0$, $W(y, h_i(x))$ can have a different range in R. There are two key issues for optimizing Eq. 3. The first one is the calculation of $W(y; h_i(x))$. As mentioned above, $P(y|h_i, x)$ is the distribution of describing the correlation between decision y and $h_i(x)$. Thus, $W(y, h_i(x))$ can be derived from y, $h_i(x)$ and the distance between y and $h_i(x)$. Here, $W(y, h_i(x))$ is assumed to be computed as

$$W(y, h_i(x)) = I(y = h_i(x)) + U(y) * V(y, h_i(x)) \qquad (2)$$

where both $U(y)$ and $V(y, h_i(x))$ are functions. $I(y = h_i(x))$ returns 1 when $y = h_i(x)$; otherwise, $I(y = h_i(x)) = 0$.

In this work, statistics the frequencies of classes at the same location from the label and the hypothesis on the validation set, and calculate the cost of two different classes (a and b) as

$$Cost(a, b) = 1 - P(a|b) \qquad (3)$$

Note that if both y and $h_i(x)$ are from the given CKD sample, then they will have a competitive relationship with eachother. Thus, V (y, $h_i(x)$) can be calculated with

$$V(y, h_i(x)) = \begin{cases} F(-CLD(y, h_i(x)))h_i(x) \in positive \\ F(CLD(y, h_i(x)))h_i(x) \in negative \end{cases} \qquad (4)$$

where F is a function of the Cost Levenshtein Distance (CLD) between y and $h_i(x)$. By a heuristic approach, the values of F can be empirically assigned at the multiple integral points, and the values at other points can be calculated by the piecewise linear interpolation. The second issue is about generating voting candidates (more probable labels of the hypotheses). Obviously, the ground truth doesn't always appear in the decisions made by H. It is compulsory to uncover an efficient way to generate high-quality candidates from all the decisions, i.e., to find a more probable label $y_i(x)$ from the existed initial label $y_i^0(x)$ of hypothesis h_i. Generally speaking, a good candidate means it has a miniature edit distance with most of the hypotheses. Following this idea, propose an algorithm too semantically generate voting candidates (see Algorithm 1).

Algorithm 1: Generating Voting Candidates

Input- H = (h₁, h₂,,,,,,, hₗ): the base classifier set, |H|= L.

Y_0-the initial decisions made by H.

ED- the measurement function of the pairwise distance.

θ-the upper bound of the distance between the candidate and the hypothesis.

Output:Y- the voting candidate set.

Parameter:H^* − a subset of H, $\forall h_i^*, h_j^* \in H^*, ED(h_i^*, h_j^*) \leq 20$

Procedure:

1. Y =0

2. For each $H^* \subset H$;

3. For each $y \in Y_0$:

4. If $\max_{h_i^* \in H^*} ED(y, h^*(x)) \leq \theta$

5. $Y = Y^i \cup \{y\}$

6. End

7. End

In Algorithm 1, the searching process of H^* is an implicit computational way for P(D|hᵢ). In experiments, a special simple case of algorithm 1 is used, where during the voting candidates generation process, Y0 is initialized only by H, the upper bound is set from θ to inf, and P(D|hᵢ) is assumed to be a constant.

EEAw-DNN

Within the above framework, the procedure of P(D|hᵢ) for CKD includes three major steps, i.e., base classifiers generation, classifier combination, and ensemble pruning.

Age Ensembles work best if the base models have high expectation results and don't cover in the arrangement of precedents they misclassify. Profound Neural Networks (DNNs) are normally utilized as a base classifier generator for groups. From one viewpoint, EEAw-DNN being made out of numerous preparing layers to learn por-trayals of information with different dimensions of deliberation. Then again, amid the preparation period of one individual profound neural system, two depictions with various minibatch orderings will merge to various arrangements. Those previews fre-quently have the comparative blunder rates, however commit distinctive errors. This decent variety can be misused by ensembling, in which numerous snapnots are normal inspecting and after that joined with dominant part casting a ballot yield layer, which gauges the arrangement likelihood adapted on the information picture, for example P(h|x), where x is the info CKD tests and h speaks to a class arrangement The center of

EEAw-DNN is to ascertain y* (by Eq. 3), i.e., the figuring of F, which is a component of separation among y and hi(x). Here, F is spoken to by the arrangement of qualities at the different vital focuses. These qualities are allocated with the most noteworthy forecast rate on the approval set.

In classifier ensemble, pruning can generally improve the ensemble performance. The attribute weight function referred to as entropy is given by

$$G(aw) = 1 + \left\{ \sum_j p(aw, c) \log p(aw, c) \right\} / \log n \text{ samples} \tag{1.6}$$

Actually, this equation represents some entropy ratio, i.e.

$$G(aw) = 1 - H(c|aw)/H(c) \tag{1.7}$$

where H(c) is the entropy of the distribution (uniform) of the classes and H(c|aw) is the entropy of the conditional distribution given that the attribute 'aw' appeared and more belongs to classes either positive or negative. The last tested global function is the real entropy of the conditional distribution:

$$G(aw) = H(c|aw) = - \sum_c p(aw, c) \log p(aw, c) \tag{1.8}$$

In EEAw-DNN pruning, initially, a population of binary weight vectors which is created from entropy function, where 1 means the classifier is remained. Secondly, the population is iteratively evolve where the fitness of a vector Aw is measured on the validation set V, i.e., $f(Aw) = PV \, Aw$ (P stands for the prediction rate). Finally, the ensemble is correspondingly pruned by the evolved best weight vector Aw.

The detail procedure of the EEAw-DNN ensemble is shown in Algorithm 2.

Algorithm 2: EEAw-DNN (classifier combination)

Input- $H = (h_1, h_2, \ldots h_L)$: the base classifier set, $|H| = L$.

F a function of distance between y and $h_i(x)$, attribute weight Aw

Parameter- Y - the voting candidates set generated by Algorithm 1.

Output- Y*- the label of CKD prediction

Procedure

1. Initialize Y by H.

2. Generation of the attribute weight by eq. 1.6

3. For y ∈ Y :

4. Calculate P(y|H, x) through Eq.1. 2.

5. End

6. Calculate y^* through Eq.1. 3.

3.2 Experimental Results

Classification Accuracy–shows the potential of the classifier algorithm to predict the dataset classes.

$$Accuracy = \frac{TP + TN}{TP + FP + TN + FN} \times 100 \qquad (1.9)$$

Sensitivity–shows the accuracy level of target class's occurrence (Eq 1.10).

$$Recall = Sensitivity = \frac{TP}{TP + FN} \times 100 \qquad (1.10)$$

Specificity associates with the ability of the test exactly to detect the patients without any condition.

$$Specificity = \frac{TN}{TN + FP} \times 100 \qquad (1.11)$$

Precision can also term as positive predictive value which is considered to be the fraction of relevant instances within the instances which are retrieved.

$$Precision = \frac{TP}{TP + FP} \times 100 \qquad (1.12)$$

F-measure is considered to be the mean of precision and recall is measured as follows:

$$F - measure = 2 \times \frac{Recall \times Precision}{Recall + Precision} \times 100 \qquad (1.13)$$

Figure 2 represents the comparison of performance outputs of sensitivity with the relevant classifiers like ANN, SVM, SVM-HWFFS, SVM-HFWE-FS and formulated EEAw-DNNNB classifier. The formulated EEAw-DNNNB algorithm yields greater results of sensitivity of 97.30% whereas other classifiers like ANN, SVM, SVM-HWFFS, and SVM-HFWE-FS yields results of sensitivity of 55.56%, 62.50%, 87.501% and 95.45%.

Figure 3 represents the outcomes of the proposed EEAw-DNNNB technique produces higher specificity results of 87.50% whereas other classifiers such as ANN, SVM, SVM-HWFFS, SVM-HFWE-FS produces specificity results of 55.56%, 62.50%, 87.50% and 87.50%.

Methods	Results(%)					
	Sensitivity	Specificity	Precision	F-measure	Accuracy	Error rate
ANN	76.19	55.56	80.00	78.05	70.00	30.00
SVM	77.27	62.50	85.00	80.95	73.33	26.67
SVM-HWFFS	90.91	87.50	95.24	93.02	90.00	10.00
SVM - HFWE-FS	95.45	87.50	95.45	95.45	93.33	6.67
EEAw-DNN	97.30	87.50	95.58	96.43	94.70	5.30

Fig. 2. Sensitivity performance comparison vs. classifiers

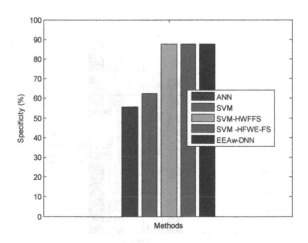

Fig. 3. Specificity performance comparison vs. classifiers

Figure 4 represents the results of the formulated EEAw-DNNNB algorithm produces higher precision results of 95.58% whereas other classifiers like ANN, SVM, SVM-HWFFS, SVM-HFWE-FS produces precision results of 80%, 85%, 95.24% and 95.45%.

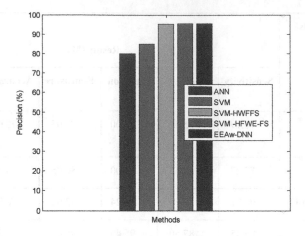

Fig. 4. Precision performance comparison vs. classifiers

Figure 5 represents the output of the proposed EEAw-DNNNB technique yields a greater f-measure output of 96.43% whereas other classifiers such as ANN, SVM, SVM-HWFFS, SVM-HFWE-FS produces f-measure results of 78.05%, 80.95%, 93.02% and 95.45%.

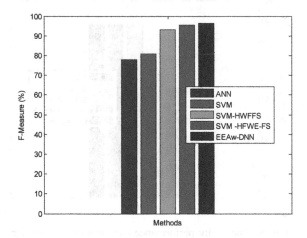

Fig. 5. F-measure performance comparison vs. classifiers

Figure 6 represents the output of the proposed EEAw-DNNNB technique yields greater results of accuracy of 94.70% whereas other classifiers like ANN, SVM, SVM-HWFFS, SVM-HFWE-FS produces accuracy results of 70%, 73.33%, 90% and 93.33%.

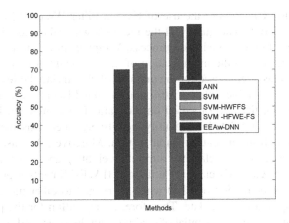

Fig. 6. Accuracy performance comparison vs. classifiers

Figure 7 represents the results of the formulated EEAw-DNNNB algorithm produces minimized error rate output of 5.30% whereas other classifiers like ANN, SVM, SVM-HWFFS, SVM-HFWE-FS produces accuracy results of 30%, 26.67%, 10% and 6.67%.

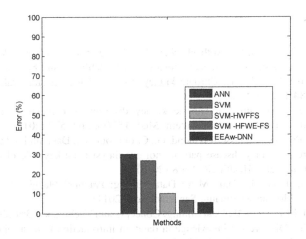

Fig. 7. Error rate performance comparison vs. classifiers

4   Conclusion and Future Work

Chronic kidney disease (CKD) is considered to be a serious problem of health aspect. Prediction of CKD needs to be accurate which is essential for the minimization of treatment cost and mortality. The classification algorithms' accuracy depends on the usage of exact feature extraction techniques leads to the classification algorithms accuracy in o5rder to reduce the data set dimension. The feasible feature set is recognized by the CKD under the use of Hybrid Filter Wrapper Embedded (HFWE) based feature selection. The quality of the methods was validated with the help of the features selected. In CKD prediction process, Wrapper, filter and embedded based FS algorithms are utilized in the minimization of dimensionality of features with the help of EEAw-DNN classifier is used for the purpose of classifying the features. The current study formulates Ensemble Entropy Attribute Weighted Deep Neural Network (EEAw-DNN) for detecting the CKD focused on the data. The formulated EEAw-DNN prediction system is formulated to identify adaptively deep classifier components at several repetitions from the entire neural system. Moreover, the ensemble is proposed as a Bayesian framework for classifier attribute weighting and combination. The output proves that the EEAw-DNN classifier utilizing HFWE-FS method produces high level of accuracy in the prediction of CKD when co-related to other methods. This dataset consists of noisy and missing data. Therefore, an algorithm for the purpose of classification is essential with the potential of dealing with the noisy and missing data, which is considered to be the future scope.

Compliance with Ethical Standards
  ✓ All authors declare that there is no conflict of interest.
  ✓ No humans/animals involved in this research work.
  ✓ We have used our own data.

References

1. Chen, P., Zhang, Q.-L., Rothenbacher, D.: Prevalence of chronic kidney disease in population-based studies: systematic review. BMC Public Health **8**(1), 117 (2008)
2. Baumgarten, M., Gehr, T.: Chronic kidney disease: detection and evaluation. Am. Fam. Physician **84**(10), 1138 (2011)
3. Moyer, V.A.: Screening for chronic kidney disease: us preventive services task force recommendation statement. Ann. Intern. Med. **157**(8), 567–570 (2012)
4. Frimat, L., Pau, D., Sinnasse-Raymond, G., Choukroun, G.: Data mining based on real world data in chronic kidney disease patients not on dialysis: the key role of early hemoglobin levels control. Value Health **18**, A508 (2015)
5. Witten, I.H., Frank, E., Hall, M.A.: Data Mining: Practical Machine Learning Tools and Techniques. Morgan Kaufmann, San Francisco (2011)
6. Jiawei Han, M.K., Jian, P.: Data Mining: Concepts and Techniques. 3rd edn. Elsevier (2012)
7. Larose, D.T.: Discovering knowledge in data: an introduction to data mining (2009)
8. Rodriguez, M., Salmeron, M.D., Martin-Malo, A., Barbieri, C., Mari, F., Molina, R.I., et al.: A new data analysis system to quantify associations between biochemical parameters of chronic kidney disease-mineral bone disease. PLoS One **11**, e0146801 (2016). https://doi.org/10.1371/journal.pone.0146801
9. Levey, A.S., Coresh, J., Balk, E., et al.: National kidney foundation practice guidelines for chronic kidney disease: evaluation, classification, and stratification. Ann. Intern. Med. **139**(2), 137–147 (2003)
10. Keane, W.F., Zhang, Z., Lyle, P.A., et al.: Risk scores for predicting outcomes in patients with type 2 diabetes and nephropathy: the RENAAL study. Clin. J. Am. Soc. Nephrol. **1**(4), 761–767 (2006)

11. Taal, M.W., Brenner, B.M.: Predicting initiation and progression of chronic kidney disease: developing renal risk scores. Kidney Int. **70**(10), 1694–1705 (2006)
12. Akay, M.F.: Support vector machines combined with feature selection for breast cancer diagnosis. Expert Syst. Appl. **36**, 3240–3247 (2009)
13. Özçift, A., Gülten, A.: Genetic algorithm wrapped Bayesian network feature selection applied to differential diagnosis of erythemato-squamous diseases. Digital Signal Process. **23**, 230–237 (2013)
14. Anandanadarajah, N., Tharmarajah, T.: Identifying important attributes for early detection of chronic kidney disease. IEEE Rev. Biomed. Eng. 1–9 (2017)
15. Varma, B.P., Raman, L.K., Ramakrishnan, L.S., Singh, L., Varma, A.: Prevalence of early stages of chronic kidney disease in apparently healthy central government employees in India. Nephrol. Dial Transplant. 3011–3017 (2010)
16. Salekin, A., Stankovic, J.: Detection of chronic kidney disease and selecting important predictive attributes. In: IEEE International Conference on Healthcare Informatics (ICHI), pp. 262–270 (2016)
17. Luck, M., Bertho, G., Bateson, M., Karras, A., Yartseva, A., Thervet, E., et al.: Rule-mining for the early prediction of chronic kidney disease based on metabolomics and multi-source data. PLoS One **11**, e0166905 (2016)
18. Huang, M.-J., Chen, M.-Y., Lee, S.-C.: Integrating data mining with case-based reasoning for chronic diseases prognosis and diagnosis. Expert Syst. Appl. **32**, 856–867 (2007)
19. José, N., Rosário Martins, M., Vilhena, J., Neves, J., Gomes, S., Abelha, A., Machado, J., Vicente, H.: A soft computing approach to kidney diseases evaluation. J. Med. Syst. **39**, 131 (2015)
20. Di Noia, T., Claudio, V., Ostuni, F.P., Binetti, G., Naso, D., Schena, F.P., Di Sciascio, E.: An end stage kidney disease predictor based on an artificial neural networks ensemble. Expert Syst. Appl. **40**, 4438–4445 (2013)
21. Polat, H., Mehr, H.D., Cetin, A.: Diagnosis of chronic kidney disease based on support vector machine by feature selection methods. J. Med. Syst. **41**(4), 55 (2017)
22. Norouzi, J., Yadollahpour, A., Mirbagheri, S.A., Mazdeh, M.M., Hosseini, S.A.: Predicting renal failure progression in chronic kidney disease using integrated intelligent fuzzy expert system. Comput. Math. Methods. Med. (2016)
23. Tangri, N., Stevens, L.A., Griffith, J., Tighiouart, H., Djurdjev, O., Naimark, D., Levin, A., Levey, A.S.: A predictive model for progression of chronic kidney disease to kidney failure. JAMA **305**(15), 1553–1559 (2011)
24. Levey, A.S., Coresh, J.: Chronic kidney Dis. Lancet **379**, 165–180 (2012)
25. National Kidney Foundation: K/DOQI clinical practice guidelines for chronic kidney disease: evaluation, classification, and stratification. Am. J. Kidney Dis. **2002**(392 Suppl 1), S1–S266 (2002)
26. Anderson, J., Glynn, L.G.: Definition of chronic kidney disease and measurement of kidney function in original research papers: a review of the literature. Nephrol. Dial. Transplant. **2011**(26), 2793–2798 (2011)
27. Novakovic, J., Strbac, P., Bulatovic, D.: Toward optimal feature selection using ranking methods and classification algorithms. Yugoslav J. Oper. Res. **21**(1), 119–135 (2011)
28. Komarasamy, G., Wahi, A.: An optimized K-means clustering technique using bat algorithm. Eur. J. Sci. Res. **84**(2), 263–273 (2012)
29. Kumar, M.: Prediction of chronic kidney disease using random forest machine learning algorithm. Int. J. Comput. Sci. Mob. Comput. **5**(2), 24–33 (2016)
30. Yang, X.S.: A new metaheuristic bat-inspired algorithm. In: Gonzalez, J.R. et al. (eds.) Nature Inspired Cooperative Strategies for Optimization (NISCO 2010), vol. 284, pp. 65–74. Springer Press (2010)

31. Yang, X.S.: Bat algorithm for multi-objective optimization. Int. J. Bio Inspired Comput. **3**(5), 267–274 (2011)
32. Nakamura, R., Pereira, L., Costa, K., Rodrigues, D., Papa, J., Yang, X.S.: BBA: a binary bat algorithm for feature selection. In: Proceedings 25th SIBGRAPI Conference on Graphics, Patterns and Images (SIBGRAPI), 22–25 August 2012, pp. 291–297 (2012)
33. .Zhou, X., Xie, L., Zhang, P., Zhang, Y.: An ensemble of deep neural networks for object tracking. In: IEEE International Conference on Image Processing (ICIP), pp. 843–847 (2014)

Swarm Intelligence Based Feature Clustering for Continuous Speech Recognition Under Noisy Environments

M. Kalamani[1](✉), M. Krishnamoorthi[2], R. Harikumar[1],
and R. S. Valarmathi[3]

[1] Department of ECE, Bannari Amman Institute of Technology,
Coimbatore, Tamilnadu, India
kalamani.mece@gmail.com, harikumarrajaguru@gmail.com
[2] Department of CSE, Dr.N.G.P. Institute of Technology, Coimbatore,
Tamilnadu, India
drkrishnamoorthim@gmail.com
[3] Department of ECE, Vel Tech Rangarajan Dr.Sagunthala R&D Institute
of Science and Technology, Chennai, Tamilnadu, India
atrmathy@gmail.com

Abstract. Swarm Intelligence based Feature Clustering using Artificial Bee Colony (ABC) technique is proposed and implemented in this research work. It is used to group and label the features of continuous speech sentence. This algorithm is unsupervised classification which classifies the feature vectors into different clusters and it will enhance the quality of clustering. Simulation is carried out for various number of clusters for different speech recognition algorithms under clean and noisy environmental conditions of NOIZEUS database signals. Experimental results reveal that the proposed hybrid clustering provides substantial enhancements in various performance measures compared with the existing algorithms and found as number of cluster as 5 produces the optimal result. For this optimal value, ABC clustering technique provides optimal enhancement in the recognition accuracy of Continuous Speech Recognition compared with the existing clustering techniques under different speech signal environments.

Keywords: Feature clustering · Labelling · K-means · Fuzzy c-means · Artificial Bee Colony · Centroid · Continuous Speech Recognition

1 Introduction

Today, the major challenging problem faced for human-machine interface through Continuous Speech Recognition (CSR). Recently, the more sophisticated recognizers are required to short out these problems. In CSR before modelling, the extracted features are labelled using efficient clustering. Recently, clustering algorithm has plays a major role in the various signal processing applications. Clustering methods mentioned in the literature are used the different principles to classify the objects into clusters. During clustering process, the properties of whole clusters are used instead of

© Springer Nature Switzerland AG 2020
S. Smys et al. (Eds.): ICCVBIC 2019, AISC 1108, pp. 1248–1255, 2020.
https://doi.org/10.1007/978-3-030-37218-7_130

individual feature and this might be useful when handling large amounts of feature vectors. Conventional clustering is mostly used to handle low dimensional features. It is mainly categorized into two methods. One is the partitional clustering in which the feature vectors are grouped into some specified number of clusters. Second is the Hierarchical clustering in which the feature vectors are clustered with a structure of partitions [1].

The partitional clustering techniques are classified into two classes. In hard clustering, each feature vectors gathered into single group. But in soft clustering, feature vectors are assigned into more than one cluster, and it have diverse membership levels. Soft clustering is used to assign the membership values and data values to one or more than one clusters. The example for most widely used partitional clustering are fuzzy c-means (FCM) and k-means (KM) techniques. KM method is a famous and approximate algorithm which is used to optimize the convergence. The objective of this method is to find K-mean vectors with K-cluster centroids. In this, global minimum is not achieved due to discrete values. Also, it depends on initial value [2].

KM with unsupervised approach is the simple clustering technique. For many applications, this clustering technique is mainly used because it is simple, efficient with low cost. Hence, it is produce the clusters with uniform size [3]. FCM clustering technique is frequently used for feature clustering applications. In this, the intra-cluster distance is minimized but it depends on initial values. FCM is the unsupervised and most widely used soft clustering method. This approach is used to label and cluster the extracted features in speech recognition system with less error rate. Fuzzy clustering algorithm is successfully applied in segmentation applications. The limitations of FCM algorithm as: Cluster number required in advance and mostly converge to a local minima or saddle point. Hierarchical clustering is broadly divided into agglomerative and divisive. This clustering does not required the number of cluster in advance. Due to its complexity, this has low execution efficiency than partitional clustering. Both clustering has pros and cons such as number and shape of clusters, degeneracy and overlapping of clusters [4, 5].

In order to overcome these issues in conventional clustering techniques, nowadays a hybrid clustering algorithms are used. The conventional techniques failed to produce efficient results when the size of the dataset increased. So clustering with swarm-based optimization algorithms emerged as an alternative to the traditional clustering algorithms. Many traditional clustering algorithms combined with optimization and this hybrid clustering are developed recently. The nature of optimization techniques is population based and can find the near optimal solutions. Optimization techniques will avoid being trapped in a local optimum when combined with partitional clustering algorithms and perform even better [6–8].

ABC clustering technique is implemented in this research work. This hybrid clustering used to group the extracted feature vectors into centroids and label it before modelling in CSR. This paper deals with: Sect. 2 provides the FCM feature clustering techniques and Sect. 3 demonstrate the proposed ABC clustering technique for Continuous Speech Recognition. Section 4 demonstrate the performance measures of proposed and existing clustering under different environments of speech signal. Finally, Sect. 5 conclude this research work.

2 Fuzzy C-Means Clustering for Continuous Tamil Speech Recognition

FCM clustering is an unsupervised and most commonly used clustering techniques. In FCM clustering, the data points for all groups have various membership values between 0 and 1. Based on the distance, this membership values are assigned. In this, all values are updated iteratively [4] and it is summarized as follows:

1. Let xi, i = 1, 2, ..., NT and it is going to clustered.
2. Assume cluster number as CL and its lies in the range $2 \leq CL \leq NT$.
3. Choose the fuzziness of the cluster (f) and it is greater than one.
4. Initialize membership matrix randomly, such that Uij $\in$ [0,1] and
$\sum_{j=1}^{CL} U_{ij} = 1$ for each 'i'
5. Compute the center of the jth cluster as follows:

$$cc_j = \frac{\sum_{i=1}^{N_T} U_{ij}^f x_i}{\sum_{i=1}^{N_T} U_{ij}^f} \tag{1}$$

6. Compute the Euclidean distance, Dij between ith feature and jth centroid as:

$$D_{ij} = \left\| x_{im} - CC_j \right\| \tag{2}$$

7. Based on Dij, update fuzzy membership matrix U. If Dij > 0, then

$$U_{jij} = \frac{1}{\sum_{C=1}^{CL} \left(\frac{D_{ij}}{D_{iC}}\right)^{\frac{2}{f-1}}} \tag{3}$$

8. If Dij = 0, then Uij = 1 because data points are having same value.
9. Continue the Steps 5–7 until termination criterion is reached.

3 Proposed ABC Clustering for Continuous Tamil Speech Recognition

The proposed ABC algorithm aims to locate the cluster centers of feature vectors based on minimum objective function. The clever searching characteristics of honey bee swarms are used to cultivate the new approach as Artificial Bee Colony (ABC). This approach contains three phases such as onlooker bee, employed bee and scout bee.

Among this three, first two phases are predictable and third phase is a random one. This ABC clustering algorithm contains four phases and step by step process in each phase is defined in detail in the subsequent sections [8].

3.1   Initialization Phase

In this phase, population size 'S' is assumed and position for each swarm is generated randomly. The entire population is divided into two with equal size and it is namely as onlooker and employed bees. Initially, the count of current scout bee is initialized to zero. After this initial process, the solution is the repeated cycles of all three phases. Consider the feature vector D consist of 'n' elements and select two features from D as centroids C1, C2 and has the distance value as i, j respectively.

After the distance calculation between the centroids and feature vectors, the feature value is stimulated to the cluster with minimum distance value. Finally, the feature vectors are classified into these two initial clusters based on their minimum distance value. Then, fitness value $f(z_i)$ for each employed bee is calculated by adding all the respective distance values.

3.2   Employed Bee Phase

Here based on fitness value, the new position of all employed bees is reformed and it is described as:

$$y_{i,j} = z_{i,j} + \Phi_{i,j}(z_{i,j} - z_{k,j}) \tag{4}$$

Where, $y_{i,j}$ is the new position and $z_{i,j}$ is the old position of employed bee; random number $\Phi_{i,j}$ and its value assumed between $[-1, 1]$; k is arbitrarily nominated neighborhood for solution i.

The optimal position of the employed bee is selected based on greedy selection strategy. In this, value of new position is greater than the previous position, then the value of new position is taken or otherwise the value of previous position is taken for further processing. After evaluating the new position, the value of the fitness is calculated for all employed bees as:

$$fit(z_i) = \begin{cases} \frac{1}{1+f(z_i)} & \text{if } f(z_i) \geq 0 \\ 1 + \text{abs}(f(z_i)) & \text{if } f(z_i) < 0 \end{cases} \tag{5}$$

The algorithm proceeds to the onlooker phase only when all the employed bees have been processed.

3.3   Onlooker Bee Phase

Here, onlooker bee position is selected based on the fitness likelihood which is supplied from the previous phase. The likelihood of position for all employed bee is calculated as follows:

$$P_i = \frac{fit(z_i)}{\sum_{i=1}^{s/2} fit(z_i)} \qquad (6)$$

The position of onlooker bee selected based on highest fitness value and finds the suitable new position in the neighborhood. Updated using Eq. (4) for quality improvement of new position for the onlooker bee. Finally, this bee compares the old position with new one. This bee will retain the new position value when it is greater than old position value or otherwise the old position value is retained.

3.4 Scout Bee Phase

This phase is the last step in the conventional ABC method. If the solution of the first phase is not improved with finite number of cycles, then solutions of final phase is uncontrolled one. Then, this scout bee searches the solution in the new space randomly. With the new populations created in this iteration, continued the next iterations. For each iteration, all the phases of the proposed approach are executed until the stopping criteria is reached. Then, final centroid is updated based on the new position.

4 Results and Discussions

The speech sentences with different noise environment are obtained from the NOI-ZEUS database and it is used throughout the evaluation process of existing and proposed algorithms. In this research work, the KM, FCM, and ABC approaches are implemented for labeling the extracted speech features before the modeling phase under different speech signal environments. For ABC algorithm, swarm size is assumed as 30 and number of iterations is fixed as 100 with Upper Bound value of 5.

The performance measures for the existing and proposed clustering techniques for sp01 speech sentence in NOIZEUS database under clean and 5 dB airport noise are described in Tables 1 and 2. From these experimental results, it is found that the proposed ABC clustering technique reduces the average Intra cluster distance, DB index and distance index from 9.4–12.9%, 14.3–20.9%, 34.3–50% respectively and increases the average Beta index from 9.6–25.1% when compared existing KM and FCM algorithms under clean signal environments.

In addition, proposed ABC clustering technique reduces the average Intra cluster distance, DB index and distance index from 3.35–6%, 13.6–22.5%, 28.9–44.9% respectively and increases average the Beta index from 17.3–31.4% when compared all existing algorithms under 5 dB airport noise signal environments. Hence, it is perceived that the proposed ABC clustering provided signification improvement in Intra-cluster distance, Beta Index, DB index, and Distance Index when compared to the existing clustering techniques.

The Recognition Accuracy (%) versus cluster number for the Expectation-Maximization with Gaussian Mixture modeling (EM-GMM) technique with K-means, FCM and proposed ABC clustering algorithms sp01 speech sentence in NOIZEUS database under various speech signal environments shown in Figs. 1 and 2. From these experimental results, it is reveal that all the clustering techniques for various modeling techniques provide the considerable enhancement in recognition accuracy for the optimal cluster number as 5 compared with other values.

Table 1. Performance comparison of existing and proposed clustering for sp01 speech sentence (Clean)

Performance measures	Existing clustering technique			Proposed ABC-FCM
	K-Means	FCM	ABC	
Intra-cluster distance	356.46	342.75	310.48	286.15
Beta index	7.23	8.72	9.65	11.43
DB index	0.91	0.84	0.72	0.59
Distance index	0.46	0.35	0.23	0.12

Table 2. Performance comparison of existing and proposed clustering for sp01 speech sentence (5 dB airport noise)

Performance measures	Existing clustering technique			Proposed ABC-FCM
	K-Means	FCM	ABC	
Intra-cluster distance	412.57	401.29	387.84	314.73
Beta Index	6.15	7.42	8.97	10.26
DB index	0.98	0.88	0.76	0.63
Distance Index	0.49	0.38	0.27	0.18

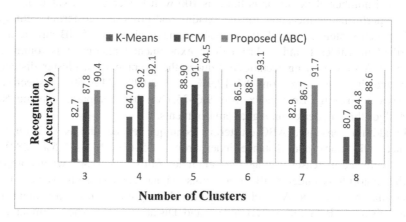

Fig. 1. Average recognition accuracy (%) versus number of clusters of the existing and proposed clustering algorithms used in EM-GMM modelling for sp01 speech sentence (Clean)

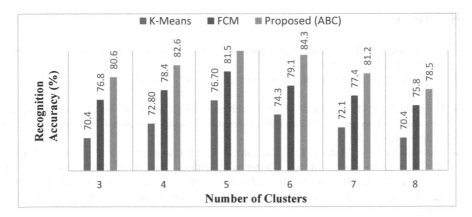

Fig. 2. Average recognition accuracy (%) versus number of clusters of the existing and proposed clustering algorithms used in EM-GMM modelling for sp01 speech sentence (5 dB Airport Noise)

In addition, proposed ABC clustering algorithm provides major improvement in Recognition Accuracy compared with existing clustering under clean and noisy environments. Hence, the ABC clustering with this optimal cluster is used for labeling the features of speech signal under the real time Continuous Speech Recognition under noise environments.

5 Conclusion

The extracted feature vectors are labelled through clustering technique before modelling in CSR. For this, ABC clustering is implemented in this research work. From the simulation results, the proposed clustering provides improvement in all the performance metrics compared with existing clustering techniques under clean and noisy environments. All clustering approaches will give better performance improvement for the cluster number '5' and it is considered as optimal one for further processing. From these experimental results, it is perceived that the ABC clustering with optimal cluster number as 5 yields the better improvement in recognition accuracy when compared to existing clustering algorithm for EM-GMM modelling technique under various signal environments of continuous speech recognition.

Compliance with Ethical Standards

✓ All authors declare that there is no conflict of interest.
✓ No humans/animals involved in this research work.
✓ We have used our own data.

References

1. Xu, R., Wunsch, D.C.: Clustering algorithms in biomedical research: a review. IEEE Rev. Biomed. Eng. **3**, 120–154 (2010)
2. Jain, A.K.: Data clustering: 50 years beyond K-means. Pattern Recogn. Lett. **31**(8), 651–666 (2010)
3. Li, X.G., Yao, M.F., Huang, W.T.: Speech recognition based on k-means clustering and neural network ensembles. In: 7th IEEE International Conference on natural computation, vol. 2, pp. 614–617 (2011)
4. Chattopadhyay, S., Pratihar, D.K., Sarkar, S.C.D.: A Comparative study of fuzzy C-means algorithm and entropy-based fuzzy clustering algorithms. Comput. Inf. **30**(4), 701–720 (2011)
5. Kalamani, M., Krishnamoorthi, M., Valarmathi, R.S.: Continuous tamil speech recognition technique under non stationary noisy environments. Int. J. Speech Technol. **22**(1), 47–58 (2019)
6. Fister Jr., I., Yang, X.S., Fister, I., Brest, J., Fister, D.: A brief review of nature-inspired algorithms for optimization. Elektrotehniski Vestnik. **80**(3), 1–7 (2013)
7. Zhang, C., Ouyang, D., Ning, J.: An artificial bee colony approach for clustering. Expert Syst. Appl. **37**(7), 4761–4767 (2010)
8. Karaboga, D., Ozturk, C.: A novel clustering approach: artificial bee colony (ABC) algorithm. Appl. Soft Comput. **11**(1), 652–657 (2011)

Video-Based Fire Detection by Transforming to Optimal Color Space

M. Thanga Manickam[(✉)], M. Yogesh, P. Sridhar,
Senthil Kumar Thangavel, and Latha Parameswaran

Department of Computer Science Engineering, AmritaVishwa Vidhyapeetham,
Coimbatore, India
{cb.en.u4.cse17161, cb.en.u4.cse17168,
cb.en.d.cse17011}@cb.students.amrita.edu,
{t_senthilkumar, p_latha}@cb.amrita.edu

Abstract. With the increase in number of fire accidents, the need for the fire detection system is growing every year. Detecting the fire at early stages can prevent both material loss and loss of human lives. Sensor based fire detection systems are commonly used for detecting the fire. But they have drawbacks like time delay and close proximity. Vision based fire detectors are cost efficient and can potentially detect fire in its early stages. Here we propose a lightweight pixel based fire detection model to extract frames from videos and identify frames with fire in it. We use matrix multiplication to transform our input frame to a new color space in which separation of fire pixels from non-fire pixels is easier. The optimal value for the matrix to be multiplied is obtained using fuzzy-c-means clustering and particle swarm optimization. Otsu thresholding is applied on transformed image in new color space to classify the fire pixels in the frame. Our result shows high accuracy on forest fire videos with very less inference time.

1 Introduction

In recent years various techniques have been used in fire detection. Sowah et al. [1] proposed a multiple sensor based fire detection method. Fuzzy logic is applied there on data received from smoke sensor, flame sensor and temperature sensor to raise fire alarm. The main drawback of these sensors is time delay [2] and high false positive rates. The distance from source of fire and sensors is also an important factor. Several pixel based fire detection methods have been proposed. Toreyin et al. [3] proposed a method based on spatial and temporal analysis. He detected the moving pixels using the difference between the frames. The fire-colored pixels are identified from moving pixels using Gaussian mixture. Then he applied temporal and spatial wavelet analysis to detect frames with fire. This method is not fast enough to identify the fire in early stages. Chen et al. [4] proposed a fire detection method based on the chromatic and dynamic features of a video. First moving regions of image sequences are identified using image differencing. Then fire and smoke pixels are extracted by chromatic features of image. Dynamic features are then used to validate the results. This is a complete rule based model and it can lead to high false positive rate. Celik et al. [5]

© Springer Nature Switzerland AG 2020
S. Smys et al. (Eds.): ICCVBIC 2019, AISC 1108, pp. 1256–1264, 2020.
https://doi.org/10.1007/978-3-030-37218-7_131

proposed a pixel based fire detection using CIELAB [6] color space. The original image in RGB color space is transformed to CIELAB color space. Then heuristic rules are applied to detect fire. This method works well only when the fire grows gradually with time. This might not be the case with sudden explosions. Demirel et al. [7] proposed a fire pixel classification method using statistical color model and fuzzy logic. The RGB image is transformed to YCbCr color space. The set of fuzzy rules is applied in YCbCr color space to detect fire in the image. This method may not work well when the intensity of the fire changes. We are proposing an efficient method that can detect fire at early stages. We also aim to ensure the proposed model fits into memory to deploy it in real time. This led us in choosing color space transformation followed by thresholding, as it requires only a 3 × 3-matrix multiplication, which is computationally less expensive and has less inference time. To measure the optimality of the matrix we use Fuzzy C Means and the matrix is optimized through particle swarm optimization.

2 Methodology

The speed of the model is very important in fire detection as early identification of fire can prevent the material and life loss. Hence, we propose a lightweight model that can classify the fire pixels in a faster way. We extracted frames from the videos and used it for training the model. We took frames with fire and non-fire as training examples. We extracted 25 × 25 pixels from the positive and negative images and flattened them into a matrix with three columns containing RGB values of the fire and non-fire pixels. The pixel values now are in RGB color space. It is the common color space used in our daily life. There are other color spaces like YIQ, HSV [8], grey-scale etc. They can be converted by performing matrix multiplication with the original pixel values. Let M be a 3 × 3 matrix with which the original pixel values are multiplied. Then the color space transformation is obtained by: **New color space pixel value = original pixel value * M,** where * is matrix multiplication. Our aim is to find the optimal value for M such that fire pixels and non-fire pixels are separable in new color space. We started with a random 3 × 3 matrix M. We project the original image to the new color space using the matrix M. Fuzzy C Means (FCM) technique is used to measure how separable fire and non-fire pixels are in new color space. FCM is applied to cluster the image into fire pixels and non-fire pixels. The error in this step is calculated as the total number of wrongly classified pixels. Particle Swarm Optimization (PSO) is used to obtain the matrix M such that error in FCM technique is minimal. After obtaining the best M value that gives the minimal cost, test image frame is matrix multiplied with optimal M. During prediction time, Otsu's thresholding takes advantage of the fact that fire pixels are more separable from non-fire pixels in the new optimal colorspace. Otsu's thresholding is more suitable here as it works well on clustered pixels (Figs. 1 and 2).

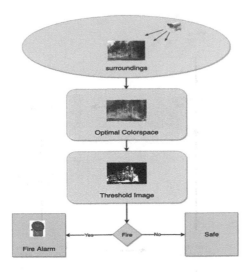

Fig. 1. Proposed system

2.1 Fuzzy C Mean Clustering (FCM)

Fuzzy C mean clustering is a soft clustering method [9]. It is used to assign each data point to one of the clusters. Each data point is assigned membership degrees to belong in each cluster. The data points on boundaries between two classes are not fully assigned to a single class through this [10]. The data points which are closer to centroid of a particular class have higher membership degree for that class. Hence it handles the outliers effectively. Here FCM is used to cluster fire pixels from non-fire pixels in new color space.

Algorithm for FCM:
Input: Fire image converted to new color spaces. Here C=2.
1. Let $x_1, x_2, x_3 \ldots x_n$ be the pixel values to be clustered.
2. Initialize C random clusters c_j and membership degree w_{ij} for each data point x_i to belong in cluster j.

3. Find cost to be minimized through following equation. $\text{cost}(w) = \arg\min \sum_{i=1}^{n} \sum_{j=1}^{C} w_{ij}^{m} \left\| x_{i\text{-}} c_j \right\|^2$
4. Update centroids and membership degrees through the following equations.

$$w_{ij} = \cfrac{1}{\sum_{k}^{c} \left(\cfrac{\|x_{i\text{-}} c_j\|}{\|x_{i\text{-}} c_k\|} \right)^{\frac{2}{m-1}}}$$

$$c_k = \frac{\sum_{x} x w_k(x)^m}{\sum_{x} w_k(x)^m}$$

Here m is the fuzziness index. Repeat step4 till the cost function converges.
   Output: Pixels clustered into fire and non-fire. We have used skfuzzy toolbox [11] to apply fuzzy c means. We used parameters as m = 2 and random seed=1.

2.2 Particle Swarm Optimization (PSO)

We have used particle swarm optimization to find the optimal value for M such that cost of the algorithm is minimal. Particle swarm Optimization is a useful optimizing technique [12] in which gradient of the cost function to be optimized cannot be

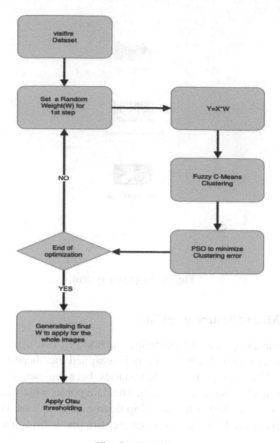

Fig. 2. Workflow

determined. It is initialized with random solutions called particles and in each of the iteration, the particles move towards the best solution with a velocity v [13]. We are trying to minimize the cost function for a matrix M given by the following expression.

The cost function of the algorithm for the matrix M is given by: Cost (M) = no of actual fire pixels incorrectly classified as non fire + no of non-fire pixels incorrectly classified as fire.

Algorithm for PSO
Input: Images, ground truth
1. Initialize each particle M_i with the velocity v_i.
2. Find cost function for each of the particles in the swarm. Maintain the local best solution for each particle in the swarm and global best solution of all the particles in the swarm.
3. Velocity of each particle is updated by the following equation,
$v_i = w * v_i + c1 * local\ best_i + c2 * global\ best$, where c1 and c2 are random numbers and w is velocity weight.
At the end of each of the iteration global best and local best solutions are updated.
4. The step3 is repeated until the cost function converges. The global solution at the end provides the optimal matrix M.

Output: Optimal matrix M
We have used w =2, random seed = 1 while applying particle swarm optimization. The optimal solution obtained by the PSO is used to transform the input image to new color space.

2.3   Otsu Thresholding

Otsu is a thresholding technique to binarize an image [14]. Otsu thresholding works well on clustered values [15]. This thresholding assumes the image to be bimodal (foreground pixels and background pixels). It then calculates the optimal threshold value using the minimal within-class variance. The within-variance is calculated using the following formula. $\sigma_\omega^2(t) = \omega_0(t)\sigma_0^2(t) + \omega_1(t)\sigma_1^2(t)$, where weights $\omega_{0,1}(t)$ are the probabilities of the two classes separated by a threshold t and $\sigma_{0,1}^2(t)$ are variances of these two classes. The class probabilities $\omega_{0,1}(t)$ are calculated from L histogram.

$$\omega_0(t) = \sum_{i=0}^{t-1} p(i)$$

$$\omega_1(t) = \sum_{i=t}^{L-1} p(i)$$

Variance $\sigma_0^2(t)$ and $\sigma_1^2(t)$ are calculated using class mean μ,

$$\mu_0(t) = \sum_{i=0}^{t-1} i * \frac{p(i)}{\omega 0}$$

$$\mu_1(t) = \sum_{i=t}^{L-1i} * \frac{p(i)}{\omega 1}$$

Then we find the minimal value from the with-in class variances of different classes of the image. The class, which has minimal with-in class variances, is the threshold value (t) to be used for Otsu thresholding. We apply Otsu's thresholding to find threshold value for each of the three channels in new color space. We have used skimage library [16] to apply Otsu thresholding. A pixel is marked as fire pixel only if it is greater than threshold in each channel. By this way, we achieve a faster way to classify the fire pixels. After training on 150 images we got the optimal value for M as:

$$\begin{matrix} 0.5905 & 1.7695 & 2.2408 \\ -1.2407 & 2.803 & 1.0292 \\ 0.483 & -2.719 & -2.5853 \end{matrix}$$

3   Dataset

We used Cair's [17] dataset of fire images for training. It consists of a variety of scenarios and different fire situations. We extracted 150 patches of 25 $\times$ 25 pixels from this dataset for the training process. We used Visifire's fire dataset [18] for testing our model. It consist videos of forest fire. Our algorithm works well on images with dark background. To verify this we used this dataset. To test our method on open environment we used fire dataset collected by smart space lab in our institution. It had 15300 frames with fire and 6300 non-fire frames. We tested this model on 480 non-fire images to evaluate the performance of the method in reducing false positives. This dataset consisted of different images in it under different lighting conditions. It had many fire-like images in it.

4 Result

Our results on 5 video clips from Visifire dataset showed high accuracy in detecting frames with fire. This shows the algorithm works well in forest fire datasets.

Here are our results on this dataset:

Video	Resolution	True positives	False positives	Accuracy
(1) Forest1	400 × 256	200	0	1
(2) Forest2	400 × 256	245	0	1
(3) Forestfire1	400 × 256	213	0	0.977
(4) Forest4	400 × 256	219	0	1
(5) Forest5	400 × 256	216	0	1

To test the method's performance in bright background we used smart space lab dataset. Results showed high accuracy with less false positive rate on this dataset. Here are our results on this dataset.

Video	Resolution	True positive rate	False positive rate	Accuracy
Smart space lab	640 × 480	0.943	0.061	0.941

Our method showed good accuracy in identifying the non-fire images. It predicted the fire-like objects like glowing lamp as non-fire with less number of false positives. This reduces the false alarms significantly.

Total number of Images	False positive rate	Accuracy
480	0.085	0.9145

The above results show the model can detect the fire efficiently with low false positive rate.

Here are the examples of our results (Fig. 3):

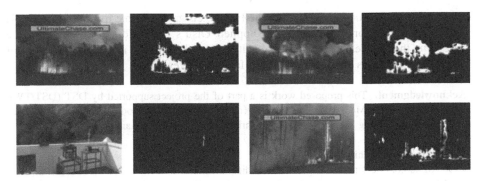

Fig. 3. (a)-Forest1, (b)-Forest2, (c) – Smart space lab, (d) - Forestfire1. The images at the left shows original images and binary images at the right shows the output of our model. The following results in forest fire images shows that our model is capable of detecting the forest fire accurately [11]

Results on Non-fire Images:

Figure 4(a)–(d) shows our sample results on the non fire images dataset prepared by us. This shows that our model can potentially avoid false alarms.

Fig. 4. Sample results on non-fire images. The images at the left shows original images and binary images at the right shows the output of our model. The output shows that fire-like images are correctly classified as non-fire

For inferring the decision on single image, our method took 0.15 s. This show our method can detect the fire at earlier stages with high speed. The above results are obtained using python version 3.6,numpy version 1.15.4 [19] and opencv version 3.4.1 [20].

5 Conclusion

We have proposed here a fast and accurate method to detect fire. We projected the Input image from RGB color space to optimal color space using linear transformation with matrix M. We used Fuzzy C means clustering and Particle Swarm optimization to find the optimal value of M such that the Fire pixels are separable from non-fire pixels in transformed color space. Then we applied Otsu's thresholding to detect the fire pixels in the image. Experimental results showed that our method is capable of detecting forest fires efficiently with less false positives rate.

Acknowledgment. This proposed work is a part of the project supported by DST (DST/TWF Division/AFW for EM/C/2017/121) titled A framework for event modelling and detection for Smart Buildings using Vision Systems to Amrita Vishwa Vidyapeetham, Coimbatore.

Compliance with Ethical Standards
   ✓ All authors declare that there is no conflict of interest
   ✓ No humans/animals involved in this research work.
   ✓ We have used our own data.

References

1. Sowah, A., Ofoli, R., Fiawoo, K.: Hardware design and web-based communication modules of a real-time multisensor fire detection. IEEE Trans. Ind. Appl. **53**(1), 559 (2017)
2. Paresh, P.A., Parameswaran, L.: Vision-based algorithm for fire detection in smart buildings. In: Lecture Notes in Computational Vision and Biomechanics, vol. 30. Springer, Netherlands, pp. 1029–1038 (2019)
3. Töreyin, B., Dedeoglu, Y., Gudukbay, U., Cetin, A.: Computer vision based method for real-time fire and flame detection. Pattern Recogn. Lett. **27**, 49–58. http://doi.org/10.1016/j.patrec.2005.06.015
4. Chen, T., Wu, P., Chiou, Y.: An early fire-detection method based on image processing. In: Proceedings - International Conference on Image Processing, ICIP, vol. 3, pp. 1707–1710 https://doi.org/http://doi.org/10.1109/ICIP.2004.1421401
5. Celik, T.: Fast and efficient method for fire detection using image processing. ETRI J. **32**. https://doi.org/http://doi.org/10.4218/etrij.10.0109.0695
6. Buckley, R.R., Giorgianni, E.J.: CIELAB for Color Image Encoding. Encyclopedia of Color Science and Technology. Springer, New York (2016)
7. Celik, T., Ozkaramanlt, H., Demirel, H.: Fire pixel classification using fuzzy logic and statistical color model. In: 2007 IEEE International Conference on Acoustics, Speech and Signal Processing - ICASSP 2007, Honolulu, pp. I-1205–I-1208 (2007)
8. Raval, K., Shah, R.S.K.: Color image segmentation using FCM clustering technique in RGB, L*a*b, HSV, YIQ Color spaces. Eur. J. Adv. Eng. Technol. **4**(3), 200 (2017). www.ejaet.com
9. Yuhui, Z., Byeungwoo, J., Danhua, X., Jonathan Wu, Q.M., Hui, Z.: Image segmentation by generalized hierarchical fuzzy C-means algorithm. J. Intell. Fuzzy Syst. **28**, 961–973 (2015)
10. Senthil Kumar, T., Suresh, A., Kiron, K., Chinnaswamy, P.: Survey on predictive medical data analysis. J. Eng. Res. Technol. **3**, 2283–2286 (2014)
11. Warner, J.D.: Scikit-Fuzzy, Zenodo. http://doi.org/10.5281/zenodo.1002946
12. Marini, F., Walczak, B.: Particle swarm optimization (PSO), a tutorial. Chemometrics Intell. Lab. Syst. **149**, 153 (2015)
13. Aerospace Design Laboratory Stanford, Gradient-Free Optimization, Aerospace Design Laboratory Stanford, April 2012

14. Kalavathi, P.: Brain tissue segmentation in MR brain images using multiple Otsu's thresholding technique. In: 2013 8th International Conference on Computer Science & Education, Colombo, pp. 639–642 (2013)

15. Bangare, S.L., Dubal, A., Bangare, P.S., Patil, S.T.: Reviewing Otsu's method for image thresholding. Int. J. Appl. Eng. Res. **10**(9) (2015)

16. van der Walt, S., Schönberger, J.L., Warner, J.D.: Scikit-image: image processing in Python. https://doi.org/10.7717/peerj.453

17. Cair, Fire-detection-image-dataset, Github. https://github.com/cair/Fire-Detection-Image-Dataset

18. Bilkent SPG, Visifire Fire dataset. http://signal.ee.bilkent.edu.tr/VisiFire/Demo/FireClips/

19. van der Walt, S., Chris Colbert, S., Varoquaux, G.: The NumPy array: a structure for efficient numerical computation. Comput. Sci. Eng. **13**, 22–30 (2011)

20. Bradski, G.: The OpenCv_library. Dr. Dobb's J. Softw. Tools, 2236121 (2008)

Image Context Based Similarity Retrieval System

Arpana D. Mahajan[✉] and Sanjay Chaudhary

Madhav University, Pindwara, Rajasthan, India
mahajan.arpana@yahoo.com, schaudhary00@gmail.com

Abstract. The rapid development in multimedia and imaging technology, the numbers of images uploaded and shared on the internet have increased. It leads to develop the highly effective image retrieval system to satisfy the human needs. The content-text image retrieval (CTIR) system which retrieves the image based on the high level features such as tags which are not sufficient to describe the user's low level perception for images. Therefore reducing this semantic gap problem of image retrieval is challenging task. Some of the most important notions in image retrieval are keywords, terms or concepts. Terms are used by humans to describe their information need and it also used by system as a way to represent images. Here in this paper different types of features their advantage and disadvantages are described.

Keywords: Context retrieval · High level features · Distances · WordNet · Flickr · CISS

1 Introduction

In this section to measure the semantic similarity between tags associated with images various semantic similarity measures are described such as WordNet Distance, Flickr distance, and Context based image semantic similarity based distance and Fuzzy string matching approach. Above all methods are called as high level features. High level features are not extracted from mathematical expression its require context data. At last the comparison between these methods is given using advantages and disadvantages of all the methods [1, 2].

Tags of the images in query class are required to extract from its respective annotation files. Tags of the images of query class is extracted as these tags are used by context based similarity techniques to measure the similarity among tags of the query pictures and tags of the pictures which are in query's class (Fig. 1).

© Springer Nature Switzerland AG 2020
S. Smys et al. (Eds.): ICCVBIC 2019, AISC 1108, pp. 1265–1272, 2020.
https://doi.org/10.1007/978-3-030-37218-7_132

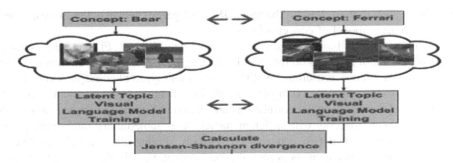

Fig. 1. Tag system

2 Related Works

2.1 Word-Net Distance (WD)

WordNet Distance is similarity measure between two keywords or concepts which are available in WordNet database. WordNet Database [6] is developed by cognitive science laboratory of Princeton University is a large semantic lexicon for English language. WordNet is a lexical database for the English language. It is an online lexical database designed for use under program control. English nouns, verbs, adjective and adverbs are organized into sets of synonyms that are in turn linked through semantic relations that determine word definitions [6]. It groups English words into sets of synonyms which provides short definitions and usage examples, and records a number of relations among these synonym sets or their members. WordNet is a combination of dictionary and thesaurus.

WordNet "includes the following semantic relations: Synonymy, Antonym, Hyponymy, Meronym, Troponymy, and Entailment. WordNet lists the alternatives from of words in different context from which choices must be made. Semantic Similarity between two concepts C1 and C2 lexicalized in WordNet is given by following equation. It is known as WordNet Distance [11].

$$sim(c_1, c_2) = \frac{2 \times \log P(C_0)}{\log P(C_1) + \log P(C_2)} \tag{1.1}$$

Where, C0 is most specific class that subsumes both C1 and C2. C0 is the most specific concept which is an ancestor of both C1 and C2 [7]. Using WD between keywords or tags of two images, we can measure the semantic relatedness between two images for retrieving the similar images in image retrieval system.

2.2 Flickr-Distance (FD)

Flickr distance is novel measurement of the relationship between different concepts in visual domain [4]. In this method first of all for each concept, a collection of images are obtained from Flickr, based on which the latent topic visual language model is built to capture the visual properties of the concept then FD between two concepts is measured by the square root of Jensen-Shannon (JS) divergence between the corresponding visual language model of concepts [4, 5].

Flickr distance is based on the visual information of images. FD can easily scale up with increasing of concepts on the web. FD is also more consisting to human cognition therefore gives better measurements. FD measures four types of conceptual correlations between given concepts: synonymy, similarity, metonymy and concurrence. The examples of these conceptual correlations are shown in Figure. Synonymy denotes the same semantic" concepts with different names, like "table-tennis" and "ping-pong". Similarity denotes that two concepts are visually similar, like "table" and "desk". Meronym denotes one concept is part of other concept, i.e., "building" and "window". Concurrence denotes situation when two concepts appear simultaneously in daily life, like "desk" and "chair" [5] (Fig. 2).

Fig. 2. Different four mutual intangible associations [7]

Following figure shows the illustration of Flickr distance calculation. Although FD is effective there are also some limitations of it. First limitation is, we do not know which topic the user is considering for each concept. Second limitation is that FD's accuracy mainly depends on the quality of web images. If there are limited numbers of images related to one concept, then the concept may not be well captured and FD is not estimated right [12].

2.3   Context-Based Image Semantic Similarity (CISS)

CISS "is a scheme for discovering and evaluating image similarity in terms of the associated group of concepts [8, 9]. The main idea in this method is to extract the text based group of concept which characterizes the image. In CISS all the concepts related to image Ii is compared with all the concepts related to image Ij then appropriate scheme Mutual Confidence (CM) is used to combine the elementary distances D of the elements in the first and second group of concepts of images. Let us we have two images image Ii and Ij respectively as shown in Figure. Ii and Ij are pair of images to be compared. Ti1, Ti2... Tim is a group of concepts related to image Ii, while Tj1, Tj2... Tin are group of concepts related to Ij. Then D Iij, the context based image semantic similarity of image Ii and Ij is defined as" [10].

$$DIij = AG2 \{AG1 [DEL (dTim \rightarrow jn)], AG1 [DEL (dTjn \rightarrow im)] \qquad (1.2)$$

Where DEL = MAX, is the most accurate group combination schema and d = Mutual Confidence (CM) is used as elementary metric. And AG1, AG2 = average metric.

CISS method, for measuring the similarity considers that the co-occurrence between tags takes place when both the tags are used with the nearly similar resources. It gives more effective and accurate similarity measurements than similarity methods based on image low level features. Also required very less computation cost. It gives more effective and accurate result similarity measurements than similarity based on image low level features both in terms of deep concept similarity as well as in terms of computational cost [13].

2.4 Fuzzy String Matching

Popular Fuzzy String matching, the pattern in words are matched approximately rather than exactly for measuring the similarity between words. It overcomes the drawback of hit/miss type search for text based image retrieval system [9]. Levenshtein approach is used for fuzzy sting matching in which string matching is done with fuzzy string matching with edit distance.

First of all number of alters or edits such as insertions and deletions are calculated to transform one string into the other string. The distance of matching is measured in terms of number of basic operations are required to convert one string to its exact match. This number is called as edit distance [9]. The number of edits obtained from fuzzy string matching is converted to percentage of matching using following equation.

$$StDist = 1.0 - (LDist(K1, K2)) / \max(K1, K2) \qquad (1.3)$$

Where LDist (K1, K2) = edit distance between K1 and K2 and K1 = length of string 1, K2 = length of string 2.

2.5 Proposed Distance Method

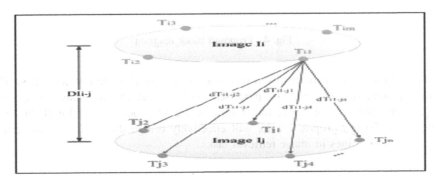

Fig. 3. Proposed distance method

We have applied context base similarity techniques between these high level features of identified query class and the query tags given by the user. Here from the .mat file which containing whole dataset annotations are used but only the tags which are of the query class are used. CISS and WordNet distance methods are applied between query tags and tags of the query class using group similarity algorithm given in [3]. Following Figure shows the concept of group similarity algorithm. In which apiece label of unique image is compared with all the tags of second image and semantic distance is calculated (Fig. 3).

3 Proposed Context Retrieval Method

Tags of the images in query class are required to extract from its respective annotation files. Tags of the images of query class is extracted as these tags are used by context based similarity techniques to measure the similarity between tags of the query image and tags of the images which are in query's class.

After extraction of tags, semantic similarity between tags of query image and tags of images in query class are calculated. For finding out the semantic similarity CISS and WordNet distance have decided to use in our system. Here CISS only calculate the co- occurrence of tags so for calculating other relationships such as hypernym, synonym etc. WordNet distance is used (Fig. 4).

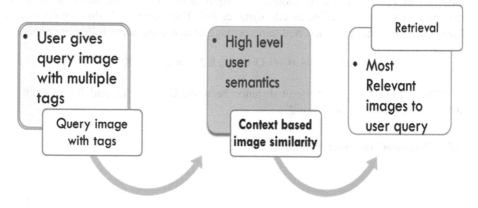

Fig. 4. Proposed block diagram

The most relevant images from database are retrieved using considering smallest semantic similarity measure. Our proposed work will retrieve the image using image Tags high level features so resultant images are very accurate to fulfill the user's perception. So our proposed work will effectively bridge the semantic gap between human and machines in image retrieval tasks.

Here we have apply context base similarity methods on the images of the query class only so our system required less calculation since we have not decided to apply context similarity methods on whole image database. Therefore our approach can also use for large database efficiently.

4 Results and Discussion

Using classification we have identified the class of the query image now next step is high level features extraction of query image. Then high level features of images in the identified query class are also extracted. Following Figures shows the examples of high level feature extraction of query images (Fig. 5).

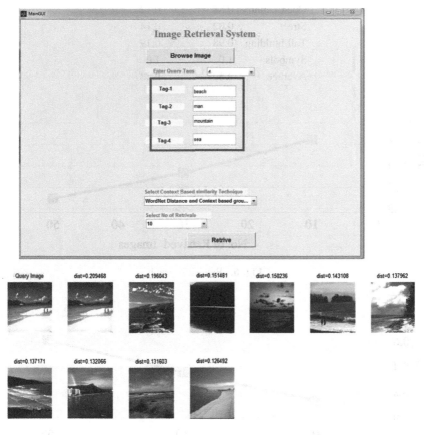

Fig. 5. GUI proposed system [12]

For calculating WordNet Distance we have installed maximum current gaps form of Word-Net i.e. Word-Net 2.1. For calculating this semantic distance we have used Java WordNet Similarity, it is open source project based on java and Hybrid Approach for Context based Image Retrieval (Table 1 and Figs. 6, 7).

Table 1. Analysis

Categories	Precision$_{N=10}$	Recall$_{N=10}$
Cost	0.82	0.12
Forest	0.94	0.11
Highway	0.82	0.13
Inside city	0.80	0.13
Mountain	0.86	0.14
Open country	0.88	0.12
Street	0.82	0.14
Tall building	0.98	0.19
Symbols	1.0	1.0
Average	**0.867**	**0.227**

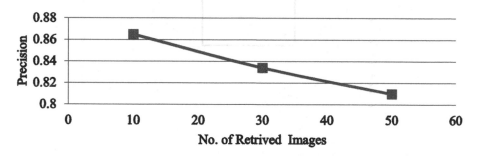

Fig. 6. Precision graph

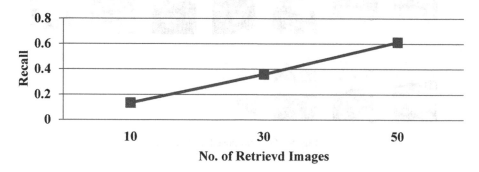

Fig. 7. Recall graph

5   Conclusion

The aim of this paper is to discuss all the phases of automatic image retrieval features. This research provide, a widespread study for feature extraction methods, classification methods and comparison measurement distances. Automatic image retrieval is a process for perceive scene feature after a photo and apportion to the corresponding classes. The feature extraction, classification and similarity measurement are used based on the size of the dataset. To overcome this issue and provide accurate retrieval of images we have proposed context image retrieval system which is based on image high levels of features.

Compliance with Ethical Standards
  ✓ All authors declare that there is no conflict of interest.
  ✓ No humans/animals involved in this research work.
  ✓ We have used our own data.

References

1. Cao, X., Wang, S.: Research about image mining technique. In: ICCIP, pp. 127–134. Springer, Heidelberg (2012)
2. Franzoni, V., Milani, A., Pallottelli, S., Leung, C.H.C., Li, Y.: Context-based image semantic similarity. In: Proceedings of IEEE Twelfth International Conference on Fuzzy Systems and Knowledge Discovery (FSKD), pp. 1280–1284 (2015)
3. Franzoni, V., Leung, C.H.C., Li, Y., Mengoni, P., Milani, A.: Set similarity measures for images based on collective knowledge. In: ICCSA, pp. 408–417. Springer, Heidelberg (2015)
4. Zarchi, M.S., Monadjemi, A., Jamshidi, K.: A concept-based model for image retrieval systems. Comput. Electr. Eng. (2015). https://doi.org/10.1016/j.compeleceng.2015.06.018
5. Goel, N., Sehgal, P.: Weighted semantic fusion of text and content for image retrieval. In: Proceedings of IEEE International Conference Advances in Computing, Communications and Informatics (ICACCI), pp. 681–687 (2013)
6. Li, X., Uricchio, T., Ballan, L., Bertini, M., Snoek, C., Del Bimbo, A.: Socializing the semantic gap: a comparative survey on image tag assignment, refinement and retrieval. ACM Comput. Surv. 49(1), 14 (2016)
7. Wu, L., Hua, X., Yu, N., Ma, W., Li, S.: Flickr distance: a relationship measure for visual concepts. IEEE Trans. Pattern Anal. Mach. Intell. 34(5), 863–875 (2012)
8. Wu, L., Hua, X.-S., Yu, N., Ma, W.-Y., Li, S.: Flickr distance. In: Proceedings of 16th ACM International Conference on Multimedia, pp. 31–40 (2008)
9. Miller, G.A.: WordNet: a lexical database for English. Commun. ACM 38(11), 39–41 (1995)
10. Budanitsky, A., Hirst, G.: Semantic distance in Wordnet: an experimental, application-oriented evaluation of five measures. In: Proceedings of WordNet and Other Lexical Resources (2001)
11. Wu, L., Jin, R., Jain, A.K.: Tag completion for image retrieval. IEEE Trans. Pattern Anal. Mach. Intell. 35(3), 716–727 (2013)
12. Chen, L., Xu, D., Tsang, I.W., Luo, J.: Tag-based image retrieval improved by augmented features and group-based refinement. IEEE Trans. Multimedia 14(4), 1057–1067 (2012)
13. Lu, D., Liu, X., Qian, X.: Tag-based image search by social re-ranking. IEEE Trans. Multimedia 18(8), 1628–1639 (2016)

Investigation of Non-invasive Hemoglobin Estimation Using Photoplethysmograph Signal and Machine Learning

M. Lakshmi[1]([⊠]) [ID], S. Bhavani[1], and P. Manimegalai[2] [ID]

[1] Department of ECE, Karpagam Academy of Higher Education,
Coimbatore, India
lakshmimuthukumar2@gmail.com,
bhavanisridharan7@gmail.com
[2] Department of BME, Karunya Institute of Technology and Sciences,
Coimbatore, India
manimegalai.vairavan@gmail.com

Abstract. Non-invasive haemoglobin (SpHb) estimation using Photoplethys-mograph signal has gained enormous attention among researches and can provide earlier diagnosis to polycythemia, anaemia, various cardiovascular diseases etc. The primary goal of this study investigation is to evaluate the efficiency of SpHb monitoring using PPG and generalized linear regression technique. PPG signal was acquired from outpatients and SpHb was calculated from the characteristic features of PPG. Haemoglobin value obtained through venous blood sample was compared with SpHb. The absolute mean difference between the SpHb and Hb_{ref} was 0.735 g/dL (SD 0.41). For statistical analysis of correlation between SpHb and Hb_{lab}, IBM SPSS statistics software was used. Bland-altman analysis, T-test, linear regression analysis was further used for finding the agreeability limits.

Keywords: Photoplethysmography · PPG · Non-invasive · Haemoglobin · Machine learning · Linear regression

1 Introduction

Haemoglobin (Hb) is a complex protein molecule in red blood cells (RBC) that transports oxygen from lungs to the rest of the body. Haemoglobin measurement is one of the most frequently performed laboratory tests. Haemoglobin test is performed during usual physical examination or when there is a sign of RBC disorder person such as anaemia or polycythaemia [1, 2]. Haemoglobin test is one of the mandatory steps to make decisions during blood transfusions. Haemoglobin measurement is generally performed by the traditional "fingerstick method" i.e., by invasively drawing blood from the body. Although the conventional laboratory measurement is accurate, it has its own limitations such as time delay, inconvenience of the patient, exposure to bio-hazards and the lack of real time monitoring in critical situations. The above said limitations can be overcome by Non-invasive haemoglobin (SpHb) monitoring.

© Springer Nature Switzerland AG 2020
S. Smys et al. (Eds.): ICCVBIC 2019, AISC 1108, pp. 1273–1282, 2020.
https://doi.org/10.1007/978-3-030-37218-7_133

SpHb monitoring has gained enormous attention as a point of care testing that allows the ability to monitor haemoglobin concentration in a continuous, accurate, and non-invasive fashion. Various technologies and methods are employed by researches all over the globe to develop a system/device for SpHb monitoring [4–12]. Among various methods adopted, measurement of SpHb from the statistical features of pho-toplethysmograph signal and adopted machine learning technique has shown an excellent correlation with the Haemoglobin (Hb) measured using invasive method [3]. The study [3] was carried out on 33 healthy subjects using various machine learning techniques. The main goal of the conducted investigation is to evaluate the said effi-ciency of calculating SpHb among the outpatients of a hospital using photoplethys-mograph (PPG) and regressive machine learning technique.

2   Methodology

2.1   PPG Signal Collection

The subject database acquisition was done in Sree Abirami Hospitals, Coimbatore after obtaining the ethical clearance for collecting the PPG signals of outpatients. A formal consent was obtained from the subjects before enrolment. Subjects aged between 18 and 55 years were enrolled. The subject data was stored in a spreadsheet, Microsoft Excel. The IR Plethysmograph transducer and Labchart software (version 7) of ADInstruments were used for signal acquisition.

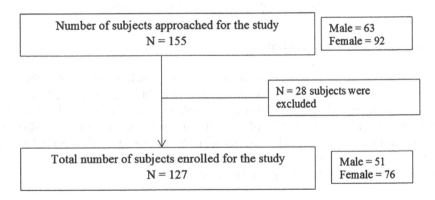

Fig. 1. Study flow

The sensor was placed in the forefinger of left arm of the subjects. While the venous blood sample was collected by the trained professionals for calculating Hb_{lab}, corre-sponding PPG signal was acquired for a 15 period sample. The study flow of the work is presented in Fig. 1. In total 155 subjects were approached. Out of which 127 subjects were enrolled for the study with prior consent received. It included subjects aged between 21 to 50 with a mean (m) age of 36 and standard deviation (SD) of 7.8. The remaining 28 subjects were excluded due to lack of co-operation and inability to receive proper PPG signal.

2.2   SpHb Calculation

The original PPG signal acquired using Labchart was saved along with a spreadsheet containing subject details. Along with original signal, derivatives of the signal were also recorded in the Labchart software, which were exported to MATLAB for subsequent signal processing. Although the signals obtained using Powerlab kit are mostly clean *i.e.*, without baseline wandering and added noise, denoising using wavelets was performed to eliminate the insignificant noises. Seven time domain features as described in [3] were extracted from the PPG signal and its derivatives. The selected features were trained using curve fitting tool in MATLAB. The selected seven features of PPG signal were given as input to the regression model with Hb_{lab} as target output. Linear model type as used in [3] was incorporated. The SpHb values predicted using the linear regression model and the Hb_{lab} were stored for further analysis. For the statistical analysis, Bland-altman plot, Regression analysis, t-test, dispersion plots, Pearson product moment correlation coefficient were done on IBM SPSS statistics package. Figure 2 depicts the flow diagram of the work performed in the study.

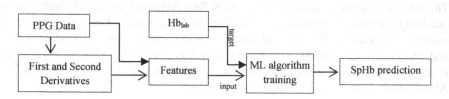

Fig. 2. Acquisition of principal components

3   Results and Discussion

Total of 127 outpatient volunteers were enrolled which includes 76 female subjects and 51 male subjects. Mean age of female subjects was 34 ranging from 21 to 48 with standard deviation (SD) of 8.9. Mean age of male subjects was 37 ranging from 23 to 50 with standard deviation (SD) of 7.7. The mean Hb_{lab} recorded was 14.09 g/dL ranging from 10.1 to 16.6 g/dL with SD of 1.75. The mean SpHb recorded was and 13.81 g/dL ranging from 10.1 to 16.4 g/dL with SD of 1.55. The Hb_{lab} recorded and the predicted SpHb values are given in Table 1.

Table 1. Hb_{lab} recorded and predicted SpHb

Subject	Hb_{lab} (g/dL)	SpHb (g/dL)	Hb_{bias} (g/dL)	Absolute error (g/dL)	Squared error
1	12	12.4	−0.4	0.4	0.16
2	12.8	13.5	−0.7	0.7	0.49
3	11.8	11.3	0.5	0.5	0.25
4	13.8	12.2	1.6	1.6	2.56
5	10.2	11	−0.8	0.8	0.64

(*continued*)

Table 1. (*continued*)

Subject	Hb$_{lab}$ (g/dL)	SpHb (g/dL)	Hb$_{bias}$ (g/dL)	Absolute error (g/dL)	Squared error
6	12.1	11.3	0.8	0.8	0.64
7	12.2	13	−0.8	0.8	0.64
8	11.4	12.4	−1	1	1
9	13.9	12	1.9	1.9	3.61
10	12.8	11.3	1.5	1.5	2.25
11	12	11	1	1	1
12	14.1	12	2.1	2.1	4.41
13	13	12.5	0.5	0.5	0.25
14	11	12.3	−1.3	1.3	1.69
15	11.1	11.5	−0.4	0.4	0.16
16	10.1	10.3	−0.2	0.2	0.04
17	13.1	13.7	−0.6	0.6	0.36
18	11.7	11.2	0.5	0.5	0.25
19	12	13	−1	1	1
20	13.7	13.3	0.4	0.4	0.16
21	12	12.2	−0.2	0.2	0.04
22	10.7	13	−2.3	2.3	5.29
23	10.6	10.2	0.4	0.4	0.16
24	10.2	11	−0.8	0.8	0.64
25	11.9	11.3	0.6	0.6	0.36
26	11.1	10.8	0.3	0.3	0.09
27	13.7	13.4	0.3	0.3	0.09
28	13.5	14	−0.5	0.5	0.25
29	12.7	13.1	−0.4	0.4	0.16
30	13.1	14	−0.9	0.9	0.81
31	10.4	10.1	0.3	0.3	0.09
32	12.2	13	−0.8	0.8	0.64
33	12.9	11	1.9	1.9	3.61
34	12.6	12	0.6	0.6	0.36
35	13.3	13	0.3	0.3	0.09
36	11.8	11.4	0.4	0.4	0.16
37	11.8	11.3	0.5	0.5	0.25
38	12.8	12.5	0.3	0.3	0.09
39	11.2	12	−0.8	0.8	0.64
40	12.1	11.8	0.3	0.3	0.09
41	11	11.3	−0.3	0.3	0.09
42	13.9	14.3	−0.4	0.4	0.16
43	12.3	11.9	0.4	0.4	0.16
44	11.5	11.9	−0.4	0.4	0.16
45	11.7	11.3	0.4	0.4	0.16
46	13.1	14	−0.9	0.9	0.81

(*continued*)

Table 1. (*continued*)

Subject	Hb$_{lab}$ (g/dL)	SpHb (g/dL)	Hb$_{bias}$ (g/dL)	Absolute error (g/dL)	Squared error
47	12.6	13	−0.4	0.4	0.16
48	11.8	14	−2.2	2.2	4.84
49	14.1	14.3	−0.2	0.2	0.04
50	11.6	11.3	0.3	0.3	0.09
51	10.5	10.1	0.4	0.4	0.16
52	10.5	11.4	−0.9	0.9	0.81
53	10.5	10.4	0.1	0.1	0.01
54	10.9	10.4	0.5	0.5	0.25
55	13.2	12	1.2	1.2	1.44
56	14.7	15	−0.3	0.3	0.09
57	14.3	14.1	0.2	0.2	0.04
58	11.4	11	0.4	0.4	0.16
59	12.1	13	−0.9	0.9	0.81
60	12.7	11.9	0.8	0.8	0.64
61	14.9	14.8	0.1	0.1	0.01
62	15.5	14	1.5	1.5	2.25
63	14.3	15	−0.7	0.7	0.49
64	13.3	14	−0.7	0.7	0.49
65	10.7	10.4	0.3	0.3	0.09
66	12.6	12	0.6	0.6	0.36
67	13.1	14	−0.9	0.9	0.81
68	12.4	13.2	−0.8	0.8	0.64
69	11.6	12	−0.4	0.4	0.16
70	10.8	11.3	−0.5	0.5	0.25
71	13.4	12	1.4	1.4	1.96
72	13.3	13	0.3	0.3	0.09
73	10.2	11	−0.8	0.8	0.64
74	10.1	12	−1.9	1.9	3.61
75	14.2	13	1.2	1.2	1.44
76	12.4	11.3	1.1	1.1	1.21
77	14.1	13	1.1	1.1	1.21
78	14	13.2	0.8	0.8	0.64
79	10.9	11.7	−0.8	0.8	0.64
80	14.9	14	0.9	0.9	0.81
81	14.1	13.8	0.3	0.3	0.09
82	16.2	15.5	0.7	0.7	0.49
83	15.8	14	1.8	1.8	3.24
84	16.1	15.4	0.7	0.7	0.49
85	15.9	15.1	0.8	0.8	0.64
86	14.5	15	−0.5	0.5	0.25

(*continued*)

Table 1. (*continued*)

Subject	Hb$_{lab}$ (g/dL)	SpHb (g/dL)	Hb$_{bias}$ (g/dL)	Absolute error (g/dL)	Squared error
87	15.3	14.6	0.7	0.7	0.49
88	13.3	13	0.3	0.3	0.09
89	15.4	15.9	−0.5	0.5	0.25
90	14.7	14.2	0.5	0.5	0.25
91	15.1	16	−0.9	0.9	0.81
92	14.2	14.5	−0.3	0.3	0.09
93	14.7	14	0.7	0.7	0.49
94	15.2	15.6	−0.4	0.4	0.16
95	15	14.6	0.4	0.4	0.16
96	15.6	16.3	−0.7	0.7	0.49
97	14.3	14	0.3	0.3	0.09
98	15.8	14.8	1	1	1
99	14	13.5	0.5	0.5	0.25
100	13.6	12.3	1.3	1.3	1.69
101	15.6	14.6	1	1	1
102	14.8	14.4	0.4	0.4	0.16
103	15.3	14	1.3	1.3	1.69
104	14.7	14	0.7	0.7	0.49
105	15	14.9	0.1	0.1	0.01
106	15.8	15.1	0.7	0.7	0.49
107	16.4	16	0.4	0.4	0.16
108	15.1	15.6	−0.5	0.5	0.25
109	16.3	15.8	0.5	0.5	0.25
110	16.5	16	0.5	0.5	0.25
111	16	16.4	−0.4	0.4	0.16
112	11	12.6	−1.6	1.6	2.56
113	13.7	13.9	−0.2	0.2	0.04
114	15.1	14.6	0.5	0.5	0.25
115	15	14	1	1	1
116	15.3	13.3	2	2	4
117	15.9	14.4	1.5	1.5	2.25
118	12.6	13.5	−0.9	0.9	0.81
119	13.1	12.2	0.9	0.9	0.81
120	16.6	16	0.6	0.6	0.36
121	15.9	15	0.9	0.9	0.81
122	13.1	13.7	−0.6	0.6	0.36
123	15.7	16	−0.3	0.3	0.09
124	14.2	14.1	0.1	0.1	0.01
125	14.9	14.8	0.1	0.1	0.01
126	15.6	14	1.6	1.6	2.56
127	16.3	15	1.3	1.3	1.69

For evaluating the performance, Mean Absolute Error (MAE), Mean Squared Error (MSE), Root Mean Squared Error (RMSE) were used as formulated below. In the following equations, Bj denotes the Hb_{lab} value and B_j' denotes the predicted SpHb value.

$$MAE = \frac{1}{n}\sum_{j=1}^{n} |B_j - B_j'|$$

$$MSE = \frac{1}{n}\sum_{j=1}^{n} (B_j - B_j')^2$$

$$RMSE = \sqrt{MSE}$$

MAE, MSE and RMSE value of 0.737 g/dL, 0.738 g/dL and 0.86 g/dL were obtained between Hb_{lab} and SpHb. Coefficient of determination-R2 value of 0.768 was obtained along with Pearson coefficient of 0.876 in SPSS. Dispersion diagram with R^2 is plotted in Fig. 3. The low values of MAE, MSE and a good R^2, Pearson coefficient shows good correlation between SpHb and Hblab. Performance evaluation of SpHb is shown in Table 2.

Table 2. Performance evaluation

Performance error	n = 127	Female subjects (n = 76)	Male subjects (n = 51)
MAE	0.737 (SD 0.44)	0.733 (SD 0.51)	0.735 (SD 0.44)
MSE	0.738 (SD 0.85)	0.803 (SD 1.17)	0.731 (SD 0.866)
RMSE	0.859	0.896	0.86
R^2	0.768	0.564	0.634
Pearson coefficient	0.876	0.751	0.796

To determine if there is a significant difference between the two measurements, one sample t-test is performed considering null hypothesis which shows the confidence interval which is tabulated in Table 3.

Table 3. One sample test

Test value = 0						
	t-static	Degrees of freedom	Sig	Difference (Mean)	95% Confidence interval	
					Lower	Upper
Hb_{bias}	2.15	126	0.03	0.165	0.013	0.032

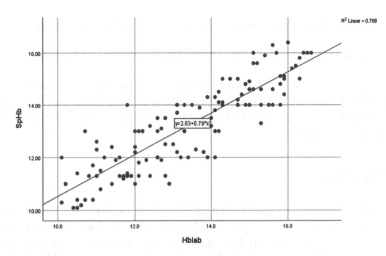

Fig. 3. Dispersion diagram

The above figure, Fig. 3 depicts the Dispersion diagram drawn between SpHb and Hb_{lab}. A regression value of 0.768 was obtained which shows satisfactory results on prediction of SpHb.

To compare the correlation and agreeability between two Hb_{lab} and SpHb, Bland-Altman analysis was performed with mean on x-axis and Hb_{bias} on y-axis which is plotted in Fig. 4.

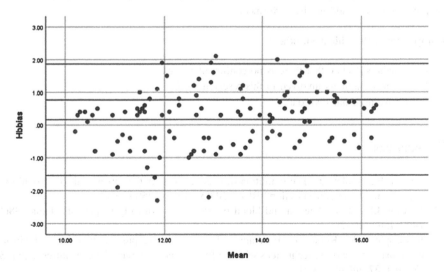

Fig. 4. Bland altman plot

To analyse whether the data fits within the expectation of agreement, the upper and lower limit of agreement is plotted which are calculated as follows.

$$\text{Upper limit of agreement} = \text{Mean Hb}_{\text{bias}} + (1.96 * \text{SD of Hb}_{\text{bias}})$$

$$\text{Lower limit of agreement} = \text{Mean Hb}_{\text{bias}} + (1.96 * \text{SD of Hb}_{\text{bias}})$$

The bland-altman plot is shown in figure for which the upper and lower limit of agreement is 1.8664 and -1.5357 respectively which indicates the 95% confidence limits. It can be seen that most of the data points are clustered around the mean difference line of 0.165.

4   Conclusion

Coefficient of Determination R^2 value of 0.77 shows satisfactory results with mean difference of 0.74 g/dL. R^2 value slightly differs for female and male population. However the study was conducted on outpatient subjects and not on healthy volunteers. One limitation of the study is that it did not include any anaemic subjects. The lowest recorded Hb value in the study is 10.1 g/dL. So, it can be concluded that SpHb measurement using PPG and generalised linear regression shows good correlation for Hb values greater than 10 g/dL. Further investigation of the method for anaemic population is needed for further validation of the method. Although the method has many positive attributes, it has to be validated in varied population set-up with larger sample size.

Acknowledgements. The authors of this paper wish to thank the help and support from the nursing and medical staff at the Sree Abhirami Hospitals, Coimbatore without whose kind support, the study could not be performed.

Compliance with Ethical Standards

✓ All authors declare that there is no conflict of interest
✓ No humans/animals involved in this research work.
✓ We have used our own data.

References

1. Beutler, E., Waalen, J.: The definition of anemia: what is the lower limit of normal of the blood hemoglobin concentration? Blood **107**(5), 1747–1750 (2006)
2. Vincent, J.L., et al.: Anemia and blood transfusion in critically ill patients. Jama **288**(12), 1499–1507 (2002)
3. Kavsaoğlu, A.R., Kemal, P., Hariharan, M.: Non-invasive prediction of hemoglobin level using machine learning techniques with the PPG signal's characteristics features. Appl. Soft Comput. **37**, 983–991 (2015)
4. Kraitl, J., Ewald, H., Gehring, H.: An optical device to measure blood components by a photoplethysmographic method. J. Opt. A: Pure Appl. Opt. **7**(6), S318 (2005)

5. Kraitl, J., Ewald, H., Gehring, H.: Analysis of time series for non-invasive characterization of blood components and circulation patterns. Nonlinear Anal. Hybrid Syst. **2**(2), 441–455 (2008)
6. Timm, U., et al.: Sensor system concept for non-invasive blood diagnosis. Procedia Chem. **1** (1), 493–496 (2009)
7. Saltzman, D.J., et al.: Non-invasive hemoglobin monitoring during hemorrhage and hypovolemic shock. Beckman Laser Inst and Medical Clinic, Irvine, CA (2004)
8. Suner, S., et al.: Non-invasive determination of hemoglobin by digital photography of palpebral conjunctiva. J. Emerg. Med. **33**(2), 105–111 (2007)
9. Kim, O., et al.: Combined reflectance spectroscopy and stochastic modeling approach for noninvasive hemoglobin determination via palpebral conjunctiva. Physiol. Rep. 2.1 (2014)
10. Esenaliev, R.O., et al.: Continuous, noninvasive monitoring of total hemoglobin concentration by an optoacoustic technique. Appl. Opt. **43**(17), 3401–3407 (2004)
11. Herlina, A.R., Fatimah, I., Mohd Nasir, T.: A non-invasive system for predicting hemoglobin (Hb) in dengue fever (DF) and dengue hemorrhagic fever (DHF). In: Sensors and the International Conference on new Techniques in Pharmaceutical and Biomedical Research, IEEE Asian Conference (2005)
12. Lbrahim, F., Ismail, N.A., Taib, M.N., WanAbas, A.B., Taib, M.N.: Modeling of hemoglobin indengue fever and dengue hemorrhagic fever using bioelectrical impedance. Physiol. Meas. **25**(3), 607–616 (2004)

Sentiment Analysis on Tweets for a Disease and Treatment Combination

R. Meena[1($\boxtimes$)], V. Thulasi Bai[2], and J. Omana[1]

[1] Department of CSE, Prathyusha Engineering College, Chennai, India
meenarajeswaran@gmail.com, omanajayakodi@gmail.com
[2] Department of ECE, KCG Engineering College, Chennai, India
thulasi_bai@yahoo.com

Abstract. The proposed work has retrieved tweets on a particular disease and treatment combination from twitter and they were processed to extract the sentiments. Initially the polarity values were set up in a range from weakly negative to strongly positive and the tweets were analyzed. The overall sentiment of the tweets related to breast cancer and chemotherapy was weakly positive. Naïve Bayes algorithm was applied on the tweets retrieved on the same disease and treatment combination. Nearly 10,000 tweets were analyzed using the Pubmed and Google book search engine as a training corpus. The sentiments were plotted in the graph which shows that the sentiments were neutral. Lastly, to find the most occurred word in the tweets, bigrams were used and co-occurrence of words were plotted using Natural Language Tool Kit in python.

Keywords: Cancer · Chemotherapy · Polarity · Naïve Bayes · Bigrams

1 Introduction

The idea of extracting health data of the public and surveillance of disease outbreaks using Internet and social media as a source has a decade of history. [1] Social networks generate a humongous amount of shared content on all the areas and the users of social media are ever increasing which is predicted to be 3.02 billion in 2021. [2] The average time spent on the social media by the users varies from places around the world though it coarsely comes around 2 h and 22 min per day. [3] Twitter is one of the popular micro blogging sites. Presently, a considerable amount of growth conquers in the study of twitter data for public health monitoring in various ways such as disease monitoring, disease outbreak and prediction using public responses and trends, etc.., [4] There is a wide range of tools and various methods are available to perform the health oriented analysis on the tweets extracted from the popular blogging service twitter. [5] Tweets posted by the users can be carefully analyzed for the traces of keywords of a particular disease outbreak and prevalence in a particular location which is used for identifying

© Springer Nature Switzerland AG 2020
S. Smys et al. (Eds.): ICCVBIC 2019, AISC 1108, pp. 1283–1293, 2020.
https://doi.org/10.1007/978-3-030-37218-7_134

seasonal diseases like influenza, flu and dengue. [6–8] Public health monitoring in twitter can be done by segregating the sentiments of users on many issues like usage of e-cigarette. [9] Twitter forums are also a great source of information which can be used to find the adverse drug effects by applying sentiment analysis and other text mining approaches. [10] NLP is applied on the twitter posts by the individual users in the discussion forum and it is used to understand the mental health of users using the sentiment scores thereby predicting the various depression conditions which actually is a less light shed analysis in healthcare. [11] Geo location features and tweets retrieved from the twitter can even be the best sources for identifying and managing the situation during a natural disaster like earthquake. [12] Twitter analysis is also a unique source of resource with tweets and images which can also be used for understanding the communication pattern, sentiment and severity of the disasters like hurricane. [13] The proposed work in this paper deals with the sentiment analysis of tweets retrieved real time from the users on a particular disease and its widely used treatment. Cancer is a dreadful disease and has over 17 million people around the world is affected by the disease as on 2018. [14] We have examined the sentiment distribution on the disease 'Breast cancer' which is of the most prevailing cancer around the world. [15] Other than surgery, chemotherapy and radiation therapy are the most recommended cancer treatments which is used to increase the life expectancy and helps them to lead a healthy life of the patients. So, chemotherapy was taken as a treatment and sentiment classes were discovered for the treatment. For both disease and treatment, tweets were retrieved and naïve bayes algorithm was applied on the data and text analysis was performed. The most common network of words were also found along with the co-occurrence of words were plotted for the disease and treatment.

The paper is structured as, Sect. 2 has literature review related to the work, Sects. 3, 4 and 5 has the proposed work and Sect. 6 has the conclusion.

2 Literature Review

Sentiment analysis is one of the major areas in text mining where opinions are categorized based on the data and used to find the users perspective. Its application in health care yields promising results and better understanding of users views in terms of medications, drugs and treatments. There is a lot of related work carried on previously in sentiment analysis of cancer patients using the social media data shared by the users. Here is the list of related works tabulated along with the data mining or machine learning technique used by the researchers (Table 1).

Related Works in Literature Review:

Table 1. Related works

Research work	Problem addressed	Techniques used
[16]	Sentiment analysis – happiness of cancer patients	Hedonometric analysis
[17]	Sentiment analysis, Temporal casuality, Probabilistic Kripke structure – Cancer Survivor Network	Machine learning – Adaboost sentiment classifier
[18]	Sentiment analysis tool (SentiHealth-Cancer) for analyzing the mood of cancer patients – online communities	Senti strength, F1, Accuracy, Harmonic mean, Recall and Precision
[19]	Psychosocial Benefits for Women with Breast cancer in Internet community	Content analysis, Thematic analysis
[20]	Cancer information dissemination in social media	Social network analysis, content analysis, sentiment analysis, and multivariate analysis of variance
[21]	Analyzing breast cancer symptoms	Data mining classification algorithms – Decision tree and Bayesian classification
[22]	Sentiment analysis – Cancer Survivors Network	Machine learning
[23]	Sentiment analysis on cancer treatments Pap smear, mammogram	Content analysis
[24]	Informatics and future research support on online social media cancer communities	Social support, member engagement
[25]	Web groups for reducing stress of Breast cancer patients	Mean scores and covariance comparing of web based blog
[26]	Sentiment analysis and subjective attitude - Cancer Survivors Network	Lexical approach
[27]	Clustering of cancer tweets	NLP, feature selection and extraction, soft clustering algorithm
[28]	Breast cancer prediction	Data mining algorithms
[29]	Ranking, sentiment analysis of cancer drugs	Machine learning, SVM

3   Sentiment Analysis of Disease (Breast Cancer) and Treatment (Chemotherapy)

3.1   Objective and Applications of Sentiment Analysis

Sentiment analysis in health care is a way to understand patient interaction in social media. It paves a path in medical diagnosis, treatment feedbacks, drug effects and mental health of patients. Social media interactions are the best way to predict the disease outbreaks.

For complex diseases such as cancer, sentiment analysis on the social media interactions which has rich source of data on health conditions, medications and treatment details provides meaningful insights from the data. People spending larger time on the social media will be benefited from the analysis on their conversations which can give them positive effect on their mental health. Sentiment analysis and tools can be the source of inspiration for taking up medical procedures such as chemotherapy. These analyses will be useful to bring up more people on the discussion in social media related to their treatments and medications which can be a mode of helping new patients to understand the positive impacts of the treatments and can boost their mental health.

3.2   Analysis of Tweets

The tweets related to the keywords cancer and chemotherapy was retrieved. Since there are many language tweets are available in twitter, only English tweets were concentrated. After retrieval of tweets, the URLs or links, special characters are removed from

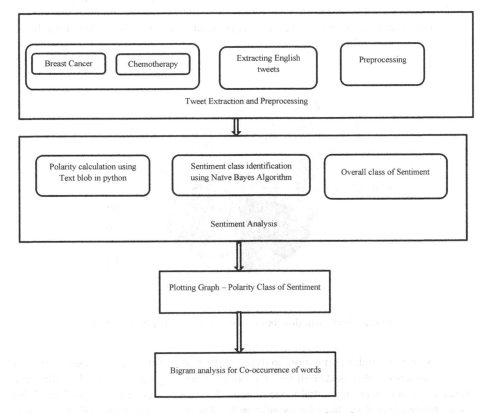

Fig. 1. Block diagram of the proposed work

Table 2. Table showing the range of the polarity value for sentiment classes

Polarity range	Category	Class
0	N	Neutral
>0 and <0.3	W_p	Weakly Positive
>0.3 and <=0.6	P	Positive
>0.6 and <=1	S_p	Strongly Positive
>−0.3 and <=0	W_N	Weakly Negative
>−0.6 and <=−0.3	N	Negative
>−1 and <=−0.6	S_N	Strongly Negative

Table 3. Sample tweets retrieved for the keyword 'Breast Cancer'

Working night shift does not increase likelihood of breast cancer WHO study says in report countering claim

She is looking at data from breast cancer studies and examining the genetic factor in breast cancer and the variants

Breast cancer touches many lives Australia and New Zealand have some of the highest rates in the world is looking

More Biocell breast implants pulled from sale in Canada Textured implants linked to rare for \xe2\x80\xa6',b'Health Canada suspends Biocell breast implants citing increased cancer risks

Just got my first peer reviewed publication on the use of Nanomaterials for Personlised breast cancer management Glad

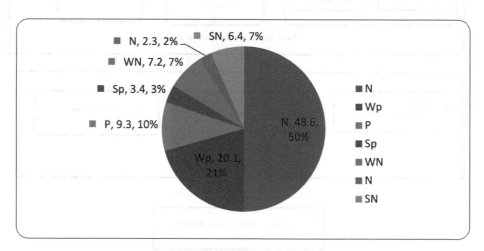

Fig. 2. Sentiment distribution for the keyword 'Breast cancer'

them using the regular expression analysis. Before analyzing the sentiment, the scale for the polarity value is determined for positive and negative tweets. Initially, nine parameters were set to the initial value of zero. The tweets were analyzed and graph has been plotted based on the classes of tweets. The total percentages of tweets were

Table 4. Sample tweets retrieved for the keyword 'Chemotherapy'

Is there a role for neoadjuvant #chemotherapy in localized #sarcoma?
Traditionally we've treated cancer with chemotherapy radiation and surgery But #immunotherapy now offers another
A variety of therapies to treat these tumors are surgery radiation therapy chemotherapy and many more
Returned to my Oncologist today to check my blood recovery from chemotherapy. All levels were back up to normal!
Please help us share It's been a really difficult year with me starting chemotherapy"
okay let's go The 8th cycle of Chemotherapy #fightmamafight #daughterduties

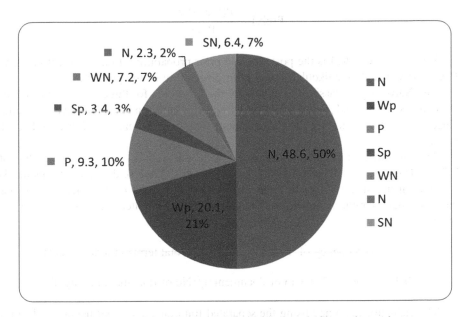

Fig. 3. Sentiment distribution for the keyword 'Chemotherapy'

calculated and the general class of report was calculated by finding the average of the results. For instance, recent tweets (1000 tweets) were retrieved for the key word 'breast cancer' for disease and 'chemotherapy' for treatment. Their reports were plotted as graph and analyzed (Figs. 1, 2 and 3) (Tables 2 and 3).

Overall average of tweet sentiment:

'Breast Cancer' – Weakly Positive

'Chemotherapy' – Weakly Positive

4   Naive Bayes Algorithm for Sentiment Analysis

The tweets related to the keyword 'chemotherapy' was retrieved and stored in a file using json format. Before scraping for the tweet, twitter posts from specific location can be mentioned and taken. The particular location of the tweets needs to be searched by using the geo code and geo fence. They are used to specify the location and the distance it has to cover for tweet search. For our search, we have specified the geocode as USA = '39.8, −95.583068847656,2500 km' which includes nearly all American states and large area of Canada. The current tweets were extracted by mentioning the date from which the data has to be taken. Nearly 10,000 tweets were extracted and stored in a file for applying the naive bayes algorithm for analyzing their sentiments. Tweets were stored separately in the two files for the keyword Breast cancer and chemotherapy respectively (Figs. 4 and 5) (Table 4).

The naïve bayes algorithm uses Bayes theorem which can be used to solve a classification problem. Bayes theorem is,

$$P(c|x) = \frac{P(x|c)\ P(c)}{P(x)}$$

Where, $P(c|x)$, $P(c)$ is the posterior and prior probability of class respectively and $P(x|c)$, $P(x)$ stands for likelihood and prior probability of the predictor.

The Naïve bayes classifier is trained by the keywords for three sentiments such as Positive, Negative and Neutral. We have used the multimodal naïve bayes classifier since it is a better option for the classifications with discrete features (Table 5) (Figs. 6 and 7).

The raw data, which here is the tweets retrieved, was converted in to a matrix of TF-IDF features. TF-IDF mentions the term frequency −inverse document frequency is a statistical measure used to find the importance of a word in the document. It has various types of ranking functions. TF-IDF was used to filter the stop-words in the tweets stored.

$$TF = (\text{Frequency of } 't' \text{ in a document})/(\text{Total terms in a document})$$

$$IDF = \log\_e\ (\text{Total no of documents})/(\text{No of document having } 't')$$

The estimator is trained using the separated train data and fitted using the data set which will be used to classify the emotions into classes. The corpus of positive, negative and neutral words were fed into the matrix representation as a train data for TF-IDF.

Words used in PubMed search queries and Google books search engine were used as the sentiment training corpus for the tweets. Nearly 25 words were given as the training data set in each class namely positive, negative and neutral and their values were assigned to the dense matrix as,

Table 5. Matrix values assigned for sentiment corpus in training data set

Positive	1	0	0
Negative	0	1	0
Neutral	0	0	1

Four classes of tweets were retrieved using the training data set such as bad, neutral, positive and negative tweets. Non English and irrelevant tweets which contain links were eliminated as bad tweets. The appropriate classes of tweets were tested against with the sentiment lexicons in the train data (Table 6).

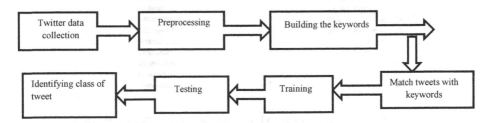

Fig. 4. Sentiment analysis of tweets using Naïve Bayes

Table 6. Result of tweets in class

Keyword	Bad tweets	Neutral tweets	Positive tweets	Negative tweets
Breast cancer	614	1536	42	31
Chemotherapy	881	2665	74	15
Chemo	467	1897	54	18

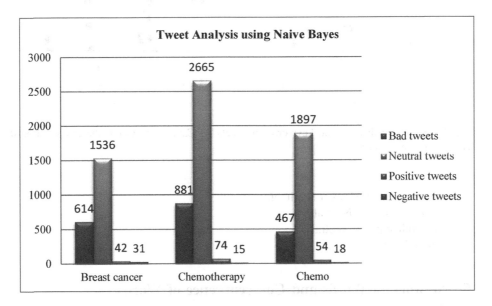

Fig. 5. Graph showing the polarity distribution of tweets using Naïve Bayes Algorithm

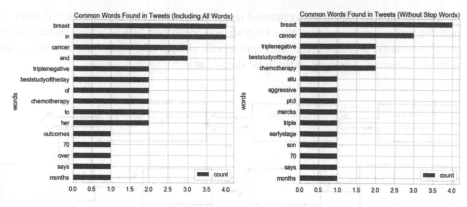

Fig. 6. Common words found in tweets with and without stop words

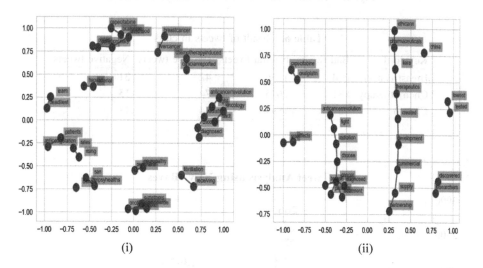

(i)   (ii)

Fig. 7. Bigram analysis of the keywords in tweets (i) Cancer and Chemotherapy (ii) Breast cancer and Chemotherapy

Overall average of tweet sentiment:
'Breast Cancer' – Neutral
'Chemotherapy' – Neutral
'Chemo' – Neutral

5   Network of Words and Co-occurrence of Words

The network of words from tweets was collected, which is counting the most popular words in the tweets. Influential words from the tweets were found and mapped on two ways such as with and without stop words. The 1000 recent tweets for the disease and generic treatment for the particular key words 'Breast cancer' and 'Chemotherapy' which has been retrieved on the particular date has been stored and the network of words has been analyzed. Initially the URLs and the stop words have been removed. To analyze the co-occurrence of words between the two keywords, corpus in tweets need to be plotted. Bigrams which are the English words occurring together very often will be used for sentiment analysis. The frequency of the bigram pairs were plotted by the natural language tool kit in python.

6 Conclusion

The results after evaluation shows that most of the discussions in the social media twitter on the disease and treatment combination (Breast cancer and Chemotherapy) were either weakly positive or neutral. The network of words and bigram analysis is used to discover the most discussed words on the terms Breast cancer and Chemotherapy. Thus sentiment analysis provides a proven impact on the health care.

Compliance with Ethical Standards
✓ All authors declare that there is no conflict of interest.
✓ No humans/animals involved in this research work.
✓ We have used our own data.

References

1. Brownstein, J.S., Freifeld, B.S.C.C., Madoff, M.D.L.C.: Digital disease detection—harnessing the web for public health surveillance. N. Engl. J. Med. Mass Med. Soc. **360** (21), 2153 (2009)
2. https://www.statista.com/statistics/278414/number-of-worldwide-social-network-users/
3. https://www.digitalinformationworld.com/2019/01/how-much-time-do-people-spend-social-media-infographic.html
4. Jordan, S E., Hovet, S.E., Fung, I.C.-H., Fu, H.L.K.W., Tse, Z.T.H.: Using Twitter for public health surveillance from monitoring and prediction to public response. MDPI, Data **4**(1) (2019)
5. Gohil, S., Vuik, S., Darzi, A.: Sentiment analysis of health care tweets: review of the methods used. JMIR Public Health Surveill. **4**, 2 (2018)
6. Signorini, A., Segre, A.M., Polgreen, P.M.: The use of Twitter to track levels of disease activity and public concern in the US during the influenza A H1N1 pandemic. PLoS ONE **6** (5), e19467 (2011)
7. Gomide, J., Veloso, A, Meira, W., Almeida, V., Benevenuto, F., Ferraz, F., Teixeira, M.: Dengue surveillance based on a computational model of spatio-temporal locality of Twitter. In: Proceedings of the 3rd International Web Science Conference, Koblenz, Germany, 15–17 June 2011
8. Santos, J.C., Matos, S.: Analysing Twitter and web queries for flu trend prediction. Theor. Biol. Med. Model. **11**, S6 (2014)
9. Chu, K.H., Allem, J.P., Unger, J.B., Cruz, T.B., Akbarpour, M.: Strategies to Find Audience Segments on Twitter for e-cigarette Education Campaigns. Elsevier, Addictive (2019)

10. Korkontzelos, I., Nikfarjam, A., Shardlow, M., Sarker, A., Ananiadou, S., Gonzalez, G.H.: Analysis of the effect of sentiment analysis on extracting adverse drug reactions from tweets and forum posts. J. Biomed. Inform. **62**, 148–158 (2016)
11. Kumar, J., Bakshi, S., Williams, K.L.: A model for sentiment and emotion analysis of unstructured social media text. Electron Commer. Res. **18**, 181–199 (2018)
12. Bhavya, S., et al: Twitter analysis for disaster management. In: 2018 Fourth International Conference on Computing Communication Control and Automation (ICCUBEA). IEEE (2019)
13. Understanding communication dynamics on twitter during natural disasters: a case study of Hurricane Sandy. Int. J. Disaster Risk Reduction, 101176 (2019)
14. https://www.cancerresearchuk.org/health-professional/cancer-statistics/worldwide-cancer
15. https://www.wcrf.org/dietandcancer/cancer-trends/breast-cancer-statistics
16. Crannell, W.C., Clark, E., Jones, C., James, T.A., Moore, J.: A pattern-matched twitter analysis of US cancer-patient sentiments. J. Surg. Res. **206**, 536–542 (2018)
17. Bui, N., Yen, J., Honavar, V.: Temporal causality analysis of sentiment change in a cancer survivor network. IEEE Trans. Comput. Soc. Syst. **3**, 75–87 (2016)
18. Rodrigues, R.G., et al.: SentiHealth-Cancer: a sentiment analysis tool to help detecting mood of patients in online social networks. Int. J. Med. Inf. **85**(1), 80–95 (2016)
19. Rodgers, S., Chen, Q.: Internet community group participation: psychosocial benefits for women with breast cancer. J. Comput. Mediated Commun. **10**(4), July 2005
20. Wang, X., Chen, L., Shi, J., Peng, T.-Q.: What makes cancer information viral on social media? Comput. Hum. Behav. **93**, 149–156 (2019)
21. Sri, M.N., Priyanka, J.S.V.S.H., Sailaja, D., Ramakrishna Murthy, M.: A comparative analysis of breast cancer data set using different classification methods. Smart Innov. Syst. Technol., 175–181 (2018). https://doi.org/10.1007/978-981-13-1921-1_17
22. Ofek, N., et al: Improving sentiment analysis in an online cancer survivor community using dynamic sentiment lexicon. In: 2013 International Conference on Social Intelligence and Technology. IEEE (2013)
23. Lyles, C.R., et al.: 5 mins of uncomfyness is better than dealing with cancer 4 a lifetime: an exploratory qualitative analysis of cervical and breast cancer screening dialogue on Twitter. J. Cancer Educ. **28**(1), 127–133 (2013)
24. Zhang, S., et al.: Online cancer communities as informatics intervention for social support: conceptualization, characterization, and impact. J. Am. Med. Inform. Assoc. **24**(2), 451–459 (2017)
25. Winzelberg, A.J., et al.: Evaluation of an internet support group for women with primary breast cancer. Cancer Interdisc. Int. J. Am. Cancer Soc. **97**(5), 1164–1173 (2003)
26. Ofek, N., et al.: The importance of pronouns to sentiment analysis: online cancer survivor network case study. In: Proceedings of the 24th International Conference on World Wide Web. ACM (2015)
27. Lavanya, P.G., Kouser, K., Suresha, M.: Efficient Pre-processing and feature selection for clustering of cancer tweets. In: Intelligent Systems, Technologies and Applications. Springer, Singapore, pp. 17–37 (2020)
28. Kumar, V., et al.: Prediction of malignant & benign breast cancer: a data mining approach in healthcare applications. arXiv preprint arXiv:1902.03825 (2019)
29. Mishra, A., Malviya, A., Aggarwal, S.: Towards automatic pharmacovigilance: analysing patient reviews and sentiment on oncological drugs. In: 2015 IEEE International Conference on Data Mining Workshop (ICDMW) (2015)

Design and Performance Analysis of Artificial Neural Network Based Artificial Synapse for Bio-inspired Computing

B. U. V. Prashanth$^{(\boxtimes)}$ and Mohammed Riyaz Ahmed

School of Electronics and Communication Engineering, REVA University,
Bengaluru 560064, India
{prashanthbuv, riyaz}@reva.edu.in

Abstract. A variety of capacitating technologies are being surveyed for the microelectronics era. Spintronics technologies exemplify notable potential due to an abundance of features that can find applications in neuro-inspired computing. The aim is to capacitate fully resilient next-generation neuromorphic hardware and to replace platitudinous CMOS microelectronics by unique devices used in neuromorphic (Bio-Inspired) computing systems. In this research we designed, planned and initiated the study along with growing the films. Further devices are fabricated and initial device characterizations are performed. The device fabrication was supported using electron beam lithography. The paper depicts the supervised pattern recognition computing on bio-inspired platform using magnetic skyrmion-based synapses on hand written recognition digit dataset. The deep learning strategies such as Artificial Neural Network (ANN) modelling is elucidated to develop and simulate an electronic synapse. This research paves a way for a bio-inspired computing based on exotic topological spin textures.

Keywords: Bio-inspired engineering · Pattern recognition · Modelling · Simulation · Skyrmion · Spintronics

1 Introduction

By the advent of mixed-mode signal processing and as scaling approaches 10 nm, the paradigm has shifted from miniaturization to learning and adaptive computing. The exceptional observation made by Gordon Moore about shrinking the size of the component on the chips has improved the VLSI industry at a dizzying rate. Though Moore's law is considered as cornerstone it will meet shuddering halt as 2D lithography approach atomic scaling. The silicon industry has doubled the computing power in tandem with Moore's law.

This remarkable thumb-rule of on-chip shrinking can be applied only for the system board [1, 2]. Neural network algorithms when run on traditional von Neumann computing systems cannot take advantage of the memory-computing intertwinement which these algorithms inherently need [3, 4]. So a new architecture has to be thought of involving artificial neurons (computing) connected through artificial synapses (memory) inspired by the human brain. Spintronic devices such as magnetic skyrmions owing to their non-volatility can make good synaptic devices and owing to their low

© Springer Nature Switzerland AG 2020
S. Smys et al. (Eds.): ICCVBIC 2019, AISC 1108, pp. 1294–1302, 2020.
https://doi.org/10.1007/978-3-030-37218-7_135

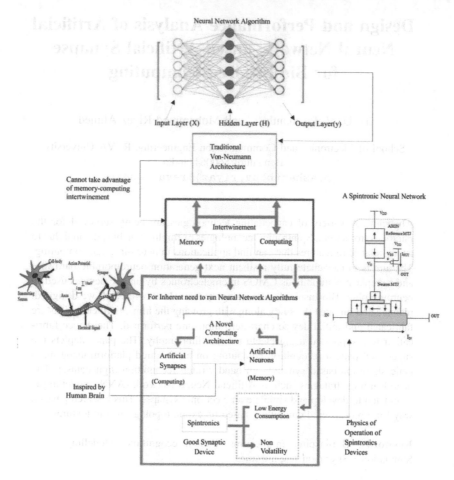

Fig. 1. Spintronics based neuromorphic computing flow diagram design

energy consumption may also make good synaptic devices [5]. In this paper as depicted in Fig. 1 we analyze the physics of operation of these spintronic devices and then how they need to be connected with each other in a neural network to do certain classification and recognition tasks with the characteristic of neural networks [6]. The magnetic skyrmions are found in unit crystals of magnetic compounds at ultra low temperatures with a non-centrosymmetric lattice and it is evident that Dzyaloshinskii-Moriya interaction (DMI) exits and is induced by spin-orbit coupling in the absence of inversion symmetry in the crystal lattice [7]. The skyrmions are also observed in ultra thin magnetic films epitaxially grown on heavy metals [8, 9], which are subjected to large DMIs induced by the breaking of inversion symmetry at the interface and to the strong spin-orbit coupling of the neighboring heavy metal. The basis for highly energy efficient data storage, transmission and processing devices features the magnetic skyrmions which are non-trivial topologically swirling spin textures. The rest of the paper is organized in the following manner: The Sect. 2 depicts the experimental setup followed by discussions on analysis of magnetic skyrmion based artificial synapse.

The ANN based Modeling and Simulation is depicted in Sect. 3 and the Sect. 4 presents the results of supervised pattern recognition simulation using skyrmion synapse with conclusions in Sect. 5 and references at the end of the manuscript.

2 The Experimental Setup

The application of electrical pulse and measurement of the device resistance is set up in the two circuits. The micrograph of the measured device with scanning electron microscope (SEM) is depicted where blue colored region describes the effective synaptic area. In this synaptic regions with the reversal of magnetization the variation of resistances are observed in Hall cross geometry.

The electrical pulse applications in a horizontal stripe in 'x' direction of the Hall bar geometry, while the successive resistance calculations are along the 'y'-axis Hall bar contacts. The developed the schematic of experimental setup for simulating the domain structures for simultaneous resistance measurements and domain imaging is as illustrated in Fig. 2 in the following page.

The following algorithm 1 depicts the steps carried out for carrying out the experiment.

Algorithm1:- Algorithm Steps for Experimental Setup depicting the design of
ANN Based Artificial Synapse for Bio-Inspired Computing
Start (Artificial Synapse Design)

Step 1:- Calculate Resistance (R) and Domain Imaging (I_{DOMAIN})

Step 2:- Calculate the X-Ray Intensity Transmitted $(X-ray)_{TX}$

Step 3:- Hall Resistance $(R_{X,Y})$ and Saturated State Hall Resistance $(R_{X,Y})_{SAT}$ as

the function of out of field magnetic field B_Z with magnetic domain configuration.

Step 4:- Calculate Normalized Hall Resistance $= (R_{X,Y})/(R_{X,Y})_{SAT}$

Step 5:- Calculated Normalized out of plane magnetization $= M_{Z,Fe}/M_{Z,Fe,SAT}$.

End

Due to the topological nature of the material, the propagation path of the skyrmion is along the axis of the current flow and opposite to the flow of electrons. The working principle of artificial synapse based on magnetic skyrmions is described as illustrated in Fig. 3(a) in the next page. The magnetic moments are pointing towards +z and −z axis in the skyrmions. The direction of electron flow opposite to the direction of the charge current pulse are defined by the enclosed electrical pulses [10]. At each electrical pulse state is simulated by the standard deviation of the resistivity measurements as elucidated in Fig. 3(b). The dissipation and accumulation of skyrmions controlled by electric current with in an enclosed active synaptic region can mimic the linearly varying synaptic weights during potentiation and depression respectively. The experimentally simulated electrical operation of skyrmion with the each resistance having the relative magnetic configuration. With the current induced spin orbit torques the single polar pulses are applied during both potentiation and depression were optimized further to delete or generate the controlled number of skyrmions [11].

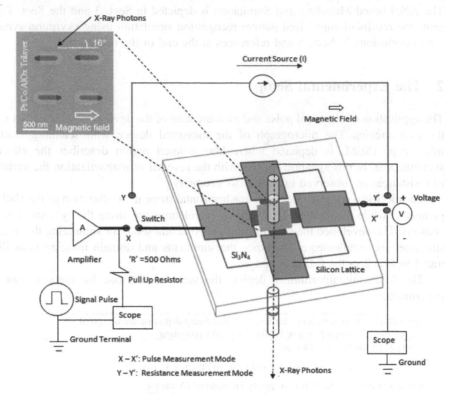

Fig. 2. An experimental setup

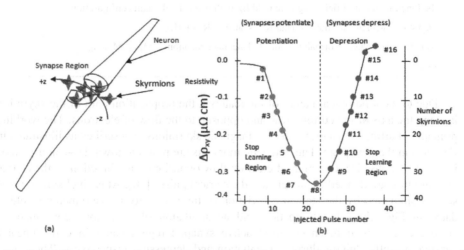

Fig. 3. (a) Skyrmion based artificial synapse (b) Shift in calculated hall resistivity and injected pulse number as function of calculated skyrmion number

The skyrmions are systematically generated with continuously applying the electrical pulse and are driven into active region of synapse area with a drop in linear resistance and the skyrmions are accumulated. During potentiation there is a drop in the average resistivity energized by each skyrmion in addition to overall acquired states. During depression the polarity of linearly increasing pulses is reversed with small increase in magnetic field and maintaining the same amplitude and the switching rate of potentiation. This leads to systematic annihilation of the skyrmions with increase in the linear resistance depicting a key characteristic of a skyrmion based artificial synapse as the analog synaptic modulation of weight is induced by electrical current [12]. The measured characteristics of artificial synapse based on skyrmion is as represented in the Fig. 4(a)–(c) respectively in the next page illuminating the simulation of the voltage measured in milli Volts (mV), current measured in nano Amperes (nA) and magnetic field is measured in terms of Tesla versus the % switching probability. The measured switching probability is the probability of switching time from skyrmion dominant state to ferromagnet dominant state. This probability of deviation is caused by the statistical distribution in the diameter of the skyrmion affected by the mutual interactions [13].

3 Artificial Neural Network (ANN) Based Modelling and Simulation

In this section the simulation of supervised pattern recognition with skyrmion synapse is presented. The input neurons are $28 \times 28 = 784$, Hidden neurons are 100, and output neurons are 10 in three stage ANN structure simulated to perform the supervised pattern recognition where the weight training is carried out with quantization-aware algorithm is utilized to learn the Modified National Institute of Standards and Technology (MNSIT) handwritted pattern dataset with the preprocessing steps with Multi-Layer Perceptron (MLP) is depicted in Fig. 5(a).

The skyrmion based artificial synapses are used during the training illustrating the simulation platform [14] to perform supervised pattern recognition presented in Fig. 5 (b) in the following page. The ANN with the softmax units is implemented using Keras with a Tensor Flow backend and the loss function and optimizer are Adam with categorical cross entropy.

The method for implementing a neuromorphic computing consists of the following:

1. Input: Consists of a network structure and the number of layers, the network input data, the number of neurons in layer, layers parameters, along with pre-synaptic neuron type and a computing window.
2. Output: Consists of Interfacing Platform with forward inference.

A neural network is constructed on the neuromorphic computing platform according to network structure and the number of layers [15]. The neuron activation takes place when the pre-synaptic neuron fire spikes. Update the weights parameters of layers according to pre-synaptic neuron types. The convolution operations by can be performed with developed neuromorphic computing platform.

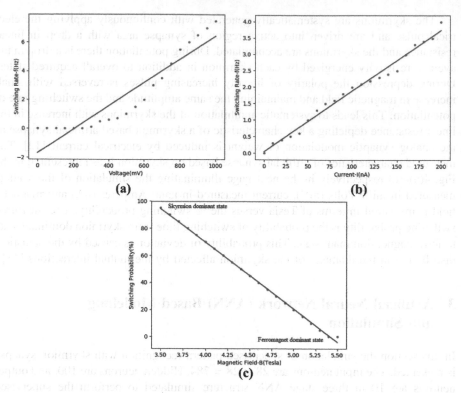

Fig. 4. (a) Measured Voltage (mV) versus Switching rate in f(Hz) (b) Measured Current (nA) versus Switching rate in f(Hz) (c) measured Magnetic Field versus % Switching Probability

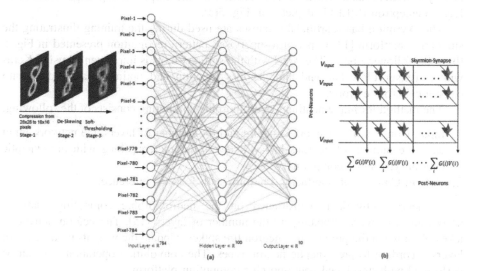

Fig. 5. (a) Artificial neural network (ANN) (b) Skyrmion based artificial synapses are used during the training illustrating the simulation platform to perform pattern recognition.

4 Results

The simulated supervised pattern recognition accuracy depends on training iteration times for skyrmion synapse-based ANN along with the result of software based neural network is also characterized for comparison as depicted in the following Fig. 6.

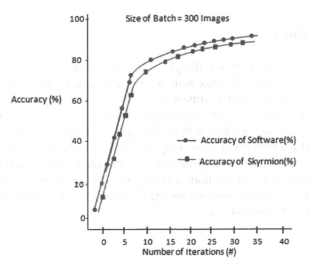

Fig. 6. Iterative training results of pattern recognition accuracy for both software based and skyrmion based devices with batch size of 300 images

Further the algorithm 2 is as illustrated below describes the quantization aware algorithm of the training process of the neural network.

Algorithm 2:- Quantization Aware Algorithm of Neural Network Training Process

---

Start (While Target == number of iterations) do
Step 1:- Assume L2 loss function, Sigmoid Activation and No bias terms
Step 2:- if normalize features then
Step 3:- Create cross-validation folds;
Step 4:- Train/evaluate the model on each fold;
Step 5:- Collect training and test data from folds;
Step 6:- Build neural network classifier model and train;
end of While loop;
Step 7:- Make predictions for training and test data; *(Compute train/test % error);*
Step 8:- Calculate and compare the % Accuracy of software and hardware
Stop

This work is based on simulations which are holistically simulated with parameters calibrated to experimental result obtained in Fig. 6 for the skyrmion based device and the circuit level device along with the software training of ANN where the size of the batch is of 300 images is utilized for each training iterations. The inference accuracy of skyrmion based synapse for hand written digit recognition task is reported to be equal to 87.1% with slight variation of 4.4% in comparison to 91.5% accuracy of ANN. This indicates a small degrade compared with that of the ANN. The input layer receives the input data and hidden layers and output layers process the data. Each neurons process sequentially and integrate each synapse.

5 Conclusions

The current induced skyrmion design with observation and deletion techniques is incorporated by a single ferromagnetic design technique to demonstrate a fully functional artificially developed skyrmion based electronic synapse device. The various convolution operations can be successfully built on the developed neuromorphic computation platform therefore it holds the great potential to map the convolutional neural networks, which is our future scope of work. In the active region of operation of the skyrmion device can be implemented for simulation of neuromorphic computing platform gaining a higher learning accuracy of supervised pattern recognition. This research work gives a new direction aiming towards neuromorphic computing based on exotic topological spin textures.

References

1. Fert, A., Reyren, N., Cros, V.: Nat. Rev. Mater. 2, 06 (2017)
2. Wiesendanger, R.: Nat. Rev. Mater. 1(7), 16044 (2016)
3. Kang, W., Huang, Y., Zhang, X., Zhou, Y., Zhao, W.: Proc. IEEE 104(10), 2040–2061 (2016)
4. Woo, S., Litzius, K., Krüger, B., Im, M.-Y., Caretta, L., Richter, K., Mann, M., Krone, A., Reeve, R.M., Weigand, M., Agrawal, P., Lemesh, I., Mawass, M.-A., Fischer, P., Klaui, M., Beach, G.S.D.: Nat. Mater. 15(5), 501–506 (2016)
5. Moreau-Luchaire, C., Moutafis, C., Reyren, N., Sampaio, J., Vaz, C.A.F., Horne, N.V., Bouzehouane, K., Garcia, K., Deranlot, C., Warnicke, P., Wohlhüter, P., George, J.-M., Weigand, M., Raabe, J., Cros, V., Fert, A.: Nat. Nanotechnol. 11(5) (2016)
6. Nagaosa, N., Tokura, Y.: Nat. Nanotechnol. 8(12), 899–911 (2013)
7. Muhlbauer, S., Binz, B., Jonietz, F., Peiderer, C., Rosch, A., Neubauer, A., Georgii, R., Boni, P.: Skyrmion lattice in a chiral magnet. Science 323(5916), 915–919 (2009)
8. Soman, S., Suri, M., et al.: Big Data Anal. 1(1), 15 (2016)
9. Panda, P., Sengupta, A., Sarwar, S.S., Srinivasan, G., Venkataramani, S., Raghunathan, A., Roy, K.: Proceedings of the 53rd Annual Design Automation Conference, p. 98. ACM (2016)
10. Sengupta, A., Banerjee, A., Roy, K.: Phys. Rev. Appl. 6(6), 064003 (2016)
11. Sengupta, A., Panda, P., Wijesinghe, P., Kim, Y., Roy, K.: Sci. rep. 6, 30039 (2016)
12. Sengupta, A., Panda, P., Raghunathan, A., Roy, K.: 2016 29th International Conference on VLSI Design and 2016 15th International Conference on Embedded Systems (VLSID), pp. 32–37. IEEE (2016)

13. Kim, H., Hwang, S., Park, J., Yun, S., Lee, J.-H., Park, B.-G.: IEEE Electron Device Lett. **39** (4), 630–633 (2018)
14. Sengupta, A., Han, B., Roy, K.: 2016 IEEE in Biomedical Circuits and Systems Conference (BioCAS), pp. 544–547. IEEE (2016)
15. Sengupta, A., Roy, K.: 2015 International Joint Conference on in Neural Networks (IJCNN), pp. 1–7. IEEE (2015)

Sentiment Analysis of Amazon Book Review Data Using Lexicon Based Analysis

Fatema Khatun[✉], S. M. Mazharul Hoque Chowdhury,
Zerin Nasrin Tumpa, SK. Fazlee Rabby, Syed Akhter Hossain,
and Sheikh Abujar

Department of Computer Science and Engineering,
Daffodil International University, Dhaka, Bangladesh
fatemamou995@gmail.com, zerinnasrin@gmail.com,
sk.fazleerabby@gmail.com,
aktarhossain@daffodilvarsity.edu.bd,
{mazharul2213,seikh.cse}@diu.edu.bd

Abstract. In this digital era, people are much more interested to buy and sell things in E-Commerce websites. Book is one of the mostly sold products on online by different online stores. Amazon is one of the leading online stores, which has million dollar transactions every year with its rich variety of products. In particular, book is one of the best-selling products of Amazon. Sentiment is therefore holding a very strong position in market analysis and future business development. Based on sentiment now it is possible to predict everything about business based on the user and other NLP related fields where opinion comes from someone. This research focuses on determining the quality of books as well as authors through the analysis of review comments and rating provided by the user. To determine that, lexicon based analysis is used here and using bag of word, the positivity and negativity of the review was determined. So that people can find the best book they need within a short time and without facing difficulty.

Keywords: Lexicon · Amazon · Online book reviews · Polarity · Sentiment analysis

1 Introduction

In this modern century, all young generation are highly dependent on online such as online books, products, and machines. People are become lazy to buy from the shop where they feel comfort to buy from the online store. Also can utilize any online books by analyzing the reviews of the customer. Amazon is the premier online book sites with so much popularity. This research work presents the polarity of online book reviews data by using both sentiment and lexicon analysis. As Sect. 2 has literature review, where demonstrated a lots of reviews of in different research work based on the same field. So the main motives on this research paper, the Decoration of online book reviews from Amazon where we easily got the point of human's emotion, which includes with anger, happiness, joy, sadness; by doing polarity observation of the online reviews or comments of the book. So in Sect. 3 a proposed model of this research work has been

© Springer Nature Switzerland AG 2020
S. Smys et al. (Eds.): ICCVBIC 2019, AISC 1108, pp. 1303–1309, 2020.
https://doi.org/10.1007/978-3-030-37218-7_136

presented in a data flow with explanation. The main part of this paper is Sect. 4 where the whole process described with collecting, preprocessing and analyzing of research data. In Sect. 5 has explanation of an expected result of this research work. Where shown that desire output of this experiment and some description of future work on this same platform what conferred in Sect. 6. Moreover In this experiment we achieved a convenient result of online book reviews established on Sentiment Analysis.

2   Literature Review

Besides there is another research experiment on sentiment analysis done by Al-Smadi, Qawasmeh, Talafha and Quwaider [1, 7]. Their experiment came out as kind of boring because they only focused on arabic text research which is not internationally done. If we do compare with our research work we experimented books review data which is based on an international website amazon book store. A text sentiment analysis work developed on clustering with support vector machine by Li and Wu [2, 8]. They tried to do it only hotspot forums. They could do it more on too many forums which could take their work on so far, otherwise that was a average research work on text sentiment analysis. Öztürk and Ayvaz completed their paper on syrian refugee crisis [3, 9], which is uncommon topic though analysed from twitter text about syrian refugee crisis based on sentiment analysis. Yoo and his team examined on latest deep learning technique for developing the accuracy for prediction and analyzing the sentiment of their work [4, 10]. Another work on survey executed by Medhat and her group [5, 11]. They did Opinion Mining extracts and analyze people's opinion; the target of their survey was to find opinions of peoples established on sentiment analysis. Haddi and his whole gang tried to find out pre- processing on text analyzing [6, 12]. After all it based on sentiment analysis accuracies using vector machine in specific area for text pre-processing.

3   Proposed Model

This research experiment introduces sentiment analysis on book reviews from Amazon.com. Where it shows in the below Fig. 1.

The first Process of the experiment is collecting data which has been completed from Amazon book website. In next step this research went through on preprocessing step. Where the data has to be preprocessed and removed some unwanted dirty data. In next step using sentiment analysis where we have done polarity accesses from the large collection book reviews where this research work displayed a result with positive either negative or even neutral. Conditions are working according to the following Fig. 2.

We collected all data from Amazon book online store. Then selected book reviews for our final output. There is a lots of platform or field that we can use for web data scraping as python, R and so on. In our work we used R platform. By writing R code we achieved our scrap data. Also we have done the emotion analysis that a customer's review depends on their comment expression or on using sign, after calculating and visualizing the output, we got our final result.

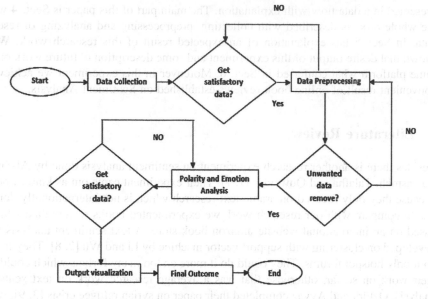

Fig. 1. Data flow diagram

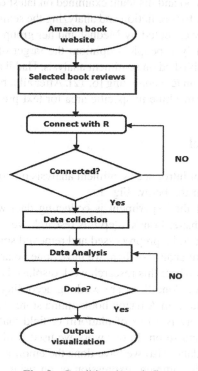

Fig. 2. Conditional work flow

4 Proposed Method

4.1 Data Collection and Preprocessing

For this system data is the most important thing and as we are collecting data from Amazon particularly for book review and comments of the user. As data analysis depends on data and attributes. So three major attributes were selected for this analysis and those are author name, customers review and rating. Though Amazon sells different types of products, but this research focuses only on the book category. There are different types of books as well as book authors. But not every author is good enough to write a good book. So this research will figure out which books are good enough and as well as which book writers are writing quality books. First of all few books data are scraped from the Amazon website with the attribute book title, author name and book rating. In this research proposed system will work with only those three types of data. Book data can be collected using an author's name, where the search result will provide all the books of the author with previously mentioned attributes. Like that several others writers book data will also be collected. A simple python program will make this easy. But every time we have to provide the book information on Amazon website separately. So that the program can collect the data and store it in a csv files. When data collection is complete it is important to preprocess the collected data. Author name and book name will not have that much effect on the analysis as our priority is sentiment analysis of review data. So the reviews and rating will get the preference. On the other hand, as there is a very low possibility of rating value being mixed with unnecessary data, so only preprocessing of review comments will be enough for the research work. During pre-processing a computer program made by R language will find different types of unnecessary data, for example- web links, spam, special signs, email address, images, etc. So when the analysis is complete it will provide a clean and simple analyzable data set.

4.2 Data Analysis and Output

For the analysis comments of each book will be responsible for the sentiment analysis result separately. If one book have around three hundred comments then each comment will own a sentiment value, either positive or negative. In this research lexicon based analysis was used to calculate the sentiment values of the comment. Therefore two separate bag of word were created. One for the positive words and another for the negative words. In the analysis part each positive and negative word will get 1 point. So the final score of a comment will be subtraction of negative words from positive words. According to that it can be written as Eq. 1 given below,

$$\text{Review comment score}, RC_s = P_s - N_s \tag{1}$$

Here, P_s is the positive number of words in the comment and N_s is the negative number of words in the comment. Therefor the final outcome will be positive sentiment if the total is positive number otherwise negative. Using the same process as Eq. (1) if all the positive and negative comments are scored separately as 1 point each it is possible to find out if the final outcome of total review comment (TRC_s) is a positive number or a negative number. If positive the book will be considered as a good book. On the other hand if negative values it will be a bad book. If any book have the score 0 then it will be considered as neutral. For this section output will be visualized as given below in Fig. 3.

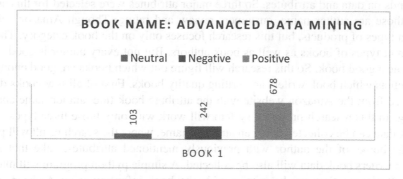

BOOK NAME: ADVANACED DATA MINING

■ Neutral   ■ Negative   ■ Positive

Fig. 3. Books review comments sentiment analysis

To measure the quality of the book all the rating will be considered and will be calculated. For a particular book all individual ratings will be added, so a final sum will be generated, where the rating score is, R_s. Now the total rating value will be added with the score found from the review comment analysis and the final score for the quality will be generated. This can be written as Eq. 2 given below

$$\text{Quality score, } Q_s = R_s + (TRC_s/\text{number of comments}) \qquad (2)$$

For this section output will be visualized as given below in Fig. 4.

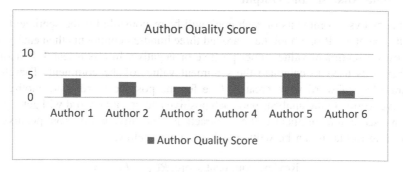

Author Quality Score

■ Author Quality Score

Fig. 4. Author quality score bar chart

Each book in the database will be calculated according to the process explained before. From this it can be predicted that if the book is a good book or it is a bad book.

5 Conclusions and Future Work

The major challenges of this research work to predict the online book reviews trough the sentiment analysis. This is unbelievable that today we are totally express our emotions on social site, because the prediction result showed us how much it is essential to analyze human's reaction on social product. I hope our experiment made it easier to find it out the online book reviews of Data mining books. Our effort shows that a customer can give any kind of reviews depending on their emotions, not all reviews are same, and not all are different. But the thing is does this review really affects to customer for purchasing the books from website, or it only depends on customer's wishes that what he wants to learn or explore. Sentiment Analysis made this easier for our work, investigation the online reviews of books from amazon.com and set up the final scoring and chart bar and finally the output observations. Our future work will be making this experiment more larger based on big data with a huge amount of online books and may other products reviews and purchasing amount of books.

References

1. Mäntylä, M.V., Graziotin, D., Kuutila, M.: The evolution of sentiment analysis - a review of research topics, venues, and top cited papers. Comput. Sci. Rev. **27**, 16–32 (2018). https://doi.org/10.1016/j.cosrev.2017.10.02
2. Piryani, R., Madhavi, D., Singh, K.V.: Analytical mapping of opinion mining and sentiment analysis research during 2000–2015. Inf. Process. Manag. **53**, 122–150 (2016). https://doi.org/10.1016/j.ipm.2016.07.001. ISSN: 0306-4573
3. Zhou, Q., Zhang, C., Zhao, S.X., Chen, B.: Measuring book impact based on the multi-granularity online review mining. Scientometrics **107**(3), 1435–1455 (2016). https://doi.org/10.1007/s11192-016-1930-5
4. Chevalier, A.J., Mayzlin, D.: The effect of word of mouth on sales: online book reviews. J. Market. Res. **43**(3), 345–354 (2006). http://www.jstor.org/stable/30162409. Accessed 06 Mar 2014
5. Kochkina, N., Shenkman, A.E., Gordienko, S.A.: The elasticity of demand for books: the research of the north-western federal district book market, vol. 67, no. 1, pp. 73–91, January 2015. https://doi.org/10.1007/bf03396924. Vestnik of Saint Petersburg University
6. Droba, D.D.: Methods used for measuring public opinion. Am. J. Sociol. **37**, 410–423 (1931)
7. Al-Smadi, M., Qawasmeh, O., Talafha, B., Quwaider, M.: Human annotated Arabic dataset of book reviews for aspect based sentiment analysis. In: 3rd International Conference on Future Internet of Things and Cloud, Date of Conference, Rome, Italy, pp. 24–26. IEEE, August 2015. https://doi.org/10.1109/ficloud.2015.62, Date Added to IEEE Xplore: 26 October 2015, Electronic ISBN: 978-1-4673-8103-1, USB ISBN: 978-1-4673-8102-4, INSPEC Accession Number: 15556781
8. Li, N., Wu, D.D.: Using text mining and sentiment analysis for online forums hotspot detection and forecast. Decis. Support Syst. **48**(2), 354–368 (2010). https://doi.org/10.1016/j.dss.2009.09.003
9. Öztürk, N., Ayvaz, S.: Sentiment analysis on Twitter: a text mining approach to the Syrian refugee crisis. Telematics Inform. **35**(1), 136–147 (2018). https://doi.org/10.1016/j.tele.2017.10.006

10. Yoo, S., Song, J., Jeong, O.: Social media contents based sentiment analysis and prediction system. Expert Syst. Appl. **105**(1), 102–111 (2018). https://doi.org/10.1016/j.eswa.2018.03.055

11. Medhat, W., Hassan, A., Korashy, H.: Sentiment analysis algorithms and applications: a survey. Ain Shams Eng. J. **5**(4), 1093–1113 (2014). https://doi.org/10.1016/j.asej.2014.04.011

12. Haddi, E., Liu, X., Shi, Y.: The Role of text pre-processing in sentiment analysis. Procedia Comput. Sci. **17**, 26–32 (2013). https://doi.org/10.1016/j.procs.2013.05.005

MicroRNA-Drug Association Prediction by Known Nearest Neighbour Algorithm and Label Propagation Using Linear Neighbourhood Similarities

Gill Varghese Sajan$^{(\boxtimes)}$ and Joby George

Mar Athanasius College of Engineering, Kothamangalam,
Ernakulam 686666, Kerala, India
gillv522@gmail.com, jobygeo@hotmail.com

Abstract. The identification of disease related microRNAs will be helpful in exploring the underlying pathogenesis of the diseases. Significant information regarding the drug discovery and repositioning are obtained from drug-disease associations. The paradigm about drug discovery has changed from finding new drugs that exhibit therapeutic properties for a disease to reusing existing drugs for a newer disease. Experimental determination of diseases associations with microRNAs and drugs could be time-consuming and costly. Computational method can be an efficient alternative to identify potential drugs related to each microRNAs. Here we present an approach which computes the missing associations that exists between microRNAs and drugs by keeping diseases as an intermediary element which could lead to the prediction of the possible chemicals that could be used for their treatment.

Keywords: Association · Disease · Drug · miRNA · Prediction

1 Introduction

Growing evidences shows that microRNAs play important functions in several biological processes. Regulating expression of disease genes is the way, the microRNAs exhibit their functions. The abnormality of microRNA biogenesis might cause different diseases. Therefore, the identification of microRNAs related to each diseases will be useful for exploring the disease pathogenesis and in planning appropriate treatments. MicroRNA-disease associations furnish significant information for drug repositioning and discovery. Drugs are chemicals used for detecting and treating diseases. Drug repositioning is the process of identifying new clinical applications for existing chemicals. It is an essential procedure for the process called drug discovery. Researches in pharmacogenomics have indicated that drugs targets microRNAs and regulate the expressions. Drug developers are becoming increasingly innovative in discovering new functions for currently available chemicals. Selecting the main restorative impacts for a chemical to perform clinically potential experiments is the major challenge lying in the

© Springer Nature Switzerland AG 2020
S. Smys et al. (Eds.): ICCVBIC 2019, AISC 1108, pp. 1310–1318, 2020.
https://doi.org/10.1007/978-3-030-37218-7_137

repositioning of drugs. Blind screening and serial testing of creature models are some of the traditional repositioning efforts. Accumulation of biomedical information suggests that computational methodologies can be a viable option in anticipating the most favorable drug-disease associations for new indication experiments for currently available drugs.

2 Related Works

Utilizing random-walk with restart algorithm a method named RWRMDA [1] was implemented for microRNA-disease associations. Later on, Xuan et al. [2] also used random-walk with restart in similar type of network for predictions. During the iterative procedure in their technique, various transition-weights were assigned to both obscure and known microRNAs of a given ailment. Additionally, they fixed a microRNA-disease bilayer network and expanded strolling in the arrangement for the prediction of new diseases. By combining the semantic similarity between diseases, functional similarity between microRNAs, Gaussian communication profile kernel similarity the existing microRNA-disease interactions into a heterogeneous chart, a system named HGIMDA [3] was introduced. Similarly many computational procedures were designed for drug-disease association predictions as well. For example, a method was implemented by Chiang and Butte [4] which incorporated a guilt by association procedure for suggesting chemical use by a common treatment profiles considering a set of diseases. The basic concept behind this method was that if some similar therapies were shared by 2 diseases, at that point different medications that were restorative for just a single of the two illnesses likewise be utilized for the other. Gottlieb et al. [5] introduced a new method dependent on the principle that comparative chemicals were connected to comparable diseases. The method make use of multiple features to compute disease-disease and drug-drug similarities and defined a categorization algorithm called PREDICT, for deducing chemical signs. Luo et al. [6] upgraded the likeness estimates utilized in the previous method and developed a method called MBiRW which uses random walk strategy in predicting potential drug-disease associations. Satisfied performance can be received from the above mentioned prediction models. But it's of considerable significance to examine how medications and illness phenotypes can be related based on molecular levels. Researches in pharmacogenomics discovered that microRNAs can be targeted by drugs and their expressions can be regulated. It gives specialists another viewpoint for drug repositioning. Hence the development of computational methods to compute drug-disease association prediction using the target microRNAs has become significant.

3 Proposed Method

Disease-disease similarities, drug-drug similarities and microRNA-microRNA similarities are key components in the construction of microRNA-disease and drug-disease association prediction models.

3.1 Disease–Disease Similarity

Using the MeSH dataset, every diseases could be represented as directed acyclic graph. If necessary the gene-gene interactions can be extracted from HumanNet dataset [7]. In the DAG of each disease, each nodes represents diseases and links represent node relationships. To define the location of a disease in the MeSH graph, each diseases have at least one address in the DAG, called as codes. The parent node's code is appended behind the child's addresses and are used to define the codes of the child node. An illness 'd' could be depicted as a directed acyclic graph, $DAG_d = (d, T_d, E_d)$, where T_d represents the group of all predecessor nodes of 'd' including 'd' itself and E_d speaks to the gathering of every relating connection. The semantic commitment esteem representing the connection of a disease t to disease 'd' is represented by $D_d(t)$, which can be calculated as follows:

$$D_d(t) = \begin{cases} 1, & \text{if } t = d \\ \{0.5 \times D_d(t')|t' \in children\ of\ t\}, & \text{if } t \neq d \end{cases} \tag{1}$$

Using Eq. (1), disease d's semantic value can be calculated as:

$$SemVal(d) = \sum_{t \in T_d} D_d(t) \tag{2}$$

By applying the basic principle that the diseases sharing bigger piece of their directed acyclic graphs will in general possess a higher semantic comparability [8], the similitude between different ailments can be calculated. The relative locations of two diseases in the MeSH disease directed acyclic graph are used for the calculation of the semantic similarity between them. closeness esteem between any two illnesses is determined as pursues:

$$S^d(A, B) = \frac{\sum_{t \in T_A \cap T_B} D_A(t) + D_B(t)}{SemVal(A) + SemVal(B)} \tag{3}$$

where $D_A(t)$ and $D_B(t)$ are the semantic values representing relationship of illness 't' to illness 'A' and illness 'B' respectively.

3.2   MicroRNA–MicroRNA and Drug–Drug Similarity

The commitments from comparable diseases which are related with two microRNAs are used to precisely calculate the functional similarity [9] between them. Therefore semantic similarity among a disease and a group of disease is to be defined first. For instance, let 'd' be one disease and let 'D' be one disease group, e.g. $D = \{D_1, D_2, \ldots, D_k\}$. The closeness among d and D is calculated as:

$$S(d, D) = max_{1 \leq i \leq k} S(d, D_i) \tag{4}$$

For clarity, to compute the useful similitude among two microRNAs, m_1 and m_2 assume D_1 represents the related diseases of m_1 and D_2 represents the related diseases of m_2. D_1 and D_2 have m and n distinct diseases respectively. The computation of functional similarity of two microRNAs will need to consider each and every diseases in D_1 as well as in D_2. Hence the similarity between two microRNAs is computed using:

$$S^m(m_1, m_2) = \frac{\sum_{1 \leq i \leq m} S(d_{1i}, D_2) + \sum_{1 \leq i \leq n} S(d_{2j}, D_1)}{m + n} \tag{5}$$

Similarly to calculate the similarity between two drugs or chemicals c_1 and c_2, assume D_1. represents the related diseases of c_1 and D_2 represents the related diseases of c_2. D_1 and D_2 have m and n distinct diseases respectively. While computing the similarity value between two drugs, it is necessary to consider each and every diseases in both D_1 and D_2. Hence the similarity of two drugs is computed using:

$$S^c(c_1, c_2) = \frac{\sum_{1 \leq i \leq m} S(d_{1i}, D_2) + \sum_{1 \leq i \leq n} S(d_{2j}, D_1)}{m + n} \tag{6}$$

Now using the values computed by Eqs. (5) and (6) both microRNA and drug similarity network can be constructed. The similarity for all pair of microRNAs and all pair of drugs are computed. Then a threshold value for both coefficients are decided. Now microRNA pairs and drug pairs with similarity value which is more noteworthy than or equivalent to the limit esteem will have an immediate connection between them.

3.3   Heterogeneous Network Construction

The drug-drug, microRNA-microRNA and disease-disease similitude network structures which are developed utilizing the process of fixing a threshold for the similarity values in the previous sections acts as the subnets for the computation of the heterogeneous networks. The two subnets namely, the disease relatedness network and microRNA comparability network, could now be connected using an experimentally proved microRNA-disease interactions from HMDD [10] dataset to form a

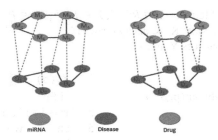

Fig. 1. MicroRNA-disease and drug-disease heterogeneous network

heterogeneous system of microRNAs and diseases. Similarly experimentally proved associations between chemicals and ailments that is contained in the CTD [11] dataset is utilized to connect the drug and disease subnets to form the drug-disease heterogeneous network. An example of the microRNA-disease and drug-disease heterogeneous network are given in Fig. 1.

3.4   MicroRNA-Disease Association and Drug-Disease Association

In this section the forecast of unfamiliar microRNA-disease interactions and drug-disease interactions utilizing the realized associations is the main aim, hence a new feature vector called "interaction profile" for both ailments and microRNAs and drugs are introduced. Let $Y = R^{m \times n}$ and $Z = R^{l \times n}$ represent a adjacency matrices constituting the microRNA-disease and drug-disease affiliations respectively whose rows have m microRNAs or l drugs and columns have n diseases, where $Y_{ij} = 1$ if microRNA m_i have a connection with sickness d_j; otherwise $Y_{ij} = 0$. Similarly $Z_{ij} = 1$ if drug c_i have a connection with sickness d_j; otherwise $Z_{ij} = 0$. The ith row vector of Y, spoken to by $Y(m_i)$, is the interaction profile for microRNA m_i. The ith row vector of Z, is the interaction profile for drug c_i and the jth column vector of both Y and Z, represented by $Y(d_j)$ and $Z(d_j)$, are the interaction profiles for disease d_j. A microRNA, drug or disease being known implies that their profiles have no less than one connection.

3.5   Known Nearest Neighbour (KNN) Algorithm

Majority of the non-connections or 0's in matrices Z and Y are obscure cases which can actually be genuine interactions, i.e. false negative values. The non-interfacing microRNA-disease matches in Y and drug-disease combines in Z are really missing edges. Hence a matrix updation process called KNN could be utilized as a pre-processing step to calculate the interaction likeliness values for these non-interfacing pairs by making use of their known neighbours, i.e., the procedure tries to replace the values of the Y_{ij} or Z_{ij} equals 0, by a value from 0 to 1 using the algorithm given below:

Input : Adjacency Matrices $Y \in R^{m \times n}, Z \in R^{l \times n}, S^m \in R^{n \times n}, S^c \in R^{l \times l}$ and $S^d \in R^{m \times m}$, neighbourhood size 'K' & decay term 'T'.

Output : Modified matrices Y and Z.

Algorithm:

1: For p=1 to n do

1.1: mn = knownNeighbour(p, S^m)

1.2: k = length(mn)

1.3: For i=1 to K do

1.3.1: $w_i = T^{i-1} S^m (p, mn_i)$

1.4: end for

1.5: $Q_m = \sum_{i=1}^k S^m(p, mn_i)$

1.6: $Y_m(p) = (\sum_{i=1}^k w_i Y(mn_i))/Q_m$

2: end for

3: For q=1 to m do

3.1: dn = knownNeighbour(q, S^d)

3.2: k = length(dn)

3.3: For j=1 to K do

3.3.1: $w_j = T^{j-1} S^d(p, dn_j)$

3.4: end for

3.5: $Q_d = \sum_{i=1}^k S^d(p, dn_j)$

3.6: $Y_d(p) = (\sum_{j=1}^k w_j Y(dn_j))/Q_d$

3.7: $Z_d(p) = (\sum_{j=1}^k w_j Z(dn_j))/Q_d$

4: end for

5: For r=1 to l do

5.1: cn = knownNeighbour(r, S^c)

5.2: k = length(cn)

5.3: For i=1 to k do

5.3.1: $w_i = T^{i-1} S^c(p, cn_i)$

5.4: end for

5.5: $Q_c = \sum_{i=1}^k S^c(p, cn_i)$

5.6: $Y_c(r) = (\sum_{i=1}^k w_i Z(cn_i))/Q_c$

6: end for

7: $Y_{md} = (Y_m + Y_d)/2$

8: $M_{cd} = (Z_c + Z_d)/2$

9: Y = max(Y, Y_{md})

10: Z = max(Z, M_{cd})

11: return Y and Z

The updated Y and Z matrices acts as new interaction profile, i.e., nearest neighbour interaction profile for microRNAs, drugs and diseases. These features are used to construct the prediction models.

3.6   Linear Neighbourhood Similarity and Label Propagation

Linear neighbourhood information [12] could be used to reconstruct each data point. MicroRNA data points are taken as an instance, denoting each of them as feature vectors x_i where i = 1,2, ...,m and they are considered to be the data objects. Then the problem is optimized using:

$$\varepsilon_i = \left\| x_i - \sum_{i_j : x_{i_j} \in N(x_i)} w_{i,i_j} x_{i_j} \right\|^2 \tag{7}$$

$$s.t. \sum_{i_j : x_{i_j} \in N(x_i)} w_{i,i_j} = 1, w_{i,i_j} \geq 0, j = 1, 2, \ldots, k$$

where $N(x_i)$ denotes the set of k neighbours nearest to x_i where $(0 < k < m)$ and w_{ii_j} speaks to the commitment of x_{i_j} to the remaking of x_i and is viewed as their relatedness. Hence a weight matrix $W_M \in R^{n \times n}$ that can be considered as the microRNA pairwise linear neighbourhood similarities is obtained. Same steps are repeated for drugs and a weight matrix $W_C \in R^{l \times l}$ that can be considered as drug pairwise linear neighbourhood relatedness is obtained. Both microRNA pairwise linear neighbourhood similitudes and drug pairwise similarities could be represented as a directed graphs. Once both the graphs are built, iteratively spread the label information of microRNAs and drugs on the microRNA and drug based graphs respectively, to reveal in secret disease-microRNA and disease-drug affiliations. The initial labels are the known associations of microRNAs and drugs with a disease d_j. In each iteration, every microRNAs and drugs receives label information from its neighbours. Each of them takes α proportion from the received information and retains 1-α portion of its initial state. The process can be described as:

$$Y_j^{t+1} = \alpha W_M Y_j^t + (1 - \alpha) Y_j^0$$
$$Z_j^{t+1} = \alpha W_C Z_j^t + (1 - \alpha) Z_j^0 \tag{8}$$

where Y_j^0 is the initial labels of all microRNAs for disease d_j, Z_j^0 is the initial labels of all drugs for disease d_j, $Y_j^t = \left(y_{1j}^t, y_{2j}^t, \ldots, y_{mj}^t\right)$ and $Y_j^t = \left(Z_{1j}^t, Z_{2j}^t, \ldots, Z_{lj}^t\right)$ represents the predicted labels at iteration t. The label matrices are then updated until convergence. Y* and Z* are the updated matrices which represents the predicted microRNA-disease and drug-disease association score matrices respectively. The Y* and Z* matrices can be used for the construction of resultant prediction model. The matrix Y* has its rows as microRNAs and columns as diseases and the values represents microRNA-disease associations. i.e., it specifies how likely is the microRNA in that row would be associated with the disease in each column. Many of the zero values in the matrix (formed using known associations) have been replaced with values within the range 0 and 1, i.e., more diseases will be associated with each microRNAs. Now using the Z* matrix whose rows represents drugs and columns represents diseases, the corresponding drugs associated with each of the predicted diseases can be retrieved. The main point to be noted is that, the number of diseases in both the Y and Z matrix has to be same while performing the above operations. In this method the nearest neighbour algorithm applied to the matrices actually acts as a pre-processing procedure which makes both the matrices more suitable for performing label propagation by linear neighbourhood similarities. The two matrices Y* and Z* together act as the microRNA-drug association model with disease acting as an intermediary to actually link between microRNAs and drugs. An example for the structure of the generated prediction model is given in Fig. 2.

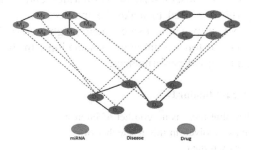

Fig. 2. MicroRNA-drug prediction network

4 Experiments and Results

Initially we computed the various drugs that are associated with each microRNAs. Then to check the prescient exhibitions of our system we applied leave one out cross validation. The anticipated signs for the test medicate were positioned by the last outcome gotten by Eq. (8). For every particular ranking threshold, if the weight of construed microRNA-disease or drug-disease affiliation was over the edge, it was viewed as a true positive. Otherwise, it was regarded as a false positive. The prediction capability of the method was represented using area under the curve (AUC) value. MicroRNAs were important biomarkers in medication repositioning. So by using this study we can confirm that computational predictions of microRNA-disease and drug-disease affiliations can provide significant help in developing microRNA-drug association prediction model. Experimental outcomes exhibit that a normal AUC estimation of 0.705 was obtained in microRNA-drug association forecasts when KNN algorithm was not applied. And when KNN was introduced, the average AUC value increased to 0.720.

5 Conclusion

The process of repositioning the available chemicals for new purposes could be beneficial to both pharmaceutical organizations just as people. The predictions of drug-disease relationship using computational methods are productive approaches to give chemicals with the most sign contender for further biomedical tests. Hence, deriving effective algorithms to gather sedate infection affiliations is of high centrality and incalculable endeavors are being made to this field. Non coding RNAs, such as microRNAs, go about as fundamental control elements of gene expressions. Expanding confirmations describes that non coding RNAs have cozy association with different diseases. All the more as of late, progresses in pharmacogenomics demonstrated that microRNAs work as basic factors in drug response and ailment treatment. Foreseeing obscure disease-microRNA associations translates malady pathogenesis. In our work, we use a method named label propagation by linear neighbourhood similarities to foresee unobserved drug-disease associations by utilizing known drug-disease and microRNA-disease affiliations. Known Nearest Neighbour algorithm is applied as a pre-processing step to improve performance. Initially we calculated drug-drug, microRNA-microRNA and disease-disease similarities and formed three adjacency matrices. KNN algorithm is used to make the matrices more suited for predictions. Then utilized the label propagation using linear neighbourhood similarities to consolidate these outcomes to foresee unknown associations. In the computational trials, our method can produce good performances.

Compliance with Ethical Standards

✓ All authors declare that there is no conflict of interest.
✓ No humans/animals involved in this research work.
✓ We have used our own data.

References

1. Chen, X., Liu, M.X., Yan, G.Y.: RWRMDA: predicting novel human microRNA-disease associations. Mol. BioSyst. **8**, 2792–2798 (2012)
2. Xuan, P., Han, K., Guo, Y., Li, J., Li, X., Zhong, Y., Zhang, Z, Ding, J.: Prediction of potential disease-associated microRNAs based on random walk, pp. 1–2 (2010)
3. Chen, X., Yan, C.C., Zhang, X., You, Z.H., Huang, Y.A., Yan, G.Y.: HGIMDA: heterogeneous graph inference for miRNA-disease association prediction. Oncotarget **7**, 65257 (2016)
4. Chiang, A.P., Butte, A.J.: Systematic evaluation of drug-disease relationships to identify leads for novel drug uses. Clin. Pharmacol. Ther. **86**(5), 507–510 (2009)
5. Gottlieb, A., Stein, G.Y., Ruppin, E., Sharan, R.: PREDICT: a method for inferring novel drug indications with application to personalized medicine. Mol. Syst. Biol. **7**, 496 (2011)
6. Luo, H., et al.: Drug repositioning based on comprehensive similarity measures and Bi-Random walk algorithm. Bioinformatics **32**(17), 2664–2671 (2016)
7. Lee, I., Blom, U.M., Wang, P.I., Shim, J.E.: Prioritizing candidate disease genes by network-based boosting of genome-wide association data. Genome Res. **21**, 1109–1121 (2011)
8. Cheng, L., Li, J., Ju, P., Peng, J., Wang, Y.: SemFunSim: a new method for measuring disease similarity by integrating semantic and gene functional association. PLoS One **9**(6), e99415 (2014)
9. Wang, D., Wang, J., Lu, M., Song, F., Cui, Q.: Inferring the human miRNA functional similarity and functional network based on miRNA-associated diseases. Bionformatics **26**, 1644–1650 (2010)
10. Bartel, D.P.: MicroRNAs: genomics, biogenesis, mechanism, and function. Cell **116**(2), 281–297 (2004)
11. Li, Y., Qiu, C., Tu, J., Geng, B., Yang, J., Jiang, T.: HMDD: a database for experimentally supported human microRNA and disease associations. Nucleic Acids Res. **42**, 1070–1074 (2014)
12. Wang, F., Zhang, C.: Label propagation through linear neighbourhoods. IEEE Trans. Knowl. Data Eng. **20**, 55–67 (2008)

RF-HYB: Prediction of DNA-Binding Residues and Interaction of Drug in Proteins Using Random-Forest Model by Hybrid Features

E. Shahna[(✉)] and Joby George

Department of Computer Science and Engineering,
Mar Athanasius College of Engineering, Kothamangalam, Kerala, India
shahnaelamchery@gmail.com, jobygeo@hotmail.com

Abstract. Here, we aimed to propose a computational approach named as RF-HYB, for forecasting the presence of the DNA-binding remains in the protein from the sequences of the amino acids and also predict the interaction of drugs in DNA. Contrast with different works here utilizing the random forest (RF) calculation, since it is quick and has powerful execution for various parameter esteems. A cross breed feature is introduced which incorporate transformative data of the amino corrosive sequence (position-specific scoring matrices (PSSMs)), auxiliary structure information (predicted solvent accessibility (PSA)) which mirrors the qualities of 20 sorts of amino acids for two physical– substance properties. After extracting the features given into a classifier. Here we are using Random forest algorithm for classification. Along with this prediction we also predict or discover the interaction of drug in the DNA and in proteins. The drug like methotrexate.

Keywords: DNA binding residues · Random-Forest · Position specific scoring matrix · Predicted solvent accessibility · Methotrexate

1 Introduction

Protein DNA coordinated efforts control an arrangement of essential characteristic systems, for instance, quality heading, DNA repetition and fix, re-combination and other fundamental walks in cell headway [1]. Changes of DNA-confining stores, for instance, those on the tumor repressor protein P53, might straightforwardly engaged by human sicknesses [2]. In this manner, the ability to distinguish the amino corrosive deposits that perceive DNA can fundamentally enhance our comprehension of these natural procedures and influence the essential for managing site coordinated mutagens ponders for the practical portrayal of DNA restricting proteins, and can additionally add to propels in drug revelation, for example, supporting the structure of fake translation factors [3]. An assortment of computational techniques have been created to identify DNA-restricting interface deposits from 3D structures [6].

Drug discovery is a response in which substance influences the movement of a drug, for example the impacts of expanded or DE wrinkled, to deliver impact that genius duce's all alone, rely on the dynamic destinations in an atom. Drug protein

© Springer Nature Switzerland AG 2020
S. Smys et al. (Eds.): ICCVBIC 2019, AISC 1108, pp. 1319–1326, 2020.
https://doi.org/10.1007/978-3-030-37218-7_138

restricting is the reversible association of drug with proteins. Proteins are vast natural elements comprises various types of AA masterminded as a direct structure, combined by peptide bonds between the carboxyl, amino gatherings of amino corrosive buildups. Here, propose a computational methodology named as RF-HYB, for anticipating DNA restricting deposits in proteins from amino corrosive groupings. A half breed highlight is introduced which incorporate evolutionary data of the amino corrosive succession (position-explicit scoring frameworks (PSSMs)), auxiliary structure data (anticipated dissolvable availability (PSA)) which mirrors the qualities of 20 sorts of amino acids for two physical synthetic properties. After extracting the features given into a classifier. Here we are using Random forest algorithm for classification. Along with this prediction we also predict or discover the interaction of drug in the DNA and in proteins. The drug like methotrexate (Fig. 1).

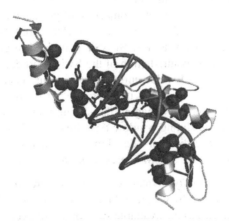

Fig. 1. Interaction of residues in DNA

2   Prediction of DNA-Binding Residues

2.1   Dataset

Here developed a set of data of 7,186 DNA chains framed by the protein bindings, which had clear target explanations in the (PDB) [8]. In the wake of evacuating the excess groupings, an aggregate of 584 non redundant protein successions were acquired with the end goal that no two arrangements had over 30% character. At that point, partitioned the non redundant successions into two sections, the preparation dataset (PDNA 543) and the autonomous test dataset (PDNA TEST). PDNA 543 comprises of 543 protein groupings, altogether discharged into the PDB. PDNA TEST incorporates 41 chains of protein, all re-rented into the PDB. All the more explicitly, there are 9,549 bindings of DNA buildups (i.e., positive examples) and 134,995 deposits for non-bindings (i.e., negative examples) in PDNA 543. PDNA TEST comprises of 734 positive examples and 14,021 negative examples. Table 1 condenses the point by point organizations of PDNA 543 and PDNA TEST.

Table 1. Creation of the preparation and autonomous approval datasets

Dataset	No. of sequences	Num P	Num N	Ratio
PDNA-543	543	9549	134995	14.137
PDNA-TEST	41	734	14021	19.102

Num P: No of +ve samples
Num N: No of −ve samples
Ratio: Num P/Num N

2.2   Protein Features

Position Specific Scoring Matrix. PSSM profiles have been displayed to be a fruitful component for imparting buildup securities and have been associated with various bioinformatics issues, for instance, the conjecture of protein work, protein discretionary structure, and protein nucleotide confining stores. Here, used the PSSM incorporate for foreseeing DNA binding deposits. PSI BLAST is utilized in generating the profile for the PSSM [12] to pursue in contradiction of the Swiss Port database [13] through three emphases, with 10^{-3}) as the E value cutoff for numerous succession arrangement. The score of the each part was rescaled utilizing the standard, the development of the feature of PSSM were expelled engaging the sliding window process followed by the rescaling. By then use the sliding window technique in light of the way that speculate that the binding of the DNA limit of a development depends upon its own one of a kind PSSM scores similarly as the PSSM scores of its neighboring buildups.

Predicted Solvent Accessibility. Waters can contact stores at the outside of a protein. Each particle can be reached by water, and the zone of a molecule externally that can be reached by water is known as the available subatomic surface, or dissolvable uncovered zone. Additionally compute the region verified by the center of a water molecule moving over the outside of a revealed protein particle. This locale is greater, clearly, and is known as the open surface. Clearly extensive polar buildups will on affirm age have a huge available surface, and little hydrophobic deposits will by and large have a little open surface. Accessibility of amino acids (AA) in various known protein representations are describe in Fig. 2. Solvent accessibility openness is especially critical in that it is firmly identified with the spatial game plan and the pressing of buildups amid the procedure of protein collapsing [14]. In addition, there is an indistinguishable connection between dissolvable availability and protein ligand co-operations, proposing that dissolvable openness data can be utilized to discourage mine protein capacities. Acquired the predicted element accessibility (PSA) attributes are every buildup the encouraging and comparing arrangement to the independent SANN program [14].

Amino acid	SEA > 30 Å²	SEA < 10 Å²	30 > SEA > 10 Å²
S	0.70	0.20	0.10
T	0.71	0.16	0.13
A	0.48	0.35	0.17
G	0.51	0.36	0.13
P	0.78	0.13	0.09
C	0.32	0.54	0.14
D	0.81	0.09	0.10
E	0.93	0.04	0.03
Q	0.81	0.10	0.09
N	0.82	0.10	0.08
L	0.41	0.49	0.10
I	0.39	0.47	0.14
V	0.40	0.50	0.10
M	0.44	0.20	0.36
F	0.42	0.42	0.16
Y	0.67	0.20	0.13
W	0.49	0.44	0.07
K	0.93	0.02	0.05
R	0.84	0.05	0.11
H	0.66	0.19	0.15

Fig. 2. Solvent accessibility of amino acids (AA)

2.3   Random Forest–Classification Algorithm

The RF is encompassed of multitudes of decision tree classifiers for the purpose of classification [15]. The RF calculation is executed the irregular R bundle [16]. To guarantee that model age and the determination of the parameter of RFs are totally autonomous of the test information, a settled cross-validation technique is performed. Settled cross-approval implies that there is an external cross-validation circle for model evaluation and an internal circle for model determination. In this examination, the first examples are irregularly isolated into k = 5 sections in the external circle. Every one of these parts is picked in a sequence for appraisal, and the rest of the 4/5 of the examples are for model determination in the internal circle are a method of cross-approval utilizing the purported out-of-bag (OOB) tests is achieved.

Random Forest Calculation

1. Randomly picks "k" high-lights in "m" high-lights.
   1. with k less than m
2. Amongst the "k" high-lights, determine the hub "d" identifying the best part point.
3. Split the hub as little girl hubs utilizing the best split.
4. Repeat 1 to 3 steps up to "l" hub has been reached.
5. Construct woodland by repeating steps 1 to 4 for n times and trees.

2.4   Random Under Sampling

The issue of imbalanced datasets happens in numerous reasonable order occasions where the objective is to recognize an uncommon however vital case. All responses for this issue can by and large be gathered into two essential arrangements [17] the fundamental involves preparing of the data to reestablish balanced class (either by resizing the minority class with higher values or down estimating the larger part of class); while the second incorporates changing the calculations involved in learning itself to adjust to data that are imbalanced. [18] Presented a calculation for the uneven determination of models by cutting back the dominant part class. Instances of the lion's share class can generally be isolated into four gatherings: class-name commotion, marginal, excess and safe models, and the calculation endeavors to make a subset holding the protected precedents [18].

2.5   Measurements of Classifier's Execution

The general forecast exactness (ACC), affectability (SE), accuracy (PR), particularity (SP) [19] and Matthew's connection coefficient (MCC) are for appraisal of the expectation framework.

3   Interaction Between Drug- DNA and Protein

Drug like methotrexate is chosen as a result of double importance to the enthusiasm of clinicians just as to the fundamental researchers. The coupling action of methotrexate drugs with different atoms including DNA and protein (enzyme)were contemplated by utilizing distinctive virtual products of bioinformatics apparatuses, SPDB Viewer, Rasmol, EMBOSS win and furthermore utilization with uninhibitedly accessible instruments, CASTp, EXPASY to choose discretionary structure and desire for structure.

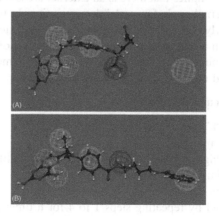

Fig. 3. Mapping of methotrexate with dynamic destinations (red)

3.1   Swiss PDB Viewer

Swiss Pdb Viewer is an app which gives an easy to understand interface permitting dissecting a few proteins in the meantime. The proteins can be superimposed so as to conclude basic arrangements and think about their dynamic destinations or some other important parts. Amino corrosive changes, H-bonds, edges and separations between iotas are anything but difficult to acquire for the unconstrained realistic and menu interface. Different displaying devices are incorporated and direction documents for well known vitality minimization bundles can be created. Figure 3 represents the interaction or mapping of methotrexate with active sites, shown in red colours.

4 Results and Analysis

4.1 RF Method Using Various Features for Analysing the Prediction Performance

See Table 2.

Table 2. The comparison of performance of the proposed model based on different features.

Features	ACC (%)	SE (%)	PR (%)	SP (%)	MCC
1	87.95	77.39	64.34	92.44	0.644
1 + 2	88.68	77.57	65.52	93.32	0.661
1 + 3	90.44	76.63	69.63	93.42	0.677
1 + 2 + 3	91.42	76.58	72.17	94.39	0.70

1: PSSMs; 2: PSA; 3: SS.

4.2 Discovery of Drug Interaction

In this research, the 3 Dimensional representation uncovers restricting locales of methotrexate (red in appearance) is appeared assistance of programming projects as depict in the table for under-stand the quantity of strings, gatherings, holding, helices and turns were seen amid examination, information of methotrexate structure which tend to cooperate with different particles covalently. The further examination of the meditation restricting was completed to the deoxyribose nucleic acid particle demonstrating the association with number of amino acids as appeared in Fig. 4.

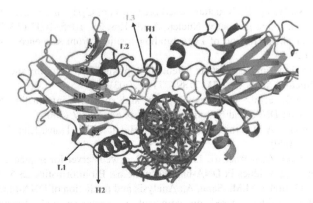

Fig. 4. Interaction of methotrexate (orange colour) in DNA (blue-green) and protein.

5   Conclusion

Here we proposed another Random-Forest based methodology called RF-HYB joining a cross breed include for forecast of DNA-restricting buildups from amino corrosive succession information. The outcomes of the RF dependent models has a forecast precision of 91.41% with MCC is 0.70. To know the best, up to this point, the RF technique joined with a mixture highlight has been the best strategy for foreseeing restricting residues reside in proteins from protein successions without the use of 3D auxiliary data, and forecast outcomes are valuable to know how the proteins interface with DNA atoms. The outcomes exhibit that the half and half component contributes most to the brilliant expectation execution. Along with the prediction of DNA binding residue we also discover the presence of drug in the DNA.

Acknowledgment. We would like to thank our faculty, Mar Athanasius College of Engineering (MACE), APJ Abdul Kalam Technological University (KTU), for their support in conducting our project.

Compliance with Ethical Standards
   ✓ All authors declare that there is no conflict of interest.
   ✓ No humans/animals involved in this research work.
   ✓ We have used our own data.

References

1. Luscombe, N.M., et al.: An overview of the structures of protein-DNA complexes. Genome Biol. **1**, reviews001 (2000)
2. Bullock, A.N., Fersht, A.R.: Rescuing the function of mutant p53. Nat. Rev. Cancer **1**, 68–76 (2001)
3. Ho, S.Y., et al.: Design of accurate predictors for DNA-binding sites in proteins using hybrid SVM-PSSM method. Biosystems **90**, 234–241 (2007)
4. Siggers, T.W., Honig, B.: Structure-based prediction of C_2H_2 zinc-finger binding specificity: sensitivity to docking geometry. Nucleic Acids Res. **35**, 1085–1097 (2007)
5. Ofran, Y., et al.: Prediction of DNA-binding residues from sequence. Bioinformatics **23**, i347–i353 (2007)
6. Jones, S., et al.: Using electrostatic potentials to predict DNA-binding sites on DNA-binding proteins. Nucleic Acids Res. **31**, 7189–7719 (2003)
7. Kuznetsov, I.B., et al.: Using evolutionary and structural information to predict DNA-binding sites on DNA-binding proteins. Proteins **64**, 19–27 (2006)
8. He, H., Garcia, E.A.: Learning from imbalanced data. IEEE Trans. Knowl. Data Eng. **21**(9), 1263–1284 (2009)
9. Hwang, S., Gou, Z., Kuznetsov, I.B.: DP-Bind: a web server for sequence-based prediction of DNA-binding residues in DNA-binding proteins. Bioinformatics **23**(5), 634–636 (2007)
10. Ahmad, S., Gromiha, M.M., Sarai, A.: Analysis and prediction of DNA-binding proteins and their binding residues based on composition, sequence and structural information. Bioinformatics **20**(4), 477–486 (2004)
11. Wong, K.C., Li, Y., Peng, C.B., et al.: Computational learning on specificity determining residue nucleotide interactions. Nucleic Acids Res. **43**(21), 10180–10189 (2015)

12. Schaffer, A.A., Aravind, L., Madden, T.L., et al.: Improving the accuracy of PSI BLAST protein database searches with composition based statistics and other refinements. Nucleic Acids Res. **29**(14), 2994–3005 (2001)
13. Bairoch, A., Apweiler, R.: The SWISS PROT protein sequence database and its supplement TrEMBL in 2000. Nucleic Acids Res. **28**(1), 45–48 (2000)
14. Joo, K., Lee, S.J., Lee, J.: Sann: solvent accessibility prediction of proteins by nearest neighbor method. Proteins Struct. Funct. Bioinf. **80**(7), 1791–1797 (2012)
15. Breiman, L.: Random forest. Mach. Learn. **45**, 5–32 (2001)
16. Liaw, A., Wiener, M.: Classification and regression by random Forest. R News **2**, 18–22 (2002)
17. Cohen, G., et al.: Learning from imbalanced data in surveillance of nosocomial infection. Artif. Intell. Med. **37**, 7–18 (2006)
18. Kubat, M., Matwin, S.: Addressing the curse of imbalanced training sets: one sided selection. In: Fisher, D. (ed.) Machine Learning: Proceedings of the Fourteenth International Conference (ICML 1997), pp. 179–186. Morgan Kaufmann Publishers, San Francisco (1997)
19. Wang, L., Brown, S.J.: Prediction of DNA-binding residues from sequence features. J. Bioinform. Comput. Biol. **4**, 1141–1158 (2006)

Investigation of Machine Learning Methodologies in Microaneurysms Discernment

S. Iwin Thanakumar Joseph[1]([⊠]), Tatiparthi Sravanthi[2],
V. Karunakaran[1], and C. Priyadharsini[1]

[1] Department of Computer Science and Engineering,
Karunya Institute of Technology and Sciences, Coimbatore, India
{iwinjoseph, karunakaran, priyadharsini}@karunya.edu
[2] Department of Computer Science Engineering,
Sree Chaitanya College of Engineering, Thimmapur, Karimnagar, India
sravanthi1424@gmail.com

Abstract. Diabetic retinopathy is one of the causes of major complication in the smallest blood vessels of diabetes that ends in a loss of eye vision if the affected candidate is not taken proper medical treatment. The early symptom of this diabetic retinopathy is microaneurysms. This research article gives the complete overview of various machine learning methodologies used for the suitable automated detection of microaneurysms in the earlier stage to avoid medical complications.

Keywords: Machine learning · Deep learning · Detection of microaneurysms · Fundus images

1 Introduction

The World Health Organization (WHO) report 2016 says that one in eleven members got affected due to diabetes. It is approximately over 400 million members all over the world. Diabetic retinopathy leads to vision blindness and WHO predicted that out of three diabetes, one is affected with diabetic retinopathy and also out of ten diabetic patient, one having the severe eye vision problem. So, initial diabetic retinopathy detection prevents the worst category of eye vision loss upto 90%. Diabetic retinopathy is categorized into two main cases:

1. Non proliferative diabetic retinopathy (NPDR)
2. Proliferative diabetic retinopathy

The screening process of diabetic retinopathy is performed in person by opthamologists and trained peoples who categories the level of disease through a appropriate visual checkup of fundus photos. But in practical, the process of grading the fundus images are mostly error prone, monotonous and complex one. Due to the increase in the number of diabetic retinopathy patient every year, it is highly needed of automatic screening methods for diabetic retinopathy disease. The World Health

S. Smys et al. (Eds.): ICCVBIC 2019, AISC 1108, pp. 1327–1334, 2020.
https://doi.org/10.1007/978-3-030-37218-7_139

Organization also pointed out that the number of diabetic affected patient may go beyond 640 million by the year 2040 (Fig. 1).

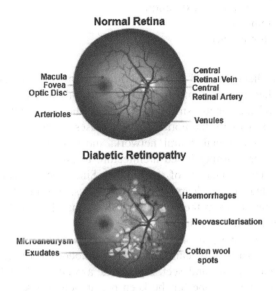

Normal Retina

Macula
Fovea
Optic Disc

Arterioles

Central
Retinal Vein
Central
Retinal Artery

Venules

Diabetic Retinopathy

Haemorrhages

Neovascularisation

Microaneurysm
Exudates

Cotton wool
spots

Fig. 1. Normal retina image and diabetic retinopathy affected retina image Source: Singapore Research Institute

In common, microaneurysms are spherical bulge of capillaries mainly due to the diminishing of the strength in vascular walls. Microaneurysms will be visualized as a small red point in the eye. It is the first or preliminary symptom or sign of diabetic retinopathy. The main advantage of automated microaneurysms detection reduce the major complexity of screening process drastically. But there are lot of challenges arises in the detection of microaneurysms such as

- Uneven image illumination
- Reflections
- Limited resolution
- Media opacity

In general, there is no well defined boundary for microaneurysms and even in high resolution images it is difficult to separate with the background. The microaneurysms diameter ranges from 10 m to 100 m which is even lesser than the size of major optic veins. Sometimes, the possibility of wrong prediction of microaneurysms with visually similar structures like

- Haemorrhages
- Junctions in thin vessels
- Separated vessel segments
- Small dark patches on blood vessels
- Patches in background pigmentation
- Reflection of dust particles present in lenses of camera.

In general, the detection of microaneurysms includes the following five steps:

1. Pre-processing
2. Microaneurysms candidate extraction
3. Blood vessel removal
4. Candidate feature extraction
5. Classification

Several researchers do well in the automation process of detection of microaneurysms using their proposed methodologies in the stipulated five steps given above for the detection of microaneurysms symptoms. But the complexity of classification of affected fundus images with the normal one increases as the dataset increases. Due to the advancement of artificial neural network, most of the researchers utilized the methodology of deep learning based architecture to predict the symptom of microaneurysms for the tested data set of the patient. Since the manual feature extraction process is eliminated in deep learning methodologies, it is highly convenient for the researchers to come out from the complexity of wrong classification of tiny particles which are similar in like characteristics.

The rest of the article is proceeded in the following manner: Sect. 2 discusses about the overall view of various machine learning methodologies used in the automated detection of microaneurysms and Sect. 3 gives the available public datasets and Sect. 4 concludes with the future scope for the keen researchers to take forward further step.

2 Overall View of Machine Learning Methodologies in Microaneurysms Detection

The overall structure of method proposed by Chudzik et al. [1] consists of the following three steps:

1. Image pre-processing
2. Patch extraction
3. Pixel wise classification

Here, pre-processing is done mainly to discard the non-uniform illumination and data which are redundant from the images. The second step is used to make the data ready for analysis and the third step, pixel wise classification is done by convolutional neural network. Chudzik et al. implemented it using keras deep learning framework and tensor flow numerical computation library.

Performance of the Algorithm

1. FROC – Free response (ROC) curve

This metrics is used widely in medical field to determine abnormality in the given medical image.

2. FPI – False positive per image

Positive Side of the Algorithm

- Algorithm performs high at low FPIs and is highly useful for applications of screening.
- Algorithm takes 220 s as a time for processing single image.
- Algorithm yields better FROCs when compared with state of art method.

Future Scope

- Efficiency of the algorithm needs to be concentrated.
- Reduction of processing time per image is further implemented by providing parallel operation of forward propagation step.

Wu et al. [2] used inverted green channel of fundus images as an input for their automated method. The reason behind the usage of green channel images are the best background contrast of microaneurysms regions are provided by these images when compared to red channel and blue channel. The overall structure of Wu et al. method is presented in the following four steps:

1. Image Pre-processing
2. MA candidate region marking
3. Feature Extraction
4. Classification

In general, retinal fundus images are normally noisy, low contrast and often non-uniform illuminated images. It is harder to detect microaneurysms in low contrast and poor brightness regions. So proper pre-processing steps are essential to make the images in more suitable form to detect the affected regions. Here pre-processing is done in three steps

1. Illumination equalization
2. Contrast limited adaptive histogram equalization (CLAHE)
3. Smoothing

The candidate extraction process provides a significant role in the complete process of detection of microaneurysms. The important function of this step is to discard the maximum region of objects that are not provides similar characteristics to microaneurysms. But if any true microaneurysms region discarded in this step cannot be get back in the further steps. Wu et al. used peak detection concept for extracting the microaneurysms candidate region.

In feature extraction step, concept of profile feature is used along with local features like hessian matrix based features, shape and intensity features etc. KNN classifier is used to classify the microaneurysms affected images from the dataset.

Metrics Used

1. FROC
2. FPIs

Positive Side of the Algorithm

- One of the efficient and full automated methodology
- KNN classifier utilized in this algorithm outperforms naïve bayes and gives same like performance when compared to adaboost classifier evaluated in e-ophtha MA dataset.

Habib et al. [3] implemented the enhanced version of Fleming's method of microaneurysms detection. The advantage of this method is it provides more detailed feature analysis and also had extensive evaluation. The methodology worked in three phases namely: (1) pre-processing (2) Microaneurysms Candidate detection and (3) Microaneurysms Candidate Classification.

During the first step, unwanted noises are removed from the image and also non uniform illumination is discarded using the concept of background subtraction. Here second step is performed using Gaussian matched filter. But it also produces some false positive due to

1. Vessel cross sections or non removed vessels before the MA candidate detection phase.
2. Presence of noise which looks same as like microaneurysms.

In order to identify these kinds of false positive images correctly, Habib et al. used suitable classifier i.e. ensemble classifier in this method. Ensemble classifier works on the concept of decision tree learning. Some of the following features are considered to extract the microaneurysms affected regions exactly

1. Gaussian features
2. Fleming's features
3. Shape features and Moment Invariants
4. Intensity features
5. Morphological features

Positive Side of the Algorithm

- It provides better results in discarding false positives from the given images.
- It provides good accuracy with comparison to the state of art method.

Lian et al. [7] proposed an effective deep convolutional neural network model for diabetic retinopathy classification [4] using the following four phases such as

1. Network architecture
2. Image pre-processing
3. Addressing class imbalance
4. Fine tuning

Lian tested in the unspecified data set of 3506 images and concludes that the accuracy level is quite improved by undergoing the above mentioned four steps.

Harangi et al. [5] proposed deep convolutional neural network based techniques for detecting microaneurysms in fundus image. In this paper, they organized the network as a super network and the super network is combined with a fusion based approach. The advantage of combination of super network and fusion based approach architecture will train as a single network and thus the member Deep Convolutional Neural Network is also trained. In the proposed method, a retinal image is divided into small sub images and given all the small sub images to an input for super network and fusion based approach. The proposed method predicts Microaneurysms efficiently and accurately compared to the following architectures such as Alexnet, googlenet, VGGnet. The proposed system outperformed on the following aspects such as Accuracy, sensitivity and specificity [8].

Shan and Li [6] proposed Deep learning method for microaneurysms detection in fundus image. Now a day's deep learning techniques are used for feature extraction and classification problems and most of the researchers achieved a good result. In this paper, Stacked Sparse Autoencoder an instance of deep learning technology is used for identifying a microneurysms in fundus image. In this work, take original fundus image and generate some n number of small image patches. Using Stacked Sparse Autoencoder learns high level features from pixel intensity from the small image patches. The learned high level features are given the classifiers for classifying, which are the image patches are having microaneurysmys or which are the image patches are not having microaneurysmys. The experiment is carried out with 89 images. From the 89 images, 8412 small image patches were generated. From the 8412 small patches, 2182 small patches are belongs to microneurysms and the remaining 6230 small patches are belongs to non microneurysms. The proposed system outperformed in the following parameters such as precision, recall, specificity, f-measure and accuracy [9] (Table 1).

Tan et al. [8] proposed automated segmentation for exudates, Hemorrhages and microaneurysmys using single convolution neural network. In this paper, authors proposed ten layer convolution neural network automatically, it will segment the image and discriminate for exudates, Hemorrhages and microaneurysmys. In the experimental process all the images are normalized before the segmentation. The proposed net is trained on two stages for improving the classification accuracy. The experimental result shows single convolution network provides considerable accuracy rate. The experiment is carried out with cleopatra dataset. The proposed system achieves sensitivity of 0.9572, 0.8758, 0.6257 and 0.4606 for background, exudates, haemorrhages and microaneurysms [10].

3   Sample Databases

Some of the available dataset utilized by various researchers to evaluate the performance of their proposed methodology are given below:

- DIARETDB0 dataset
- DIARETDB1 dataset
- DRIVE dataset
- MESSIDOR dataset

Table 1. Few research articles in machine learning methodology

Author	Features used	Metrics	Classifier	Database
Chudzik et al. (2018) [1]	Not mentioned	Sensitivity FROC	CNN classifier	E-ophtha ROC DIARETDB1
Wu et al. (2017) [2]	Local features Profile features	Sensitivity FROC	KNN classifier	E-ophtha ROC
Habib et al. (2017) [3]	Fleming's features Shape features Moment invariants Gaussian features Intensity features Morphological features	Sensitivity FROC	Ensemble classifier	MESSIDOR ROC DIARETDB1
Eftekhari et al. (2019) [4]	NIL	Sensitivity	CNN architecture	E-ophtha ROC
Harangi et al. (2018) [5]	NIL	Sensitivity	DCNN architecture	MESSIDOR ROC DIARETDB1 E-ophtha
Shan et al. (2016) [6]	NIL	Sensitivity FROC	Stacked sparse autoencoder + Softmax classifier	DIARETDB

- E-Ophtha dataset
- ROC dataset
- DRIONS dataset
- STARE dataset
- ARIA dataset

4   Conclusion

In this research article, the suitability of various machine learning techniques in the field of microaneurysms detection is discussed along with the advantages and limitations. The hybrid approaches are also well defined in this survey article. This research article also aids in highlighting significant contributions of different approaches in the detection of microaneurysms.

Compliance with Ethical Standards
   ✓ All authors declare that there is no conflict of interest.
   ✓ No humans/animals involved in this research work.
   ✓ We have used our own data.

References

1. Chudzik, P., Majumdar, S., Calivá, F., Al-Diri, B., Hunter, A.: Microaneurysm detection using fully convolutional neural networks. Comput. Methods Programs Biomed. **158**, 185–192 (2018)
2. Wu, B., Zhu, W., Shi, F., Zhu, S., Chen, X.: Automatic detection of microaneurysms in retinal fundus images. Comput. Med. Imaging Graph. **55**, 106–112 (2017)
3. Habib, M.M., Welikala, R.A., Hoppe, A., Owen, C.G., Rudnicka, A.R., Barman, S.A.: Detection of microaneurysms in retinal images using an ensemble classifier. Inform. Med. Unlocked **9**, 44–57 (2017)
4. Eftekhari, N., Pourreza, H.R., Masoudi, M., Ghiasi-Shirazi, K., Saeedi, E.: Microaneurysm detection in fundus images using a two-step convolutional neural network. Biomed. Eng. Online **18**(1), 67 (2019)
5. Harangi, B., Toth, J., Hajdu, A.: Fusion of deep convolutional neural networks for microaneurysm detection in color fundus images. In: 2018 40th Annual International Conference of the IEEE Engineering in Medicine and Biology Society, EMBC, pp. 3705–3708. IEEE, July 2018
6. Shan, J., Li, L.: A deep learning method for microaneurysm detection in fundus images. In: 2016 IEEE First International Conference on Connected Health: Applications, Systems and Engineering Technologies, CHASE, pp. 357–358. IEEE, June 2016
7. Lian, C., Liang, Y., Kang, R., Xiang, Y.: Deep convolutional neural networks for diabetic retinopathy classification. In: Proceedings of the 2nd International Conference on Advances in Image Processing, pp. 68–72. ACM, June 2018
8. Tan, J.H., Fujita, H., Sivaprasad, S., Bhandary, S.V., Rao, A.K., Chua, K.C., Acharya, U.R.: Automated segmentation of exudates, haemorrhages, microaneurysms using single convolutional neural network. Inf. Sci. **420**, 66–76 (2017)

Segmentation Algorithm Using Temporal Features and Group Delay for Speech Signals

J. S. Mohith B. Varma, B. Girish K. Reddy, G. V. S. N. Koushik,
and K. Jeeva Priya[✉]

Department of Electronics and Communication Engineering, Amrita School
of Engineering, Bengaluru, Amrita Vishwa Vidyapeetham, Bengaluru, India
mohithvarma@gmail.com, girish2531998@gmail.com,
Gvsnkoushik@gmail.com, k_jeevapriya@blr.amrita.edu

Abstract. Automatic Speech segmentation is a technique to segment the
speech signals automatically into phonemes or syllables which form the basic
units of speech. This plays an important role in automatic speech recognition
(ASR) systems. In this work, we have implemented a method for automatically
segmenting a speech signal using temporal features such as Short Term Energy
(STE), zero crossing rate (ZCR) and the group delay of the speech signal
computed using the energy for Hindi speech. The syllable boundaries are
determined by the group delay algorithm. Analyzing the experimental results for
comparing the manual segmentation and automatic segmentation using the
group delay, we can infer that the outcome remains similar to the manual
segmentation.

Keywords: ASR · Speech segmentation · Phonemes · Syllabes · STE · ZCR ·
Group delay

1 Introduction

Automatic Speech recognition (ASR) primarily involves recognition of a speech and
converts into equivalent text content. Human Computer Interaction (HCI) is exem-
plified with the developments in speech recognition of various languages.

The sequential steps involved in automatic speech recognition include pre-
processing of speech signals that are acquired followed by segmentation of speech
signals and labeling them. Further, conventional ASR systems have acoustic, language
and lexicon models which mathematically relate to identify a spoken word. Speech
segmentation is a fundamental task in automatic speech recognition and speech synthesis systems. With the development of large vocabulary continuous speech recognition systems (LVCSR), the role of speech segmentation is quite important.

Speech segmentation is an important task to recognize the limits between words,
syllables, or phonemes in spoken dialects. Speech segmentation techniques are two-
fold: acoustic and phonetic segmentation. Acoustic segmentation techniques involve
the determination of various time domain features such as Short term energy (STE),
Zero crossing rate (ZCR), spectral flux, and Mel frequency cepstral coefficients
(MFCC) Perceptual Linear Predictive (PLP) coefficients. Most often the temporal

© Springer Nature Switzerland AG 2020
S. Smys et al. (Eds.): ICCVBIC 2019, AISC 1108, pp. 1335–1343, 2020.
https://doi.org/10.1007/978-3-030-37218-7_140

features mentioned above were used in combinations to segment the speech signals in the form of combinational units of vowels and consonants called syllables.

The phonetic segmentation is a technique in which a forced alignment between the speech signal and its transcription. There have been relatively few works described for segmenting of speech in Indian languages, automatically. In this paper, we have tried speech segmentation of sentences spoken in Hindi using STE and group delay algorithms. The organization of the paper is as follows: In Sect. 2 a brief overview of existing algorithms and methods has been described. Section 3 demonstrates the technique for automatic Speech segmentation and its various stage. In Sect. 4 results are explained and an analysis is carried out by comparing with manual segmentation. In Sect. 5, summary of present work has been shown in order to conclude the discussion.

2   Related Works

Preliminary work on speech segmentation has been reported since 2003. Nagarajan et al. [1] have proposed subband based phase delay technique to segment speech signals into syllables. Prasad et al. [2] also have described a technique for segmenting the speech signals into syllable that used short time energy. Lakshmi et al. [3] in their work have discussed about using STE to segment speech into syllables.

For dividing the speech signals into simpler units called syllables, dependent on typical syllable model where the syllable units boundaries are recognized by determining the HMM grouping and then by SVM labeling., proposed by Hamza et al. [4] for Arabic. Zhao et al. [5] proposed a technique that uses silence detection followed by energy and spectral analysis for segmenting Mandarin language into its syllables. Sahana et al., in their work [6] for recognition of speech in Kannada have described segmentation of utterances and also in paper [7] speech segmentation in Kannada has been discussed.

Continuous Punjabi Speech signal is segmented into syllables by using high resolution properties of group delay in which Kaur et al. [8] have concluded group delay is a better representation of STE for the detection of syllable boundary. Sudhakar Kumar et al. [9] have worked on segmenting speech signals in Maithilwords are segmented by using group delay to identify syllable boundaries and achieved with an accuracy of 85.2%. Geetha [10] has proposed in their work to segment speech signal using STM and level building programming which uses the spectral variations for the boundaries of phoneme.

Automatic Segmentation of speech in Tamil for the syllables was carried out by using vowel onset Point and by finding the significant spectral changes for the utterances in the speech (Spectral Transition Measure). And has been compared it with manual segmentation by Vadivel et al. [11].

The VOP identification along with the ZCR performs the automatic speech segmentation at syllable boundaries dynamically to overcome the limitations due to time and databases of manual segmentation has been proposed by Soumya et al. [12]. In this work, speech segmentation for Hindi speech is described based on temporal features and group delay, and compared with the manual segmentation.

3 Proposed Method

The following (Fig. 1) block diagram proposes the method to segment speech signals. The preliminary work is to preprocess the acquired speech signals, which involves sampling and quantization. Further, the features of the speech signal is extracted, the word boundaries are identified. Post this, the words are divided into syllables, identified using group delay algorithm.

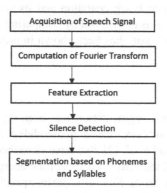

Fig. 1. Block diagram for segmentation of speech

Database Creation and Preprocessing of Speech Signals

For the experimental Speech samples have been recorded from ten speakers for the analysis of segmentation of speech signals in Hindi. The speech samples were collected from speakers in the age group of 20–25 years and the recordings were done at a sampling frequency of 16 kHz, using mono channel in.wav format, in an echo free environment using the open source tool, Praat (6.0.43). The pre-processing of speech samples involves sampling and quantization, along with the removal of unwanted noise in the speech signals, using Praat software.

The speech samples have been divided into frames of 20 ms each, by multiplying the speech signals with a Hamming window of length 20 ms and overlap of 10 ms.

Table 1 shows the ten sentences used for analysis.

Computation of FFT

The speech signal is a quasi stationary signal. The spectrum of the signal cannot be taken in the entirety, hence the speech signal is segmented into smaller frames and the spectrum that is computed is the short time FFT spectrum (STFT) The phase spectrum extracted from the STFT is used to determine the group delay.

Feature Extraction

Speech feature extraction is important for defining the speech signals into set of vectors coefficients Features of speech can be either be observed from time or spectral domain.

In this proposed technique for segmenting the speech signal, a combination of time domain and spectral domain features have been used. For demarcating voiced and unvoiced regions of the signals the temporal features such as ZCR and STE have been

Table 1. Sentences used for segmentation

Sentence 1	मेंएकडॉक्टरकोदेखनाचाहताहूँ
Sentence 2	क्याआपमेरीमददकरसकतेहैं?
Sentence 3	मैंबीमारहूँ
Sentence 4	मुझेहस्पतालमेंबरतीकरायाजानाहैं
Sentence 5	कैसाचलरहाहैं
Sentence 6	तुमकितनेसालकेहो?
Sentence 7	मुझेपानीचाहिए
Sentence 8	मेराभाईकैसाहै?
Sentence 9	मुझेएकनर्सकीज़रुरतहैं
Sentence 10	आपकास्वागतहै

used and further to segment the speech with syllables, group delay functions on the spectrum of the signal is implemented.

Zero Crossing Rate (ZCR) depicts the number of zero crossings of the speech signal. The voiced component of the speech signal has lower zero crossings than the unvoiced component and the amplitude of voiced segments is higher than that of unvoiced segments. Based on Eq. (1), ZCR is computed for a speech signal s(n), framed with windows, w(n).

$$Z(n) = \frac{1}{2N} \sum_{m=0}^{N-1} s(m).w(n-m) \tag{1}$$

ZCR is computed for all the frames of the speech signal and we compute the threshold value by considering the average Zero crossing rate. Short Term Energy (STE) is yet another method to segment a speech signal based on the energy content of the signal, given by Eq. (2).

$$e(n) = \sum_{m=-\infty}^{\infty} (s(m).w(n-m))^2 \tag{2}$$

The sentence, "क्याआपमेरीमददकरसकतेहैं" has been plotted in Fig. 2(a–d)

After phonemes, syllables have been identified as basic unit of representation of speech signal in ASR system. Syllables have three distinct parts viz., onset, nucleus and coda. Though the syllables can have all the three parts or many syllables should have any two and in rare situations syllables should have the nucleus part. The high energy regions of the speech signal represents the nucleus of the syllable and the valleys at the ends of the nuclei represents the syllable boundaries.

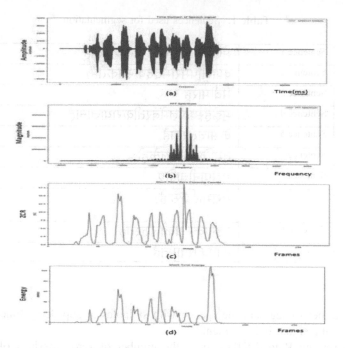

Fig. 2. (a) Speech signal (b) Spectrum of the signal (c) ZCR (d) STE of the speech signal

The following steps describe the procedure to segment the speech signals based on syllables:

Step 1: The digitized version of the speech signal is considered.

Step 2: The STE of each frame of the speech signal is computed where e_1, e_2, e_3 ..... are the energies of the frames.

Step 3 : The windowed speech signal is passed through a Hibert transformer (all pass filter) to detect the phase change in the given speech signal. The inverse of the energies are determined.

Step 4: The inverse Fourier transform of the energy signal is computed, which would be the root cepstrum and is depicted as

$$Z(t) = Z_r(t) + jZ_i(t) \qquad (3)$$

The phase is extracted using above equation by using

$$\emptyset(w) = arc(Z(w)) = tan^{-1}\left(\frac{z_i(t)}{z_r(t)}\right) \qquad (4)$$

Step 5: The phase for each frame is differentiated to obtain group delay

$$\tau_g(w) = -\frac{d\emptyset(w)}{dw} \qquad (5)$$

4 Result

The experimental analysis to segment the speech signals for automatic speech recognition was done for the sentences mentioned in Table 1. The steps mentioned in the Sect. 3 has been done in Python to segment the speech signals into voiced and unvoiced regions and further the words have been segmented into syllables.

The sentence "क्या आप मेरी मदद कर सकते हैं" has been used for the analysis. Figure 3 shows the speech signal overlapped with the energy computed. As seen from the Fig. 3, the signal is segmented into voiced and unvoiced regions based on the energy content. The Silence is detected in the speech signals when the energy content is very minimal.

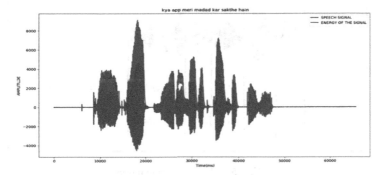

Fig. 3. Automatic speech segmentation using STE

The group delay of the speech signal is shown in Fig. 4. The syllable segmentation of each word is analysed from the peaks that are observed in the group delay function as shown for the words 'meri' and 'madad' in Figs. 6 and 7.

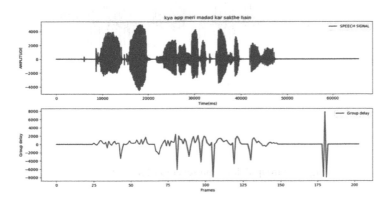

Fig. 4. Speech signal and group delay

Table 2. Comparison of time frames

	Kya	Aap	Me	Ri	Mad	Ad	Kar	Sak	The	Hain
Manual segmentation	29–44	48–62	72–82		90–96	98–104	109–117	121–124	130–138	140–147
Group delay	30–43	49–61	72–80	82–87	92–96	98–101	110–115	120–123	128–127	148–154

The manual segmentation of the speech signal is shown in Fig. 5. Table 2 shows the time frames for the words taken for a sample and we can observe that the time frames are more similar for manual and automatic segmentation described above.

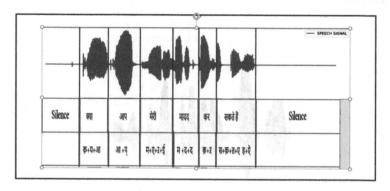

Fig. 5. Manual segmentation

Whenever there is syllable break in the spoken words the group delay goes to negative peak as observed in the Figs. 6 and 7.

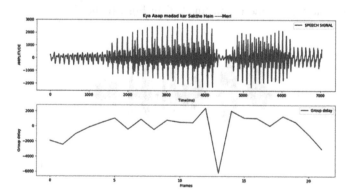

Fig. 6. Group delay representation for the word 'Meri'

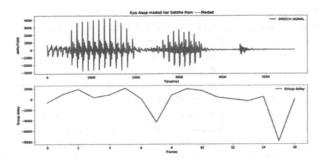

Fig. 7. Group delay representation for the word 'Madad'

5 Conclusion

We have discussed a technique to segment the Hindi speech signals based on time and spectral domain features and compared with manual segmentation techniques. The experimental analysis is carried out for 10 sentences collected from 5 speakers. The proposed method segments the speech signals automatically based on group delay and shows an improvement over manual segmentation. Further, this technique can be applied to continuous speech signals and the recognition rate of these segmented speech signals can be compared with manual segmented speech signals in any ASR system.

Compliance with Ethical Standards
✓ All author states that there is no conflict of interest.
✓ We used our own data.
✓ Animals/human are not involved in this work.

References

1. Nagarajan, T., Murthy, H.A.: Sub band-based group delay segmentation of spontaneous speech into syllable-like units. EURASIP J. Appl. Signal Process. 2614–2625 (2004)
2. Prasad, V.K., Nagarajan, T., Murthy, H.A.: Automatic segmentation of continuous speech using minimum phase group delay functions. Speech Commun. **42**(3), 429–446 (2006)
3. Sarada, G.L., Lakshmi, A., Murthy, H.A., Nagarajan, T.: Automatic transcription of continuous speech into syllable-like units for Indian languages. Sadhana **34**, 221–233 (2009)
4. Frihia, H., Bahi, H.: HMM/SVM segmentation and labelling of Arabic speech for speech recognition applications. Int. J. Speech Technol. **20**, 563–573 (2017)
5. Zhao, X., Shaughnessy, D.O.: A new hybrid approach for automatic speech signal segmentation using silence signal detection, energy convex hull, and spectral variation. In: Canadian Conference on Electrical and Computer Engineering, pp. 145–148 (2008)
6. Sahana, T., Srilasya, N., Priya, K.J., Gupta, D., Vinay, S.: Comparison of different Acoustic Models for Kannada language using Kaldi Toolkit. In: 2018 International Conference on Advances in Computing, Communications and Informatics (ICACCI), pp. 2415–2420 (2018)

7. Priya, K.J., Sree, S.S., Navya, T., Gupta, D.: Implementation of phonetic level speech recognition in Kannada using HTK. In: 2018 International Conference on Communication and Signal Processing (ICCSP), Chennai, pp. 0082–0085 (2018)
8. Sharma, A., Kaur, A.: Automatic segmentation of Punjabi speech signal using group delay. Global J. Comput. Sci. Technol. **13** (2013)
9. Kumar, S., Phadika, S.: Modified segmentation algorithm based on short term energy & zero crossing rate for Maithili speech signal, pp. 21–42 (2016)
10. Geetha, K., Chandra, E.: Automatic phoneme segmentation of Tamil utterances. In: International Conference on Advanced Computing and Communication, pp. 05–07 2015
11. Geetha, K., Vadivel, R.: Syllable segmentation of Tamil speech signals using vowel onset point and spectral transition. Autom. Control Comput. Sci. **52**, 25–31 (2018)
12. Panda, S.P., Nayak, A.K.: Automatic speech segmentation in syllable centric speech recognition system. Int. J. Speech Technol. **19**, 9–18 (2016)

Efficient Ranking Framework for Information Retrieval Using Similarity Measure

Shadab Irfan[✉] and Subhajit Ghosh

Galgotias University, Greater Noida, Uttar Pradesh, India
shadab710@gmail.com,
subhajit.ghosh@galgotiasuniversity.edu.in

Abstract. The information on the web is increasing day by day and to manage such vast amount of information is really a difficult task. The user finds it really hard to capture the desired information as per their need and maximum amount of time is spent in framing proper query and filtering the resultant web pages. The search engine plays a major role in filtering the information and ranking the desired result. The quest for accurate information is still a dream and in this regard this paper presents an approach that tries to optimize the ranking algorithm by employing document clustering and similarity measures. In this paper we present an outline of different ranking algorithms and proposed an approach where PageRank algorithm is optimized by using document clustering. It also employs content mining along with structural mining that help to reduce the computational complexity of the algorithm and thereby diminish the time in performing the ranking of the web pages.

Keywords: Information retrieval · Web mining · Ranking algorithms · Document clustering · Similarity measure

1 Introduction

The explosion of data that is emerging from social networking sites is creating a major hindrance in the searching and retrieval of information. The users are faced with a challenging task at every step to find the relevant information that quest their thirst for accurate information. Though the World Wide Web is flooded with voluminous data but still the lame users have to struggle hard in order to get the needed information. The time required to retrieve the information may take several minutes on the basis of query entered by the users. The role of Information Retrieval system is to facilitate the user in finding the desired information as per the need of the query. The core objective that should be fulfilled by the system is to save the user time in searching for the desired information. Precision and Recall are two important measures that help to estimate the relevance of the Information Retrieval System.

The search engine plays a major role in satisfying the requirements of the user where the information is extracted from the repository on the basis of some matching function which targets the query. The information provided to the user is normally ranked using various ranking algorithm so that the user can get their information easily. Using a single measure for ranking is not sufficient to achieve better results so different

© Springer Nature Switzerland AG 2020
S. Smys et al. (Eds.): ICCVBIC 2019, AISC 1108, pp. 1344–1354, 2020.
https://doi.org/10.1007/978-3-030-37218-7_141

measures can be incorporated for obtaining good results. The approach is to inculcate content mining and structural mining together. Filtering data on the basis of content similarity is being employed which are clustered together. After clustering, on basis of the similar pattern data PageRank algorithm is employed. The approach being used helps to reduce the computation complexity of the process thereby decreasing the processing time.

The basic outline of the paper is specified as. Section 2 provides a brief outline about Information Retrieval while Sect. 3 covers an overview of Web Mining. Section 4 provide a glimpse of various ranking algorithm that are used for ranking web pages and Sect. 5 gives information about similarity measure and clustering approach. Section 6 presents the proposed approach for ranking the web pages while Sect. 7 is conclusion part.

1.1   Related Work

Various methods are being proposed in order to increase the efficiency of the ranking algorithm in the information retrieval process. Some of the works undertaken in this respect are reviewed properly.

Alam [1] et al. throw light on search result clustering techniques and define various ways in which the documents can be clustered together like graph based and hybrid methods. Steinbach [2] et al. specifies various measures by which the quality of the clusters can be measured like entropy and F measure. LaTorre [3] et al. present different criteria for clustering like structure, usage and semantic based approach. Rafi [19] et al. proposed an approach which is based on topic maps which are used for linking relevant knowledge. The method is quite effective for clustering the documents.

2   Information Retrieval

The World Wide Web is heavily loaded with vast amount of information which is normally unstructured. The search engines perform the task of returning the highly relevant pages to the user on the basis of query entered by the user. The queries entered by lame user are normally vague and the pages which are retrieved are not relevant in most of the cases and the user has to formulate the query again and again in order to get the much needed information. The users normally traverse the pages to find the most relevant matter. In order to ease the task of information retrieval document clustering can be employed which help to group relevant information together thus helping the users in getting the much needed result [1].

In the area of information retrieval and text mining document clustering plays a major role by analyzing and grouping documents on the basis of similarity so that the most similar documents can be found at one place and overall help in increasing the precision and recall of the information retrieval system [2].

One of the major roles of Information Retrieval is to find the relevant information. Normally IR accepts query from the user and by employing some matching function provide the user with relevant information. The information that is provided is ranked by using some ranking algorithm but still it comprises of non-relevant data that create hindrance for the user in filtering the needed information. Clustering in Information Retrieval play a major role and help in associating the documents in specific order and assist in extracting similar interesting patterns [3].

3 Web Mining

Web Mining help in processing the information on the basis of content and links. It performs various processing tasks in order to retrieve desired information from the web. On the basis of the process used by it in extracting relevant information it is basically characterized into Web Content Mining (WCM), Web Structure Mining (WSM), and Web Usage Mining (WUM). On one hand WCM make use of content in filtering the information, on the other hand WSM employs link structure among the web pages which are connected with each other to retrieve information. WUM normally make use of log files to study the pattern of the user and retrieve information accordingly [4]. Data can be analyzed on the basis of browsing pattern and other similar measure which help in the analysis and enhance the security of the data [5, 6].

4 Ranking Algorithm

Ranking Algorithms are considered as the backbone of the Information Retrieval System. A number of measures are used for ranking web pages like content of the page, structural links among pages and log information. The links between the pages play an important role in ranking web pages. More important pages have high link within it and signify the ranking of the pages [7]. In order to extract relevant information from the web pages, ranking is done for easy retrieval. Similarity measure should not be the only criteria for ranking so links among web pages are also used for ranking process. Log information extracted from query logs is also helpful in ranking process [8]. Ranking play an important role in the Information Retrieval System and is one of the major driving force for ranking the web pages. Not only links help in deciding the importance of a page but the content is also an important factor to be considered [9].

Various Ranking algorithms came into existence that plays a major role in ranking the web pages. PageRank is one of the important algorithms which is based on hyperlink structure and capable of managing huge volume of data [10]. The PageRank not only take into account local topology but also consider global topology of the web. It work by counting the links to a particular page and consider it important on basis of link attached to it [11]. Hypertext Induced Topics Search (HITS) [12] consider web pages as hub and authority and calculate the score on the basis of weight assigned to them. Weighted Page Rank Algorithm [13] consider the popularity of the web page and distribute the rank score on basis of it. The EigenRumor Algorithm [14] is designed for blogging site and eigenvector calculation is performed for the blog entries. The Distance Rank Algorithm devised [15] work on logarithmic distance that is shortest and ranking is done on them The Time Algorithm [16] work on the time frame which user normally spent on visiting a web page.

5 Document Clustering and Similarity Measure

It has been found that web clustering is one of the important measures which help to group related web pages on basis of some similarity measure. The pages in the clusters are tightly coupled within each other that help in finding similar pattern objects and overall help in the retrieval and analysis process. Various clustering methods are used like semantic clustering, structure and usage based clustering [17]. It has been found out that clustering of documents can help to boost the retrieval process, the precision and recall of the documents can be enhanced if for retrieval of documents clustering is used. Cosine, Pearson correlation and extended Jaccard are some of the common similarity measure which can help in clustering of the documents [18].

Similarity measure is one of the important task for clustering. Different measures can be employed for grouping similar web pages into clusters which can then be used for pattern recognition and other retrieval task. Accuracy of the clustering is based on the similarity measure. Euclidean and cosine similarity are some of the measures that can be used for clustering objects [19]. For ranking the document, the query is represented as a vector, the search engine adjusts the term on the basis of importance and ranking with respect to the query is given to the users. The query and document can be represented by TFIDF vectors and later cosine similarity is considered [20]. The tf-idf score is represented as

$$\text{tf} - \text{idf}_{t,d} = \text{tf}_{t,d} \times \text{idf}_t. \tag{1}$$

After calculating the tf-idf score, cosine similarity between query vector and document vector is calculated as given [21].

$$\text{score}(q, d) = \frac{\vec{V}(q) \cdot \vec{V}(d)}{|\vec{V}(q)||\vec{V}(d)|}. \tag{2}$$

6 Proposed Approach

The proposed approach help to ease the task of information retrieval using both link structure and content of the nodes, as single measure is not sufficient for ranking of the web pages. In the present work we are incorporating content similarity measure approach before applying PageRank algorithm. On the basis of the query Q entered by the user N amount of pages are retrieved which are then filtered by applying different conditions and after applying the cosine similarity measure are clustered into groups. The PageRank algorithm is then applied on the clustered web pages that help in reducing the computational complexity and final ranked nodes are retrieved. The overall flow of the proposed approach is depicted in Fig. 1.

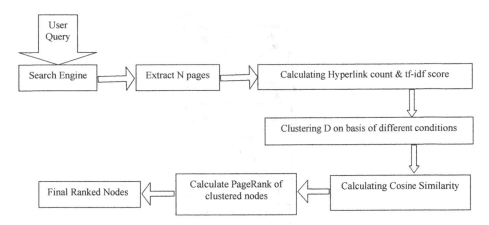

Fig. 1. Proposed approach model

The proposed algorithm accept the user query and present ranked web pages by following proposed steps of the algorithm and is depicted as -

Input: User Query
Output: Ranked Web Pages/documents
Steps:
1. Enter query Q consisting of 'n' terms that extract 'N' pages by removing stopwords.
2. Count number of terms 'n' of query 'Q'.
3. Calculate hyperlink count.
4. Calculate tf-idf score.
5. Webpages/ documents are clustered on basis of following criteria-
    Case I: If all term 'n' of 'Q' appear in Di where i=1, 2, 3,...M group them in Ci, where i=1, 2, 3.
    Case II: If ' n-1 > = D < = n/2 ' group them in Ci, where i=1, 2, 3.
    Case III: If no term 'n' match group them in cluster Ci, where i=1, 2, 3.
6. Calculate Cosine similarity between Di where i=1, 2, 3,...M and Q which are grouped into clusters.
7. Calculate PageRank of the web pages grouped in clusters.
8. Finally ranked web pages are retrieved.

To implement the proposed approach an example is taken which consist of seven web pages A, B, C, D, E, F, and G having 500, 300, 200, 400,300, 300 and 200 words respectively. The directed graph which is shown in Fig. 2 presents seven web pages which are interlinked among each other.

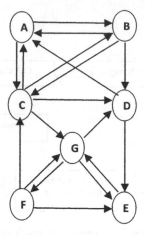

Fig. 2. Hyperlink structure of web pages

The frequency of the terms of the specified query in the various web pages is shown in Table 1.

Table 1. Term frequency of web pages

	A	B	C	D	E	F	G
Information	20	8	8	10	20	0	25
Retrieval	10	12	13	15	10	0	0
Evolutionary	12	0	6	8	5	0	0
Computation	8	0	2	6	2	6	0

The hyperlink count for the respective nodes on the basis of hyperlink structure is calculated and shown in Table 2.

Table 2. Hyperlink count for the nodes

Node	Inlink	Outlink	Hyperlink count
A	3	2	5
B	2	3	5
C	3	4	7
D	3	2	5
E	3	1	4
F	1	3	4
G	3	3	6

For all the nodes the tf-idf score is calculated and the nodes are clustered following the rules of the algorithm. It is found that node A, D, C and E contain all the words of the query and hence Case I is applied and they are clustered in one group and on node B Case II is applied which is placed in second cluster while the two nodes F and G are left out according to Case III. After clustering the nodes, cosine similarity is calculated for Cluster 1 and Cluster 2 as shown in Tables 3 and 4.

Table 3. Cosine similarity of cluster 1

Cluster 1	
cos(Q,A)	0.97755
cos(Q,D)	0.925222
cos(Q,C)	0.88167
cos(Q,E)	0.842342

Table 3 present the cosine similarity for nodes A, D, C and E clustered together. Table 4 present the cosine similarity for node B.

Table 4. Cosine similarity of cluster 2

Cluster 2	
cos(Q,B)	0.48662

The Fig. 3 represents the cosine similarity of the final nodes clustered together.

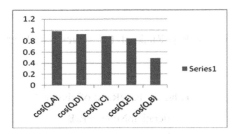

Fig. 3. Cosine similarity of clustered documents

After finding the similarity among the nodes, finally the PageRank formula is applied as given in equation on the clusters and ranking of the nodes are done as shown in Tables 5 and 6.

$$PR(A) = (1 - d) + d(PR(T1)/C(T1) + \ldots + PR(Tn)/C(Tn)) \qquad (3)$$

The Table 5 shows the PageRank of the nodes which are clustered together in Cluster 1.

Table 5. PageRank of Cluster 1

Iteration\Node	PR(A)	PR(C)	PR(D)	PR(E)
0	1	1	1	1
1	1.070833333	1.141666667	0.959270833	1.124356771
2	1.0836276	1.1772084	0.9668235	1.1275666
3	1.0943901	1.1817825	0.9677954	1.1279797
4	1.0957752	1.1823711	0.9679205	1.1280329
5	1.0959534	1.1824469	0.9679366	1.1280397
6	1.0959764	1.1824566	0.9679387	1.1280406
7	1.0959793	1.1824579	0.967939	1.1280407
8	1.0959797	1.182458	0.967939	1.1280407
9	1.0959797	1.1824581	0.967939	1.1280407
10	1.09597975	1.1824581	0.967939	1.1280407
11	1.09597975	1.1824581	0.967939	1.1280407
12	1.09597975	1.1824581	0.967939	1.1280407
13	1.09597975	1.1824581	0.967939	1.1280407
14	1.09597975	1.1824581	0.967939	1.1280407
15	1.09597975	1.1824581	0.967939	1.1280407

The Table 6 shows the PageRank of the nodes which are clustered together in Cluster 2.

Table 6. PageRank of cluster 2

Iteration\Node	PR(B)
0	1
1	0.7875
2	0.7875
3	0.7875
4	0.7875
5	0.7875
6	0.7875
7	0.7875
8	0.7875
9	0.7875

(*continued*)

Table 6. (*continued*)

Iteration\Node	PR(B)
10	0.7875
11	0.7875
12	0.7875
13	0.7875
14	0.7875
15	0.7875

After calculating the PageRank of the nodes it has been found that after Iteration 10 the value are same in all the iteration so the process can be stopped. The final rank of the nodes is shown in Table 7.

Table 7. Ranked nodes

Ranked nodes
C
E
A
D
B

It has been found that the following strategy help the user in performing lesser calculation for computing the PageRank of the nodes thus reducing the computational complexity of the process and overall time. We are getting the result in less number of iteration thus minimizing the cost and time also.

Table 8. PageRank of all nodes

Iteration \Node	PR(A)	PR(B)	PR(C)	PR(D)	PR(E)	PR(F)	PR(G)
0	1	1	1	1	1	1	1
1	1.070833333	0.817604167	1.120092014	0.9030074	1.100444812	0.433333333	1.446175421
2	1.003452212	0.814486743	0.930016212	0.988149391	1.102490972	0.559749703	1.44334152
3	0.99836318	0.771932797	0.951614393	0.979879115	1.133991137	0.558946764	1.474478775
4	0.987380975	0.771854973	0.946697406	0.987634427	1.145881868	0.567768986	1.486040666
5	0.989610073	0.77175748	0.95011678	0.99160929	1.153346683	0.571044855	1.494040539
6	0.991998383	0.773499129	0.952553442	0.994887179	1.157934579	0.573311486	1.499100253
7	0.994402744	0.775038772	0.954653739	0.99720331	1.160994733	0.574745072	1.502553879
8	0.996269645	0.776278519	0.956204617	0.998862661	1.163084667	0.575723599	1.504937134
9	0.997655692	0.77719715	0.957331215	1.000037597	1.16453652	0.576398855	1.506601934
10	0.998654721	0.77786114	0.958135255	1.000868279	1.165552575	0.576870548	1.507770086
11	0.99936675	0.77833461	0.958705664	1.001454617	1.166266392	0.577201524	1.508591819
12	0.999871306	0.778670258	0.959108977	1.001868246	1.166768785	0.577434349	1.509170524
13	1.000227902	0.778907516	0.95939372	1.002159943	1.16712269	0.577598315	1.509578308
14	1.000479604	0.779074997	0.959594604	1.002365623	1.1673721	0.577713854	1.50986573
15	1.000657159	0.779193146	0.959736276	1.00251064	1.167547904	0.57779529	1.510068343

In contrast to the above proposed method if we follow the original PageRank we are unable to get the similar value until iteration 15, while this is achieved after iteration 10 on following the proposed methodology. The Table 8 shows the result of PageRank of all the nodes by following the standard method of PageRank calculation.

By observing the computation it has also been found that the PageRank of Node G is high, though it contains only single term of the query thus showing inadequacy to capture the true value. It also signifies that alone structural links are not sufficient for calculating the rank of the nodes.

7 Conclusion

The paper presents an approach for ranking the web pages by using not only the structural links among the pages but also incorporating content measure for high precision results. It has been found out that a single measure for ranking documents is not sufficient to provide accurate result, and the precision value can be enhanced if more than one measure is used for ranking the web pages. In the proposed work we not only applied document clustering and similarity measure for ranking web pages but also used link structure among the pages. It has been observed that though the hyperlink count for a particular node is high but still due to lack of content it is inefficient for ranking purpose.

The proposed methodology helps in overall reducing the computational complexity of PageRank. As it is known that the process of information retrieval is a complex task which need sufficient time for retrieval, so by using this process we are undertaking only relevant nodes and exempt the nodes which lack content, thus decreasing the retrieval time for accessing relevant information. In future, further work can be done on this procedure by employing evolutionary computation techniques which help in clustering the documents and will help in increasing the efficiency of the proposed approach.

References

1. Alam, M., Sadaf, K.: A review on clustering of web search result. In: Advances in Computing & Information Technology, AISC, vol. 177, pp. 153–159. Springer, Heidelberg (2013)
2. Steinbach, M., Karypis, G., Kumar, V.: A comparison of document clustering techniques (2000)
3. Leuski, A., Allan, J.: Improving interactive retrieval by combining ranked lists and clustering (2000)
4. Sheshasaayee, A., Thailambal, G.: Comparison of classification algorithms in text mining. Int. J. Pure Appl. Math. 116(22), 425–433 (2017)
5. Jain, R., Purohit, G.N.: page ranking algorithms for web mining. Int. J. Comput. Appl. 13(5), 0975–8887 (2011)
6. Srivastava, J., Cooley, R., Deshpande, M., Tan, P.-N.: Web usage mining: discovery and applications of usage patterns from web data. In: ACM SIGKDD, January 2000

7. Wang, Z.: Improved link-based algorithms for ranking web pages. In: WAIM. LNCS, vol. 3129, pp. 291–302. Springer, Heidelberg (2004)
8. Yates, R.B., Hurtado, C., Mendoza, M.: Query clustering for boosting web page ranking. In: AWIC 2004. LNAI, vol. 3034, pp. 164–175. Springer, Heidelberg (2004)
9. Irfan, S., Ghosh, S.: A review on different ranking algorithms. In: International Conference on Advances in Computing, Communication Control and Networking IEEE ICACCCN (2018)
10. Brin, S., Page, L.: The anatomy of a large-scale hypertextual web search engine. In: Proceedings of the Seventh International World Wide Web Conference (1998)
11. Masterton, G., Olsson, E.J.: From impact to importance: the current state of the wisdom-of-crowds justification of link-based ranking algorithms. Philos. Technol. **31**, 593–609 (2018)
12. Kleinberg, J.M.: Authoritative sources in a hyperlinked environment. J. ACM **46**, 604–632 (1999)
13. Xing, W., Ghorbani, A.: Weighted pagerank algorithm. In: Proceedings of the Second Annual Conference on Communication Networks and Services Research. IEEE (2004)
14. Fujimura, K., Inoue, T., Sugisaki, M.: The eigenrumor algorithm for ranking blogs. In: WWW (2005)
15. Bidoki, A.M.Z., Yazdani, N.: DistanceRank: an intelligent ranking algorithm for web pages. Inf. Process. Manag. **44**, 877–892 (2007)
16. Jiang, H.: TIMERANK: a method of improving ranking scores by visited time. In: Proceedings of the Seventh International Conference Machine Learning and Cybernetics, Kunming, 12–15 July 2008 (2008)
17. LaTorre, A., Pena, J.M., Robles, V., Perez, M.S.: A survey in web page clustering techniques (2019)
18. Sandhya, N., Govardhan, A.: Analysis of similarity measures with wordnet based text document clustering. In: Proceedings of the InConINDIA. AISC, vol. 132, pp. 703–714. Springer, Heidelberg (2012)
19. Rafi, M., Shaikh, M.S.: An improved semantic similarity measure for document clustering based on topic maps (2013)
20. Markov, Z., Larose, D.T.: Data Mining the Web: Uncovering Patterns in Web Content, Structure, and Usage. Wiley, Hoboken (2007)
21. Manning, C.D., Raghavan, P.: An Introduction to Information Retrieval. Cambridge University Press, Cambridge (2008). Preliminary draft© 2008

Detection and Removal of RainDrop from Images Using DeepLearning

Y. Himabindu[(✉)], R. Manjusha, and Latha Parameswaran

Department of Computer Science and Engineering, Amrita School
of Engineering, Coimbatore, Amrita Vishwa Vidyapeetham, Coimbatore, India
hima.harshu@gmail.com,
{r_manjusha,p_latha}@cb.amrita.edu

Abstract. Dynamic climatic conditions like rain, affects the performance of vision algorithms which are used for surveillance and analysis tasks. Removal of rain-drops is challenging for single image as the rain drops affect the entire image and makes it difficult to identify the background affected by the rain. Thus, removal of rain from still pictures is a complex and challenging task. The rain drops affect the visibility and clarity of the image which makes it difficult to read and analyze the information present in the image. In this paper, we identified and restored the rain drop affected regions using a deep learning architecture.

Keywords: Resnet · Rain removal · Denoising

1 Introduction

Digital image processing is one of the thrust areas of research. Image processing techniques are used to improve images or to draw out some information from the images.

1.1 Applications of Image Restoration

A few applications of image restoration are discussed here:

1. In the field of medical imaging such as Brain tumor identification, X-ray imaging, MRI, etc.
2. Pattern Recognition.
3. One of the major areas is surveillance. Traffic signals, Shopping complex and many more places where surveillance is essential has camera deployed outside. The videos or images captured from such cameras may be affected with noise such as rain or snow.

Rain removal is similar to image enhancement. Main challenge involved is identifying the regions affected by the rain and restoring those without loss of information. Differentiating raindrops from objects is one of the challenges involved. As mentioned, removing rain drops from images is complex than removing rain drops from video. This is because the video is a collection of frames and using the frames the correlation

© Springer Nature Switzerland AG 2020
S. Smys et al. (Eds.): ICCVBIC 2019, AISC 1108, pp. 1355–1362, 2020.
https://doi.org/10.1007/978-3-030-37218-7_142

between the consecutive frames can be calculated and further the rain drops direction and regions affected by it can be identified and removed accordingly. Differentiating the rain streaks from that of background with respect to object information like orientation and structure is complex and is a major challenge while trying to restore a rain affected image.

2 Literature Survey

The indices of neighboring samples were represented using parameters p and p + 1. The slope with respect to p is calculated using mod (fp, fp + 1). The non- zero components count that satisfies the parameter p is denoted by operator {}. To extract basic data i.e., to straighten the details and sharpen the fundamental edges the strategy was joined with a general limitation, that is, there existed an auxiliary likeness between input signal g and the outcome f. The pixel splendor was preserved and managed using contrast enhancement. Histogram extending was utilized to move the pixel esteems to fill the splendor bringing about high contrast. In [2] authors have used a mixed strategy. The image that was given as input, was converted to YChCr color space to acquire precise gray component.

The algorithm began about the centroid and iterated until a neighborhood ideal was found.

$$J = \sum_{j-1}^{k} \sum_{i-1}^{n} \left\| x_i^{(j)} - c_j \right\|^2 \tag{2}$$

Equation 2 is used to measure the distance between clusters formed. The clusters were discovered by the peaks and valleys from histogram. From this bunches they acquired the sliced areas and by consolidating the boundary extraction the image highlights was obtained. To get the accurate number of clusters, elbow algorithm was considered. The peak value of the histogram was obtained from the number of clusters. After the potential raindrop regions were recognized by the saliency method, a strategy for check was put in to attest if the zone contains a raindrop or not. The details regarding the shape of the drop was acquired using Hough circle transform. K-means was used for classifying the regions. The edges and boundary information was acquired using canny edge detection. By separating the background details the rain drops were identified. The raindrops were evacuated using the median filter.

In [3], Deep CNN was the base to handle the rain drop problem. The mapping between the rainy and clean layers is learnt instinctively. To yield a better quality of the restored image, both removal of rain and enhancement of the image were performed. Low frequency layer was termed ad base layer and the high frequency layer was termed as detail layer. The input for the proposed architecture was the detail layer i.e., high frequency layer. Image enhancement operation was performed to improve the quality of image. The network was trained on the detail layer obtained after splitting the image. The image that is free from raindrops was obtained by combining the base layer with obtained detail layer.

In [4], layer decomposition was the proposed solution for rain drop removal. The image O is split into two layers. B represents rain free layer and R represents layer with rain drops.

$$O = B + R \tag{3}$$

The split layers are assigned with priors based on patch. Using the priors traced by Gaussian mixture models the rain drops of various directions and sizes were handled.

In [5], the proposed strategy does not need any kind of statistical information with respect to pixel. This strategy is reasonable for both images just as videos as it does not require any arrangement before recognition. Rain free image was obtained by determining the intensity changes formed by rain on the pixels. The acquired reference picture gave draft forms yet it did not have those edges which were influenced by raindrops. The affected regions were refurbished using a guided filter for preserving the edges and the acquired reference image.

In [6], authors introduced an approach that is free from explicit motion analysis. The technique probed the transient changes and chose pixels affected by rain.

In [7], a new model with two elements was created by collaborating binary map with existing model. The rain drops were represented by one element and another element represented various shapes and directions of rain drops.

A multi-task Convolutional network JORDER (joint rain detection and removal) was constructed. JORDER (joint rain detection and removal) identified the raindrop influenced locales and a contextualized expanded network, which was used for separating the raindrops.

In [8] authors have made use of an orientation filter for identifying the rain drops. Remaining raindrops were identified using entropy maximization followed by removal of background information. Here the deterioration depended on the direction of rain drops. A low pass filter was used to estimate the background. The backdrop information in pictures [9] was acquired by high recurrence concealment and direction. It was noticed that at least one color channel of the pixels influenced by raindrops had low intensity value. The thickness of mist was determined based on the prior and the image was reconstructed.

In [10], the image was split into two layers. One layer was represented by B (rain free layer) and another layer was represented by R (rainy layer). This separation was achieved by constituting three priors on layer B and layer R. The three priors included smoothening the rain streaks in B, preserving the object details in B and hide the non-rainy details from R. From the obtained rainy layer, the regions populated by rain drops were identified using histogram and the direction of the rain was estimated by using canny edge detector the slope of the rain drop direction is calculated. The rain patches were randomly selected and restored.

In [11] a framework constituting deep learning and layer priors was introduced.

The framework included a CNN of residual type for training the images.

Along with this they used or incorporated the concept of total variation energy which would enhance the recovered image.

From the literature survey it is observed that rain removal and restoration is still an open challenge and hence can be explored.

3 Proposed Work

The first step performed is background subtraction i.e., separating the rain drop layer and detail layer. The objective of splitting image into the two layers was partially achieved by using high-pass and low-pass filters. Image impainting is used for filling those rain affected regions by taking the mean of surrounding pixels. Further, Applying the L0 gradient smoothening gave a partial rain free image. But the result was not as expected as the images obtained as output was in a blurred state. Hence Using the ResNet Architecture, the objective of removing rain drops from images is achieved. The number of layers which produced a good result was 26 layers; ResNet Architecture helps in handling the gradient vanishing problem. Here we obtained the base layer using a low pass guided filter and take the difference of base layer and image to obtain the detail layer. The obtained detail layer has the required rain drops and object structures which make sure that background information is not lost.

$$\text{Image} = \text{Base layer} + \text{Detail layer}$$

The detail layer is taken as input to the network and the training of images is performed. The objective function can be defined as

$$\mathcal{L} = \sum\nolimits_{i=1}^{N} \left\| f(X_{i,\text{detail}}, W, b) + X_i - Y_i \right\|_F^2 \tag{5}$$

N denotes the number of training images, ResNet is denoted by f(.), the network parameters which are supposed to be learned are W and b. To obtain the detail layer, firstly the base layer is calculated using a low pass guided filter and then the image is separated as base and detail layer. As the dataset has color images, we take c i.e., the number of channels as 3. The filter size for the first layer is $c \times f1 \times f1 \times m1$, f represents the filter size and m represents the feature maps. Numbers of layers are represented by L. For the layers till $L - 1$, the filter size is $m1 \times f2 \times f2 \times m2$. The size for last layer is $m2 \times f3 \times f3 \times c$. Using this we calculate the residual value of the image.

Figure 1 gives a brief description about the workflow used in the proposed method of identifying and removing the raindrops from images.

3.1 Results and Discussions

From [15], Stochastic Gradient Descent has been used to minimize the objective function. A dataset of 14,000 pairs of rain images is trained. The testing and training is in the ratio 13: 7. The parameters are set namely, number of layers is taken as 26 and the learning rate is 0.1, the iterations are taken at 10k and 210k. The filter size is set to 3 and the mapping parameter is set to 16. The number of channels, c is set to 3. A guided filter of radius 15 is used.

Table 1 gives the SSIM, Structural similarity index for Figs. 1 and 2 obtained as a result of proposed method. It basically requires two images, one is original image and the other one is a processed image. Using the structures visible it calculates the

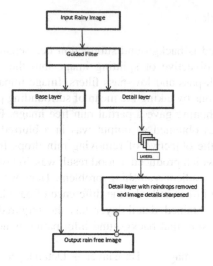

Fig. 1. Workflow of proposed method

Table 1. Table with SSIM values

Figures	SSIM value
Figures 2 and 3	0.87
Figures 4 and 5	0.926
Figures 6 and 7	0.948
Figures 8 and 9	0.932

Fig. 2. Input rainy image [7]   Fig. 3. Output normal image

perceptual difference between the images and gives the value accordingly. The SSIM value ranges between −1 and 1. Higher the SSIM value, the more similar are the images. The higher the SSIM is, the closer the image is to the ground truth. Figures 1 and 2 are the input and output images respectively. Table 2 gives the average SSIM for different network sizes. Increasing the number of hidden layers gives a good SSIM value. A good SSIM value means the images are quite similar.

Figures 1 and 2 are the input and output images respectively.

Table 2. Table for network sizes

Layers	m1 = m2 = 16	m1 = m2 = 48
12	0.876	0.88
26	0.916	0.920
36	0.918	0.924

4   Results

Figures 4 and 6 are given as input to the proposed method and the figures obtained as output are Figs. 5 and 7. In Figs. 3 and 4, we can see one rainy image which is given as input to the proposed method and the other which is obtained as output.

The above results are obtained for the proposed approach with 26 layers at a learning rate of 0.1.

When the layers were 12 the obtained images were obtained as shown in Fig. 9.

Fig. 4. Input rainy image [8]

Fig. 5. Output normal image

Fig. 6. Input rainy image [1]

Fig. 7. Output normal image

Fig. 8. Input rainy image [11]   **Fig. 9.** Output normal image

5 Conclusion

Using the proposed approach, the input image is split into detail layer and base layer benefits in preserving the object details. In this paper, the pairs of clean and rainy images are used for network learning as the ground truth is not taken. Firstly, we tried to achieve the objective by using classic techniques. We calculated the mean of surrounding pixels and used the value to fill the raindrop affected regions. In the resulting image the rain drops were removed partially and the quality of image was not as expected. Hence used the deep learning architecture, where the input image has the object details and rain streaks. This approach helped us to preserve the background information of the image. Though it produced good results still the architecture can be improved for differentiating the rain streaks and object details which have similar frequency.

Compliance with Ethical Standards
   ✓ All authors declare that there is no conflict of interest.
   ✓ No humans/animals involved in this research work.
   ✓ We have used our own data.

References

1. Manu, B.N.: Rain removal from still images using L0 gradient minimization technique. In: 2015 7th International Conference on Information Technology and Electrical Engineering, ICITEE. IEEE (2015)
2. Kanthan, M.R., Sujatha, S.N.: Rain drop detection and removal using K-Means clustering. In: 2015 IEEE International Conference on Computational Intelligence and Computing Research, ICCIC. IEEE (2015)
3. Fu, X., et al.: Clearing the skies: a deep network architecture for single-image rain removal. IEEE Trans. Image Process. **26**(6), 2944–2956 (2017)
4. Li, Y., et al.: Rain streak removal using layer priors. In: Proceedings of the IEEE Conference on Computer Vision and Pattern Recognition (2016)

5. Xu, J., et al.: Removing rain and snow in a single image using guided filter. In: 2012 IEEE International Conference on Computer Science and Automation Engineering, CSAE, vol. 2. IEEE (2012)
6. Subhani, M.F., Oakley, J.P.: Low latency mitigation of rain induced noise in images, p. 13 (2008)
7. Yang, W., et al.: Deep joint rain detection and removal from a single image. In: Proceedings of the IEEE Conference on Computer Vision and Pattern Recognition (2017)
8. Dodkey, N.: Rain streaks detection and removal in image based on entropy maximization and background estimation. Int. J. Comput. Appl. **164**(11), 17–20 (2017)
9. He, K., Sun, J., Tang, X.: Single image haze removal using dark channel prior. IEEE Trans. Pattern Anal. Mach. Intell. **33**(12), 2341–2353 (2011)
10. Zhu, L., et al.: Joint bilayer optimization for single-image rain streak removal. In: Proceedings of the IEEE International Conference on Computer Vision (2017)
11. Luo, Y., Xu, Y., Ji, H.: Removing rain from a single image via discriminative sparse coding. In: Proceedings of the IEEE International Conference on Computer Vision (2015)
12. Fu, X., et al.: Removing rain from single images via a deep detail network. In: Proceedings of the IEEE Conference on Computer Vision and Pattern Recognition (2017)
13. Schaefer, G., Stich, M.: UCID: an uncompressed color image database. In: Storage and Retrieval Methods and Applications for Multimedia 2004, vol. 5307. International Society for Optics and Photonics (2003)
14. Manjusha, R., Parameswaran, L.: Design of an image skeletonization based algorithm for overcrowd detection in smart building. In: Pandian, D., Fernando, X., Baig, Z., Shi, F. (eds.) International Conference on ISMAC in Computational Vision and Bio-Engineering, vol. 30, pp. 615–629. Springer, Cham (2018)
15. Bai, Z., et al.: Algorithm designed for image inpainting based on decomposition and fractal. In: 2011 International Conference on Electronic and Mechanical Engineering and Information Technology, EMEIT, vol. 3. IEEE (2011)
16. Liu, R., et al.: Deep layer prior optimization for single image rain streaks removal. In: 2018 IEEE International Conference on Acoustics, Speech and Signal Processing, ICASSP. IEEE (2018)

A Review on Recent Developments for the Retinal Vessel Segmentation Methodologies and Exudate Detection in Fundus Images Using Deep Learning Algorithms

Silpa Ajith Kumar[⊠] and J. Satheesh Kumar

Dayananda Sagar College of Engineering, Bangalore, India
silpaajithkumar@gmail.com, jsatheeshng1@gmail.com

Abstract. Retinal image analysis is considered as a well-known non-intrusive diagnosis technique in modern opthalmology. The pathological changes which occurs due to hypertension, diabetic retinopathy and glaucoma can be viewed directly from the blood vessels in retina. The examination of the optic cup-to-disc ratio is the main parameter for detecting glaucoma in the early stages. The significant areas of the fundus images are isolated using the segmentation techniques for deciding the value of cup-to-disc ratio. The deep learning algorithms, such as the Convolutional Neural Networks (CNNs), is often used technique for the analysis of fundus images. The algorithms using the concepts of CNNs can provide better accuracy for the retinal images. This review explains the recent techniques in deep learning relevant for the analysis of exudates.

Keywords: Blood vessel segmentation · Convolutional neural networks · Deep learning · Diabetic retinopathy · Exudate detection · Fundus images · Machine learning

1 Introduction

The common cause for vision impairment in adults is due to the diabetic retinopathy (DR) [1]. The presence of the exudates on the retina plays a key role in the vision loss. The detection of exudates is an important diagnosis step for diabetic retinopathy. The automatic detection of exudates was analyzed in the literatures done before [2, 3]. This review paper discusses about the exudates which is considered as the main reason for the existence of Macular Edema and DR in retinal images.

Deep learning methods have great significance in recent years since it is able to solve complex problems efficiently [4]. The deep learning techniques use the feature learning techniques from raw data [5]. CNNs are implemented to detect the DR in fundus images which helps in the applications of deep learning in the pixel based classification and labeling. The automatic exudate detection methods are classified into two major classifications namely morphological and machine learning methods [6–8].

S. Smys et al. (Eds.): ICCVBIC 2019, AISC 1108, pp. 1363–1370, 2020.
https://doi.org/10.1007/978-3-030-37218-7_143

Sopharak et al. [9] discussed the exudate detection using the morphological reconstruction techniques in the fundus images. In the work done by Sanchezet al. [10, 11] a
statistical mixture model-based clustering method for dynamic thresholding for the
exudate detection was implemented. Fraz et al. [12] in his paper tried the bootstrapped
or bagging decision trees based ensemble classifier for the exudate detection.

The paper discusses the concepts of fundus image processing, available public
datasets and analysis of performance measures for different extraction techniques in
Sect. 2. The deep learning methods and the analysis of different methods used in
exudates detection and segmentation for DR is outlined in Sect. 3. Section 4 concludes
with the summary and enhancements for future research.

2   Fundus Image Processing

2.1   Fundus Photography

Fundus photography records the retinal images and the neurosensory tissues in our
eyes. The images are converted into electrical impulses which the brain easily
understands. Fundus photography uses three types of examination namely: the color
photography, red-free photography and angiography [13]. The techniques of adaptive
optics are used to increase the areas of optic systems. The AO retinal imaging helps in
the classification of normal and diseased images of eyes which are done using spacing,
mosaic and photography cell thickness. The fundus imaging methods are based on the
fundus fluorescein angiography and Optical Coherence Tomography [14].

2.2   Publicly Available Retinal Image Datasets

In this section, the summarization of the public domain datasets like DRIVE, STARE,
MESSIDOR, ORIGA, Kaggle and e-Ophtha is done in detail.

Table 1. Summary of datasets used for blood vessel segmentation and DR detection

Databases	Total no. of images	Fundus camera	Resolution (pixel)	Field of view FOV	Tasks
Digital Retinal Images for Vessel Extraction (DRIVE) [15]	40 [33 Normal images & 7 images contains exudates, hemorrhages and pigment epithelium changes]	Canon CR5	768×564	45°	Blood vessel extraction, exudates, hemorrhages
Structured Analysis of Retina (STARE) [16]	400 (20 images for blood vessel segmentation and 80 images ONH is localized)	Top Con TRV-50	605×700	35°	Blood vessel segmentation, exudates, MicroAneursyms (MA)

(continued)

Table 1. (*continued*)

Databases	Total no. of images	Fundus camera	Resolution (pixel)	Field of view FOV	Tasks
MESSIDOR [17]	1200	Non mydriatic 3CCD camera, (Top Con TRC NW6)	1440 × 960, 2240 × 1488, 2304 × 1536	45°	Diabetic retinopathy (DR) grading and macular edema risk analysis
Kaggle	80000	–	–	–	DR grading
ORIGA [18]	482 (Normal & 168 (Glaucomatous images)	–	–	–	Optic disk, Optic cup and cup-to-disc ratio (CDR)
e-ophtha	463 (148MAs, 233 Normal, 47 exudates, 35 Normal non-exudates)	OPHDIT Tele- medical Network	Exudates	–	Grading of DR lesions

2.3   Analysis of Performance Measures

The segmentation process produces a pixel based outcome. A pixel can be classified either as a vessel or surrounding tissue. Table 1 shows the four different distinctive areas namely (i) the true positive area (ii) the true negative area (iii) the false negative area (iv) the false positive area. In the four cases discussed above, two are considered as classifications and the other two are treated as misclassifications (Table 2).

Table 2. Performance measures for assessment of feature extraction techniques

Measurement parameters	Relevant formulas
Sensitivity or True Positive Fraction or (TPF) or Recall	$\frac{TP}{TP+FN}$
Specificity or True Negative Fraction (TNF)	$\frac{TN}{TN+FP}$
Precision or Positive Predicted Value (PPV)	$\frac{TP}{TP+FP}$
Negative Predicted Value (NPV)	$\frac{TN}{TN+FN}$
Accuracy	$\frac{TP+TN}{TP+TN+FP+FN}$
False Positive Fraction (FPF)	$\frac{FP}{TP+FN}$
False Negative Fraction (FNF)	$\frac{FN}{TP+TN}$
Receiver Operating Characteristic (ROC)	–
Intersection Over Union (IOU)	$\frac{Area\,S\cap M}{Area\,S\cup G}$ 'S' indicates the segmentation of the output and 'M' represents the manual ground truth segmentation
F-score	$2\frac{PPV\times TPF}{PPV+TPF}$

3   Deep Learning Methods

In this section a brief description of deep learning, architectures and techniques is provided for the effective detection of exudates in fundus images. A brief summary of exudate detection is described in Table 3.

3.1   Machine Learning Algorithms

The machine learning algorithms are generally classified as supervised and unsupervised learning algorithms. In supervised learning, a model is presented with a dataset $D = \{x, y\}_{n-1}^{N}$, were 'x' represents the input features and 'y' represents the label pairs. The final output value for any correct input image is predicted with the help of the regression function. The unsupervised learning algorithms can be implemented using various loss functions like reconstruction loss. The principal component analysis and clustering methods are the traditional unsupervised learning algorithms [20].

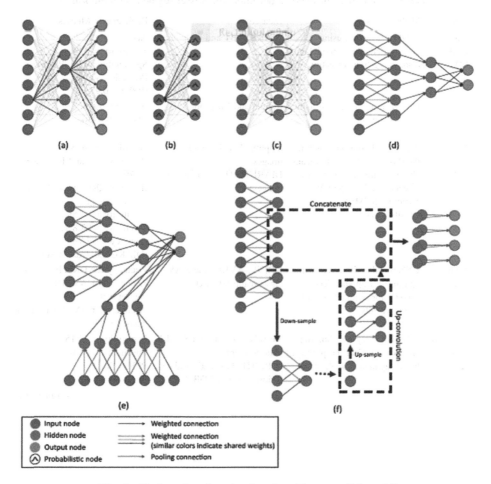

Fig. 1. Various deep learning based architectures [20, p. 64]

3.2 Neural Network Architectures

The neural network architectures (Fig. 1) used for the exudate detection are convolutional neural networks (CNNs), auto encoders (AEs), deep belief networks (DBNs) and recurrent neural networks (RNNs). The CNNs are used for various tasks and comprises mainly of three types of layers namely convolutional, pooling and fully connected layers. The AlexNet, VGGNet, GoogLeNet and ResNet are the CNN models used for the exudate detection. An auto encoder (AE) is usually a single hidden layer network having the identical input and output values. It is used to construct stacked – auto encoder (SAE). The pre-training and fine-tuning are the two phases in the training of a SAE model. The DBN architecture is constructed by cascading the restricted Boltzmann machines. DBN is a probabilistic model which is pre-trained in an unsupervised way and finally fine-tuned with the help of a gradient descent and back propagation algorithms.

Table 3. Summary of exudate detection and vessel segmentation algorithms

Authors	Methodology	Datasets	Performance Metrics
Parham et al. [21]	Exudate detection using 9 layered CNN-based model and color space transformation	DIARETDB1 and e-Ophtha	Accuracy = 98.2%, Sensitivity = 0.99, Specificity = 0.96, Positive Prediction Value (PPV) = 0.97
Chen et al. [22]	Diverse lesion detection using subspace learning	Kaggle and Messidor	AUC = 0.9956, MAP = 0.8236
Long et al. [23]	Algorithm based on dynamic threshold and Fuzzy–C means clustering for hard exudates detection and Support Vector Machine (SVM) for classification	e-ophtha EX database (47 images) DIARETDB1 database (89 images)	Sensitivity = 76.5%, PPV = 82.7% and F-score = 76.7%. (For e-ophtha EX database) overall average sensitivity = 97.5%, specificity = 97.8%, accuracy = 97.7%, (DIARETDB1 database)
Badawi et al. [24]	CNN based multi loss function optimized deep encoder–decoder for segmentation of retinal vessels	DRIVE, AVRDB, and AV classification dataset	Accuracy for DRIVE = 96%, Accuracy for AVRDB dataset = 98% Accuracy for AV classification dataset = 97%
Almotiri et al. [25]	Adaptive fuzzy thresholding and mathematical morphology based segmentation algorithms	Vessels (DRIVE and STARE datasets), optic disc (DRISHTI-GS) and exudates lesions (DIARETDB1)	Sensitivity = 93.15%, Specificity = 97.09%, PPV = 90.15%

(*continued*)

Table 3. (*continued*)

Authors	Methodology	Datasets	Performance Metrics
Fu et al. [26]	Automatic Glaucoma screening done using Disc-aware Ensemble Network (DE Net) for automatic glaucoma screening	SCES and SINDI	**SCES Dataset:** AUC = 0.9183, B-Accuracy = 0.8429, Sensitivity = 0.8479, Specificity = 0.8380 **SINDI Dataset:** AUC = 0.8173, B-Accuracy = 0.7495, Sensitivity = 0.7876, Specificity = 0.7115
Geetha Ramani et al. [19]	Methods used are image pre-processing, supervised and unsupervised learning and image post-processing techniques.	DRIVE database	Accuracy = 95.36%, Sensitivity = 70.79%, Specificity = 97.78%, Positive Prediction Value (PPV) = 75.76%

4   Conclusion and Future Work

The early detection of DR helps to avoid the vision loss. The deep learning methods has helped to implement very accurate detection and diagnosis of the diabetic retinopathy complications. The review paper gives an outline of the recent deep learning algorithms used in exudate and blood vessel segmentation. Much research has not been done in the field for the pathological images. The future work in the detection of hard exudates, microaneurysms and hemorrhages detection using deep learning can be implemented by preventing overfitting of data in the future research.

Compliance with Ethical Standards
✓ All authors declare that there is no conflict of interest.
✓ No humans/animals involved in this research work.
✓ We have used our own data.

References

1. Cheung, N., Mitchell, P., Wong, T.Y.: Diabetic retinopathy. Lancet **376**, 124–136 (2010)
2. Philips, R., Forrester, J., Sharp, P.: Automated detection and qualification of retinal exudates. Gaefes Arch. Clin. Exp. Ophthalmol. **1231**, 90–94 (1993)
3. Gardner, G.G., Keating, D., Williamson, T.H., Elliott, A.T.: Automatic detection of diabetic retinopathy using an artificial neural network: a screening tool. Br. J. Ophthalmol. **80**, 940–944 (1996)
4. LeCun, Y., Bengio, Y., Hinton, G.: Deep learning. Nature **521**(7553), 436–444 (2015)
5. Simonyan, K., Zisserman, A.: Very deep convolutional networks for large-scale image recognition. CoRR, vol. abs/1409.1556 (2014). https://arxiv.org/abs/1409.1556
6. Naqvi, S.A.G., Zafar, M.F., ul Haq, I.: Referral system for hard exudates in eye fundus. Comput. Biol. Med. **64**, 217–235 (2015)

7. Walter, T., Klein, J.C., Massin, P., Erginay, A.: A contribution of image processing to the diagnosis of diabetic retinopathy-detection of exudates in color fundus images of the human retina. IEEE Trans. Med. Imaging **21**(10), 1236–1243 (2002)
8. Niemeijer, M., van Ginneken, B., Russell, S.R., Suttorp-Schulten, M.S., Abramoff, M.D.: Automated detection and differentiation of drusen, exudates, and cotton-wool spots in digital color fundus photographs for diabetic retinopathy diagnosis. Invest. Ophthalmol. Vis. Sci. **48**(5), 2260–2267 (2007)
9. Sopharak, A., Uyyanonvara, B., Barman, S., Williamson, T.H.: Automatic detection of diabetic retinopathy exudates from non-dilated retinal images using mathematical morphology methods. Comput. Med. Imaging Graph. **32**(8), 720–727 (2008)
10. Sanchez, C.I., García, M., Mayo, A., Lopez, M.I., Hornero, R.: Retinal image analysis based on mixture models to detect hard exudates. Med. Image Anal. **13**(4), 650–658 (2009)
11. Ali, S., Sidibe, D., Adal, K.M., Giancardo, L., Chaum, E., Karnowski, T.P., Mériaudeau, F.: Statistical atlas based exudate segmentation. Comput. Med. Imaging Graph. **37**(5–6), 358–368 (2013)
12. Fraz, M.M., Jahangir, W., Zahid, S., Hamayun, M.M., Barman, S.A.: Multi scale segmentation of exudates in retinal images using contextual cues and ensemble classification. Biomed. Sig. Process. Control **35**, 50–62 (2017)
13. Cassin, B., Solomon, S.A.B.: Dictionary of Eye Terminology, 2nd edn. Triad Publishing Company, Gainesville (1990)
14. Bouma, B.E., Tearney, G.J.: Handbook of Optical Coherence Tomography, 1st edn. Marcel Dekker, New York (2001)
15. Niemeijer, M., Staal, J.J., Ginneken, B.V., Loong, M., Abramoff, M.D.: DRIVE: digital retinal images for vessel extraction (2004). http://www.isi.uu.nl/Research/Databases/DRIVE/
16. Hoover, A., Kouznetsova, V., Goldbaum, M.: Locating blood vessels in retinal images by piecewise threshold probing of a matched filter response. IEEE Trans. on Med. Imaging **19**(3), 203–210 (2000)
17. MESSIDOR: Methods for Evaluating Segmentation and Indexing techniques Dedicated to Retinal Ophthalmology (2004). http://messidor.crihan.fr/index-en.php
18. Zhang, Z., Yin, F.S., et al.: ORIGA–light: an online retinal fundus image database for glaucoma analysis and research. In: Proceedings of the Annual International Conference of the IEEE Engineering in Medicine and Biology Society, EMBC 2010, Buenoa Aires, Argentina, September, pp. 3065–3068. IEEE (2010)
19. Geeta Ramani, R., Balasubramanian, L.: Retinal blood vessel segmentation employing image processing and data mining techniques for computerized retinal image analysis. Biocybern. Biomed. Eng. **36**, 102–118 (2016)
20. Litjens, G., Kooi, T., Bejnordi, B.E., Setio, A.A.A., Ciompi, F., Ghafoorian, M., Van Der Laak, J.A.W.M., van Ginneken, B., Sánchez, C.I.: A survey on deep learning in medical image analysis. Medical Image Anal. **42**, 60–88 (2017)
21. Khojasteh, P., Aliahmad, B., Kumar, D.K.: A novel color space of fundus images for automatic exudates detection. Biomed. Sig. Process. Control **49**, 240–249 (2019)
22. Chen, B., Wang, L., Sun, J., Chen, H., Fu, Y., Lan, S.: Diverse lesion detection from retinal images by subspace learning over normal samples. Neurocomputing **297**, 59–70 (2018)
23. Long, S., Huang, X., Chen, Z., Pardhan, S., Zheng, D.: Automatic detection of hard exudates in color retinal images using dynamic threshold and SVM classification: algorithm development and evaluation. BioMed Res. Int. (2019). http://doi.org/10.1155/2019/3926930
24. Badawi, S.A., Fraz, M.M.: Multiloss function based deep convolutional neural network for segmentation of retinal vasculature into arterioles and venules. BioMed Res. Int. (2019). https://doi.org/10.1155/2019/4747230

25. Almotiri, J., Elleithy, K., Elleithy, A.: A Multi-anatomical retinal structure segmentation system for automatic eye screening using morphological adaptive fuzzy thresholding. IEEE J. Transl. Eng. Health Med. **6**, 1–23 (2018)
26. Fu, H., Cheng, J., Xu, Y., Zhang, C., Wong, D.W.K., Liu, J., Cao, X.: Disc-aware ensemble network for glaucoma screening from fundus image. IEEE Trans. Med. Imaging **37**(11), 2493–2501 (2018)

Smart Imaging System for Tumor Detection in Larynx Using Radial Basis Function Networks

N. P. G. Bhavani[1,2], K. Sujatha[2(✉)], V. Karthikeyan[3], R. Shobarani[4], and S. Meena Sutha[5]

[1] Department of EEE, Meenakshi College of Engineering, Chennai, India
[2] Department of EEE, Dr. MGR Educational and Research Institute, Chennai, India
drksujatha23@gmail.com
[3] Department of EIE, Dr. MGR Educational and Research Institute, Chennai, India
[4] Department of CSE, Dr. MGR Educational and Research Institute, Chennai, India
[5] Department of ECE, Rajalakshmi Engineering College, Chennai, India

Abstract. Tumor in Larynx is a fatal disease, which is capable to make the humans speechless. The Computed Tomography (CT) images are used to detect the benign and malignant tumors in larynx. The objective of this project is to develop an organized scheme to analyze and evaluate its probabilities with the help of a typical user friendly simulation tool like MATLAB. The innovation in this scheme has a well developed strategy for detection of malignant and benign tumors using parallel computing image related machine learning algorithms. The preliminary stage includes noise removal and feature extraction using Principal Component Analysis (PCA) from the region of interest in the larynx. This method uses images from the open source data base. The feature set extracted serves as the input for training the Radial Basis Function Network (RBFN) so as to identify the malignant and benign tumor present in the larynx. The Diagnostic Efficiency (DE) is found to be 94 to 95%.

Keywords: Larynx tumor · Principal Component Analysis · Radial Basis Function Network · Diagnostic efficiency

1 Introduction

The larynx which is otherwise called as voice-box, is a triangular shaped structure responsible for voice generation, respiration and serves as a protection to trachea from aspirations. The larynx is made up of cartilages, epiglottis, thyroid gland, cricoid, arytenoid, corniculate and cuneiform. Muscles and connective tissues help to connect the cartilages. The epiglottis is a piece of tissue which safeguards the trachea [1, 2]. Due to its strange importance, the disorders like laryngeal cancer require a meticulous

© Springer Nature Switzerland AG 2020
S. Smys et al. (Eds.): ICCVBIC 2019, AISC 1108, pp. 1371–1377, 2020.
https://doi.org/10.1007/978-3-030-37218-7_144

care. The treatment includes excision surgery, radiation, chemotherapy or even total laryngectomy. The pre-staging methods used are Computed tomography (CT) and magnetic resonance imaging (MRI) prior to surgery so as to identify vocal cord disorders. Unfortunately, the other structures like nerves and blood vessels adjacent to the larynx to may be affected because of complications such as nerve damage, inflammation, tissue breakdown, fibrosis and pain. Manual diagnostic reviews by radiologists, oncologists and surgeons are tedious, delayed and subjected to inaccuracies in determining the exact stage and offering an appropriate treatment. Therefore, this automated smart imaging diagnostics for larynx tumor identification offers best sites for radiotherapy which will reduce complications thereby producing a favorable outcome. The CT image of larynx shown in the Fig. 1 indicates normal, subglottic and supraglottic tumors [3, 4].

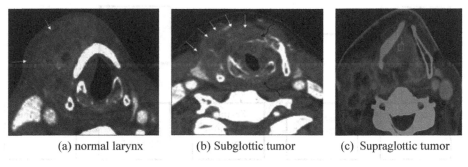

(a) normal larynx   (b) Subglottic tumor   (c) Supraglottic tumor

Fig. 1. CT images of larynx [4]

The entire manuscript contains four sections with Sect. 1, introducing the readers to Larynx tumors, Sect. 2, work carried out so far related to laryngeal cancer, Sect. 3, describing the various steps involved in identifying the laryngeal cancer, Sect. 4, depicting the results and their connected discussion and finally Sect. 5, with concluding remarks.

2 Literature Survey Related to Laryngeal Cancer

Cancer includes the chondrosarcoma, lymphoma and the paraganglioma to be the commonest types of laryngeal cancers. The endoscopy may not reveal such sub-mucosal types except for a bulge. The presence of sub-mucosal mass may be confirmed if boundaries are detected by the imaging techniques. This will guide the endoscopist to the most appropriate site for biopsy [5].

In pretherapeutic staging of SCC in larynx, CT and MRI play an important role in diagnosis. The extent to which, the cancer is spread is determined by 'T' staging method for treatment of laryngeal cancer in patients. The tumor spread in the different regions of larynx can be interpreted from MRI and CT scan using standardized image processing algorithms which identify those using key anatomical features and characteristic patterns [6, 7].

3 Methodology

Images of the larynx with benign and malignant tumor are taken from the open source data base for diagnosis. This data base consists of images which are nearly 25 benign and 25 malignant accounting to a total of 50 images of the larynx. The noise removal is done using median filtering. The overall schematic is represented in Fig. 2.

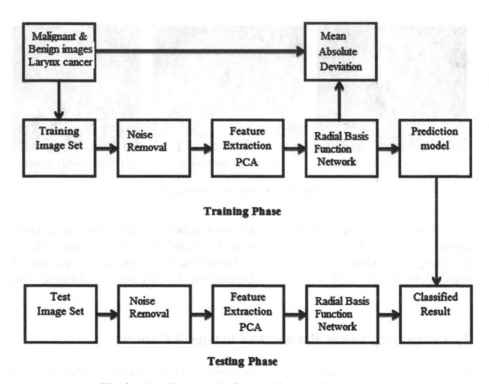

Fig. 2. Overall schematic for identification of larynx tumor

3.1 Feature Extraction

For the detection to take place the multi-dimensional image should be transferred to a two dimensional feature space which is done using Principal Component Analysis (PCA). The PCA computes Eigen values and Eigen vectors followed discriminant vectors which facilitates the mapping of N-dimensional feature set on to a two dimensional space thereby reducing the computational complexity. The Eigen values and Eigen vectors are unique values for each and every image of larynx, hence it helps to discriminate the images of larynx which are affected by benign and malignant tumors [8, 9].

3.2   Detection of Skin Lesion Using Radial Basis Function Networks (RBFN)

RBFN approach is more spontaneous than the Multi Layer Perceptron (MLP). The similarity measure is used by RBFN to classify the input samples used for training the RBFN. Every node of RBFN stores a sample as an example from the training set. When the real time classification has to take place with a new and unknown input, each node computes the Euclidean distance between the input and its sample. If the input resembles the first sample type then the new input is assigned to benign category else it is assigned to malignant category. The network architecture for RBFN is shown in Fig. 3. The flowchart in Fig. 4, depicts how the RBFN is used for detecting the benign and malignant tumors in larynx

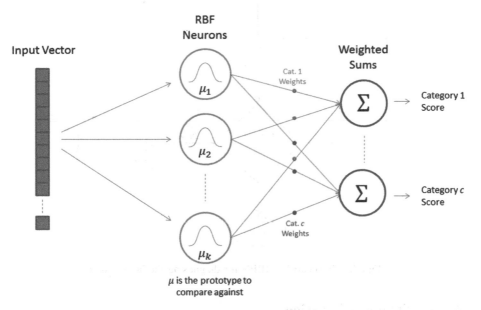

Fig. 3. Architecture for RBFN

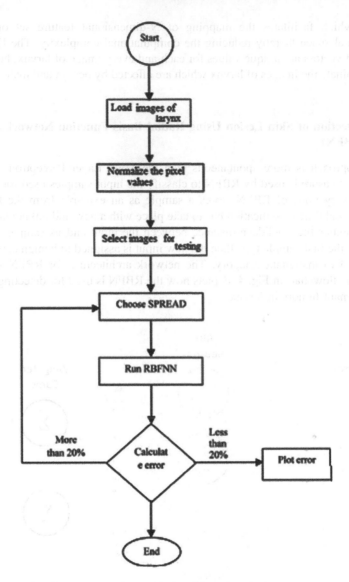

Fig. 4. Flowchart for RBFN for diagnosing the larynx tumor

4   Results and Discussion

The RBF network uses discriminant vectors from PCA as inputs. The number of neurons in hidden layer is called as RBF neurons which uses Gaussian function as the activation function and finally the output layer uses threshold function as the activation function.

The input vector is the two-dimensional vector which is the output of the PCA which is subjected to classification. The entire input vector is connected to each of the

RBF node. Each RBF node stores a sample vector which is one of the discriminant vectors from the training set and compares the same with sample to produce the outputs in the range of the similarity index so that the benign and malignant tumors present in the larynx are identified as shown in Fig. 5. Those images marked with red circles indicate the images of the larynx with benign tumors and those indicated by a blue cross denotes the images of the larynx with malignant tumors. The Diagnostic Efficiency is illustrated in Table 1.

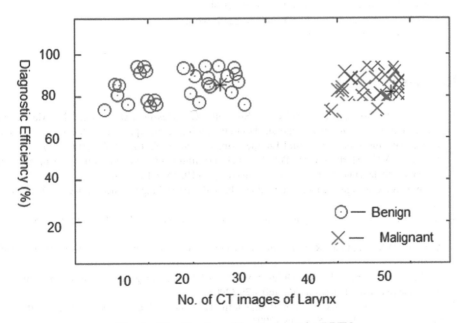

Fig. 5. Classification of larynx tumor by RBFN

Table 1. Determination of diagnostic efficiency for RBFN

S. No	Total no. images for larynx		Diagnostic efficiency (%)	
	Benign	Malignant	Benign	Malignant
1.	25	25	94.2	95.3

5 Conclusion

The RBFN when used to develop an automated larynx tumor detection system yielded a diagnostic efficiency of nearly 94.2% and 95.3% to identify the benign and malignant tumors. On par with the performance, these outputs are tested by taking the opinion of experts for demonstrating an artificially intelligent scheme for classifying the larynx tumor with a competence level comparable to the radiologists. Automated medical imaging system using RBFN, can also be developed as a mobile application so that the devices can efficiently comply with the requirements of the radiologists outside the hospital. It is projected that nearly six billion smart phone subscriptions will exist in a couple of years and can therefore potentially provide a smart detection low cost universal access to vital diagnostic care. In future the classification efficiency can be improved by using advanced filtering techniques and robust classifiers like deep learning neural networks.

Compliance with Ethical Standards
   ✓ All authors declare that there is no conflict of interest.
   ✓ No humans/animals involved in this research work.
   ✓ We have used our own data.

References

1. Hashibe, M., Boffetta, P., Zaridze, D., Shangina, O., Szeszenia-Dabrowska, N., Mates, D., et al.: Contribution of tobacco and alcohol to the high rates of squamous cell carcinoma of the supraglottis and glottis in central Europe. Am. J. Epidemiol. **165**, 814–820 (2007)
2. Nikolaou, A.C., Markou, C.D., Petridis, D.G., Daniilidis, I.C.: Second primary neoplasms in patients with laryngeal carcinoma. Laryngoscope **110**, 58–64 (2000)
3. Connor, S.: Laryngeal cancer: how does the radiologist help? Cancer Imaging **7**, 93–103 (2007)
4. Hermans, R.: Staging of laryngeal and hypopharyngeal cancer. Eur. Radiol. **16**, 2386–2400 (2006)
5. Yousem, D.M., Tufano, R.P.: Laryngeal imaging. Magn. Reson. Clin. N. Am. **10**, 451–465 (2002)
6. Becker, M., Burkhardt, K., Dulguerov, P., Allal, A.: Imaging of the larynx and the hypopharynx. Eur. J. Radiol. **66**, 460–479 (2008)
7. Curtin, H.D.: The larynx. In: Som, P.M., Curtin, H.D. (eds.) Head and Neck Imaging, 4th edn, pp. 1595–1699. Mosby, St Louis (2003)
8. Zbaren, P., Becker, M., Läng, H.: Pretherapeutic staging of laryngeal carcinoma: clinical findings. CT MRI histopathol. Cancer **77**, 1263–1273 (1996)
9. American Joint Committee on cancer: Larynx. In: Greene, F.L., (ed.) American Joint Committee on Cancer: AJCC Staging Manual, 7th edn., pp. 57-68. Springer, New York (2010)

Effective Spam Image Classification
Using CNN and Transfer Learning

Shrihari Rao(✉) and Radhakrishan Gopalapillai

Department of Computer Science and Engineering, Amrita School
of Engineering, Bengaluru, Amrita Vishwa Vidyapeetham, Bengaluru, India
Shrihari.rao@gmail.com,
g.radhakrishanan.new@gmail.com

Abstract. Machine Learning is emerging as one of the most active research areas in the field of Artificial Intelligence [AI]. It is significantly beneficial for artificial systems to be able to find reusable characteristics that capture structure in the setting in order to enable a more flexible and general learning. In this work, we have developed an Image Classifier model, which can identify spam images such as "good-morning", "good-night" and "few quotes message with image" from the group of images and classify it accordingly. To accomplish our objective, together with pre-trained information set such as VGG16, VGG19 and MobileNet model, we retrained the model with fresh information dataset. In the assignment of large-scale image classification, supervised learning using profound convolutionary neural network has demonstrated its promise. It is now well placed as a construction block to be component of a bigger scheme that addresses multimedia duties in real life. Our model is capable of handling dynamic data sets and delivers an excellent accuracy.

Keywords: Deep learning · Incremental model · Spam image classification · Fine tuning model · CNN model

1 Introduction

Object classification is a simple job for human beings, but for machine it has been viewed as a complicated issue. It has received an increased attention due to an increase in high-capacity pcs, accessibility of high-quality, low-priced video cameras, and a growing need for automatic video or image analysis. Currently, the Internet has several effects on our daily life cycle. One of the researchers multifarious duties is the accurate classification of the field-captured photographs, social media pictures, whatsapp/stored pictures, etc. Human can do manual surveillance can be performed using field surveys. This method's constraints are high labor costs and it takes a long time.

Convolution Neural Networks have been successful in classifying images. CNN fundamentally has a layer of convolution, a pooling layer and a completely connected layer. Filter weights are defined in the convolutionary layers. While preparing the CNN model, the uneven seated weights are refreshing for subsequent epochs. Relu function is provided in order to stay away from the vanishing gradient issues. In the interim, the CNN adds non-linearity, enabling the scheme to portray non-linearities [1, 5].

© Springer Nature Switzerland AG 2020
S. Smys et al. (Eds.): ICCVBIC 2019, AISC 1108, pp. 1378–1385, 2020.
https://doi.org/10.1007/978-3-030-37218-7_145

In addition, it is essential to create incremental teaching techniques in an epochs in which information sets are developing frequently, with fresh courses and samples. A typical idea to mitigate problem of building of model from scratch we can utilize pre-prepared CNNs for work on a careful arrangement of information and adjust them to contemporary data sets or assignments, rather than train the full system starting with scrach [2].

One strategy to adapting a network to fresh information or assignments is fine tuning [3]. Here, the initial network's output layer is adapted either by replacing it with classes that correspond to the fresh assignment, or by adding fresh courses to current ones. At that point the loads in this layer are introduced haphazardly, and all system parameters are acclimated to the objective for the new assignment. While this structure on the crisp classes is exceptionally viable, its proficiency on the antiquated classes will endure drastically if the system isn't prepared together on all classes [4]. This issue, wherein a neural system overlooks prior got the hang of understanding when it is changed in accordance with a new task, is called heartbreaking obstruction or over-looking. This is also called overfitting.

2 Related Work

Our study endeavors to carry the guarantee of profound figuring out how to make it practical and increasingly broad worldview of gradual learning. We sort out the exchange of the related work according to studies in computer vision fields.

Classification of Using CNN: Characterization step orders identified objects in pre-defined classes [6, 7] using reasonable techniques that contrast image designs and target patterns. The classification can be visualizing by below picture Fig. 1(a).

The key structure of CNN comprises of two layers, the first is for extraction of the capacity, and every one of the neurons in a single layer is associated with the past layers nearby window. Other layer is the mapping layer of highlights. CNN is utilized as it permits changes and bending in the image to be strong, uses less memory requests, is easier, and offers the better preparing model [6].

Convolution layer that converts with a tiny filter through the input picture. This weight may be random at the beginning, and subsequently the filter's co-efficient will be used to convert. Dot weight product resulting in comparative figures with specified inputs. Using the RELU layer to activate the next layer [7]. For effectiveness and precision purpose zero padding and striding will be used.

Classification Using Deep Learning and Incremental Model: Noteworthy advancement has been made in sorting multi-class pictures over the previous century, both regarding classification amount and exactness. ImageNet, for instance, has around 22 K categories [8].

An unrefined meaning of gradual learning is that learning with crisp data is a progressing technique. Earlier research managed two critical circumstances, the second is appropriate to this exploration. The first is idea float in the information stream of training, this helps the classifier to learns in a non-stationary condition [6, 10]. The second is when current classifiers are related with the crisp courses to be educated.

Along these lines proposed the inquiry about in the case of inclining the nth classifier is simpler (than examining the first), and from that point forward researchers have started to handle this issue utilizing move learning instinct, utilizing strategies that utilization less examples to get new models or better speculation [11].

In the context of incremental learning or in transfer learning, the task with DCNN is to mix in one architecture function extractor and classifier. Goodfellow et al. [8] studied in gradient-based neural networks the disastrous issue of forgetting. However, the analysis has been done on the tiny networked MNIST data set, and recommend to use available deep convolution neural network configuration. Nevertheless, it reaffirms qualitatively the trouble of achieving brilliant execution all the while on old and new assignments. Numerous works have focused on domain adjustment, for example, one-shot learning by modifying attributes learned ImageNet 1 K information set to different datasets [12]. The latest zero-shot learning work assumes that a semantic embedding room is available to which DCNN outputs are projected. Our job varies in the objective as model has to transfer the learning with a bigger dataset in the same assignment. There were 2 different incremental model i.e. clones and a flat increment [13].

Calibrating among order and division offers reasonable projections of the system. Indeed, even the most exceedingly terrible model accomplished almost 75% of the past best results. VGG16 is by all accounts 56.0 superior to past techniques. Despite the fact that VGG and GoogleNet are as precise as classifiers, FCN-VGG19 is more exact than other pre-prepared models [12].

3 Fine Tuned Incremental Model

Its most necessary to use a pre-prepared model for all of ours own needs, we start by evacuating the first classifier, at that point we include another classifier that accommodates our motivations, lastly it should tweak our model to one of three methodologies:

Train the Whole Model: Its utilization the pre-prepared model design for this situation and train it as indicated by your informational index. You're taking in the model sans preparation, However its required huge data set for training from scratch for better accuracy (and higher computational power PC).

Train Certain Layers and Keep Other Layer Solidified: Lower layers suggests general properties (free issue), while higher layers insinuate express characteristics (subordinate issue). As a general rule, if a little course of action of data and various parameters, to sidestep over fitting, required to leave more layers hardened. Of course, if the instructive file is huge and the amount of parameters is close to nothing, then required to improve the model by means of getting ready more layers for the new task as over fitting isn't an issue.

Convolution Foundation Frozen: Major idea is to keep up the base of the convolution in its unique structure and utilize its yields to encourage the classifier. Once utilized pre-prepared model as a fixed extraction component, then it helpful for less chance of computers computational power. This aids the advantage that informational collection is little and additionally pre-prepared model takes care of an issue that is fundamentally builds the model. Below Fig. 2(b) provides a schematic presentation of above mentioned three approaches.

Fine tuning algorithm with second method is updated below in Fig. 1.

Input:	1 **repeat**
$\mathcal{U} = \{C_i\}, i \in [1, n]$ {$\mathcal{U}$ contains labelled data } 2	**for** each $C_i \in \mathcal{U}$ **do**
$C_i = \{x_i^j\}, j \in [1, m]$ {C_i has m frozen layer } 3	$p_i \leftarrow P(C_i, \mathcal{M}_{t-1})$
$\mathcal{M}_0$: pre-trained CNN 4	**if** $mean(p_i) > 0.5$ **then**
b: batch size 5	$S_i' \leftarrow$ top α percent of the patches of C_i
α: patch selection ratio 6	**else**
Output: 7	$S_i' \leftarrow$ bottom α percent of the patches of C_i
$\mathcal{L}$: labeled Images 8	**end**
$\mathcal{M}_t$: fine-tuned CNN model at Iteration t 9	
Functions: 10	**end**
$p \leftarrow P(C, \mathcal{M})$ {outputs of $\mathcal{M}$ given $\forall x \in C$} 11	Sort $\mathcal{U}$ according to the numerical sum of R_i
$\mathcal{M}_t \leftarrow F(\mathcal{L}, \mathcal{M}_{t-1})$ {fine-tune $\mathcal{M}_{t-1}$ with $\mathcal{L}$} 12	Query labels for top b candidates, yielding $\mathcal{Q}$
$a \leftarrow mean(p_i)$ {$a = \frac{1}{m}\sum_{j=1}^{m} p_i^j$} 13	$\mathcal{L} \leftarrow \mathcal{L} \cup \mathcal{Q}; \quad \mathcal{U} \leftarrow \mathcal{U} \setminus \mathcal{Q}$
Initialize: 14	$\mathcal{M}_t \leftarrow F(\mathcal{L}, \mathcal{M}_{t-1}); t \leftarrow t + 1$
$\mathcal{L} \leftarrow$ Null $t \leftarrow 1$ 15	**until** classification performance is Best

Fig. 1. Fine tuning algorithm using Keras

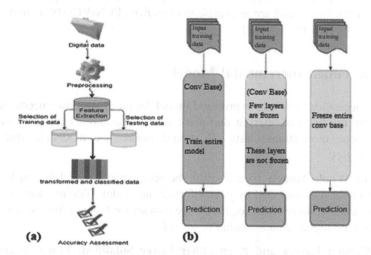

(a)   Accuracy Assessment   (b)

Fig. 2. (a) Basic steps for followed for CNN (b): Various way of incremental model building

Indeed, only optimizing in a subgroup of the feature space has not been updated with new weights for the network. If our training dataset is similar to any image dataset, there wont be concern about accuracy if datasets are very distinct from the image set, freezing will mean a reduction in precision. Unfreezing everything will allow us to optimize the entire space of the feature and receive a better optima. However computational time will be huge for this.

Selection of New Samples for Training the Incremental Model: We propose two distinctive memory arrangements to test the conduct of our gradual learning approach in various situations. The principal arrangement thinks about that there is restricted limit of K tests in the agent memory. The more classes are put away, the less examples per class are held, as the memory limit is free of the quantity of classes. The quantity of tests per class "m" is in this manner characterized by m = K/c. Here c is the quantity of classes put away in the memory. Here assumption is memory limit is K. The subsequent establishment retains a many duplicates for each class. Sample training pictures are shown below Fig. 3 for incremental model.

Fig. 3. Sample image used for spam detection [8]

4 Implementation Details

Our models are actualized on Keras, tensor flow and python, 30 epochs and an additional 30 epochs are performed for adjusted calibrating for each steady stage. Our initial 30 epochs learning rate starts at 0.1 and is divided by 10 each on 10 epoch. In case of tweaking, a similar abatement is utilized, then again, actually the beginning recurrence is 0.01. We train the systems utilizing traditional stochastic inclination decent with around 500 examples, which is smaller than expected groups, 0.0001 weight decay and 0.9 energy. This limit has been set to handle over fitting. We apply L2-regularization and irregular clamor with parameters $\pi = 0.5$, $\pi = 0.55$) on the slopes. However all these parameter we could configurate in Keras while building the model.

Since class-steady learning benchmarks are not effectively available, we are following the standard setup [9, 14] of partitioning the classes of a conventional multiclass accumulation of data into gradual groups.

VGG 16 and VGG 19 Dataset [4]: In the 2014 [8] article, profound convolution neural system Simonyan and Zisserman propelled the VGG arrange design. This system is characterized by its straight forwardness, utilizing just 33 convolution layers stacked in developing profundity over one another. Max pooling is utilized to decrease volume size. A SoftMax classifier pursues two fully connected layers, each with 4,096 nodes. On top of it we added few spam images Fig. 4(a) depicts the accuracy against epochs for incremental model with frozen and partially frozen layer. It can be observed that all *frozen conv layer* has less accuracy Since its holds only predefined images. Red color graphs depicts accuracy for the few layers frozen network with weights transferred to next layer with new set of data.

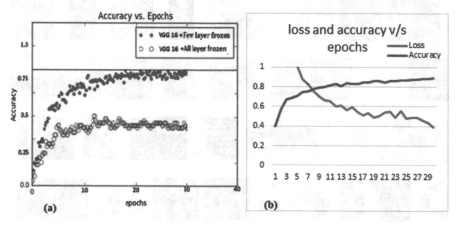

Fig. 4. (a) Comparison of partially and fully frozen incremental model, (b) Loss and Accuracy against epochs for partially frozen VGG16

High accuracy is accomplished after 35 epochs in the case of few layers frozen network. The accuracy does not enhance in later epochs even reduces in 25–45 epochs intervals.

The graph in Fig. 4(b) shows the intersection between accuracy for new data and validation loss. The loss of validation propagates the model's capacity to simplify fresh information. If our training information accuracy ("acc") continues to improve as our validation information loss ("val loss") deteriorates, For our data set on nearly 45 epoch validation accuracy is deteriorating as training accuracy is increased. Around 32nd epoch onwards, it is likely to be in an overfitting scenario, i.e. our model just begins memorizing the information.

This means the model can show the same result after the 35th epoch, but it won't be better. This model is therefore enough to train on 45 epochs. However, to receive the best accuracy, its mandatory to have more datasets for training. My current experiment yields result as tabulated below. Epochs vs loss and epoch vs accuracy plots captured while building the model for vgg16 of partially frozen model.

As we observe the graphs, loss is less against than accuracy. İdeally loss would be zero if system understood image clearly. Practically its impossible in practical.

However specific to respective task model can yields better results Detailed numerical information of loss and accuracy has been tabulated Table 1 below.

Table 1. Accuracy and loss for 45 epochs on incremental learning

Epochs	Loss	Accuracy	Epochs	Loss	Accuracy	Epochs	Loss	Accuracy
1	3.0904	0.3962	16	1.5459	0.5587	31	1.3981	0.6522
2	2.4273	0.3982	17	1.5134	0.5575	32	1.3988	0.6825
3	2.1312	0.4025	18	1.5376	0.5624	33	0.3987	0.6477
4	2.0602	0.4058	19	1.4886	0.5928	34	0.388	0.6492
5	2.0085	0.4568	20	1.5021	0.5728	35	0.3851	0.6583
6	1.8754	0.4275	21	1.5417	0.5792	36	0.3746	0.6588
7	1.8271	0.4712	22	1.5534	0.5876	37	0.3758	0.6592
8	1.7587	0.4952	23	1.4928	0.5881	38	0.3801	0.6631
9	1.7025	0.5002	24	1.5577	0.5993	39	0.3325	0.6691
10	1.6631	0.5093	25	1.4779	0.6025	40	0.3672	0.6694
11	1.6595	0.5174	26	1.4881	0.6482	41	0.2564	0.6718
12	1.6051	0.5237	27	1.492	0.6132	42	0.3456	0.6718
13	1.6069	0.5329	28	1.4561	0.6285	43	0.3782	0.6757
14	1.5683	0.5391	29	1.4365	0.6297	44	0.3473	0.6773
15	1.5974	0.5486	30	1.3926	0.6359	45	0.3684	0.6774

On analyzing the above table, it clearly indicates that accuracy is increased as epochs increased, and loss is travelling opposite to accuracy. As we increase data set accuracy will reduce. Overall accuracy has been listed in Table 2 for various pre-prepared model. However, with small data set it achieved better results.

Table 2. Accuracy for various incremental learning

Models	Accuracy (45 epochs)	Validation accuracy (45 epochs)	Loss (45 epochs)
New data scratch model build	31.15%	9%	12.78%
New data model merged with VGG19	65.25%	52.41%	38.42%
New data model merged with VGG16	67.44%	58.53%	36.84%

5 Conclusion

In this paper the convolution neural system-based methodology alongside with a steady methodology that has been used for developing a binary classification of spam pictures for example sort of "hello", "goodbye" picture messages and couple of other spam pictures with statements. It tends to make a progressively summed up model by gathering more spam pictures for information collection. It will likewise improve the precision of the model. In future study, CNN and Text classification techniques can be combined to make the model to be much more precise in recognizing the spam pictures.

Compliance with Ethical Standards
  ✓ All authors declare that there is no conflict of interest.
  ✓ No humans/animals involved in this research work.
  ✓ We have used our own data.

References

1. Sowmya, V., Soman, K.P., Deepika, J.: Image classification using convolutional neural networks. Int. J. Sci. Eng. Res. **5**(6), 1661–1668 (2014)
2. Wu, Y., Qin, X., Pan, Y., Yuan, C.: Convolution neural network based transfer learning for classification of flowers. College of Computer and Information Engineering, Guangxi Teachers Education University Nanning, China
3. Girshick, R., Donahue, J., Darrell, T., Malik, J.: Rich feature, hierarchies for accurate object detection and semantic segmentation. In: CVPR (2014)
4. Neena, A., Geetha, M.: Image classification using an ensemble-based deep CNN. In: Advances in Intelligent Systems and Computing, vol. 709, pp. 445–456 (2018)
5. Suresh, R., Prakash, P.: Deep learning based image classification on amazon web service. J. Adv. Res. Dyn. Control Syst. **10**, 1000–1003 (2018)
6. Kavitha, D., Hebbar, R., Harsheetha, M.P.: CNN based technique for systematic classification of field photographs. In: International Conference on Design Innovations for 3Cs Compute Communicate Control (2018)
7. Korshunova, K.P.: A convolutional fuzzy neural network for image classification. Moscow Power Engineering Institute in Smolensk, Smolensk, Russia
8. Simonyan, K., Zisserman, A.: Very deep convolutional networks for large-scale image recognition. arXiv preprint arXiv:1409.1556 (2014)
9. Shmelkov, K., Schmid, C., Alahari, K.: Incremental learning of object detectors without catastrophic forgetting. Univ. Grenoble Alpes, Inria, CNRS, Grenoble INP, LJK, 38000 Grenoble, France
10. He, K., Zhang, X., Ren, S., Sun, J.: Deep residual learning for image recognition. In: CVPR (2016)
11. Thrun, S.: Is learning the n-th thing any easier than learning the first? In: Advances in neural information processing systems, pp. 640–646 (1996)
12. Deng, J., Dong, W., Socher, R., Li, L.-J., Li, K., Fei-Fei, L.: ImageNet: a large-scale hierarchical image database. In: Computer Vision and Pattern Recognition (2009)
13. Shelhamer, E., Long, J., Darrell, T.: Fully convolutional networks for semantic segmentation. IEEE Trans. Pattern Anal. Mach. Intell. **39**(4), 640–651 (2017)
14. Rebuffi, S.A., Kolesnikov, A., Sperl, G., Lampert, C.H.: iCaRL: incremental classifier and representation learning. In: CVPR (2017)

Unsupervised Clustering Algorithm as Region of Interest Proposals for Cancer Detection Using CNN

Ajay K. Gogineni[1]([⊠]), Raj Kishore[2], Pranay Raj[1], Suprava Naik[3], and Kisor K. Sahu[4]

[1] School of Electrical Sciences, IIT Bhubaneswar, Bhubaneswar 752050, India
{akg11,prl6}@iitbbs.ac.in
[2] Virtual and Augmented Reality Centre of Excellence, IIT Bhubaneswar, Bhubaneswar 752050, India
rkl6@iitbbs.ac.in
[3] Department of Radiodiagnosis, All India Institute of Medical Sciences, Bhubaneswar, Bhubaneswar 751019, India
radiol_suprava@aiimsbhubaneswar.edu.in
[4] School of Minerals, Metallurgical and Materials Engineering, IIT Bhubaneswar, Bhubaneswar 752050, India
kisorsahu@iitbbs.ac.in

Abstract. Deep learning methods are now getting lot of attention due to their success in many fields. Computer-aided bio-medical image analysis systems act as a tool to assist medical practitioners in correct decision making. Use of Deep Learning algorithms to predict Cancer at early stage was highly promoted by Kaggle Data Science Bowl 2017 competition and Cancer Moonshot Initiative. In this article, we have proposed a novel combination of unsupervised machine learning tool which is modularity optimization based graph clustering method and Convolutional Neural Networks (CNN) based architectures for lung cancer detection. The unsupervised clustering method helps in reducing the complexity of CNN by providing Region of Interest (ROI) proposals. Our CNN model has been trained and tested on LUNA 2016 dataset which contains Computed Tomography (CT) scans of lung region. This method provides an approximate pixel-wise segmentation mask along with the class label of the ROI proposals.

Keywords: Machine learning · Graph clustering · Modularity · CNN · ROI · CT scan · Cancerous nodules

1 Introduction

The current research in cancer detection using deep learning methods is increasing exponentially. There are several software based firms which are spending huge amount of money on machine learning based researches including Google and Microsoft [1]. The main reason behind this growth is the availability of a huge amount of labelled data as open source in various competitions such as LUNA 2016 and Kaggle Data Science Bowl 2017 [2, 3] and also the development of fast computing tools like GPUs which

© Springer Nature Switzerland AG 2020
S. Smys et al. (Eds.): ICCVBIC 2019, AISC 1108, pp. 1386–1396, 2020.
https://doi.org/10.1007/978-3-030-37218-7_146

makes computation several times faster than CPUs [4]. Cancer Moonshot Initiative [5] promoted the idea to accelerate cancer research and predict cancer at an early stage. This increased the number of researchers working on it. Around 9.6 million people died in 2018 due to cancer [6] and around 33% of the deaths were caused by smoking and use of tobacco, which directly affects the lung. The chance of survival of a lung cancer patient decreases with increase in the size of the cancerous nodule. So the early detection of a cancerous nodule in the patient lung helps in increasing his/her survival probability. Due to early detection, doctors can take timely therapeutic intervention and can reduce mortality. Several deep learning models have already been developed for computer-aided lung cancer detection. Many deep learning algorithms, for both segmentation and classification are being used to analyse CT scans of lung region by taking advantage of the labelled data. Approach to Kaggle Data Science Bowl 2017 competition of a particular team is described in [7]. Use of 3D Dense U-Net for Lung Cancer segmentation from 3D CT scans is proposed in [8]. The purpose of the present article is to introduce a novel combination of unsupervised machine learning tool, which is a modularity optimization based graph clustering method and deep learning method for classification of cancerous nodules inside the lung. The clustering algorithm was developed for spatially embedded networks [9, 10].

2   Method

This section consists of three parts: A, B and C. In Part A, we will discuss about the type of data used and its pre-processing, In Part B, we will discuss the unsupervised machine learning method which is used to detect the nodules in the lung. In Part C, we will discuss the CNN architecture that is used for classification of cancerous nodules which are detected in part B and how to obtain a rough segmentation mask of cancerous nodules.

2.1   Part A: Data Type and Pre-processing

The data is obtained from LUNA 2016 challenge [2]. It has 880 CT scans of clinically proven lung cancer cases. A CT scan is stored in a dicom format which is converted to a 3D matrix of size $512 \times 512 \times z$. z is the number of 2D slices in the 3D volume, which lies in the range 150 to 280. The 2D images are obtained by taking a particular slice along the z direction which has a shape of 512×512 and contains grey scale values which indicate the electron density obtained from the CT Scans. A significant fraction of the scan does not contain lung region. Since we can find cancerous nodules only in the lung region, an initial pre-processing step is to mask the lung region from the entire scan (explained in the next section). The masked lung region is shown in Fig. 1(b). Further processing is done on the masked image. A separate annotation file contains the locations of cancer in all the scans. This information is used to extract the 2D slices containing cancer.

2.1.1   Lung Segmentation

Lung segmentation is done using marker controlled Watershed segmentation [11], which is a morphological operation. Sobel filter [12] is applied on the Image along x and y directions which are later used to compute the Sobel gradient. The pixel values of the image lie in the range -1024 to $+1000$. The portion surrounding the lung region has a very high positive value of intensity as compared to the lung region. Inside the lung, there are blood vessels, cancer nodules, etc. which has high positive intensity value. The remaining portion of the lung region has intensity value around -750. Using these intensity values, we obtain internal and external markers corresponding to inside and outside the lung region. Watershed segmentation algorithm is then applied to the Sobel gradient image along with the markers. Markers are used to avoid over segmentation.

2.2   Part B: Unsupervised Machine Learning Method

The segmented portion is then converted into a network. This network is then clustered using unsupervised clustering method discussed in Sect. 2.2.2 to detect the different nodules present. These steps are discussed in details in following sections.

2.2.1   Network Generation

A network is defined in terms of nodes and edges between these nodes. The segmented lung region is represented into network by considering its pixels (2D) centre as node and an edge is formed between each node and its first nearest neighbouring pixels. In 2D, there may be either four or eight neighbours are possible based on the neighbourhood definition. Here, we have considered edge and diagonal pixels as neighbours. Thus eight neighbours are possible for inner pixels. The generated network is unweighted, since weight of each edge is unity.

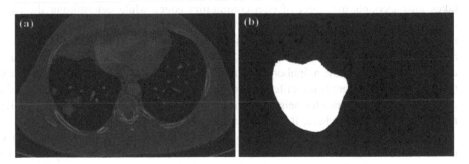

Fig. 1. Lung portion segmentation (a) the Dicom image and (b) the segmented left lung portion (white region).

2.2.2   Network Clustering

The generated network of lung portion is partitioned into "communities" which can also be called as nodules in this case by maximizing a modularity function given in Eq. 1. The modularity function was developed by Raj et al. for spatially embedded networks [6]. They have successfully applied this function on granular networks obtained from granular assemblies [7] for identifying the community structure which has highest associability with the naturally identifiable structures. The modularity function reads as:

$$Q(\sigma) = \frac{1}{2m} \sum_{i,j} \left(a_{ij}A_{ij} - \theta(\Delta x_{ij}) |b_{ij}| J_{ij} \right) \left(2\delta(\sigma_i, \sigma_j) - 1 \right) \tag{1}$$

Here m is the total number of edges in the network, aij and bij are the strength of connected and missing edges between ith and jth nodes respectively. The edge strengths, aij and bij, can also measures the similarity between the ith and jth nodes. The connection matrix Aij contains the information of edge between nodes. The elements of connection matrix is either zero or one. The parameters aij and bij is the difference between the local average grey scale value (a similarity parameter) of node i and j and the global average grey scale value <k> of the given network as shown in Eq. 2.

$$a_{ij} = b_{ij} = \left(\frac{k_i + k_j}{2} - <k> \right) \tag{2}$$

Where, $<k> = \frac{1}{N} \sum_{r=1}^{N} k_r$ and N is the total nodes and k is the node degree.

$$\{2\delta(\sigma_i, \sigma_j) - 1\} = \left\{ \begin{array}{ll} 1 & \sigma_i = \sigma_j \; (i.e \text{ in same community}) \\ -1 & \sigma_i \neq \sigma_j \; (i.e. \text{ in different communities}) \end{array} \right\} \tag{3}$$

The modularity function in Eq. 1 is not comparing the actual connections (Aij) with some predefined null model, so it is independent of any null model comparison. It also takes into accounts the effect of inter-community edges while determining the most modular structure as depicted in Eq. 3. It has another distinction over other existing methods of modularity calculation. It uses Heaviside unit step function θ (Δxij) to restricts the over penalization for missing links between two distant nodes within same community. It defines the neighborhood and penalizes only for the missing links inside the neighborhood. Here is the difference in Euclidian distance between nodes i, j and a predefined cutoff distance for neighborhood xc, which is chosen as xc = 1.05 (Ri + Rj) for the present study (where, R is the distance between two touching pixels).

The unsupervised learning algorithm consists of three main steps. First step is initialization. Here, each node is considered to be in a different community. Later we merge the nodes to form a community. Each node is selected iteratively and checked for possible merging sequentially with all its connected neighbor communities. Modularity is calculated after each merge and the merge which has highest positive change is selected. Node merging continues until no further merging increases the modularity of the system. Final step is Community merging. The node merging often converges at local maxima in the modularity curve. In order to remove any local maxima trap, we merge any two connected community. If this merging increases the modularity then it is selected. The output of community merging is the final clustered network which has the highest value for modularity.

2.3   Part C: Supervised Machine Learning

Various semantic segmentation algorithms exist [13] out of which Regional based Convolutional Neural Network (RCNN) related algorithms [14] use a two steps process: first step is to segment the image into various ROI's and the second step to perform semantic segmentation on the proposed ROI's. RCNN proposes ROI's from an image by using selective search [15]. Initially the image is segmented into multiple small regions and later these small regions are combined (based on colour similarity, texture similarity, size similarity, etc.) to form a larger region. These regions are the Region of Interest Proposals (ROI's). These ROI's are warped or resized to a specific size which is fed to a ConvNet where the output of ConvNet is used as a Bounding Box Regressor and a SVM classifier. Around 2000 ROI's are generated per image and requires the CNN to be run 2000 times on a single image. FastRCNN [16] is built to share the computation across all the 2000 regions. The ROI's are generated using a CNN. ROI's are resized to a specific shape using ROI pooling and the resized regions are fed into a Fully Convolutional Network (FCN). A softmax layer is used on top of the FCN to predict output classes.

The RCNN type architectures perform very well when we have a precise bounding box coordinates and labels of each bounding box which are hard to obtain and expensive. An end-to-end learning process using MaskRCNN [17] to perform all the tasks of interest at once such as ROI proposal, Bounding Box Regression, Classification and semantic segmentation of all the ROI's is very well suited when the dataset is well labelled which is not always the case. Some domains such as medical does not have datasets with rich labels for all the problems of interest. In such sparsely labelled datasets, dividing the training process into sub-blocks can help in understanding the data better. We can deal with individual blocks separately and combine the results when required.

Instead of using CNN or selective search, we propose a clustering based algorithm to find regions in the 3D volume and in 2D images that are similar to each other based on the pixel values. Reducing the amount of data fed to CNN reduces computation time and learns features that are more relevant. The clustering method also stores cluster ID's which are serial numbers where pixels with same cluster ID belong to a cluster which can be used to fine-tune the segmentation masks obtained as outputs from U-Nets [18].

2.3.1   Proposed Model

Using the clustering algorithm, we obtain less than 100 ROI's of fixed size and fed as input to CNN which acts as a classifier and predicts whether the ROI is cancerous or not. The ROI's of fixed size are fed to CNN where binary classification is done. The data consists of 8050 Images of size 50×50 generated from LUNA dataset. Out of these 8050 images, 5187 images are used for training, 1241 for validation and 1622 for testing. Training of CNN was done on K80 GPU, provided for free by Google Colaboratory.

We used Transfer Learning, where the initial weights of CNN are same as that of pre-trained CNN model weights trained on ImageNet [19]. These weights are then fine tuned to perform classification on a different dataset. We trained the network using fit_one_cycle and learning rate schedulers techniques [20]. The main idea is to train different regions (hidden layers) of the CNN with different learning rates. The initial layers of the CNN learns features that are general to any image classification task such as horizontal and vertical edges, so we use a higher learning rate for the initial layers. As we go deeper into the network, the layers learn features that are more specific to the target dataset and also learn the discriminative features between various classes [21]. So, these layers are trained at a lower learning rate. The learning rate is varied between a specific upper and lower bound that are selected using learning rate finder.

Using a lower learning rate results in very less loss and hence very minor changes in the gradient which takes lot of time to converge. Higher value of learning rate results in large gradients and we might overshoot from attaining the global minima. We should pick a learning rate that is neither too high nor too low to get a better generalisation and quicker training [20]. Using a single learning rate might raise the problem of getting stuck at local minima or at saddle points where the gradient is not enough to push the loss to a global minima or away from local minima in the loss landscape. We select an upper and lower bound on learning rates which are 10 times different in magnitude and train the CNN. Typically the upper bound is 1e−4 and lower bound is 1e−5. The higher value of learning rate helps in getting out of local minima or saddle points. This helps in moving to a stable point in the loss landscape.

The algorithm consists of the following steps. Initially, the Lung region is masked from the image. Later, we perform clustering on the lung region which divides the image into a number of smaller regions (clusters). A 50×50 region is cropped around all the clusters. These 50×50 regions are used to train the CNN to perform binary classification.

2.4 Different CNN Architectures

We have used various pre-trained CNN architectures *e.g. SeResNext50, ResNet50* and *Densenet161* and among them *Densenet161* provided better results for classification. These different architectures are discussed below.

ResNet50: Deep Neural networks were initially harder to train and diverge from local minima frequently which resulted in a poor training loss. Typically the mapping of inputs *x* to outputs *y* is done through a single neural network, represented by function *H(x)*. The idea of *ResNet* [22] was to modify the mapping by adding an extra skip connection to the network, represented by *H(x) + x*. They result in a smooth loss landscape, due to which we do not get stuck at local minima and saddle points and get a better training. We used a variant of *ResNet* named *Resnet50* which has 50 convolutional layers [22].

SeResNext50: *SeNets* are proposed in [23]. Unlike many neural network architectures that focus on finding informative features along the spatial dimension, *SeNets* focus also on the channel-wise features. The basic *SeNet* block, SE block maps the input *X* to a feature map, *U* by convolution. In *SeResNext* [23, 24], a squeeze-and-excitation block is applied at the of each non-identity branch of *ResNext* Block. *ResNext* block is an extension to *ResNet* block. *ResNet* block has a single skip connection and a single convolution layer. *ResNext* block has multiple convolution layers along with a single skip connection [24].

DenseNet161: DenseNet architecture is similar to *ResNet* architecture, where each layer is connected to every other layer in a feed forward fashion [22]. The skip connections are not just added but are concatenated after each layer. The features computed at initial layers can be used at later layers. The individual blocks of various CNN architectures are stacked on top of one another with some simple convolution layers in between to maintain the dimensionality while going from one layer to the next. It contains various Dense Blocks as shown in Fig. 2(d).

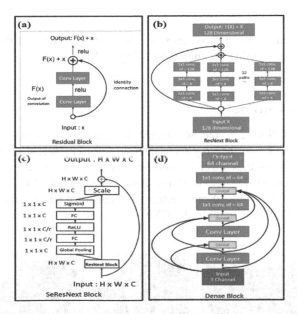

Fig. 2. Building blocks of CNN based architectures (a) *ResNet50* (b) *ResNext* (c) *SeResNext50* (d) *DenseNet161*

3   Segmentation Mask of Cancerous Nodules

The cluster ID's can be used to get rough object localisation similar to object localisation in [25]. Segmentation masks of various classes in an image is obtained using a CNN trained to perform classification. Saliency map of a particular class captures the most discriminative part of an image responsible for CNN to predict that class as output. Applying GrabCut segmentation algorithm [26] on these Saliency Maps provides a corresponding mask of the particular class. Class Activation Maps (CAM) [27] can also be used to find the discriminative regions in the image used by the CNN to perform classification. CAM can be obtained by performing global average pooling over the final activation layer of CNN. The feature map obtained after global average pooling is extrapolated to the size of original image. These maps show which regions in the image was highly activated to produce a certain class. Instead of computing CAM, saliency maps and performing GrabCut segmentation for semantic segmentation, we can use a combination of clustering algorithm and CNN to localise objects and get a segmentation mask. The clustering algorithm produces bounding boxes and cluster ID's. The bounding boxes are classified using a CNN and the corresponding cluster ID's act as a Segmentation map of a particular class predicted by the CNN.

4   Results

The result section is also divided into two parts. Part A contains the result of unsupervised machine learning technique and Part B has the result of proposed neural network architecture.

4.1   Part A: Clustering

As discussed in Sect. 2.2, the segmented lung portion is clustered into different "communities" based on the grey scale intensity of each pixel (Fig. 3). For the given system, these communities are called nodules. We have drawn a 50×50 square box to surround each nodule. The region inside these boxes are called ROI and later used as input to different CNN based architectures for classification purpose. We have generated ~ 8050 ROI's for training the CNN.

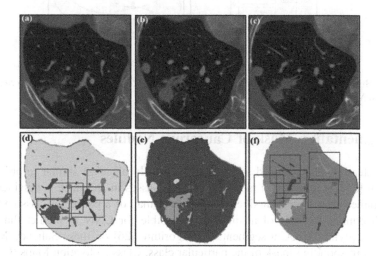

Fig. 3. Nodule detection in 2D. (a–c) Different CT scan slices at different height (Z locations) and (d–f) the clusters/nodules detected using clustering algorithm. The colour assigned to each nodule and the background is randomly selected. The red square boxes are the ROI's which are used in different CNN architectures.

4.2   Part B: Cancerous Nodule Detection

For cancerous nodules detection, we have used approximately 8050 ROI's for training different CNN architectures. These ROI's (Fig. 5(a)) are classified as either cancerous (labelled as 1) or non-cancerous (labelled as 0) as shown in Fig. 4(b). Based on these classifications the cancerous ROI are selected and the cancerous nodule region (detected by clustering algorithm) is masked by assigning the pixel values as 1 and the rest pixels as 0. This creates a binary image as shown in Fig. 4(c).

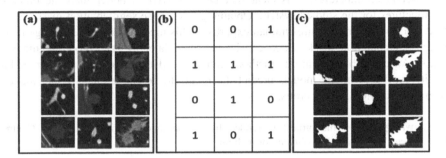

Fig. 4. Cancer segmentation (a) Different ROI's generated by image clustering (b) Classification labels on each corresponding ROI's in panel (a), obtained by CNN based architectures (c) The cancer mapping of the nodule area detected by clustering algorithm.

The confusion matrix obtained from a CNN based architecture (Fig. 5) is used for accuracy measurement. The matrix has four elements named 'True positive (TP)', 'True negative (TN)', 'False positive (FP)' and 'False negative (FN)'. The TP and TN are the correctly detected nodules whereas FP and FN are the wrongly detected nodules respectively. An accurate model should have very low value of FN and high value of TP.

(a)

		Predicted Label	
Label	Non Cancerous	Cancerous	
Non Cancerous	1280 (TN)	60 (FP)	
Cancerous	48 (FN)	234 (TP)	

(b)

		Predicted Label	
Label	Non Cancerous	Cancerous	
Non Cancerous	1305 (TN)	35 (FP)	
Cancerous	42 (FN)	240 (TP)	

(c)

		Predicted Label	
Label	Non Cancerous	Cancerous	
Non Cancerous	1298 (TN)	42 (FP)	
Cancerous	50 (FN)	232 (TP)	

Fig. 5. Confusion matrices of different CNN based architectures: (a) *SeResNext50*, (b) *DenseNet161* and (c) *ResNet50*.

5 Conclusions

We have proposed a novel combination of unsupervised machine learning tool with CNN based architectures for lung cancer detection/classification. Complexity of a CNN can be reduced by using the techniques such as ROI proposals. We have shown that this method is helpful in providing an approximate pixel-wise segmentation mask along with the class label of the ROI proposals. We have tried different CNN based architectures like *SeResNext50, DenseNet161* and *ResNet50*. From the false negative value of different architectures, it is clear that *DenseNet161* is better than the other two architectures for the studied system. Obtaining pixel-wise segmentation masks is often time consuming and requires human expertise. This method can be used to generate a rough segmentation mask which can be modified later to get a good segmentation mask. An ensemble of various methods can be used to reduce the number of False Positives. If any of the models predict the image to be cancerous, we can make the overall prediction as cancerous.

Acknowledgement. This project is funded by Virtual and Augmented Reality Centre of Excellence (VARCoE), IIT Bhubaneswar, India.

Compliance with Ethical Standards
   ✓ All authors declare that there is no conflict of interest.
   ✓ No humans/animals involved in this research work.
   ✓ We have used our own data.

References

1. https://www.techrepublic.com/article/report-the-10-most-innovative-companies-in-machine-learning/#ftag=CAD-00-10aag7f
2. LUNA. https://luna16.grand-challenge.org/
3. Kaggle Data Science Bowl (2017). https://www.kaggle.com/c/data-science-bowl-2017
4. https://medium.com/@andriylazorenko/tensorflow-performance-test-cpu-vs-gpu-79fcd39170c
5. Cancer Moonshot. https://www.cancer.gov/research/key-initiatives/moonshot-cancer-initiative
6. Cancer Research UK. https://www.cancerresearchuk.org/health-professional/cancer-statistics/worldwide-cancer. Accessed January 2019
7. Kuan, K., Ravaut, M., Manek, G., Chen, H., Lin, J., Nazir, B., Chandrasekhar, V.: Deep learning for lung cancer detection: tackling the Kaggle Data Science Bowl 2017 challenge. arXiv preprint arXiv:1705.09435 (2017)
8. Kamal, U., Rafi, A.M., Hoque, R., Hasan, M.: Lung cancer tumor region segmentation using recurrent 3D-DenseUNet. arXiv preprint arXiv:1812.01951 (2018)
9. Kishore, R., Gogineni, A.K., Nussinov, Z., Sahu, K.K.: A nature inspired modularity function for unsupervised learning involving spatially embedded networks. Sci. Rep. **9**(1), 2631 (2019)

10. Kishore, R., Krishnan, R., Satpathy, M., Nussinov, Z., Sahu, K.K.: Abstraction of meso-scale network architecture in granular ensembles using 'big data analytics' tools. J. Phys. Comm. **2**(3), 031004 (2018)
11. Sankar, K., Prabhakaran, M.: An improved architecture for lung cancer cell identification using Gabor filter and intelligence system. Int. J. Eng. Sci. **2**(4), 38–43 (2013)
12. Sharma, D., Jindal, G.: Identifying lung cancer using image processing techniques. In: International Conference on Computational Techniques and Artificial Intelligence (ICCTAI), vol. 17, pp. 872–880 (2011)
13. Guo, Y., Liu, Y., Georgiou, T., Lew, M.S.: A review of semantic segmentation using deep neural networks. Int. J. Multimedia Inf. Retrieval **7**(2), 87–93 (2018)
14. Girshick, R., Donahue, J., Darrell, T., Malik, J.: Rich feature hierarchies for accurate object detection and semantic segmentation. In: Proceedings of the IEEE Conference on Computer Vision and Pattern Recognition, pp. 580–587 (2014)
15. Uijlings, J.R., Van De Sande, K.E., Gevers, T., Smeulders, A.W.: Selective search for object recognition. Int. J. Comput. Vision **104**(2), 154–171 (2013)
16. Girshick, R.: Fast R-CNN. In: Proceedings of the IEEE International Conference on Computer Vision, pp. 1440–1448 (2015)
17. He, K., Gkioxari, G., Dollár, P., Girshick, R.: Mask R-CNN. In: Proceedings of the IEEE International Conference on Computer Vision, pp. 2961–2969 (2017)
18. Ronneberger, O., Fischer, P., Brox, T.: U-Net: convolutional networks for biomedical image segmentation. In: International Conference on Medical Image Computing and Computer-Assisted Intervention, pp. 234–241. Springer, Cham (2015)
19. Simon, M., Rodner, E., Denzler, J.: ImageNet pre-trained models with batch normalization. arXiv preprint arXiv:1612.01452 (2016)
20. Smith, L.N.: Cyclical learning rates for training neural networks. In: IEEE Winter Conference on Applications of Computer Vision (WACV), pp. 464–472 (2017)
21. Zeiler, M.D., Fergus, R.: Visualizing and understanding convolutional networks. In: European Conference on Computer Vision, pp. 818–833. Springer, Cham (2014)
22. He, K., Zhang, X., Ren, S., Sun, J.: Deep residual learning for image recognition. In: Proceedings of the IEEE Conference on Computer Vision and Pattern Recognition, pp. 770–778 (2016)
23. Hu, J., Shen, L., Sun, G.: Squeeze-and-excitation networks. In: Proceedings of the IEEE Conference on Computer Vision and Pattern Recognition, pp. 7132–7141 (2018)
24. Xie, S., Girshick, R., Dollár, P., Tu, Z., He, K.: Aggregated residual transformations for deep neural networks. In: Proceedings of the IEEE Conference on Computer Vision and Pattern Recognition, pp. 1492–1500 (2017)
25. Simonyan, K., Vedaldi, A., Zisserman, A.: Deep inside convolutional networks: visualising image classification models and saliency maps. arXiv preprint arXiv:1312.6034 (2013)
26. Rother, C., Kolmogorov, V., Blake, A.: GrabCut: interactive foreground extraction using iterated graph cuts. ACM Trans. Graph. (TOG) **23**(3), 309–314 (2004)
27. Zhou, B., Khosla, A., Lapedriza, A., Oliva, A., Torralba, A.: Learning deep features for discriminative localization. In: Proceedings of the IEEE Conference on Computer Vision and Pattern Recognition, pp. 2921–2929 (2016)

Assistive Communication Application for Amyotrophic Lateral Sclerosis Patients

T. Shravani[✉], Ramya Sai, M. Vani Shree, J. Amudha,
and C. Jyotsna

Department of Computer Science and Engineering, Amrita School of
Engineering, Bengaluru, Amrita Vishwa Vidyapeetham, Bengaluru, India
shravani_2194@yahoo.com, ramineniramya1@gmail.com,
vanishri.mallampati@gmail.com,
{j_amudha, c_jyotsna}@blr.amrita.edu

Abstract. Eye tracking is one of the advanced techniques to enable the people with Amyotrophic lateral Sclerosis and other locked-in diseases to communicate with normal people and make their life easier. The software application is an assistive communication tool designed for the ALS Patients, especially people whose communication is limited only to eye movements. Those patients will have difficulty in communicating their basic needs. The design of this system's Graphical User Interface is made in such a way that it can be used by physically impaired people to convey their basic needs through pictures using their eye movements. The prototype is user friendly, reliable and performs simple tasks in order the paralyzed or physically impaired person can have an easy access to it. The application is tested and the results are evaluated accordingly.

Keywords: Amyotrophic lateral sclerosis · Eye tracking · Eye tracker · Gaze data · Fixation value · Graphical user interface

1 Introduction

In patients with ALS, the brain loses the ability to initiate and control muscle movement, as the motor neurons die and the patient become paralyzed. He also loses his ability to speak. This condition is very frustrating to the patient. One of the Eye Tracking applications is, using this mechanism for assistive communication for physically disabled people.

In this paper we are describing the assistive application we have developed for ALS patients which helps them to convey their basic needs through Human Computer Interaction (HCI) [1, 2]. Eye Tracker is a device which tracks the movement of eyes and determines the exact positions where the user is looking at the screen. Our application works on the data collected by the eye tracker.

The easiest way to communicate is to use an application on a traditional computer where the mode of understanding Graphical User Interface (GUI) is done using pictures and conveying his needs is done using his eye movements which are being tracked by an eye tracking device. It is uncomplicated and painless way for the patients to convey the basic needs rather than typing the message by the patient with eye movements

© Springer Nature Switzerland AG 2020
S. Smys et al. (Eds.): ICCVBIC 2019, AISC 1108, pp. 1397–1408, 2020.
https://doi.org/10.1007/978-3-030-37218-7_147

which is a stressful task. Pictures are also comfortable way of understanding the needs of patients by the care taker.

An advantage of it is once the user or the patient gets used to the application, he needn't search for the position of the particular block of need on the screen. This makes his job easier to convey his need to the care taker.

Section 2 discuss the related work on establishing communicative system with the help of eye gaze features. Section 3 explains the system architecture which comprises the creation of GUI and its usage. Section 4 presents the results obtained. Heat map is used to visualize various results obtained. Section 5 draws the conclusion and future scope of this research study.

2   Related Work

Various interaction methods are being invented based on eye-tracking technology with the information concerning the patient's needs and limitations of eye trackers combined [3–5]. There are various approaches which use eye tracking as input to computers. The first approach uses eye gaze as a pointing device and combined with touch sensor to work as a multimodal input [6]. The second approach uses eye gestures as input to computer. The third approach is providing a user interface which assist the user with eye tracking technology.

An eye communication system is a graphical user interface with the icons of basic needs of the patient and alphabets for entering text and it works with the help of an eye tracker [7]. Functionality of the communication system is accessed based on the usability tests.

A cheap, reliable and portable communication support system for the disabled developed using Electro-Oculography. Four directional eye movements and blink are tracked using 5 electrode EOG configuration for communication using a visual board for reference [8]. The board was specifically designed based on modelling of English language to increase the speed by reducing the number of eye movements required. Using these system users has achieved an average speed of 16.2 letters/min.

An eye tracking application has been developed using the infrared sensors and emitters as the input device. The screen was divided into no one grids and hierarchical approach is used for selection of those grids [9]. The advantages of this system are low cost and user friendliness.

An assistive communication tool has been developed which consists of a camera and a computer. The images will be captured by the camera and the system can interpret the data contained in the images [10]. Thus, this tool focuses on camera-based gaze tracking. Various eye tracking applications built using different technologies, their advantages and limitations are shown in Table 1.

Table 1. Comparison of different applications

S. No	Application	Eye tracking technology	Communication	Advantages	Limitation
1	Eye communication system for disabled people	Eye tracker - eye tribe	Communication via symbols or icons, texts	Can specify needs through images. If the desired option is not available, then user can convey it by texts	Very small icons
2	EOG based text communication	Electro Oculography	Communication with texts	Average typing speed 16.2 Letters/min Easy and invasive method for recording large eye movements	Not suitable for daily use due to close contact of electrodes
3	Eye gaze tracking for people with locked in diseases	Developed using infrared Sensors	Communication with texts	Low cost, user friendliness, and eye strain reduction.	Eye movement are limited to certain degrees
4	Eye - GUIDE (eye gaze user interface design)	Built using a computer and a camera	Communication through texts	Low cost	Limitation is due to saccades. fixations and noise that contaminates the gaze data

3   System Architecture

Our system comprises a GUI which can help an ALS patient to convey his basic needs. Our model helps the communication using pictures. There is no limitation in eye movements and the care taker can easily understand patient's need.

3.1   Application Program/Interface (API) Overview

The Fig. 1 shows hardware and software components of the eye tracking system. Application Program Interface (API) layer overview is given below.

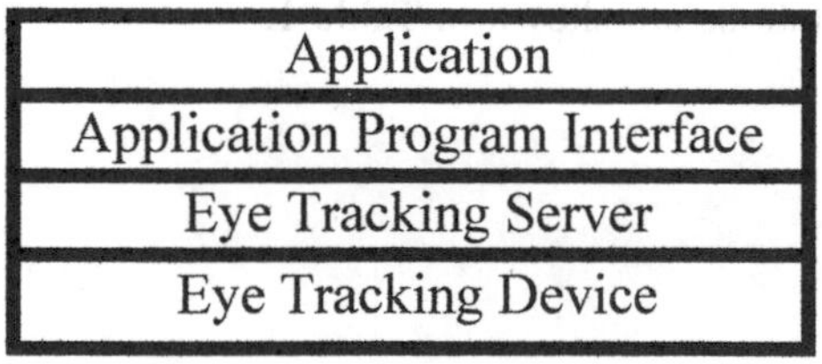

Application
Application Program Interface
Eye Tracking Server
Eye Tracking Device

Fig. 1. API layers

Application: The first screen of the API is the welcome page which is divided into nine blocks as shown in the Fig. 2. Each block provides with the user's basic needs which are Thirst, Food, Switch on/off the lights, Switch on/off the fan, Washroom, Entertainment, Emergency and Exit the Application.

Our application is made in such a way that if the user gazes for 0.5 s or above at a particular block, event gets triggered and a new page will be opened in order to specify the desired option as described below.

Fig. 2. GUI of the welcome page [4, 6, 8]

When you gaze at the Thirst block for 0.5 s or above it will lead you to the next screen which has specific options on what a person wants to drink as shown in Fig. 3. It has options like tea, coffee, milk, coconut water, soft drinks, water, and a back option which will take the application to the welcome page.

Assume the user desires to drink milk and selects the bottom right block with his eye gaze. Then the GUI of milk block shown in Fig. 4 will be opened. He can go back to welcome page by selecting top right back block in case it is a wrong option selected or he desires to convey his other requirement.

Other blocks also perform the similar functionalities. On gazing at the Exit block which is at the bottom right corner of the welcome page, the user means that he has nothing to convey and the application will end and the eye tracker will stop recording the gaze data.

API: Programmable interface to provide access to eye tracking device. iView X™ API which we have used is part of the iView X™ SDK (Software Development Kit) which can be downloaded from SMI website. We have used Python 2.7 programming language to build eye tracking application.

Fig. 3. GUI of thirst block [6, 8, 9]

Eye Tracking Server: Eye tracking server application collects the data from the eye tracking device and provides the data via the API [9, 11]. Eye tracking Server is used as a generic name for software component.

Eye Tracking Device: The eye tracker used is SMI RED-m eye tracker. The use of a laptop allows conceiving a mobile, practical and economic solution, applicable in the person's everyday life. Laptop with USB 2.0 port is used to connect the eye tracker.

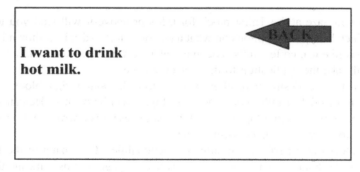

Fig. 4. GUI of milk block

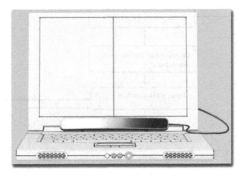

Fig. 5. Positioning the eye tracker

Place the RED-m Eye Tracking Device at the center point (located in step 1 above) and in the hinge area of the laptop as shown in Fig. 5. Angle the RED-m Eye Tracking Device upwards towards the eyes of the user.

3.2   Working of the Application

The Fig. 6 shows the activities performed by the application. When the helper opens the application calibration and validation functions are performed and when they get the appropriate accuracy the application is opened.

If the user gazes the intended block for more than 0.5 s, then the corresponding action is performed and repeated unless the user gazes at the exit block. Once the user gazes at the exit block the eye tracker stops collecting the data and the application is closed.

3.3   Steps Involved

Eye Tracker Set Up: Now the Eye Tracker device needs to be connected to the Eye Tracking Server. Start the iView RED-m application which automatically starts the eye tracking server and establish connection with the eye tracking device.

Calibration and Validation: Once the application is started the calibration and the validation processes will take place and the result of it which is accuracy and it is displayed on the command window. Make sure that accuracy value is less which means deviations happened are less. After the completion of calibration and validation the application is started. It will take the gaze data and perform the preferred action until the user gazes at the exit block.

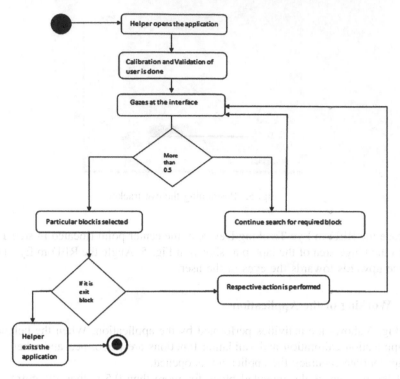

Fig. 6. Activity diagram

4   Results and Analysis

4.1   Heat Map

Several test cases have been applied to the application and the results were analyzed using heat maps. The user was asked to follow a specified path and using the gaze data which consists of timestamp, gaze x position, gaze y position and the duration of the event the following heat maps were generated as shown in Fig. 7.

Test Case 1: The path asked to follow: thirst → coffee → back → exit

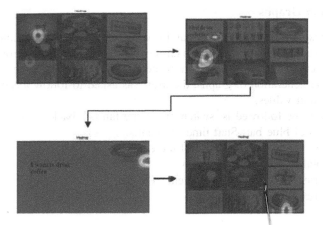

Fig. 7. Heat maps drawn with Test case 1 data

Test Case 2: The path to follow: emergency → back → food → snacks → back → exit is shown in Fig. 8.

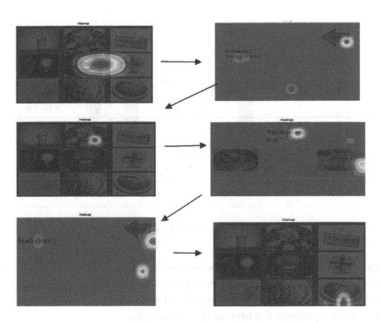

Fig. 8. Heat maps drawn with Test case 2 data

From the heat maps, we conclude that, the gaze data generated the correct flow of events. In the interface where the message is displayed only the back option block was enabled, when the user gazes at the remaining areas, no event is triggered.

4.2   Fixation Graphs

Fixation or visual fixation is the maintaining of the visual gaze on a single location. The application was implemented using fixation values with the range from 0.1 s to 1 s. Graphs are generated using event data to determine appropriate range of the fixation value. For the generation of graphs, the user was asked to follow a specific path with different fixation values.

The path to be followed is: switch on/off the fan $\rightarrow$ back $\rightarrow$ exit

NOTE: Top of blue bar: Start time of event
Top of red bar: End time of the event
Total red block: Event span
Series 1: Time taken for event to occur
Series 2: Event duration

Fixation Value of 0.1 s: As the fixation value is very small many number of events are triggered before we intend to see the desired block. As shown in Fig. 9, since the user had fixated only for 0.1 s, the corresponding page will not be opened. The y-axis shows the time in micro second.

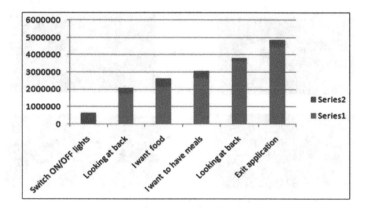

Fig. 9. Fixation graph for 0.1 s

Fixation Value 0.3 s: In this case before triggering the actual event, many events which are not specified in the path are triggered in between the actual path. But, the number of events triggered for 0.3 s are less when compared with 0.1 s. It is shown in Fig. 10. Y-axis represents the time in micro seconds.

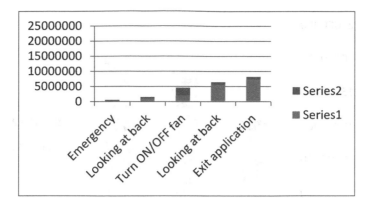

Fig. 10. Fixation graph for 0.3 s

When the fixation value is in between 0.5 s to 1 s the user is able to gaze at the specified path and generate the events appropriately.

Fixation Value 0.9 s: But, as the fixation point increases the time taken to trigger the event also increases. This may lead the user with uneasiness as he has to gaze at a desired block for more amount of time. It is shown in Fig. 11.

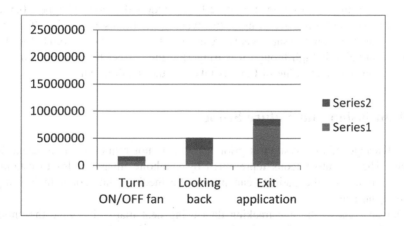

Fig. 11. Fixation graph for 0.9 s

Fixation Value 1 s: User has to gaze for a long time for the events to trigger. It is shown in Fig. 12.

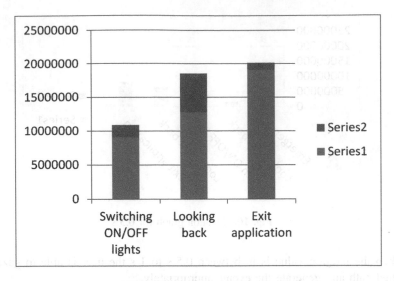

Fig. 12. Fixation graph for 1 s

When the fixation value is in between 0.5 s to 1 s the user is able to gaze at the specified path and generate the events appropriately. But, as the fixation point increases the time taken to trigger the event also increases. Our application is successful so far in the amount of time the user needs to gaze for the required action to happen. 0.5 to 0.7 s interval is precise for the occurring of the event as per the user's choice. If the fixation value is less, then unnecessary events are triggered (i.e. if value is from 0.1 to 0.4 s). If the fixation value is high though it is triggering the correct block it is taking a lot of time. So, an optimum value of 0.5 s is taken as the fixation value.

5   Conclusion and Future Scope

The advantage of this system is prior learning is not required for its usage by any patient. The figurative icons represented by symbolic images helps to establish the connection so that the patients can inform the medical staff about their symptoms, feelings and needs.

Our analysis of the eye tracking data using heat maps also gave the anticipated results. The optimal fixation value is determined to be 0.5 s upon the analysis done using graphs.

Advancements can be done to our application by incorporating text to speech conversion techniques, application can be enhanced in such a way that care takers or doctors can have the flexibility of modifying the GUI according to each patient's explicit needs, differentiating natural and intended eye movements. The application can be customized according to the need of each patient.

Acknowledgment. This study was conducted in the Eye Tracking and Computer Vision Research Lab in Amrita School of Engineering, Bengaluru.

Compliance with Ethical Standards
✓ All authors declare that there is no conflict of interest.
✓ No humans/animals involved in this research work.
✓ We have used our own data.

References

1. Kiyohiko, A., Shoichi, O., Minoru, O.: Eye-gaze input system suitable for use under natural light and its applications toward a support for ALS patients. In: Current Advances in Amyotrophic Lateral Sclerosis. IntechOpen (2013)
2. Venugopal, D., Amudha, J., Jyotsna, C.: Developing an application using eye tracker. In: 2016 IEEE International Conference on Recent Trends in Electronics, Information & Communication Technology (RTEICT). IEEE (2016)
3. Pavani, M.L., Prakash, A.B., Koushik, M.S., Amudha, J., Jyotsna, C.: Navigation through eye tracking for human–computer interface. In: Information and Communication Technology for Intelligent Systems, pp. 575–586. Springer, Singapore (2019)
4. Drewes, H.: Eye gaze tracking for human computer interaction. Dissertation LMU (2010)
5. Jyotsna, C., Amudha, J.: Eye gaze as an indicator for stress level analysis in students. In: International Conference on Advances in Computing, Communications and Informatics (ICACCI). IEEE (2018)
6. da Eira, M.S.: Eye communication system for nonspeaking patients (2014)
7. Liu, S.S., et al.: An eye-gaze tracking and human computer interface system for people with ALS and other locked-in diseases. In: CMBES Proceedings, vol. 33, no. 1 (2018)
8. TallaVamsi, S.M., Shambi, J.S.: EOG Based Text Communication System for Physically Disabled and Speech Impaired (2008)
9. Lupu, R.G., Ungureanu, F.: A survey of eye tracking methods and applications. Buletinul Institutului Politehnic din Iasi, Autom. Control Comput. Sci. Sect. **3**, 72–86 (2013)
10. Anacan, R., et al.: Eye-GUIDE (eye-gaze user interface design) messaging for physically-impaired people. arXiv preprint arXiv:1302.1649 (2013)
11. Sawhney, T., Pravin Reddy, S., Amudha, J., Jyotsna, C.: Helping hands – an eye tracking enabled user interface framework for amyotrophic lateral sclerosis patients. Int. J. Control Theor. Appl. **10**(15) (2017). ISSN: 0974-5572
12. Lule, D., Kübler, A., Ludolph, A.: Ethical principles in patient-centered medical care to support quality of life in amyotrophic lateral sclerosis. Front. Neurol. **10**, 259 (2019)
13. Lenglet, T., Mirault, J., Veyrat-Masson, M., Funkiewiez, A., del Mar Amador, M., Bruneteau, G., Lacomblez, L.: Cursive eye-writing with smooth-pursuit eye-movement is possible in subjects with amyotrophic lateral sclerosis. Front. Neurosci. **13** (2019)
14. Chipika, R.H., Finegan, E., Shing, S.L.H., Hardiman, O., Bede, P.: Tracking a fast-moving disease: longitudinal markers, monitoring, and clinical trial endpoints in ALS. Front. Neurol. **10** (2019)
15. Yunusova, Y., Ansari, J., Ramirez, J., Shellikeri, S., Stanisz, G.J., Black, S.E., Zinman, L.: Frontal anatomical correlates of cognitive and speech motor deficits in amyotrophic lateral sclerosis. Behav. Neurol. (2019)

Author Index